T0296879

Orogenesis

The Making of Mountains

Orogenesis, the process of mountain building, occurs when two tectonic plates collide – either forcing material upwards to form mountain belts such as the Alps or Himalayas or causing one plate to be subducted below the other, resulting in volcanic mountain chains such as the Andes. The relatively recent discovery of plate tectonics gave us answers to questions such as why mountains are found in particular places on Earth, how long it takes them to form and what the limitations of their formation are.

Integrating the approaches of structural geology and metamorphism, this book provides an up-to-date overview of orogenic research, and an introduction to the physico-chemical properties of mountain belts. Global examples are explored, from the Scottish Caledonides, the Alps, the Andes and the Himalayas, and other chapters examine the deep structures and nature of mountain roots, with examples from Canada, Greenland and Antarctica. It includes a review of the role of the interactions of temperature and deformation in the orogenic process, and explains important new concepts such as channel flow.

This book provides a valuable introduction to this fast-moving field for advanced undergraduate and graduate students of structural geology, plate tectonics and geodynamics, and will also provide a vital overview of research for academics and researchers working in related fields including petrology, geochemistry and sedimentology.

Michael R. W. Johnson is a Fellow of the Royal Society of Edinburgh, and taught structural geology and tectonics in the University of Edinburgh for 40 years. He has undertaken research on orogenic belts in many parts of the world – Scotland, North America, the Alps and the Himalaya – and has continued his researches since his retirement in 1997. Dr Johnson has written over 80 papers, co-edited and contributed to several books, and organised and given key-note lectures at many international conferences. For many years he was on the editorial board of *Tectonophysics*, and has served on international committees such as the International Geological Correlation Programme (IGCP).

Simon L. Harley is Professor of Lower Crustal Processes at the School of Geosciences, University of Edinburgh. He has taught metamorphism and tectonics, Earth evolution, and aspects of isotope geology at the University of Edinburgh and Oxford University since 1983. He is recognised internationally as a world authority on metamorphism at extreme temperature conditions in the crust. He has undertaken field and laboratory-based research on several mountain belts, and has a particular interest in Antarctica, its evolution and environment. Professor Harley has written more than 100 papers, co-edited several conference proceedings and special volumes, and acted on the editorial boards of several key journals in geosciences, including *Geology* and the *Journal of Petrology*. He is a Fellow of the Royal Society of Edinburgh, and a recipient of the Imperial Polar Medal for contributions to Antarctic science.

Orogenesis

The Making of Mountains

MICHAEL R. W. JOHNSON

SIMON L. HARLEY

University of Edinburgh

CAMBRIDGE
UNIVERSITY PRESS

University Printing House, Cambridge CB2 8BS, United Kingdom

One Liberty Plaza, 20th Floor, New York, NY 10006, USA

477 Williamstown Road, Port Melbourne, VIC 3207, Australia

314-321, 3rd Floor, Plot 3, Splendor Forum, Jasola District Centre, New Delhi - 110025, India

103 Penang Road, #05-06/07, Visioncrest Commercial, Singapore 238467

Cambridge University Press is part of the University of Cambridge.

It furthers the University's mission by disseminating knowledge in the pursuit of education, learning and research at the highest international levels of excellence.

www.cambridge.org
Information on this title: www.cambridge.org/9780521765565

First published 2012
Reprinted 2014

A catalogue record for this publication is available from the British Library

Library of Congress Cataloging in Publication data
Johnson, M. R. W. (Michael Raymond Walter)
Orogenesis : the making of mountains / by Michael R.W. Johnson and Simon L. Harley.
p. cm.
Includes bibliographical references and index.
ISBN 978-0-521-76556-5 (Hardback)
1. Orogeny. 2. Orogenic belts. 3. Plate tectonics. 4. Geology, Structural.
I. Harley, Simon, 1956– II. Title.
QE621.J64 2012
551.8′2–dc23
201102938–9

ISBN 978-0-521-76556-5 Hardback

Contents

The colour plates can be found between pages 342 and 343.

Preface

Mountains have attracted the attention of mankind at least since Rousseau (or did Petrarch precede him?*) who devoted much thought to nature, perhaps because the height and scale of mountains induced a sense of awe. A love of nature showed itself in the fairly recent desire to get to the top of mountains. George Mallory gave his reason for wanting to climb Everest as "because it is there", but long before that mountains were important for humankind, because they formed natural barriers for trade and the movement of armies. Perhaps the ancient Egyptians tried to simulate mountains in the pyramids of Giza. The same is true of builders of Gothic cathedrals, which were built ever higher so as to imitate mountains which reach up to heaven. The Greeks worshipped the gods on Mount Olympus, and mountains appeared often in Greek mythology; Prometheus, for example, was chained to a mountain side. The Greeks saw mountains as mysterious and frightening places, and even today for Hindus and Buddhists there are sacred mountains in the Himalaya such as Nanda Devi, Kailas and Everest – Qomolungma, the goddess mother of the Earth. Badrinath near the source of the Ganges in the High Himalaya is the home of the gods and a place of pilgrimage. Moses came down from a mountain bearing his famous tablets. Noah is supposed to have docked his ship on Mount Ararat. The Bible states "the mountains shall melt before the Lord" (Judges 5:5), but perhaps the reference was to volcanoes rather than orogenic mountains.

Many artists, too, have been fascinated by mountains. Leonardo Da Vinci realised that the fossils in the rocks of the Apennines showed that the rocks were once below sea level, and he and other painters used mountain scenes as backgrounds. Cezanne painted many pictures of Mont St. Victoire in Provence.

The word orogenesis means 'birth of mountains' (Greek, *oros*, a mountain and *genesis*, be produced, creation). From this we get orogen, which is a term for the characteristically long, narrow linear or curvilinear mountain belts, whether or not they have a marked topographic expression. So what is the definition of a mountain? In North America it is 600 m which seems very acceptable, but when we come to the definition of "high" mountains answers vary, from only a few hundred metres in Scandinavia, to 1660–1700 m in central Europe and 5500 m in central Asia. These numbers refer to the practice of defining mountains by reference to particular landscape features such as the tree line or snow line.

* We are grateful to Professor Usher of the University of Edinburgh who gave us the reference to Petrarch's *c.*1350 letter to his former confessor (Book 1V, Letter 1 of the *Familiares*). In this, Petrarch describes the ascent of Mont Ventoux in Provence, more a moral allegory than an ascent according to Usher. At the top he not only admired the view but read St Augustine's words, "Men go to admire the high mountains and the great flood of the seas and the wide-rolling rivers and the ring of the Ocean and the movement of the stars; and they forget themselves."

Our concern in this book is with the science of mountains and with the processes involved in their formation. Although the aim is not to give a comprehensive account of the orogens of the world, it is still necessary to present an account of the major features of selected orogenic belts. In Chapters 1, 2 and 3 we give brief elementary accounts of background material needed for the understanding of mountain building processes. Some readers may wish to skip these chapters or follow up the further reading given in the references.

Acknowledgments

For help with this book, Michael Johnson wishes to thank Peter Clift, Patience Cowie, Ian Dalziel, Barry Dawson, Ben Harte, Brian Horton, Linda Kirstein, Dirk Kroon, Tom Pedderson, John Platt, Brian Price, Mike Searle, Rob Strachan, Jan Wijbrans and two anonymous referees. Jan has been hugely helpful with advice on the form and content of several parts of the book. Yvonne Cooper handled the photographic and scanning work with skill and a great deal of patience.

More generally, the following have greatly influenced my work on mountains – Ian Dalziel, John Dewey, Ernst Cloos, John Crowell, Dave Elliott, Derek Flinn, Jake Hossack, Geoff Milnes, Neville Price, Hans Ramberg, John Ramsay, Alastair Robertson and Gilbert Wilson.

Last but by no means least I am very grateful to my wife Anne for unflagging support during the writing of this book. Anne processed my words and took on the daunting task of correcting and formatting the text. Only she knows how many times I had to be saved from some computing impasse or disaster.

Simon Harley wishes to thank Bas Hensen and David Green for inspiring him to take up the challenge of metamorphic geology. His approach to metamorphism has also been influenced by Mike Brown, David Ellis, Becky Jamieson, John Platt, Roger Powell, Mike Sandiford, Frank Spear, Alan Thompson and Ron Vernon, all of whom are thanked for their key contributions to the subject.

Simon Harley also thanks the following for discussions, at various times, that have informed his ideas: Bill Carlson, Damian Carrington, Geoff Clarke, Dan Dunkley, Ian Fitzsimons, Joerg Hermann, Tomokazu Hokada, Nigel Kelly, David Kelsey, Peter Kinny, Jana Kotkova, Jean-Marc Montel, Yoichi Motoyoshi, David Pattison, Daniella Rubatto and Olivier Vanderhaeghe.

Most of all, Simon Harley acknowledges and thanks his wife, Annie, for her patience and forbearance during the long evenings spent writing his contributions to this book.

Geological timescale

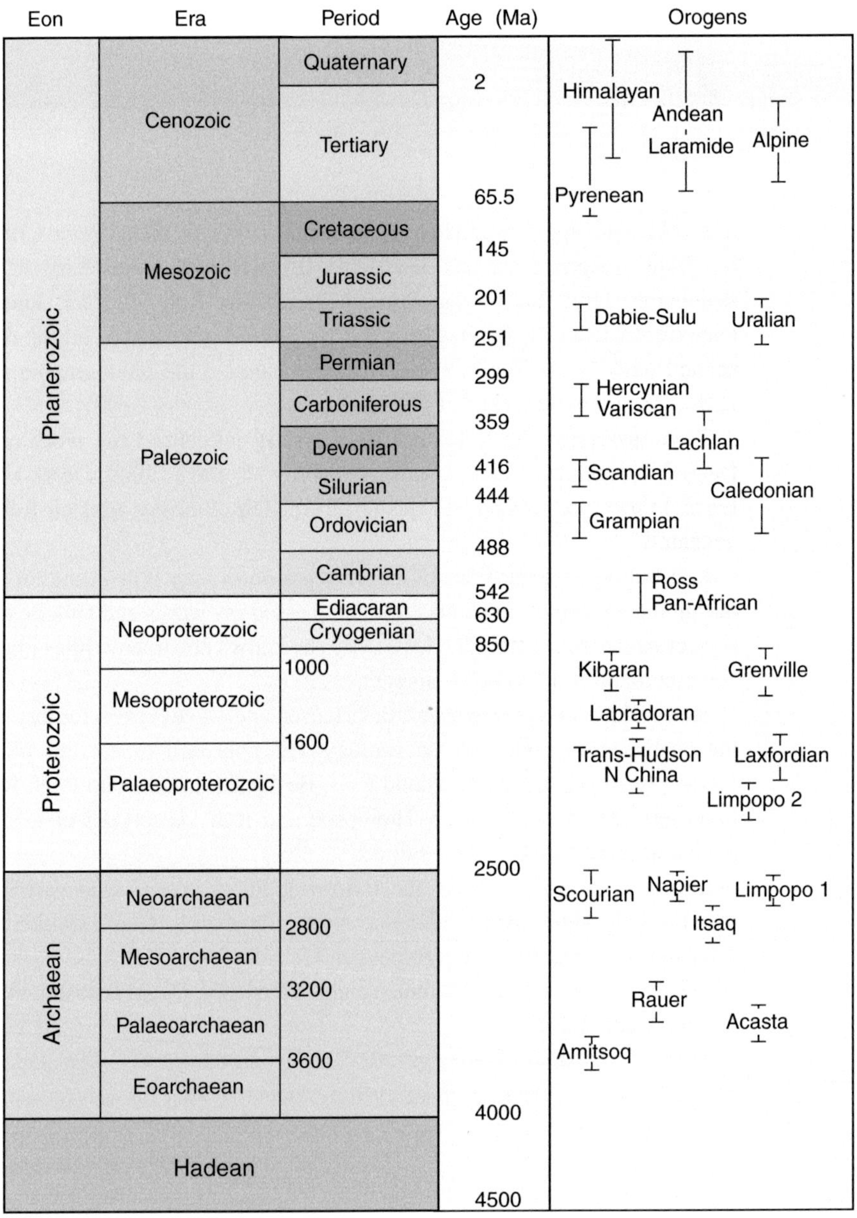

Eon	Era	Period	Age (Ma)	Orogens
Phanerozoic	Cenozoic	Quaternary	2	Himalayan, Andean
		Tertiary	65.5	Laramide, Alpine, Pyrenean
	Mesozoic	Cretaceous	145	
		Jurassic	201	
		Triassic	251	Dabie-Sulu, Uralian
	Paleozoic	Permian	299	
		Carboniferous	359	Hercynian Variscan
		Devonian	416	Lachlan
		Silurian	444	Scandian, Caledonian
		Ordovician	488	Grampian
		Cambrian	542	Ross Pan-African
Proterozoic	Neoproterozoic	Ediacaran	630	
		Cryogenian	850	
	Mesoproterozoic	1000		Kibaran, Grenville, Labradoran
	Palaeoproterozoic	1600	2500	Trans-Hudson N China, Laxfordian, Limpopo 2
Archaean	Neoarchaean			Scourian, Napier, Limpopo 1, Itsaq
	Mesoarchaean	2800		
	Palaeoarchaean	3200		Rauer, Acasta, Amitsoq
	Eoarchaean	3600	4000	
Hadean			4500	

1 Major features of the Earth and plate tectonics

The father of geology, James Hutton, in the late eighteenth century provided the insights which led nearly a century later to the first understandings of how mountains are constructed and what causes them. In the nineteenth and early twentieth centuries Lapworth, Peach and Horne, and Clough in Britain, and Argand, Bertrand, Heim and others working in the Swiss Alps, revolutionised our understanding of what the German geologist Kober called 'orogens' and of the process of orogenesis, the building of mountains.

To understand the significance of orogens, it is necessary to know something about plate tectonics, which has been remarkably successful in explaining many of the features of the Earth. In particular it deals with large-scale dynamic processes in the planet. Plate tectonics developed from the preceding ideas of continental drift, but in essence originated from an idea put forward in the 1960s by H. H. Hess. Hess postulated a surprising concept: that the ocean floor is in motion and is older as one moves away from submarine mountains known as mid ocean rises: for this reason the model became known as sea-floor spreading. The ocean floor is like a giant conveyor belt, and the interesting question is, where does it go? What this idea meant was that, for the first time in the history of geology, attention was turned on the oceans rather than the continents.

This new approach brought about a revival of the older idea of continental drift, which proposed that the continents over geological time have not been fixed in position but have drifted across the surface of the Earth. Half a century ago, many thought this idea to be nonsense. In Britain, the great Arthur Holmes, who had already provided the first quantification of geological time and the age of the Earth in his famous book of 1944 (see also Holmes 1928, 1929), made an eloquent case for continental drift, but most geologists were strongly opposed to the idea, pointing out there was no mechanism for such motion. However, in 1928 Holmes had outlined a mechanism for drift by invoking convection cells in the mantle which dragged the crust along, a view which anticipated Hess's discovery. Hess's sea-floor spreading model and the then new powerful technique of palaeomagnetism swung geological opinion round into acceptance of mobile continents which have moved over time, albeit through the creation and destruction of intervening oceans rather than by ploughing their own furrows across the Earth's mantle.

A spectacular consequence of the nineteenth-century work on continental tectonics was the discovery of thrust faults. There was the almost simultaneous discovery in the 1980s of the Moine thrust in Scotland and the Glarus thrust in the Swiss Alps, leading to the realisation that large bodies of rock can be transported along low angle faults horizontally over other rocks, sometimes for great distances. The existence of mountain belts has long been known, if only because they may be conspicuous features, but their causal origin and nature has only become clear in the past 50 years, following the discovery of plate tectonics.

In this book the term orogen is used for the narrow but long belts of crust which have been deformed mainly by compressional stresses in the process of orogenesis. It is better to avoid the term mountain belt, because many ancient orogens are rather poor mountains in the topographic sense! What follows is a brief account of the plate tectonic model, starting with a review of the Earth's major features including the crucial differences between oceans and continents.

Plate tectonics

Oceans make up 75% of the Earth's surface and they are young, say <200 million years (Ma). Their structure is simple. In contrast the continents are sometimes as old as *c*.4.0 billion years (Ga), and are structurally complex. Oceanic crust is uniform, between 3–10 km thick and mainly made of basaltic igneous rocks. Continental crust is, on average, *c*.35 km thick. Being largely composed of silica-rich, iron- and magnesium-poor rocks, it is less dense than ocean crust. On average the continental crust is regarded as intermediate in composition – not granitic, not basaltic but in between (andesitic). But in orogens continental crust can be twice the normal thickness: for instance, it is *c*.70 km thick under Tibet. On the other hand in rift zones like the Basin and Range of the western United States the crust is thinned. Compression and extensional tectonics are common on the continents and in rocks of all ages. The upper levels of continental crust are composed of sediments which are sometimes metamorphosed. The middle levels are composed of high-grade metamorphic rocks including migmatites which result from partial melting. The lower levels are composed of granulite facies gneisses with mafic and silicic plutons.

The oldest rocks on the continents are seen in so-called Precambrian Shields, specifically in cratons. These make up the central regions of most continents such as India, Africa, Asia and America. The rocks in the Shields are more than 600 Ma and stretch back to the oldest dated rocks at *c*.4 Ga. Within the Shields, cratons are those regions of continental crust that experienced their last significant crust-forming or generation in the Archaean, prior to about 2500 Ma. In these areas we see high-grade metamorphic rocks, comprising gneisses which were formed at great depths in the Earth and then exhumed at the surface, as well as magmatic belts and sedimentary–volcanic belts known as greenstone belts and their accompanying deformed ‘granitoids’ which are dominated by tonalites and related felsic intermediate rocks.

The greenstone belts are interpreted as ultramafic to mafic volcanic igneous complexes accompanied by more felsic volcanic centres. In addition, sediments such as quartzite, carbonate and ironstone are found, and these are witnesses to the presence of seas, even up to at least 3700 Ma ago in the case of the Isua greenstone belts of west Greenland. The origins and tectonic settings of these ancient rocks remain controversial, but for some geologists, applying the famous Huttonian principle that the present is the key to the past, the rocks can be interpreted in terms of plate tectonics. If so, the implication is that, as today, oceans and continents existed in the Archaean. The apparent problem is that the present oceans are very young. What, then, has happened to the ancient floors of the oceans? To answer this we need to know more about the oceans.

Mid Ocean Rises

Although much of the ocean floor is flat and featureless, there are conspicuous elevated regions. These are called Mid Ocean Rises, although not all are positioned centrally in an ocean. These ocean rises are long linear features, up to 40,000 km long, 2.5 km high and 2000 km wide. This means that they are the biggest mountain ranges on Earth. That they are special places in the oceans is shown by the fact that the heat flow at the rises is much higher than is normal for oceans. They are hot because basaltic magmas, dykes and lavas are being intruded and extruded there. The space for the magmatic rocks is provided by the normal faults which are structures associated with extension.

In the plate model, the Mid Ocean Rises are places where new ocean crust is actively being created. This continuous creation of new crust poses a problem. If the Earth does not expand to accommodate the extra crust, where does it go? This is where sea-floor spreading comes in, as does the analogy of a conveyor belt. The ocean floor moves away from the Mid Ocean Rises and disappears – where? At the margins of oceans, another link in the chain of events is provided by the ocean trenches which can be as much as 12 km deep, where the ocean floor plunges beneath the continents. The process is called subduction. Thus there is a balanced system of creation of new crust and its eventual disappearance. Long before the discovery of plate tectonics, subduction zones were recognised by study of patterns of shallow, medium and deep foci for earthquakes, and were named Benioff zones. The Benioff zone is essentially a thrust along which ocean crust slides down to great depths beneath the continents. Subduction zones are associated with curvilinear zones of volcanoes, termed island arcs, situated above the Benioff zone. Good examples are the Indonesian islands and the Japanese islands. The volcanoes show that the activity of the Benioff zone has induced melting of rocks at depth and the rise of magma to the Earth's surface.

Plate boundaries

We can now define what is meant by a plate. In the oceans a plate is the lithosphere forming ocean floor between the Rises and the subduction zones. Some plates also carry continental crust. The boundaries of a plate are subduction zones, Mid Ocean Rises or transform faults (see later). The plate boundaries are termed *divergent* where the plate motion is away from the boundary, as at a Rise, and *convergent* where the motion is towards a boundary, as at a subduction zone. At the present the Earth is covered by about seven or eight major plates and two minor plates.

Magnetic anomalies

The whole system is moving and the rate of plate motion can be measured by using the Global Positioning System (GPS). But the first timescale for plate motion and indeed the final confirmation of the reality of sea-floor spreading came from studies of rock magnetism. Earlier we indicated that palaeomagnetism provided a confirmation of continental drift. This is because certain rocks can capture the orientation of the Earth's magnetic field at the time and place of their formation and retain this as a memory. Because the magnetic inclination varies from the equator to the poles and is dependent on latitude it is possible to show that rocks have wandered across the Earth's surface. The principles of rock magnetism can also be applied to the problem of sea-floor spreading.

Mapping the magnetic intensity of the ocean floor using sea-borne magnetometers revealed a strange feature. The intensity varied in a systematic way, so that the maps of magnetic intensity showed over large areas of ocean a striped pattern of alternating high and low intensity, so-called magnetic anomalies. This was unexplained until Fred Vine and David Matthews spotted the truth in 1963: the variations in intensity reflected periodic reversals in the Earth's magnetic field, with north and south poles switching position. Rocks with reversed magnetism showed lower intensity whereas those with normal or present day polarity showed high magnetic intensity. The next step was to date the normal and reversed stripes of the ocean. At first sight the inaccessibility of ocean floors might seem to be an insurmountable problem, but there is a way round this involving the intricate correlation of the stripes from the oceans with a similar reversal pattern in dated rocks from the continents, the latter being easy to hand. From this arose a magnetic stratigraphy which dated ocean floors. The result was clear: the oceans were youngest at the Rises and aged towards the trenches. This is exactly what the conveyor belt idea predicted. From this work we can say that the rate of motion of plates varies from *c*.2 cm to *c*.18 cm *per year.* The former rate is often expressed as being about the same as the rate of growth of your fingernails.

The analysis of plate motion has been very thorough and involves consideration of plate motion on the curved surface of the Earth using Eulerian geometry. Another component of the ocean-floor spreading must be added – the transform faults that are needed to account for the interaction of the moving plates and to understand what happens at the end of a plate boundary. Motion along transforms is strike-slip. Transform faults are common features of ocean floors but they occasionally affect continental rocks, e.g. the San Andreas and Dead Sea transform faults.

Lithospheric plates

The final point is to consider the plates in three dimensions: in other words, how thick are they? The plate is composed of more than continental or oceanic crust because it includes the upper mantle. Chemically the mantle is quite unlike the crust because it is much richer in iron and magnesium and is made of distinctive rocks such as peridotites and lherzolites. The whole column of rock in a plate, that is the crust and the upper mantle, is referred to as the lithosphere. This model marks an important, indeed key, departure from the previous continental drift hypothesis which assumed that the continents moved independently, and this is crucial in the discussion of orogens. The base of the plate is the contact between the lithosphere and the underlying asthenosphere. The latter is mantle, and as we will see it is important in the study of orogens as a source of heat. Plate tectonics provides a broad framework and indeed a *raison-d'etre* for orogens.

Orogens and plate tectonics

Most orogens are sited on convergent plate boundaries, and the most familiar illustration of this is in orogens like the Alps or Himalaya, which have been formed by the collision of continents. Such collisions are inevitable in plate tectonics. Many but not all orogens mark the sites of ancient oceans which have closed completely as a result of plate motion. Examples are the Alps which formed during the closure of the Mesozoic Tethys Ocean and the Caledonian orogens which formed during the closure of the early Palaeozoic Iapetus Ocean. The history of the planet is one of ocean growth followed by ocean closure, a process called the Wilson cycle after Tuzo Wilson, the great Canadian geophysicist.

Rheological control over continental break-up

Where are the likely places in the continental crust for continental break-up? Krabbendam (2001) cited inherent strength differences between different orogens, and he emphasised that the relatively weak orogens serve as loci for continental break-up. One example is the way the split between America and Europe seems to have been controlled in part at least by the Caledonian orogen. Other stronger orogens have resisted being sites of break-up. For example the Urals, in contrast to the Caledonian belt, retain a strong mantle root, and are also strengthened by the presence of ophiolites and other mafic rocks.

Dewey and Bird classification of orogens

In 1970 Dewey and Bird gave a classification of orogens based on their settings in relation to plate boundaries:

1. The Andean-type belts, situated on continental margins above subduction zones where ocean crust is being subducted beneath continental lithosphere. The name is from the Andes of South America.
2. The continent–continent collision belts such as the Himalaya, resulting from the collision of India and Asia, or the Alps, which result from the collision of Europe and Africa.
3. The (island) arc–continent collisional belts such as in New Guinea, which are formed when an island arc collides with the adjacent continent as a result of plate reorganisation.

Examples of the three types of orogen will be given in Chapter 5. In addition a fourth type may be added – intracratonic orogens resulting from so-called far-field stresses transmitted through the continental crust from an active plate margin. The Neoproterozoic orogens in central Australia appear to have formed far away from a plate boundary and may be examples of this type.

Ancient plates and orogeny

The first three orogen types given above come from the orogenic belts formed in the past 50 Ma, but the classification has been extended to older belts. For example, the Early Palaeozoic Grampian orogen of Scotland and the Taconic orogen in North America are interpreted as arc–continent collision belts. Other parts of the Caledonian belt are viewed as continent–continent collision belts, examples being the Early Palaeozoic Scandinavian Caledonides and the thrusts in the Moines of northwest Scotland.

A famous formulation of James Hutton's theory of the Earth was that the ‘present is the key to the past’. This sounds simple and obvious, but it involves huge assumptions about whether processes we see operating now did so in the distant geological past. Were rivers, and the sediment they are carrying into the seas, always there? The Huttonian formulation receives its most severe test when we go back to the Archaean, but even Palaeozoic orogens call for geological detective work on a grand scale. In older orogenies it is as if we are dealing with a manuscript from which most of the pages have been torn. If we apply the Huttonian dictum, we need to ask: what are the keys to present day orogens?

Firstly, there is the ocean crust which is being subducted. From this we have a method of determining the rates of convergence, using the magnetic anomalies. Obviously most or all of this evidence is lost in ancient orogens where it is a case of *habeas oceanus* – produce the ocean! However, there are places where bits of ancient oceanic crust have survived because the crust has not been subducted, but instead obducted onto the continent. Secondly, much of the evidence for continental drift comes from palaeomagnetism, from which it is possible to determine the latitude but not the longitude at which a rock was formed. A sequence of such measurements through time gives us a means of tracking continental drift. This is still possible in ancient orogens unless metamorphism has destroyed the fossil magnetism. Thirdly, within orogens the original order of deposition of strata (that is, oldest at the bottom and the youngest at the top) may become inverted by either recumbent folding or thrusting. In the Alps, stratigraphic inversions were recognised by the nineteenth-century geologists using the fossil content. But the fossil assemblages which make it possible to determine a succession of strata did not exist before about 600 Ma, so this technique cannot be employed in the Precambrian orogens. Long before continental drift was accepted, palaeontologists puzzled over the fact that the Cambrian of Britain showed two distinct trilobite faunas, the Olenellids of NW Scotland and Paradoxides of Wales. The conclusion is that as we go back in time the standard techniques disappear one by one. Of course the lack of a timescale based on fossil assemblages can be countered by use of geochronological dating, but as we go back in time our geological ‘book’ loses more and more of its pages.

Going back further in time, the Grenville Province of NE Canada, which is found to extend southwards through the USA and into Mexico and South America, has been shown by Gower and others to record the accretion of crustal blocks and collision of continents in the late Mesoproterozoic. Even older collisional belts are identified: for example, St-Onge *et al.* (2006) interpreted the Palaeoproterozoic Trans-Hudson belt of Canada as a continent–continent collision zone with many similarities to the Himalaya. These and other examples from around the globe (e.g. the Irumide Belt, Limpopo Belt) provide reasonable grounds to presume that plate tectonics have operated for at least 2.0 Ga. The Hadean and Archaean, from 4.4 to 2.5 Ga, are much more problematic. Since the time of Sederholm (early 20th century), the metamorphic rocks of the Archaean have been regarded mostly as the deeply buried parts of orogenic belts now exposed at the surface as a result of erosion. Since the thickness of Archaean crust is much the same as present day continental crust outside orogens (*c.*35 km), the inference is made that there has been a great loss of cover. This is supported by the estimates of elevated pressures and temperatures identified in exposed rocks of Archaean age. Therefore a process of crustal thickening must be involved in Archaean tectonics, presumably a result of thrust stacking or folding similar to that which can be seen in the Himalaya or the Alps. Friend and Nutman (2005a) and others have argued that this was the case in the late Archaean assembly of the gneiss belts of west Greenland, which have distinct histories prior to their shared deformation events after 2700 Ma. However, the recognition of thrusts and estimation of the amount of lateral movement along them are greatly hampered by the general lack of stratigraphical information, especially way-up criteria, in many Archaean rocks.

Despite these difficulties, the conclusion is that the Dewey and Bird (1970) classification probably applies even in the Archaean, at least that part of it where continents and oceans existed and the plate tectonics model of subducting ocean floor can be applied. If Archaean oceans and ocean crust existed then they have all been eliminated, the only possible traces being the occurrences of Archaean ophiolites now incorporated into continental crust. Plate tectonics is the only process known that can eliminate oceans.

Before taking up specific problems it is useful to give a brief review of the main features of some orogenic belts in order to gain an acquaintance with them. The treatment here is on a large scale, therefore many interesting topics – for instance structural analysis, strain determination, metamorphic studies – are at best mentioned briefly. In the first part of the book, examples are confined to Phanerozoic orogens because, as mentioned above, the younger mountain belts offer a better chance of understanding evolutionary processes in orogenesis than the older deeply eroded belts.

Major features of orogens

Our aim is to examine the factors controlling the morphology and evolution of orogenic belts that are familiar features in the geological record from the Archaean. Characteristically orogenic belts are long linear or curvilinear features, as much as several thousand kilometres long, within which the continental lithosphere has undergone contraction to varying extents. Conspicuous examples are the Alpine–Himalayan belt and the Cordillera on the west coasts of North and South America.

Orogenic deformation

Calculations of orogenic strain by study on a small scale of deformed objects, such as deformed pebbles or fossils, and on a large scale using a procedure called retro-deformation (removal of the deformation structures in order to restore the rocks to their state prior to orogenesis), suggest that contractions vary greatly from belt to belt but can reach the order of 80%, meaning that the rocks in an orogen occupy only a fifth of their original horizontal dimensions.

Mechanisms of lithospheric thickening

The principal mechanisms of thickening are folding and/or thrust faulting. In the latter, thick bodies of rock are transported along a low angle fault, sometimes for distances of several hundred kilometres. The transport direction (or vergence) for thrusts leads us to another useful terminology, foreland and hinterland – the thrust vergence is usually towards the foreland and away from the hinterland of the orogenic belt. In simple terms, the contraction results in a thickened lithosphere and its topographic expression is high mountains.

One of the interesting aspects of orogeny is the mechanism whereby thickened lithosphere returns to normal thickness. The process is termed orogenic collapse and accounts for the well known fact that many older orogens are now at low elevations. However, an understanding of the process is made easier in recent orogens which show a better preserved record of the orogen, especially the early stages of collapse.

Orogenic metamorphism

An extremely important feature of many orogens is the heating of the rocks, commonly producing regional metamorphism, over a range of temperatures from about 300 °C to 750 °C and pressures ranging up to 10–12 kbar, equivalent to a depth of *c.*40 km. The resulting regional metamorphism, known as medium-pressure, medium P/T or 'Barrovian' metamorphism, is, as we shall see, typical of many orogens and as such provides an important constraint on how orogens may evolve and 'work' at depth. However, it is not only the metamorphic signature of orogeny. In many young orogens we also see mineralogical evidence for the burial of oceanic and continental rocks to depths of 100–150 km or more, producing eclogites and ultra-high-pressure rocks that in some instances contain microdiamonds as well as the high-pressure equivalent of quartz, coesite. The Triassic collision belts of China (Dabei Mountains, Shandong) and parts of the Caledonian in Norway provide spectacular examples of the extreme of metamorphism, which has also been identified in the Alps and Himalaya by Chopin (2003), O'Brien (2001) and others.

Orogeny may also produce metamorphic rocks that have experienced remarkably high temperatures of over 900 °C at depths of 20–40 km. These ultra-high-temperature (UHT) granulites are not only found in exhumed ancient belts like the *c.*2600–2500 Ma Archaean Napier Complex of Antarctica, the 100 Ma Eastern Ghats granulites of India and numerous 'pan African' (600–530 Ma) complexes preserved in the southern continents, but are also recognised in young and recent orogens. We can directly study the processes operating in the deeper parts of orogenic belts, where granulites appear to dominate, by looking at old deeply eroded belts such as the Caledonian in Scotland and Norway and the Grenvillian in Canada. However, geophysical techniques nowadays permit insights into the rheology and mode of deformation of the rocks, with an increasing weakening or softening as the temperature rises and metamorphic reactions take place. In addition to this the breakdown of hydrous minerals, through metamorphic reactions that take place as temperatures increase, results in the release of water. This provides another important control on rheology, for example by facilitating grain boundary migration of material or by raising fluid pressures in a column of rock; see Chapter 3.

Further reading

Bailey, E.B. (1935). *Tectonic Essays: Mainly Alpine*. Oxford: Clarendon Press.

Holdsworth, R.E., Buick, I.S. and Hand, M. (eds.) *Continental Reactivation and Reworking*. Geological Society of London, Special Publication, **184**, 57–75.

Le Pichon, X., Francheteau, J. and Bonnin, J. (1973). *Plate Tectonics: Developments in Geotectonics*, **6** Amsterdam, London, New York: Elsevier.

Moores, E.M. and Twiss, R J. (1995). *Tectonics*. New York: W.H. Freeman and Company.

Windley, B.F. (1995). *The Evolving Continents*, 3rd edition. New York: John Wiley and Sons.

2 Driving mechanisms for plates, slab retreat and advance, and a cause of orogenesis

In 1928 Arthur Holmes suggested that the mechanism for continental drift is cells of convection in the mantle. This was a remarkable insight, although many would now question the one-to-one connection between plate motion and mantle convection. So what is the modern view on the driving force for plate movements? There are two models in which the plates drive themselves. The first is called 'slab pull', which means that the dense ocean crust exerts a pull on the ocean floor during subduction as it plunges into hot asthenosphere. In contrast, the less dense continental crust is relatively buoyant. Sometimes the subducted slab of ocean crust breaks off and sinks into the hot asthenosphere, but if it survives it will exert a traction and in effect pull the ocean crust away from the Mid Ocean Rise. The opposite view is 'slab push', which means that the driving force for the moving ocean floor is situated at the Mid Ocean Rise which is opening under extension to allow in the new ocean crust.

Perhaps it should not be either/or here. Phillip England (1982) calculated the required stresses at the Mid Ocean Rise in the Indian Ocean if slab push were to be responsible for the northward movement of the Indian plate carrying the Indian continent. The forces acting on a plate boundary must do work against gravity during the raising of high mountains and plateaux. The force balance must take into account the Argand number, which expresses the relative magnitudes of the buoyancy forces arising from contrasts in crustal thickness and the forces required to deform the medium. England's results show that the horizontal stress arising from slab push is enough to explain not only the motion of the Indian plate before collision but also the continuation of motion after the India–Asia collision, with the result that India indents Asia, and a wave of deformation has spread across the Asian continent for over 2000 km north of the Himalaya. Using a viscous sheet model and a non-Newtonian rheology for the lithosphere, England showed that the elevated Mid Ocean Rises relative to the deep ocean basins provide a force of 2×10^{12} N per metre and the negative buoyancy of the slabs may provide about 2×10^{13} N per metre of trench. The estimated average driving force per unit length of subducting slab is 5×10^{12} N/m, enough to provide a reasonable balance for the forces resisting plate motion. The implication is that the continuing orogeny in southeast Asia requires the driving force given by 2500–5000 km of subducting slab or 12,000 km of Rise. These conditions are met in broad terms in the Indian plate, and so there is plenty of available force (slab push) to drive the Indian plate. The extensive slab system there is crucial in the Indo-Asian collision, in contrast to the smaller slab in the case of the Africa–Europe collision which was marked by slow collision and substantial strike-slip motion.

The above account firmly relates orogenesis to plate convergence, and there is a broad consensus now on that interpretation, but we should mention other models. For example, the compressional stress which generated orogens has been related to the intrusion of large batholiths which made space for themselves in the upper crust. Another suggestion is that orogens can form far from subduction zones, the compressional stress arising from the stress transmission from subduction zone far into the overlying plate. Some of the late Proterozoic orogens in central Australia have been explained by so-called far-field stresses (Dyksterhuis and Müller, 2007).

Retreat and advance of subduction zones

Royden (1993a,b) has discussed the expression of slab pull at convergent plate boundaries (Figs. 2.1, 2.2). She classified orogens according to whether the subduction zone retreated, i.e. the hinge of the down-bent slab moved away from the continent, or advanced, i.e. moved towards the continent. Orogens such as the Apennines or the Hellenides are in extension and are associated with retreating slabs, while the Alps or Himalaya, which are in compression, are associated with advancing slabs.

Advancing and retreating slabs are determined by different rates of subduction. If the overall convergence rate is less than the rate of subduction then the deformation of the overriding plate is extensional. Where the rate of plate convergence is greater than the rate of subduction the deformation in that plate is compressional. In addition, the hinge line retreat has been explained by several mechanisms: negative buoyancy of the downgoing plate with respect to the asthenosphere into which it sinks (Fig. 2.1a); pushing of the overriding plate on the hinge caused by its excess potential energy or by tectonic forces (Fig. 2.1b); or overall east-directed flow in the asthenosphere or a net westerly rotation of lithosphere which is related to the Earth's rotation and/or tidal effects (Fig. 2.1c).

Based on examples from Italy, Greece, the Alps and the Himalaya, Royden set out the differences between the orogens at the retreating and advancing boundaries, as given in Table 2.1. This table shows the rather neat way in which large-scale plate motions have had important consequences for the nature of orogens.

Schellart and Lister (2004) discussed four tectonic models for arc-shaped convergent zones and back arc basins. Of these four, three, including gravity collapse, extrusion tectonics and orogen-parallel, were discarded but one, slab rollback, was adopted. Although at retreating convergent zones, the hinge line of the subducting plate moves causing extension in the overriding plate, contraction continues near the trench owing to friction between the two plates.

Gravity differences as driving forces in orogeny

Platt and England (1993) have taken this topic further and discussed the problem of late stage extension and orogenic collapse, as seen in many orogens but best exemplified in the Tibetan Plateau where recent extension by normal faulting occurs despite the fact that plate

Table 2.1 Comparison of orogens at retreating (RSB) and advancing subducting boundaries (ASB)

	ASB	RSB
Topographic elevation	High (2–6 km)	Low (1–3 km)
Erosion	Great (<30 km)	Minimal (*c*.5–10 km)
Metamorphism	Medium to high grade	Low grade at most
Post-collisional convergence	Protracted (tens of Ma)	Short-lived (a few Ma)
Sedimentary facies in the foredeep basin	Molassic	Flysch
Compensation	Topographic	Dominated by subduction loads

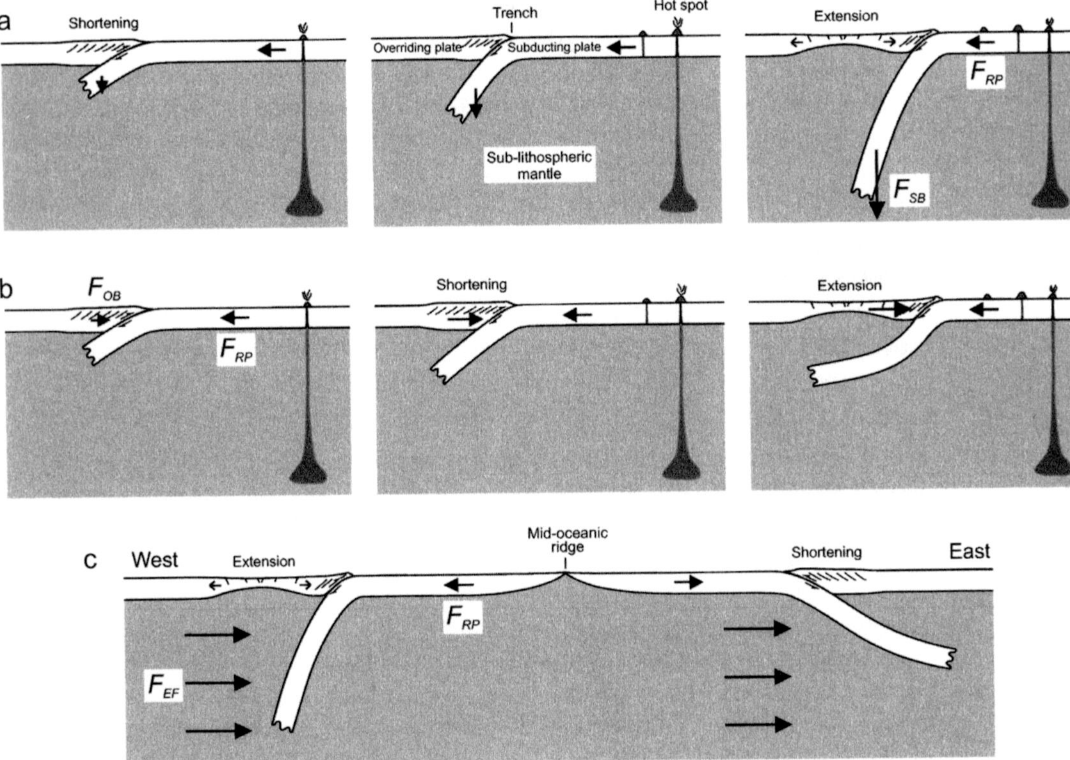

Figure 2.1 Three possible driving mechanisms for regressive hinge-line migration of a subducting slab. a, Migration due to sinking of the slab because of negative buoyancy of the slab. b, Migration due to push exerted from the overriding plate to the hinge line. c, Migration of west-dipping subduction zone due to global east-directed asthenospheric flow. F_{SB}, slab buoyancy force; F_{RP}, ridge push force; F_{OB}, overriding plate buoyancy force; F_{EF}, eastward asthenospheric flow force. From Schellert and Lister (2004), *GSA Special Paper* **383**, published by permission of the Geological Society of America.

convergence continues. In contrast to the recent extension in Tibet, recent deformation in the Himalaya is accommodated by compressional south-verging thrusts. From this contrast in the stress regimes in Tibet and Himalaya it is inferred that the horizontal stress arising from plate motion no longer supports the high plateau, or, more likely, that there have been

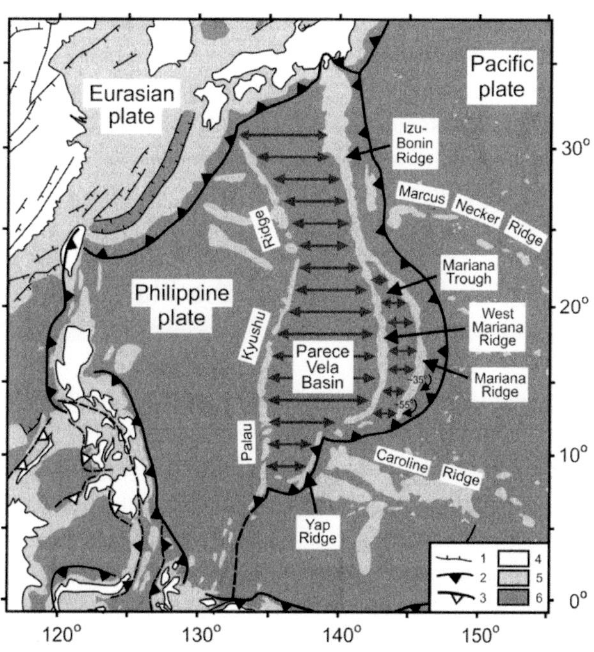

Figure 2.2 Tectonic setting of the Philippine Sea region with large E–W extension. The figure indicates the following features as shown in the key: 1 – normal fault; 2 – subduction zone; 3 – inactive subduction zone; 4 – land; 5 – continental shelf/topographic feature on ocean floor, 6 – ocean floor/back-arc basin. Curved arrows along the Mariana Ridge indicate rotation since the Late Eocene. From Schellert and Lister (2004) *GSA Special Paper* **383**, published by permission of the Geological Society of America.

density changes in the Tibetan lithosphere. Similarly in the Andes above an altitude of 3–4 km the dominant mode of deformation is normal faulting with, as in Tibet, thrusting at lower levels directed at high angles to the plate vector.

Platt and England argued that plate convergence is a result of the differences of gravitational potential energy between two columns of rock, one existing at Mid Ocean Rises and the other on the continents. The basis of the idea is that these columns achieve complete isostatic compensation (see Box 8.1 in Chapter 8) at some level. The column with the greater gravitational energy will exert a horizontal compressive deviatoric stress on the column with lower potential energy. Elevated Mid Ocean Rises, where the hot asthenosphere is near to the Rise crest, will be columns of high gravitational energy relative to orogens on continents where the thickening dense mantle root of orogens will partly or completely offset the buoyancy of the thickened crust. Sandiford and McLaren (2002) have shown that this gravitational state will evolve with time, as the thickened continental crust self-heats because of its inbuilt inventory of heat producing elements.

The important observation here is that whole lithosphere thicknesses are considered, not just crustal thickening. In consequence it shows that the thickening of the dense lithospheric mantle which may take place in an orogenic belt implies a *decrease* in gravitational potential energy. If, however, the mantle lithosphere is thinned, say by convective removal or delamination of the lower part of the mantle, called the thermal boundary layer, uplift results simply because of an increase in the gravitational potential energy of the orogen. This is an interesting model which has been advanced by England and Houseman (1988) and developed by Sandiford and Powell (1990) for variable proportions of crustal versus lithospheric thickening. The old concept, alluded to in Chapter 6, is that thickening of the crust will slow down and cease when the surface elevation is too high to be supported by the horizontal stress.

Applications of the Platt and England model

Applying this concept to the Himalayan–Tibet orogen, Platt and England considered that the crucial point is that the proposed change in Tibet from lithospheric thickening, under compressional stress, to extension and collapse can be attributed to an increase in gravitational potential resulting from the thinning of the lithospheric mantle. Under these circumstances the removal of dense lithosphere would trigger uplift. The extension is occurring because the enhanced potential energy of the Plateau is too great to be supported by the compressional stresses.

What is the time lag between the onset of compressive stress and the attainment of the large downward velocities that leads to loss of part of the dense mantle root? This depends on the Rayleigh number (a dimensionless number related to the coefficient of thermal expansion, gravity layer thickness and heat flow through the layer) of the convecting upper mantle. The time interval between the start of compression and mantle thinning decreases with increasing Rayleigh number. At low Rayleigh numbers the time may be 100 Ma, but with very high numbers the interval is only about 10 Ma. Furthermore, once convective removal reaches peak velocities it proceeds to completion in a short time.

There are important consequences to the sudden removal of the thermal boundary layer. Hot asthenosphere is brought into direct contact with the overlying crust and mantle. The result of this is the generation of melts which appear at the surface as lava flows. Another consequence is that a wave of temperature increase, enhanced by the transfer of mantle-derived melts into the crust and their crystallisation there, would move through the upper lithosphere giving rise to high-temperature (and perhaps UHT) granulite metamorphism. These points are further discussed below and in Chapter 5.

Platt and England applied their model to other regions for example the Basin and Range of the western United States (WUSA), Anatolia and the Betic cordillera in southern Spain. The Basin and Range region evolved from compressional tectonics in the late Mesozoic and early Cenozoic era to an extensional regime in the Cenozoic after the Eocene to the present. Extension was accompanied by considerable igneous activity during extension and also before it. Platt and England's contention that the evolution of the Basin and Range is related to mantle thinning, as in Tibet, has been contested. Thus an alternative mechanism to convective removal has been suggested, called slab steepening. Anatolia is another place where the Platt and England model may apply. There, a late Cenozoic volcanic province occurs in a belt running from eastern Turkey to Iran, 900 km long and <350 km wide. These volcanics were erupted at *c*.6 Ma after a period of crustal thickening.

In the Betic cordillera of southern Spain and North Africa, coeval (early Miocene) magmatism and extension have been recognised as late events in a collision type orogen with high grade metamorphic rocks which are dated as early Miocene and have been exhumed along high-temperature paths accompanied by appreciable loss of pressure, as manifested in the partial replacement of denser high-pressure metamorphic assemblages by lower-density but still high-temperature ones. These 'near isothermal decompression' P–T records, or ITD paths, provide important metamorphic constraints on the behaviour of the orogen and hence models for how orogens work. In the case of the Betics the complementary time information obtained from dating of minerals grown in the metamorphism and the cooling ages of micas in the schists shows that the exhumation took at most a few million years.

The Platt and England conclusion is controversial in two respects: firstly the postulated timescale for Tibet, especially the suggestion that a sharp uplift of Tibet was dated in the late Miocene or early Pliocene (8 Ma), and secondly the proposed link between lithospheric thinning and magmatism in Tibet, i.e. the sudden removal of the lower mantle lithosphere and the consequent heating of the lithosphere by hot asthenosphere. These points are pursued further in Chapter 5.

Finally, the England–Houseman (1988) model invoked a homogeneous thickening of the Tibetan lithosphere. This proposition is in conflict with suggestions that the crustal thickness of Tibet is due to the subduction of Indian lower crust and mantle lithosphere under Tibet. The subduction model involves no mantle thickening because Asian mantle has been pushed out of the way in order to make room for the Indian mantle. Manifestly there is much more work to be done on this topic, but that fact does not diminish the importance of the Platt and England model.

3 Physical and chemical principles: rock deformation, isostasy, geochronology and heat production in the lithosphere

The aim of this chapter is to set out some important physical and chemical properties which are highly relevant to the study of orogens and which are referred to in later chapters.

Rock deformation

The topic is covered in several excellent books, for example Ramsay and Huber (1983, 1987), Ramsay and Lisle (2000), or Jaeger and Cook (1976) (see Further Reading), and the discussion here is limited to giving a basic introduction to the subject.

The fact that some rocks have been deformed was probably first recognised by James Hutton towards the end of the eighteenth century, because at Siccar Point near Edinburgh he realised that some rock strata had been tilted so that the bedding was now dipping in a near vertical attitude. This was an amazing insight at the time: he realised that the strata had been rotated from their horizontal attitude on the sea-bed.

It was in the nineteenth century that Swiss geologists presented evidence in the form of folds and thrusts, clearly visible on mountain sides, that rocks had been subjected to compression, as well as evidence that rocks had been stretched. Faced with this evidence which has been confirmed in many parts of the world, we are forced to consider the strengths of rocks in orogenic belts. If you sit by a river and watch the intricate patterns of air bubbles that result from the complex flow patterns of the river then you get a feeling for what happens when rocks lose their strength and flow.

Folds result from shortening or contraction of the crust, and to an extent their shape shows the degree of shortening. Thus open folds are the result of minor shortening whereas tight folds result from intense shortening. Another common small-scale structure is boudinage (named because they look like boudins, a Belgium sausage) which results from extension ('necking') of rock layers which are being pulled apart. Other ductile structures include C-S fabrics and ductile shear zones. These and many others are small-scale structures that can be observed at outcrop in an orogen. They can be systematically mapped and the large-scale structures can gradually be elucidated. When that is done, it is often noted that the small structures and the large structures share similar geometries and kinematics.

Folds mostly result from a deformation which has maintained the coherence of the rock. Such a deformation is termed ductile. In contrast, brittle deformation involves loss of cohesion. Faulting is the usual expression of brittle deformation. There are three types: normal faults, which result from extensional deformation; thrust faults, which result from contractional deformation; and strike-slip or wrench faults, which are contractional with horizontal shortening but, unlike thrusts, have a vertical intermediate stress axis.

Coulomb fracture theory

Brittle behaviour is perhaps easier to understand than ductile. A theory which deals with the brittle behaviour of materials is termed the Coulomb fracture criterion, which recognised that the shear stress tending to cause failure across a plane is resisted by cohesion. Shear stress is related to a term which is the inherent strength of the rock and also a frictional term (the coefficient of internal friction). Mohr diagrams are used to depict the strain of a rock undergoing brittle failure. In addition, the Griffith criterion states that fracture is caused by stress-concentrations at the tips of minute cracks ('Griffith cracks') which pervade the rock. There is a wealth of theoretical and experimental work based on these criteria.

Stress and strain

The term deformation stress is used to refer to the force applied to a body per unit area, and strain refers to the change of shape or volume of a body resulting from the application of stress. It is easier for the geologist to work on strain simply because in older inactive orogens the stress cannot usually be measured in rocks. The deficiency can, however, sometimes be accommodated by using data from experimental rock deformation.

Strain can be illustrated by the strain or deformation ellipsoid which is derived from an original sphere of the same volume. The three principal axes (*X*, *Y*, *Z*) of the ellipsoid define directions of shortening and extension. In pure shear or coaxial strain, the principal axes do not rotate, whereas in simple shear or non-coaxial strain, they do. Three-dimensional strain can be pictured by the shapes of end-member ellipsoids. A prolate ellipsoid, like a cigar, has shortening of two principal axes (*Y*, *Z*) and elongation of the third (*X*) indicating all-round squeezing or constriction. An oblate ellipsoid (like a squashed orange) has one axis (*Z*) along which shortening has taken place and two (*Y*, *X*) along which elongation has occurred. This is the process of flattening. In between is an ellipsoid where one axis (*Y*) has neither shortened nor lengthened while the other two have changed length. The depiction of strain in terms of ellipsoids is well expressed by plotting axial ratios of the ellipsoid on a Flinn plot. Ellipsoids are assigned a *k* value between 0 and infinity, defined by axial ratios of the ellipsoids: k = 0 is the oblate ellipsoid, *k* = infinity is the prolate ellipsoid and *k* = 1 is the ellipsoid in which one axis has not changed in dimension. This is called plane strain.

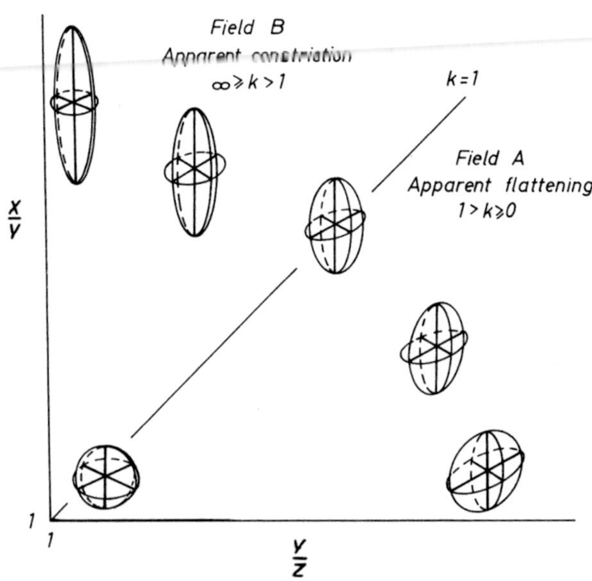

Figure 3.1 Flinn plot or diagram showing the shape of any strain ellipsoid derived by the deformation of an original sphere. There are three lines ($k=0$, $k=1$, $k=$infinity) and two fields ($0<k<1$ and $1<k<$infinity) representing finite strain ellipsoids for constant-volume deformation. Modified from Ramsay and Huber (1983), published with permission from Elsevier.

S- and L-tectonites

The existence of different strain ellipsoids in rocks is detectable in the field. L-tectonites are deformed rocks formed in the constrictional field of the strain ellipsoid; thus the ellipsoid is prolate, and accordingly the rock fabric is dominated by lineation, especially a stretching lineation. In S–L tectonites, lineation is still prominent but in addition foliation is strongly developed, and the strain ellipsoid tends towards oblateness. In S-tectonites, foliation is dominant and the strain ellipsoid approximates to an oblate ellipsoid.

The Flinn plot is an excellent way of visualising strain in three dimensions (Fig. 3.1). Firstly, it is possible to depict strain paths, that is, the succession of ellipsoids or strain states, produced during the transformation of the original sphere into an ellipsoid. This is incremental strain. It is easy to think of strain paths as straight lines, but in nature it is probably more realistic to imagine that they are curved. In rock deformation we only see the end product; nevertheless, the concept of strain paths is a valuable aid to thinking about strain. Secondly, to convert, as it were, the ellipsoid into rocks means considering the changes in length of planes such as bedding planes in sedimentary rocks or foliations in metamorphic rocks, or similar changes of the length of linear features in rocks such as lineations of fold axes. The deformation ellipsoids give us a framework with the depiction of fields of shortening or extension for planes and lines. Strain is progressive, and a material plane or line will rotate as the ellipsoid develops. Thus a plane may undergo initial shortening but as a result of rotation it may end up being elongated, which may be expressed as the unrolling of the folds.

Rotational and non-rotational strain

As mentioned above, the important difference between progressive pure shear and progressive simple shear lies in the behaviour of the principal axes of the strain ellipsoid. The principal axes of strain do not rotate in pure shear but they do in simple shear, hence the terms rotational and non-rotational strains. The difference in the behaviour of the principal axes is referred to as the vorticity of the deformation, which is a measure of the average rate of rotation of material lines of all orientations about each coordinate axis. The vorticity is zero for pure shear and non-zero for simple shear. In progressive pure shear the principal axes of finite strain are always parallel to the principal axes of incremental strain. The opposite is true of simple shear. In progressive pure shear all material lines and planes rotate toward parallelism with the principal axes of strain.

Three-dimensional strain

Now we will consider what happens in three dimensions when stress acts on a point in a rock. In the theoretical treatments, stress can be represented as maximum, intermediate and minimum stresses, labelled like this:

$$\sigma_1 \geq \sigma_2 \geq \sigma_3$$

These principal stresses are the surface stresses acting on three mutually perpendicular planes through a point which can be represented as an infinitesimal cube. In relation to the planes, then, these are normal stresses. The faces of the cube are parallel to the principal planes and perpendicular to the principal axes of stress. The deviatoric stress can be defined by subtracting the mean normal stress from each of the normal stress components in the three- or two-dimensional stress tensor. So-called hydrostatic pressure occurs where $\sigma_1 = \sigma_2 = \sigma_3 = p$. In this case all the principal stresses are compressive and equal. No shear stress exists on any plane. Shear stress, as opposed to normal stress, acts as opposite tractions on a surface, producing a shear couple.

Faults are classified according to the orientation of the stress axes. Thus for normal faults σ_2 and σ_3 are horizontal and σ_1 is vertical. For strike-slip faults: σ_1 and σ_3 are horizontal and σ_2 is vertical. For thrust faults: σ_1 and σ_2 are horizontal and σ_3 is vertical.

Rheology

Rheology is the study of flow in general, and several rheological models have been used to describe the behaviour of materials undergoing stress and also the stress–strain history. There is plastic deformation involving a yield point where elastic or impermanent strain gives way to permanent strain and ultimately to rupture. In addition there are visco-elastic,

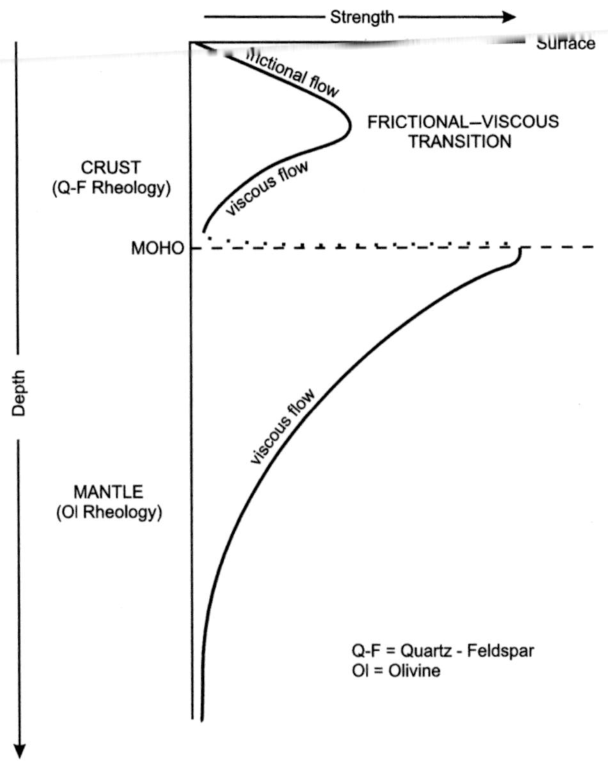

Figure 3.2 Strength profile for the lithosphere showing a strong upper crust and mantle and weak lower/middle crust.

Bingham and Newtonian viscosities. A Newtonian substance is said to be perfectly viscous and demonstrates linear viscosity, in which the flow rate is proportional to the applied stress. A more complicated model is one of non-linear viscosity, for example Bingham rheologies. These different rheological bodies can be described by different plots of stress and strain. A perfectly plastic body shows a yield stress separating fields of elastic from plastic or permanent strain. In linear Newtonian behaviour the plot is linear. Non-Newtonian viscosity is non-linear: the rate of strain is a function of stress but is not proportional to it. In structural geology, identification of these different viscosities is elusive, but experimental work has shown the relevance to natural rock deformation.

The factors which determine whether deformation is brittle or ductile are discussed later. For present purposes suffice it to say that temperature is important in the sense that as a rock is heated in the crust it becomes weaker. There is a loss of strength until on melting it changes from the solid to the liquid state.

A popular model ('jelly sandwich' model; see Chapter 10) for lithospheric rheology has a strong upper crust with dominantly frictional flow expressed by brittle fractures (Fig. 3.2). At deeper levels viscous flow predominates with thermally activated deformation mechanisms including crystal plasticity and diffusion creep. These processes control the ductile behaviour of the deeper levels, with a loss of strength and weakening. In the mid crust is a frictional–viscous regime and a transition from brittle to ductile deformation (or frictional to viscous, as some would term it). In this model the strong upper mantle

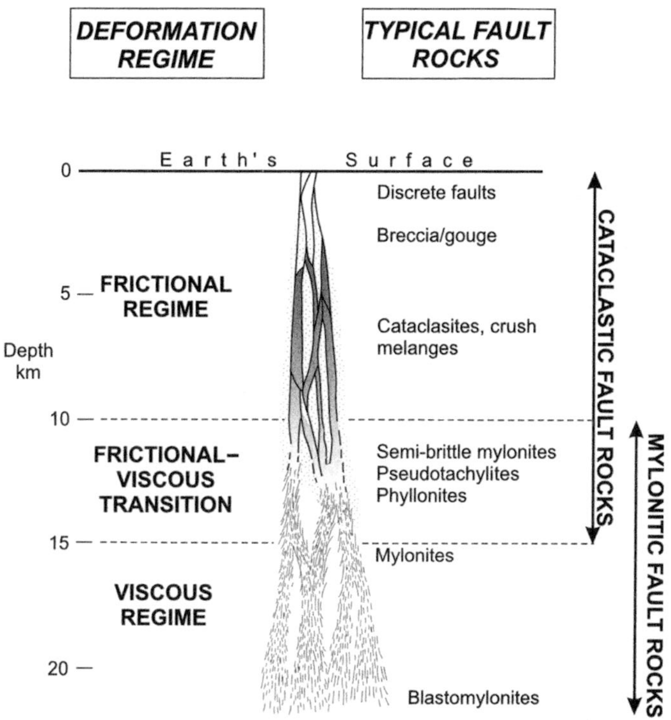

Figure 3.3 Schematic diagram to show transition from frictional to viscous or brittle to ductile as depth increases. The consequent change in the style of faulting from brittle in shallow depths to a ductile shear zone at depth where slip is probably aseismic. Modified from Figure 1 of Holdsworth *et al.* (2001), published by permission of the Geological Society of London.

represents the main load-bearing region of the lithosphere, which means that large-scale deformation is mainly determined by ductile shear in the olivine-rich upper mantle.

At this point, we should pause to define a few terms. In orogenic belts the rocks have reacted to stress in quite different ways, and we can recognise that some rocks are strong and others weak under the same physical conditions. Some materials are fragile, like glass, and the readiness of a material to fracture or rupture is called the brittle state. In orogenic processes there are transformations or changes of the physical conditions, and in this context the question is – what happens to rheological behaviour if the physical conditions change? This is not too difficult to picture, because we are familiar with the way sealing wax loses its brittleness when it is heated. Heated wax flows. So here is a clue to the behaviour of rocks, namely loss of strength caused by a rise of temperature. For this rheology we need another term, and it is ductility. Ductile deformation involves considerable deformation or flow without loss of cohesion. This is in complete contrast to brittle behaviour, in which there is loss of cohesion – glass or sealing wax will fracture as will rocks if we hit them with a hammer.

The ability of rocks to undergo large-scale flow or deformation without loss of cohesion is made clear at any outcrop of rocks in an orogenic belt by, for example, ductile folding or stretched pebbles or crystals. How is it achieved? The answer is simple: those rocks at outcrop were once deep down, which means, firstly, that temperatures were higher than at the Earth's surface, and, secondly, that the rocks were under pressure from the pile of rocks above them; this is called the confining pressure. The expectation is that going down into the Earth's crust we would encounter brittle rocks near the surface, a sort of brittle 'lid' to the crust, but on going deeper we would come across rocks which were ductile. Between the brittle and ductile regimes there is a frictional–viscous transition or brittle–ductile transition (Fig. 3.3). This is a very important concept.

The terms brittle and ductile in the sense defined here are dependent on scale. That is, we can have brittleness on a small scale but ductility on a larger scale. To picture this, take a piece of paper and make cuts in it, but be careful not to extend the cuts to the edges of the paper. Putting this into a geological context we have cuts (faults) on one scale but overall, on the largest scale, the whole piece of paper, we have maintained coherence, and that is ductility.

Controls of deformation

In discussing the different responses of rocks to stress we need to set the range of temperature and pressures that are likely in the crust. From experiments it is known that temperatures in orogenic belts are in a range from 0 to *c*.1000 °C. The increase of temperature downwards in the crust varies widely, the geothermal gradient being greater in hot places like Mid Ocean Rises or in active orogens. A temperature gradient of 25 °C per km, which is an acceptable value in an orogen, means that at the base of crust of normal thickness we would expect a temperature of several hundred degrees, but in the thickened crust of orogenic belts obviously much higher temperatures would occur in the deeper levels.

An important change in rheological behaviour starts to take place at *c*.350 to 400 °C. It is then that rocks pass through the brittle–ductile transition: they become softer and more and more ductile. The reason for this is a change in the mechanisms by which the rocks deform. The ductile mechanism involved on a microscopic scale is called crystal plasticity, which is a general term for a variety of mechanisms by which crystal aggregates – rocks – undergo deformation. It can involve change of shape of a crystal by internal gliding on slip planes, or it can involve movement of atoms by diffusion processes.

Temperature

Rising temperature promotes mineral reactions (metamorphism) which involve diffusion and mass transfer of material resulting in the growth of new minerals. Usually under these conditions rocks lose strength and undergo ductile flow rather than brittle fracture. Weakening is further enhanced at the upper temperature limits of Barrovian metamorphism (see Chapter 7) because of the partial melting of the rock mass.

Confining pressure

Another major control on rock deformation is the so-called confining pressure which increases with depth in the Earth and is measured in bars or megapascals (MPa). A rock at high confining pressure is stronger than one at low confining pressure. Thus confining pressures appear to counteract the effect of rising temperature, but this is a two-horse race and temperature wins. So rocks become softer and more ductile with increasing depth. Increases in confining pressure due to deep burial of rocks, as happens for example in subduction zones, may result in so-called phase changes. This means that the increased depth promotes the growth of denser minerals such as the high-density phase of quartz or the garnet–pyroxene mineral assemblage in high-density rocks such as eclogite. Changes of pressure may cause rheological changes in rocks as crystal lattices are reconstituted. In Chapter 10 we discuss the transformation of granulite facies rocks into denser eclogites, a change that has profound consequences for the evolving orogen.

Strain rate

A third important control in rock deformation is strain rate. In geology we are used to having plenty of time, but this does not mean that given enough time we can do anything. If a man leant against the Houses of Parliament for a thousand years he would not bring them down (at least, there is a very small chance that he would). But the time factor, meaning whether we have fast or slow strain rates, is extremely important in inducing ductile behaviour. 'Silly putty' demonstrates this beautifully. Roll it into a ball and leave it, and it will flow continuously; after about half an hour it will be a flat disc. This is long-term strain, which is called creep and is permanent strain without loss of cohesion. Solid rocks will creep given enough time even at low temperatures, as bent gravestones show. However if we hit the silly putty with a hammer it will fracture like glass: it is now a brittle substance because the strain rate is much faster. What else can we do with silly putty? Pull it out slowly and it flows plastically like hot toffee, the strain rate being intermediate between those of the first two experiments. Lastly roll it into a ball again and it bounces wonderfully: that is elastic deformation.

The classic rock deformation experiments that demonstrated the importance of strain rate in geological processes were carried out in California by Hugh Heard in the 1960s. His data show that marble loses strength as the strain rate decreases. Strain rate is written as so much strain per second, thus: 10^{-14}/s. Supposing we shorten a block of rock by 10% of its length in 1 second. This can be written as 10^{-1} (i.e. one tenth) /s. Now suppose that we change to a slower strain rate, say 10^{-8}, so that 10% shortening takes much longer, indeed millions of seconds. Realistic geological strain rates come out to be between 10^{-11} and 10^{-14}/s. The latter figure is the rate for the San Andreas fault.

Is creep at slow strain rates a reasonable proposition for processes in orogens? Orogenic belts evolve over millions of years; for instance, the development of the Himalaya started about 50 Ma ago and is still active. So we can be sure that long-term deformation creep is taking place at slow rates and is a viable process in geology.

Fluids in rock deformation

Another important controlling factor in rock deformation is the presence or absence of fluids (and, at the highest temperatures attained, melts) in rocks. Fluids, water for the most part, are common in the Earth's crust even down to deep levels. We know this from a variety of lines of evidence. Most people are familiar with the water table in the upper 1 to 2 km of the crust, below which there is saturated rock. However, the interest here is in evidence for water at deeper levels. The evidence comes from the presence of water in deep bore holes, going down as much as 12 km, also from geophysical data. The fluid, usually water, may have been trapped in sediment at the time of deposition or it may arise from certain metamorphic reactions which involve the loss of water from crystals (so-called dehydration or, more generally, devolatilisation reactions).

Geophysicists have used seismic activity in order to detect water reservoirs at depth. They also identify zones in the crust where electrical conductivity is low, probably owing to layers or pockets of water. Such zones have been found at depths of 10 to 20 km. Then there is evidence from oxygen isotopes in rocks which are taken as indicators of water in the rocks at depth. Finally, studies of metamorphic rocks show that fluids have an important role in mineral reactions which occur at depths of more than 20 km and in lower-temperature thermal regimes significantly deeper. Progressive metamorphism may involve dehydration reactions and thus the release of water into rocks; we will retun to this process later, as it is significant to the evolution of the mineral assemblage of metamorphic rocks in orogens. Furthermore, many metamorphic rocks from lower-grade slates and phyllites to high-pressure blue schist may contain veins in which typically metamorphic minerals such as garnet, kyanite and even omphacite have crystallised. This evidence makes the case for a significant presence of water in the crust.

Permeability and non-permeability of rocks

Water in the crust is mainly lodged in the spaces between grains and crystals, called pore spaces (intergranular fluid). It is also lodged within grains, for example in micro-cracks (intragranular fluid). The volume of pore spaces and cracks determines the water-carrying capacity of rocks: this is referred to as the porosity. Porosity is high in newly deposited sands but it gradually decreases as they are compacted into sandstones. The other point we should consider is the circulation of water in the crust. The term for this is permeability. Some rocks, sandstones for example, are very permeable and the passage of water through them is not difficult, whereas other rocks such as shales tend to be impermeable and therefore serve as seals for reservoirs. It is the interlayering of sandstones and impermeable shales that allows the formation of reservoirs in the crust. Water flow rates in the crust are governed by Darcy's law, which states that the volume of fluid crossing a unit area per unit time is proportional to the pore pressure gradient at that point, the viscosity of the fluid and a constant which depends on the medium and is independent of the fluid.

How do these features relate to rock deformation? Let us consider first the significance of a film of water surrounding a grain. The presence of this film assists deformation simply because it serves as a lubricant in the rotation of grains under stress. But the aqueous film has a more general role, because it aids the movement or diffusion along grain boundaries of material. Diffusion operates on an ionic or atomic scale and is insignificant at surface temperatures. However, diffusion is a significant process at temperatures above 200–300 °C and indeed even lower-temperature diffusion, so-called pressure solution, is a very important mechanism in rock deformation. Pressure solution was first recognised by Sorby in pitted pebbles in conglomerates; it is also seen in stylolites in carbonates where volume loss has occurred by solution along prominent wavy surfaces. We return to this later in this chapter.

Another factor is the water present within grains. This also weakens the grain and makes it more readily deformable. This was recognised very clearly in the example of the behaviour of quartz, silica dioxide, which is an important constituent of the continental crust. In laboratory experiments dry quartz was found to be very strong, needing high stresses to make it flow. This was completely contrary to the experience of the rheological behaviour of quartz in nature. When water was added to the quartz, however, it showed a dramatic loss of strength, a phenomenon which was called 'water or hydrolytic weakening'.

Enhanced pore fluid pressure

Lastly, water influences rock deformation in another way: by enhanced pore fluid pressure. Fluids are found in many rocks and at many levels in the crust, well below the water table, and exist in whatever spaces, such as pore spaces between grains or in micro-fractures, are available. In order to create a buoyancy effect, however, something else is required: so-called fluid pressure or, more precisely, enhanced pore fluid pressure. Water occupies spaces between grains in a sandstone; if the spaces shrink in size during compaction or by tectonic compression, the trapped water will exert a pressure on the sand grains. This is the pore fluid pressure. Since water has a lower density than rock, the effect of fluid pressure is to create a buoyant state. The essential requirement for the buoyant state is that fluid pressures must be high and approaching the value of the normal stress or lithostatic stress.

Effective stress

A column of rock exerts a stress on its base, and this is called the lithostatic stress ('litho' meaning rock). Now consider the case where the rock column contains fluids. The fluids can be of several origins: meteoric, from melts, from dehydration reactions in metamorphism and from aquathermal pressuring (pressure by hot fluids). Fluids may also come up into the crust from a subduction zone. We now have to deal with two stresses: the load on the base exerted by the rock and the load exerted by the water. The term 'Effective stress' was coined to recognise the fact that water, being less dense than the rock, will tend to counteract the load of the rock. The normal stress or load stress exerted by the rock column on its base is reduced because of the pore fluid pressure. In other words, the water acts as a buoyancy device. Effective stress means that it is necessary to rewrite the normal stress as the lithostatic pressure minus the pore fluid pressure. This is termed $\underline{\sigma}_{\mathrm{n}}$.

The water must not escape from the column but must be trapped in the pore spaces. In this way the system is closed. A seal preventing the passage of fluids may also be provided by an impermeable layer such as shale or by gouge in a fault zone. But the effectiveness of the seal may be limited if the gouge is impermeable at one stage but has fracture permeability at another. Thus the maintenance of high pore fluid pressure is likely to be intermittent.

The enhanced fluid pressures reduce the stresses acting on a rock, hence the effective stress is given by $S_{\mathrm{v}} - p$, S_{v} being the vertical stress and p the pore fluid pressure. Actually this should be written as $S_{\mathrm{v}} - p(1 - a)$ where a is a bulk modulus term. Enhanced pore fluid pressures are denoted by a ratio $\lambda = p/S_{\mathrm{v}}$, where λ can reach about 0.9. If so, the rock sheet is buoyant. From pore pressures measured in boreholes, it is known that the condition where pore pressure and hydrostatic stress are about equal does exist in nature. We will take this conclusion a little further in discussing thrusts, but for the moment we can say that in various ways the presence of water in the crust promotes what we might call the deformability of rocks.

Fluids in earthquakes and in pathways in the crust

Other examples of fluids in rock deformation can be mentioned. The idea of thrust faults as pathways for fluid flow has been proposed by several workers for the Canadian Rocky Mountains and for the famous Glarus thrust in the Swiss Alps. The Glarus thrust is a pathway simply because it is a fracture plane, and the fluids are generated by metamorphic reactions taking place in the deeper parts of the Alps.

Fluids are also important in fracturing and earthquakes. As explained above, high fluid content in a rock increases its buoyancy and reduces the vertical stress. The term effective stress is used to take note of the way fluid pressures modify the stresses acting on rocks. The lower effective stress lowers the strength and thereby promotes fracture.

A rather good illustration of the importance of rock fluids is shown by their role in rock fracture and earthquakes. The connection is termed hydraulic fracture, which essentially relates to effective stress mentioned earlier.

Failure occurs when $\sigma_3 - p = -T$, where σ_3 is the least compressive stress, p the pore fluid pressure and T the tensile strength of the rock. In igneous intrusions, this relation controls the opening of fractures in order to accommodate the intrusion of dykes and sills. The molten igneous rocks exert a magma pressure which promotes fracture. There is also a link between water content, tensile fracture and the deposition of mineral veins and lodes.

So we have many examples of hydrothermal deposition associated with intermittent fluid activity around faults. The cyclic nature of fracture opening and mineral deposition is well shown by the so-called crack seal mechanism: fractures open under the control of effective stress and then are sealed by the deposition of minerals.

Seismic pumping and valves

Two models apply these principles to earthquake activity. The first model is 'seismic pumping' (Sibson, 1977), in which stress drives the process which consists of the following sequence of events:

1. Stresses rise in the fault zone, resulting in the dilation of the rocks, and because open spaces form in the rocks, fluid pressure drops.
2. The result is tensile cracking with fractures orientated perpendicular to the minimum stress direction. If the stress system results in horizontal tensile fractures then the surface expression of this phase is a bulge.
3. Fluids migrate into the fault zone and fluid pressures rise.
4. Frictional resistance is lowered and stress is relieved by seismic failure and slip on the fault, with a consequent earthquake.
5. The cracks relax and fluids are expelled from the fault zone, sometimes in large amounts.

The second model (see Sleep & Blanpied, 1992) is one in which fluids drive the faulting. Between earthquakes, ductile creep in the fault zone results in the compaction of the zone, thereby increasing the fluid pressure and finally allowing frictional failure. A seal to the fault zone is essential because it allows at most only a small amount of fluid flow between the fault and country rock. In other words, the fault zone is isolated, thereby allowing effective stresses to operate in it. The seal is created by deposition of material from the pore water existing in cracks or pore spaces. This is a sealing of fluid pathways. The seal is not broken in the country rock. The earthquake restores porosity and fluid pressure falls below a hydrostatic level. Within the fault zone, freely circulating fluids deposit a cement between the grains. Therefore the fluid pressure rises again and the effective stresses are lowered. The process is cyclical, and if we knew the time involved then earthquake prediction would be much easier. Diffusive mass transfer (see below) creates a pressure seal at the edges of the pressure zone. The time taken for a single pore pressure cycle may be of the order of 10^4–10^6 years.

Mechanisms of rock deformation

At low temperatures, cataclasis is an important mechanism. This refers to the displacement of the grains relative to one another, leading to mechanical granulation and fracturing. Rock types such as gouge, a finely comminuted rock, or breccias occur on faults as a result of frictional sliding on the fault plane. Gouge is a response to cataclasis and the wearing away during fault slip of uneven parts of the fault called asperities. The direction of slip is shown by slickenside lines (i.e. striations on fault planes) and by steps or chatter marks on the fault.

Turning to mechanisms of ductile flow in rocks, it is useful to think of the process as a softening or weakening of rocks, and to an extent it is brought about by the interplay of deformation and metamorphism. The mechanisms are briefly reviewed as follows.

Transformation and reaction enhanced ductility

This is brought about by:

(1) Phase changes in the mineral assemblages, in which the breaking of atomic bonds results in a loss of strength. The phase change from alpha to beta quartz sets up internal stresses in the crystal lattice because of a volume change in the crystal. This lowers the yield stress and therefore increases ductility. This phase change is especially significant in the mantle where high pressures prevail, but as noted in Chapter 10, phase changes are also likely in the lower crust.

(2) Mineral reactions: for example a change in the mineral assemblage during retrograde metamorphism, as in the change from a gneiss to a schist or mylonite. The change involves diffusion and recrystallisation, and this promotes ductility. In addition, fluids in rocks promote diffusion. Water at grain boundaries weakens rocks and increases ductility. Polar interaction between water and the grain surface leads to reduced surface energy or reduced interfacial tension. Thus work hardening due to pile up of dislocations at crystal faces is reduced because there is an increased probability that dislocations will emerge onto the crystal surface. Also, water lowers the temperature at which grain boundary diffusion occurs, as in pressure solution. Dehydration reactions in serpentine result in enbrittlement.

(3) Chemical changes in rocks: these may lead to softening, as in ductile shear belts where water enters the mineral assemblage owing to retrograde metamorphic changes.

(4) Grain size reduction during deformation, such as occurs in the formation of mylonites: this increases diffusion in rocks and increases grain boundary sliding. Diffusion is related to grain size as in the expression Diffusion $\propto 1/d^2$ or d^3, where d is the grain size. Therefore as the grain size is reduced the diffusion rate increases.

These mechanisms are promoters of softening. But if the conditions change so that mineral reactions cease or grains are coarsened by recrystallisation, then softening may be replaced by work hardening which means a slowing of the strain rate.

Crystal plasticity

This may be within grains, intracrystalline, or outside them, intercrystalline plasticity. It may also involve diffusive mass transfer and superplasticity. The latter term refers to enhanced ductility due to grain boundary rotation in an aggregate. Intracrystalline plasticity involves gliding on lattice planes within crystals. Lattice planes of potential gliding are found in many common minerals – quartz, calcite, dolomite, olivine, hornblende – and so this type of plasticity is a very important deformation mechanism. From a knowledge of the crystal structure it is expected that potential glide planes will be those of atomic weakness. As a result of the operation of glide planes controlled by the stress system, mineral assemblages acquire a preferred crystallographic orientation. For example, the c-axis of quartz in deformed rocks shows patterns which may be related to stress or strain. Symmetry of the microstructure has been used to distinguish between pure and simple shear in deformed rocks. In Chapter 5, for example, when describing the evolution of the Himalaya, we refer to quartz microstructural studies which are held to support the extrusion mechanism in the Greater Himalayan Sequence. Twinning planes in calcite are also likely glide planes under stress conditions and have been used to determine shortening and extension directions in marbles.

Slip on glide planes is temperature controlled, and low- and high-temperature gliding can be recognised. For example, glide on calcite twins tends to be a low-temperature phenomenon but glide on prism faces in quartz occurs at higher temperatures. The slip on the glide planes is controlled by the movement of dislocations, tiny defects in crystals which under stress move and create the glide plane. Dislocation theory tells us that at low temperatures dislocation movement is by dislocation glide, in which the dislocation moves through the crystal in the same way as a ruck can be propagated across a rug. At higher temperatures other mechanisms are triggered, called dislocation climb. Log jams can occur in the flow of dislocations, like traffic jams, and this constitutes a work hardening, a slowing of the strain rate.

Diffusive mass transfer

Diffusive mass transfer (DMT) is the diffusional transport of matter, usually at a high temperature. The transfer can occur between the surfaces of crystals which are differently orientated with respect to the applied stress. Stress gradients may control diffusional flow from highly stressed to less-stressed crystal faces. This is stress-induced recrystallisation. In rocks we find pressure shadows in which crystals have grown around a resistant grain but only on the low-stress surfaces, hence the name pressure shadow.

A low-temperature type of DMT is called pressure solution, as we mentioned before. This is the diffusion of solute ions along grain boundaries in an aqueous intergranular film. Pressure solution is the cause of indentations observable on the surfaces of pebbles which have been pressed closer together by tectonic stress or compaction. Other pressure solution features are dissolved fossils and stylolites. Higher-temperature diffusion processes like Coble and Nabarro–Herring creep are important in crustal deformation. Nabarro–Herring creep involves lattice diffusion while Coble creep, like pressure solution, is a grain boundary diffusion but taking place at elevated temperatures.

From the preceding account it is clear that the mechanisms of rock deformation are complicated. Because most of them involve metamorphic reactions in addition to deformation, there is an intricate interplay of these controls. For that reason much of the earlier work on rock deformation dealt with the rather more tractable brittle state, fracturing and faulting.

Isostasy

The concept of isostasy (from the Greek ‘isostasis’ meaning in equipoise) provides an explanation for the support of mountains and plateaux. It is generally assumed that the excess volume in a mountain chain is compensated by lighter crustal material at depth;

otherwise either large shear stresses would be required on vertical planes in the crust or it is necessary to postulate higher pressure at depths below mountain ranges compared with adjacent lower terrains. These possibilities can be ruled out because the Earth is not strong enough to maintain the necessary shear stresses or the pressure differences over quite long geological timescales. Therefore excess mass above sea level must determine the magnitude of the force per unit length that a plateau and the surrounding lowlands must apply to one another in order to maintain the plateau at its current height.

There are two models for isostasy. The first, the Airey model, postulates that compensation for columns of rock with differing densities occurs at different depths. Thus mountains are like blocks of wood floating in water: the tallest block stands higher and goes deeper than the adjacent blocks. The second model is the Pratt model, which postulates compensation at the same level in the Earth for columns of differing density. Pratt isostasy suggests that the base of the upper layer in the lithosphere is at a constant depth and that isostatic equilibrium is achieved by allowing the upper layer to be composed of columns of constant density. In this model compensation is achieved because mountains consist of, and are underlain by, material of low density.

Geochronology and thermochronology

As noted in previous chapters, the tectonic processes leading to and involved in orogeny take time to occur and leave their imprints on the rocks affected by them. The crust and lithosphere take time to deform and undergo metamorphism, with the timescales dependent on the rates of heat transfer, plate movement and accommodation of strain, amongst other things, which may vary considerably.

Understanding the processes driving orogeny requires an intimate knowledge of the timing and timescales of the events recorded in mountain belts, from the burial of sediments to great depths during construction of the orogen, through deformation and metamorphism during orogen maturation, to the exhumation of deep-level crust and the demise of the orogen. Questions such as ‘How quickly were the materials present in the orogen accreted, assembled and taken to depth?’, ‘How long did the rocks stay hot?’, ‘What was the duration of deformation associated with thickening or thinning?’, and ‘How rapid was exhumation to expose deeper-level rocks?’ are all central to evaluating how an orogen ‘works’, and indeed to all modelling of orogen development.

Geochronology, the age dating of rocks and minerals, is therefore of singular importance to the study of mountains and their making, as it provides the age data that constrain timescales and hence rates of processes. It is also, when coupled with isotope geochemistry, critical in revealing the identity, or provenance, of the materials contributing to and caught up within a mountain belt. Collision generally involves the joining together of continental blocks that were previously distinct and may have had rather different geological histories and constituent rock units. Discrete crustal blocks, or terranes, accreted to or along an active continental margin may also become trapped within an orogen formed by collision, providing clues as to the scale and extents of horizontal tectonic movements prior to collision. Identification and distinction of these terranes, and of units within an orogen that are derived from previously separate continents, relies to a large extent on determination of the age signatures (i.e. age–event chronologies) of the protoliths in the units within the orogen, and correlation of these signatures with those of other regions. The application of geochronology to define such age-based provinces has been complemented and enhanced greatly by the use of isotope geochemistry, in which the initial isotopic compositions (e.g. the ratios $^{87}Sr/^{86}Sr$; $^{143}Nd/^{144}Nd$) of igneous rocks at the time of their emplacement can be determined and used as a fingerprint for correlation. The derivation of rock units caught up or produced in an orogen can also be ascertained by documenting the range of ages (age spectra) of robust U–Pb bearing accessory minerals, in particular the zircon ($ZrSiO_4$) which may be present as 'inherited' grains that have survived melting and metamorphism during orogenesis, and comparing these spectra with those determined for various crustal blocks. This is especially useful for ancient orogens, where the collided continental blocks may no longer be adjacent because of the ravages of subsequent plate movements.

Geochronology involves determining a radiometric date for a rock, or suite of rocks, or a mineral present within a rock. Early geochronological work, from the late 1950s, concentrated on determining the ages of igneous rock suites and in doing so often utilised what is known as the 'isochron' methodology. This is based on determining the parent/daughter isotopic compositions of a number of rocks assumed or constrained to be of a common, shared, origin, and fitting the results to a straight line, the slope of which – in an ideal case – would reflect the age of the rock suite. Examples of such 'isochron' systems include the Rb–Sr and Sm–Nd whole-rock methods. Somewhat in parallel with the development of such isochron methods applied to whole rocks was the application and refinement of mineral dating. In this case, specific minerals containing modest (e.g. 100–1000 parts per million, ppm) to appreciable (1–10 wt%) concentrations of radioactive elements are analysed for the compositions of parent and daughter isotopes in order to arrive at a radiometric date. Many of the radiogenic minerals important to defining the ages of events in orogeny are present only as trace or very minor amounts in their host rocks. These are the 'accessory minerals', and include zircon ($ZrSiO_4$), monazite (Ce–LREE–Th–U phosphate, where LREE means light rare earth elements), xenotime (Y–Th–U–phosphate), rutile (TiO_2 containing some uranium) and titanite (CaTi phase containing some uranium). The term 'accessory mineral geochronology' is commonly used to denote this approach, which is also generally based on the decay of the unstable radioactive isotopes of uranium (^{238}U, ^{235}U) and/or thorium (^{232}Th) to their final daughter products – the isotopes of lead ^{206}Pb, ^{207}Pb and ^{208}Pb respectively.

Mineral age dating is also applied to major mineral phases present in rocks from within orogens, such as schists, gneisses and magmatic rocks. In particular, the micas, potash feldspar and hornblende all contain appreciable potassium, some of which is in the form of ^{40}K that undergoes radioactive decay to ^{40}Ar. These minerals are used in 'argon–argon' dating. Garnet, a key mineral in many metamorphic mineral assemblages formed under a wide range of pressure–temperature (P–T) conditions, contains trace amounts of U, high Lu relative to Hf, and low Nd relative to Sm. As a result, garnet/whole-rock Sm–Nd isochrons have been used extensively for age determination, and garnet/whole-rock

Lu–Hf isochrons are now being applied similarly. High-precision chemical separation and mass spectrometry methods also have occasionally been applied to determine garnet U–Pb ages.

Radioactive decay of an unstable isotope is a spontaneous process that is only time-dependent. The rate of decay of the parent, radioactive, isotope to a stable, daughter, isotope is characteristic for the given parent and is defined by the decay constant λ. In all radioactive decay systems the rate of decrease in the number of parent atoms is equal to the number of parent atoms remaining, multiplied by the decay constant. This results in radioactive decay following an exponential function, formally written as:

$$N(t) = N_0 \, e^{-\lambda t} \tag{3.1}$$

where $N(t)$ refers to the number of parent atoms at time t, and N_0 to the initial number of atoms, which will equal the number of parent atoms remaining plus the number of daughter atoms produced. If both parent and daughter isotopes can be measured, we are able to evaluate the time elapsed since the daughter atoms started to accumulate in the rock or mineral being analysed, provided we can correct for the presence of any pre-existing daughter atoms and provided the system is 'closed' on the scale of examination. Whilst the decay constant formally defines the decay of a radioactive isotope in terms of time, it is generally more intuitive to compare such isotopic systems in terms of their 'half-lives'. The half-life indicates the time required for half of the radioactive isotope initially present in the rock or mineral to decay. Half-lives vary greatly; the transformation of ^{238}U to ^{206}Pb has a half-life of 4.468 billion years, whereas ^{235}U to ^{207}Pb is 704 million years. The half-life for ^{232}Th decay to ^{208}Pb is about 14 billion years, and that of ^{87} Rb to ^{87}Sr is 48.8 billion years, but the half-life for ^{14}C decay to ^{14}N is only 5730 years. That is why the latter is used for archaeological dating. It is vital to choose the radiometric system that is appropriate to the dating required; that is, its half-life must be suitable.

Given the timescales of orogeny, and the many events involved from subduction to accretion, collision and perhaps collapse, it is important to understand what a radiometric date means.

Even though radioactive decay itself is a spontaneous statistical process in which diffusion plays no role, the retention of potentially poorly fitting daughter atoms within mineral lattices may be affected by diffusion. Diffusion is a thermally activated (temperature-dependent) process that will for most radiometric isotopic systems lead to age dates that are less than the true age of a mineral, because the daughter atoms do not accumulate but instead move out of the system, either partially or completely (in the latter case the 'age' would be zero). As temperature decreases, the rates of diffusion decrease and hence the distance over which an atom or isotope will move or diffuse (the diffusion distance) decreases. If this temperature-dependent diffusion distance is smaller than the scale of our measurements, then the system being measured is effectively closed and the age will be retained. The temperature below which the diffusion distances are small enough to ensure the isotopic system is closed is the blocking temperature, or 'closure temperature' (T_c). Above the closure temperature, daughter isotopes can diffuse over distances such that they do not accumulate in the same volume in which the original parent isotope decayed, but below it the daughter isotopes accumulate within that volume.

Table 3.1 Apparent closure temperatures for commonly used geochronometers, for cooling rates in the range 2–20 °C per million years

System	Grain sizes	T_c (°C)	Comments
Zircon U–Pb	All	>1000	Fluid impacts, dissolution–precipitation
Monazite U–Th–Pb	40–500	>800–900	Fluid impacts, dissolution–precipitation
Garnet U–Pb	>1000	>750–800	Mineral inclusions concerns
Allanite U–Pb	>500	630–680	Useful in leucosomes
Titanite U–Pb	>500	570–630	Useful in metabasites
Xenotime U–Pb		500–600	Susceptible to fluid impacts
Rutile U–Pb	100–500	380–420	Useful in blueschists, eclogites
Apatite U–Pb	200–1000	350–500	Compositional effects (Cl, F)
Garnet Sm–Nd	>1000	600–700	Mineral inclusions concerns
Garnet Lu–Hf	>1000	600–700	Mineral inclusions concerns
Hornblende Ar–Ar	160	480–560	
Muscovite Ar–Ar		320–380	
Biotite Ar–Ar		250–330	
K-feldspar Ar–Ar	125–250	150–250	Microstructural considerations
Muscovite Rb–Sr		400–450	
Biotite Rb–Sr		350	Depends on exchanging phases
K-feldspar Rb–Sr	>1000	320	
Zircon FT		225–175	
Apatite FT		120–80	For recent exhumation
Zircon U–Th–He		40	For near-surface conditions

FT: fission track dating.

In his elegant exposition on closure temperatures Dodson (1973) showed that the closure temperature for an isotopic system, the 'temperature at the time corresponding to the apparent age', depends on the rate of cooling as well as the material properties and grain size. As a consequence, there is no single closure temperature that can be applied to any one mineral-isotope system for all possible cooling rates. It is important to understand that closure temperatures in rapidly cooled systems, such as shallow-level contact aureoles around igneous bodies, will be significantly higher than those relevant to slowly cooled deep crustal rocks. Those geochronological methods for which closure temperatures are less than the peak temperatures attained by the rock or the temperature at which a dated mineral crystallised will not record the age of the peak event or mineral growth. Instead, they will measure the age corresponding to the time that the rock or mineral cooled below the closure temperature for that isotopic system. In other words, these methods are best viewed as geochronometers that in general give cooling ages unless the peak thermal event lies at or below their closure temperature, T_c.

Examples of T_c for a variety of commonly used geochronometers are listed in Table 3.1. Where diffusion is the only process governing the retention of ages, it is apparent that peak metamorphic ages in rocks that have been subjected to upper amphibolite facies conditions, for example in Barrovian metamorphism, or under medium-T eclogite facies conditions of >600 °C, will potentially only be recorded by a few of the U–Pb mineral

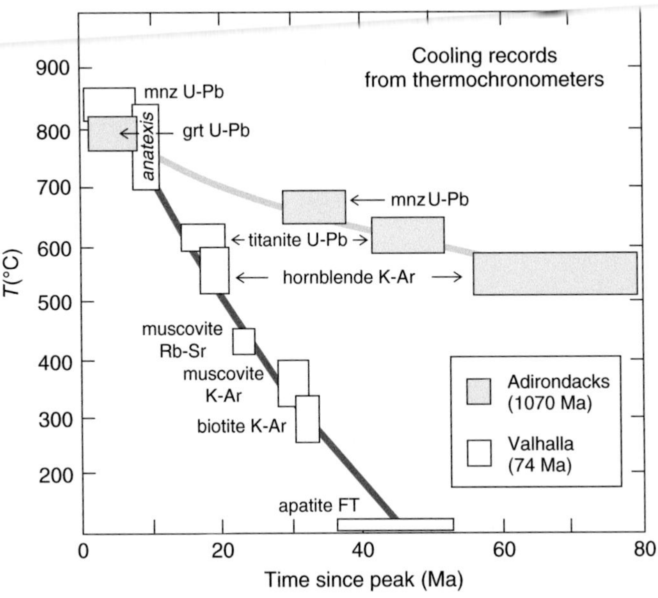

Figure 3.4 Temperature–time diagram showing examples of cooling histories for high-*T* metamorphic belts, as determined using a variety of thermochronometers. Time is shown as 'time since peak' so that the two regions can be compared. The Adirondacks of north eastern USA is a 'Grenvillian' granulite belt that experienced its metamorphic peak at 1070 Ma and then underwent slow cooling (at 3°C/Ma) at depth for a period of over 100 Ma (Mezger *et al.*, 1991). The Valhalla Complex of north west Canada is a Cordilleran granulite belt that experienced its metamorphic peak at 74 Ma and then underwent relatively rapid cooling (20–25°C/Ma) with exhumation (Spear & Parrish, 1996).

chronometers (zircon, monazite, allanite, titanite, garnet) and the REE-based garnet chronometers (Sm–Nd, Lu–Hf). Fewer of the methods will preserve the ages of peak metamorphism if the orogen is hot and dominated at depth by granulites (>800 °C) or 'hot' eclogites (>700 °C) – zircon, monazite and perhaps garnet U–Pb being the only mineral chronometers that retain the record. This is a key problem for constraining the age of peak metamorphic conditions, and the timescales of high-temperature portions of metamorphic *P–T* paths, for formerly hot rocks from the deep interiors of orogens. Conversely, the wealth of lower-*T* mineral chronometers now available, for the temperature range from 500 °C to 40 °C, means that cooling histories (temperature–time records) can be constructed in some detail, using a combination of U–Pb, K/Ar, $^{40}Ar/^{39}Ar$, Rb/Sr and fission track methods. Those mineral geochronometers with closure temperatures below 400 °C can be used to track exhumation rates by converting the temperature–time records to depth–time, by assuming typical post-orogenic geothermal gradients for the upper crust (e.g. 20–30 °C/km). Figure 3.4 provides some examples of such cooling histories, for both 'fast' and 'slow' cooled metamorphic regions that have a range of geochronometers applied to them.

The closure temperature approach to interpreting ages only addresses part of the problem of understanding what an age means in a complex orogenic system. As noted above, if a mineral grows or is recrystallised at a temperature below its closure temperature, then, ideally, it will record the age of that growth or recrystallisation event. It then becomes quite critical to be able to link the growth event as seen by the mineral to what is going on around it – in other words to metamorphic conditions, deformation and melting where these occur. This may be relatively easy to do for a major mineral like garnet, which can be linked explicitly to other minerals and structural elements through its textural relations coupled with chemical zoning features and mineral inclusion analysis. This approach has been well described for garnets and modelled in depth for many metamorphic belts by many workers (e.g. Spear *et al.*, 1984; Kohn, 2003; Carlson, 2006). In contrast, establishing the age-to-event link is not so obvious for the high-closure-temperature accessory minerals, zircon and monazite, which in principle could be used to constrain the high-temperature record but which might not have grown at the 'peak' conditions or along with the other minerals used to infer the pressure–temperature record of the rock.

Zircon is a somewhat unreliable witness of orogenic events. Its virtues of chemical inertness in many geological environments, and extremely low diffusivities for Pb and other cations, mean that zircon can record the most ancient events in Earth history, as shown by workers on the Jack Hills zircons (e.g. Wilde *et al.*, 2001; Valley *et al.*, 2005; Harrison, 2009), and be used to trace the origins of igneous rocks and sources of sediments incorporated into orogens (e.g. Kemp *et al.*, 2006). However, the virtues that render it a great survivor mean that it might not grow, recrystallise, or be modified and isotopically 'reset' on sufficient scales to provide accurate ages for metamorphism during orogeny, even up to granulite facies conditions. When it does grow, for example during partial melting or melt crystallisation at 700–1000 °C, or undergo recrystallisation facilitated by mineral–fluid interactions to form 'metamorphic' zircon, it can provide excellent age information for these events. As a consequence, experimental and empirical data are now used to link zircon formation to the growth of other minerals such as garnet using trace element (mainly rare earth element, REE) chemical signatures (Rubatto, 2002; Whitehouse & Platt, 2003; Harley *et al.*, 2007), and to temperature using the Ti contents of the zircon as calibrated by Watson and coworkers (Watson *et al.*, 2006). These approaches to evaluating what zircon ages might mean rely on the ability to microanalyse the mineral *in situ* using mass spectrometry, either secondary ion (SIMS; ion microprobes such as the SHRIMP) or laser ablation, underpinned by detailed microtextural characterisation of the mineral using electron imaging and cathodoluminescence as advocated by Hanchar and others (e.g. Hanchar and Miller, 1993).

Similar observations can to some extent be made for monazite. It too can preserve age records from events that pre-date those occurring in an orogeny, even up to high temperatures, but as it is more chemically reactive during prograde metamorphism its 'survival rate' is much lower than that of zircon, and it is more likely in most circumstances to preserve age information relevant to the metamorphic events accompanying orogen evolution. Pyle and Spear (2003) and others have developed and applied methods based on the distribution of REE and yttrium between monazite, garnet and xenotime that allow the timing of monazite growth to be related to metamorphic reactions and mineral assemblages under amphibolite facies conditions, and similar criteria are being tested and developed for higher-pressure and -temperature metamorphic rocks (Hermann & Rubatto, 2003; Kohn *et al.*, 2005). Once again, this is particularly useful when the same grains are analysed *in situ* for their age and chemistry, and for monazite this can be accomplished using SIMS and ICP-MS for isotopic analysis or electron microprobe chemical dating as developed by Suzuki and others (e.g. Suzuki & Kato, 2008; Braun *et al.*, 1998).

The final complication with respect to the interpretation of ages from zircon, monazite and potentially other higher T_c minerals relates to post-growth mineral behaviour that is not diffusion-controlled. Several workers, notably Schaltegger *et al.* (1999) and Hoskin and Black (2000), have shown that zircon is susceptible to internal recrystallisation and modification, probably via coupled *in situ* dissolution–reprecipitation, subsequent to its growth. This grain-specific modification could be fluid-activated, as argued by Giesler and others (Giesler *et al.*, 2007), and may lead to significant or total Pb loss from the affected zircon zones or domains. The ages obtained may define the time of fluid-activated reaction, and so give additional information on the thermal history of the orogen if independent temperature constraints are available, or they may scatter between this and the age of zircon growth and be extremely difficult to interpret. A good example of the effects of fluid-activated dissolution–reprecipitation of zircon is provided by the slowly cooled late Archaean UHT Napier Complex. In this relict orogen, the peak granulite event at *c.*2590 Ma and *c.*1000 °C is rarely recorded by zircon whereas abundant 2500–2480 Ma zircons occur, modified and reset because of dissolution–reprecipitation activated by fluids possibly liberated from residual melts at *c.*700–750 °C, temperatures constrained by zircon Ti contents and adjacent rutile Zr content thermometry. Dissolution–reprecipitation phenomena are also documented in monazite, and can be recognised by the formation of microporosity, the presence of trails of exotic minerals such as thorite, and localised decrease in monazite Th contents. For both zircon and monazite the effects of dissolution–reprecipitation are to update the mineral ages, leading to ages that may in some cases overlap with or even be younger than those obtained from low-T_c minerals with closure temperatures in the range 600–300 °C. Reddy *et al.* (2007) have shown that this resetting and 'updating' of zircon ages also can occur through the mobilisation and loss of Pb induced by internal crystal plastic deformation.

Geotherms and thermal structure

The importance of temperature as a controlling variable affecting rock strength, deformation mechanisms, densities, metamorphic mineralogies, extents of melting and, on the larger scale, the evolution of an orogen, has been alluded to in earlier parts of this chapter. At the simplest level the Earth can be regarded as a hot body in cold space, cooling over time but with the additional complication of also producing heat over time and so needing to dissipate this continuous heat production. This is, ultimately, why the Earth is not static: why convection currents move heat in the mantle, why deep plumes may exist, why plate tectonics occurs and why orogens happen.

It is well established that temperature in general increases with depth in the Earth. The lines of evidence for this are many and varied: temperatures in deep mines, boreholes and wells; volcanic eruptions of hot lavas coupled with seismic and satellite-based measurements locating the sites of magma storage at depth prior to eruption; measurements of heat flow at the Earth's surface; changes in velocities of seismic waves with depth; attenuation of seismic shear waves indicating a liquid outer core; thermometry of xenoliths from within basalts and kimberlites; experiments on melting of peridotite and typical crustal rock types; experiments on phase transitions involving mantle minerals coupled with observation of these minerals as inclusions in diamond; and many more.

The rate of change of temperature with depth below a point (or small area) on the Earth is known as the geothermal gradient. Typical surface geothermal gradients, measured from boreholes and mines within crust far removed from plate boundaries, are 25–30 °C/km. On the other hand, seismic wave modelling and experiments that constrain the equations of state of materials such as quartz and feldspar in the crust, olivine in the mantle and iron in the core lead to temperature estimates of about 450–550 °C at the base of 35 km thick continental crust, 1250–1330 °C at the base of the lithosphere (120 km) and 5400 ± 600 °C at the centre of the Earth (6400 km). These lead to average geothermal gradients of 13–16 °C/km for old continental crust, 10–11 °C/km for the lithosphere, and ‘whole Earth’ gradients of less than 1 °C/km, and demonstrate very clearly that geothermal gradients steepen with depth in the ‘steady state’ case.

Geothermal gradients, conventionally plotted with depth vertically downwards and temperature on a horizontal axis, dramatically steepen with depth within the conducting lithosphere primarily because radiogenic heat production is concentrated within the upper parts of the continental crust, which contains the highest concentrations of ^{40}K, ^{238}U and ^{232}Th – the principal heat-producing isotopes today. Beneath the lithosphere heat transfer is deduced, following the remarkable insights of Holmes and subsequent modelling by Turcotte and Schubert (1982) and others, to be controlled by convection and the advective transport of viscous mantle material as plumes or in cells. Geothermal gradients within the convecting mantle are therefore essentially determined by the adiabat for peridotite, enthalpy changes at phase transitions, and the adiabats of deeper mantle materials such as majorite and perovskite, and are estimated as about 0.3 °C/km. The temperature of convecting upper mantle is currently considered to be *c.*1300–1350 °C at 120 km depth.

One of the most effective ways to characterise geographic variability in how temperature increases with depth is by heat loss as measured by surface heat flow (in milliwatts per square metre: mW/m^2). The average surface heat flow from the Earth is about 60 mW/m^2, representing current global heat loss of about 4.2×10^{13} watts. In a global analysis of heat flow data, Sclater and others showed that continental interior areas, such as cratons, that have been unaffected by major plate tectonic processes and volcanism over the past several hundred million years have surface heat flows of about 46 mW/m^2, similar to that of oceanic crust between 120 and 140 Ma in age. Young oceanic crust and recently active tectonic zones on continents have much higher surface heat flows, for example greater than 120 mW/m^2 in oceanic crust less than 10 Ma old and over 70 mW/m^2 in mountain belts such as Tibet and the Andes, and 100 mW/m^2 in the Basin and Range province of western USA.

A ‘geotherm’ is a curve that depicts the relationship between temperature and depth beneath a chosen point on the Earth, generally from the surface down to the convecting mantle. Geotherms are not to be confused with ‘isotherms’. The latter are contours of equal temperature at depth in the Earth and so are at least two-dimensional, as shown on lithosphere cross-sections, and may vary in depth depending on the temperature–depth distributions (geotherms) at specific points along the cross-section. The surface heat flows referred to in the preceding paragraphs reflect these different temperature–depth distributions, or geotherms, though the relationship is a complex one as it depends on several factors. These include the distribution of heat production with depth, the relative importance of material advection versus heat conduction, and the thermal diffusivity of the rocks at depth – which is itself a complex parameter that depends on thermal conductivity, density and heat capacity of the material. Hence, though it is common practice to construct 'steady-state' geotherms labelled with their corresponding surface heat flows, these are non-unique: the same surface heat flow can result from different combinations of heat production distribution, thermal diffusivities and advection/conduction efficiencies (which will be returned to later). Nevertheless, in dynamic plate settings the rate of increase in temperature with depth varies from what might be considered to be the 'norm' – the narrow range of steady-state temperature–depth distributions, geotherms, that generate surface heat flows of about 46 mW/m^2 from old continental and oceanic lithospheres. The comparison is essentially between those geotherms in which there is long-term equilibrium between the base and top of the lithosphere column (steady-state geotherms) over hundreds of millions of years, and those in which this equilibrium is disturbed (transient or perturbed geotherms).

It is worthwhile to consider further the relationships between heat production, heat transfer and heat loss in order to develop an understanding of how orogens may evolve thermally and hence also in their mechanical behaviour. We will firstly consider heat production and then go on to examine heat transfer in the lithosphere in terms of the roles of conduction and advection.

Global heat content in the Earth is dominated by a combination of residual heat from planetary accretion (about 20%) and heat produced through decay of radioactive isotopes (80%). Further sources of heat, more 'localised' in that they do not occur throughout the whole or large parts of the Earth, include mechanical and chemical heat production (e.g. Stüwe, 2002).

Radioactive heat production is directly dependent on the amounts of radioactive heat-producing isotopes present in the Earth at any time, and the effects of this heat production in terms of raising internal temperatures is further dependent upon the disposition and distribution of the heat-producing isotopes with depth. In its simplest terms, if all of the inventory of heat-producing elements are held at depth then temperatures will be higher because the heat transfer paths to the surface are longer and slower. In contrast, if all the heat-producing elements in the Earth – both long-term and short-lived – had been concentrated in a surface layer and the heat thus allowed to dissipate into the atmosphere and space, then the Earth would truly be a dead planet, hot for perhaps only the 30 million years that Lord Kelvin deemed necessary to cool it from an infernal initial state.

Fortunately for the Earth, mountain building, and ourselves, considerable radioactive heat production is still occurring from the deeper Earth. For example, the mantle is estimated, from present day abundances of U, Th and K, to generate about 7.4 picowatts per kilogram of peridotite (Table 3.2). This may not appear to be much, but given the total volume of mantle it amounts to over 60% of global radiogenic heat production.

The continental crust has a much higher concentration of heat-producing elements – about 1.6 ppm U and 5.8 ppm Th (roughly 50 times as much as the mantle) and 2000 ppm

Table 3.2 Present day major heat-producing isotopes

Isotope	Heat release (W per kg of isotope)	Half-life (years)	Mean mantle concentration (ppm)	Heat release (W per kg mantle)
^{238}U	9.46×10^{-5}	4.47×10^{9}	0.031	2.91×10^{-12}
^{235}U	5.69×10^{-4}	7.04×10^{8}	0.00022	1.25×10^{-13}
^{232}Th	2.64×10^{-5}	1.40×10^{10}	0.124	3.27×10^{-12}
^{40}K	2.92×10^{-5}	1.25×10^{9}	0.037	1.08×10^{-12}

total K (roughly 70 times as much as the mantle). These present-day concentrations lead to total heat productions of about 390–400 picowatts per kilogram of crust, which is equivalent to 1 $\mu W/m^3$ if one takes an average crustal density. Granites typically have heat productions of about 3 $\mu W/m^3$. Some highly fractionated granitoids such as the central Australian Proterozoic granites (Tea Pot, Mount Painter) described by Sandiford and others (Sandiford *et al.*, 1998), and the alkaline-mafic 'durbachites' of the Bohemian Massif in the Czech Republic, have notably high heat productions of up to 10 $\mu W/m^3$. Indeed, the long-term burial of such 'hot' rocks beneath thick sedimentary cover has been invoked as a cause of non-magmatic, non-orogenic thermal metamorphism and belated hydrothermal circulation in the cover rocks (e.g. Sandiford *et al.*, 1998), and more broadly as a possible cause of enhanced high-T metamorphism in regional scales.

Radioactive heat production in the Earth has inevitably decreased throughout time as the amounts of parent isotopes – the heat producers – have decreased to form the stable daughter products referred to in the section on geochronology. This statement of course assumes that the flux of new radioactive material onto the Earth (via meteorites, asteroids and other collisions) is trivial compared with the mass of the Earth attained by the time of the 'big impactor event' at about 4.5 Ga. The major contributors to present-day radioactive heat production listed above – U, Th and K – have long half-lives and so still contribute heat. Even so, it is certainly the case that only half the ^{238}U initially in the Earth remains today, as its half-life is close to the age of the Earth. Only about one-seventieth (1.4%) of the initial ^{235}U and 11% of the initial ^{40}K remain, though about 80% of the original ^{232}Th still exists. Simple calculations for typical rock types, for example a granodiorite with 1 wt% K, 4 ppm Th and 1 ppm U, show that their heat production 4 billion years ago would have been three to four times their present-day production. Adding to this the heat produced in the past by short-lived unstable isotopes leads to the conclusion that internal heat production even 3 billion years ago was at least twice the present-day amounts. How this influenced geotherms within the Earth, tectonics, plume behaviour and rates of mantle convection is a continuing debate.

Heat transfer in the lithosphere

The conductive or diffusional transfer of heat is governed by *Fourier's law*:

$$q = -k(\mathrm{d}T/\mathrm{d}z) \tag{3.2}$$

where q is the heat flux at a point on the Earth's surface, k the thermal conductivity of the rocks there, and dT/dz the measured geothermal gradient. Thermal conductivities are in the range $k = 3.0$ W/(m K). For a typical surface geothermal-conducting gradient of 0.02 K/m (20 °C/km) we get that $q = 60$ mW/m^2. This estimate of the global average heat flux is corroborated by thousands of observations of heat flow in boreholes all over the world.

The heat content of a body of rock depends on the difference between heat flow into the body (q_{in}) and heat flow out (q_{out}). Energy balance in the Earth dictates that the rate of temperature change with time is proportional to the rate with which heat content changes. Hence:

$$\frac{dT}{dt} \propto \frac{dq}{dz} \tag{3.3}$$

Heat content H (in J/m^3) is a function of temperature T, density ρ and the heat capacity C_p of the material: $H = T\rho C_p$, so that

$$\rho C_p(dT/dt) = -(dq/dz) = d[k(dT/dz)]/dz \tag{3.4}$$

And if k is independent of depth z then

$$\rho C_p(dT/dt) = k(d^2T/dz^2) \tag{3.5}$$

So, defining the parameter *thermal diffusivity* κ as $k/\rho C_p$ we arrive at:

$$dT/dt = \kappa(d^2T/dz^2) \tag{3.6}$$

which basically shows that the rate of change in temperature is proportional to the curvature of the temperature profile with depth – the curvature of the geotherm (Stüwe, 2002). Typical values of the key parameters used in many modelling studies of crustal rocks are 2750 kg/m^3 for density ρ, 1–1.2 × 10^3 J/(kg K) for heat capacity C_p, and 2.75 J/(s m K) for thermal conductivity k. These lead to estimates for thermal diffusivity κ of 0.5–1.5 × 10^{-6} m^2/s.

Thermal diffusivity proves to be a very useful parameter for estimating some first order features of the thermal behaviour of the Earth. In particular, in the case of conduction the characteristic timescale of thermal equilibration proves to be a simple function of the length scale of the thermal anomaly and the thermal diffusivity: $\tau = L^2 / 2\kappa$ (Turcotte & Schubert, 1982; Stüwe, 2002) in the case of linear or layered thermal structures, where L is the length scale over which the thermal perturbation occurs and relaxes over the characteristic time such that thermal equilibration is about 84% complete. Based on the formulation given above, the characteristic timescale for conductive relaxation and thermal equilibration over a 1 km length scale is 10 000 years, 1 Ma for a 10 km length scale and 100 Ma for a 100 km length scale, such as the whole lithosphere. An alternative formulation that does not assume a linear geometry but instead adopts a spherical to elliptical geometry of relaxation ($\tau = L^2 / \pi^2 \kappa$) reduces these timescales by about 80% (e.g. for 100 km length scale the characteristic timescale is about 20 Ma).

Heat sources

There are three principal sources of heat that influence the thermal behaviour of the crust and lithosphere. These are radioactive heat sources, chemical heat production, and mechanical heat production. Following Stüwe, we will define these by their volumetric rates of heat production, S, and total S is $S_{rad} + S_{chem} + S_{mech}$.

Adding this to the rate of temperature change due to the curvature of the geotherm gives us a more general expression for change in T:

$$dT/dt = \kappa(d^2T/dz^2) + [S/\rho C_p] \tag{3.7}$$

As $\kappa = k/\rho C_p$, we can rearrange the above equation to consider the conditions for which there will be no tendency for temperature to change (i.e. $dT/dt = 0$), the steady-state case. Hence, for steady state to apply, the curvature in the geotherm must be balanced by the rate(s) of heat production divided by the thermal conductivity (or conductivities) of the materials:

$$d^2T/dz^2 = -S/k \tag{3.8}$$

Radiogenic heat production is well constrained for typical rocks and model lithospheres, as considered above, with present-day continental crustal values being in the range 1 $\mu W/m^3$ to 3 $\mu W/m^3$. What is not so well understood, despite extensive geochemical measurement and modelling of crustal composition with depth, is the distribution of this heat production with depth. This is crucial, as the deeper the heat-producing elements are stored in the crust the higher the temperatures are at depth. England and Thompson (1984), in their classic modelling of crustal doubling, used a layered crust model in which all radiogenic heat production is concentrated in the upper 10–12 km of crust. Other modellers have employed a more even distribution to 20 km, with lower heat production at a given depth, whereas still other workers have assumed distributions in which heat production decays exponentially with depth. The only constraints on these distributions is that the surface heat flows generated must match up with those observed in old, tectonically inactive crust, and the range in heat productions used must correspond to those observed in typical crustal rock compositions.

The significance of mechanical heat production has, in contrast, been controversial in terms of its contribution to overall heat budgets (Stüwe, 2002). The mechanical work done on rocks, for example via bending of lithosphere at a subduction zone or the sliding of a slab of rock past another, is largely manifested as frictional heat, also known as shear heating or viscous dissipation. Such shear heating is essentially proportional to the deviatoric stress acting on the body of rock. It is also strongly dependent on the strength of the rock, which itself is strain-rate dependent. It turns out that mechanical heat production, S_{mech}, is independent of the deformation mechanism that controls how the material behaves, but is a simple function of deviatoric stress and strain rate (where strain is ε):

$$S_{mech} = \tau(d\varepsilon/dt) \tag{3.9}$$

This means that the rate of change of temperature related to mechanical heat production is also simply related to deviatoric stress and strain rate:

$$dT/dt = [\tau(d\varepsilon/dt)]/\rho C_p \quad (3.10)$$

For strain rates of 10^{-12} to 10^{-14} /s, which result in 100% strain over time periods of 1–10 Ma, and rock strengths in the range 50–100 MPa, Stüwe has estimated that shear heating may contribute about 30–40 °C to rock temperatures at depth during crustal doubling. However, this effect may be magnified if the heating is constrained to occur in a narrow zone in which strain rates are high, and may produce a significant contribution to metamorphic temperatures attained during subduction-accretion (Peacock, 1992), as considered in detail by Gerya and co-workers in recent modelling (Gerya *et al.*, 2008; Faccenda *et al.*, 2008).

Chemical heat production, S_{chem}, may result from the crossing of exothermic chemical reactions by significant volumes of rock. This may occur during cooling or, for some phase transitions, during compression and burial. Although dehydration reactions may lead to heat transfer via liberated fluids, this is not generally regarded as significant over the long timescales of typical metamorphic events (Peacock, 1992). In contrast, even though melting is endothermic, requiring an input of heat to progress, melting can potentially be important in moderating or buffering temperatures and dT/dt in the deep crust during metamorphism. This arises because of the latent heat of fusion that is released when melts crystallise, the converse of latent heat of melting. For melt-related processes:

$$S_{chem} = L_m\rho(dV/dt) \quad (3.11)$$

where L is the latent heat of melting, estimated at about 3.2×10^5 J/kg for crustal melting, and dV/dt is the volume proportion of reaction per unit time, in s^{-1}. This leads to

$$dT/dt = L_m(dV/dt)/C_p \quad (3.12)$$

so that the contribution to heating is clearly linked to the amount of melt present, which varies with temperature depending on the intervals over which chemically complex crustal rocks undergo partial melting, and the temperature interval over which melts crystallise. In reality the effect of melting will be to moderate the rate of change in T at a given depth (Stüwe, 2002).

Heat advection

Heat is advected when warm materials move relative to other material, and is a general term applicable to all cases of heat transfer through the active transfer of mass or bodies of rock, melts or fluids. Convection is simply a special case in which the transport occurs in a closed loop. One-dimensional heat advection, the simplest scenario, could describe the transfer of heat associated with a plume or hot jet that has a very small diameter compared with its length. Although in reality one needs to consider three-dimensional heat advection, the one-dimensional case provides the essential insights into the process, particularly if there is limited lateral heat transfer on the timescale of the advective process. Following Turcotte and Schubert (1982) and Stüwe (2002), one-dimensional heat advection in the vertical direction is given by:

$$\mathrm{d}T/\mathrm{d}t = \mu(\mathrm{d}T/\mathrm{d}z) \tag{3.13}$$

where μ is the transport velocity (parallel to z, positive if upwards transport). This simplified approach can be applied to consider the effects of erosion, which is in a sense the advective movement of a body of crust and lithosphere through a surface boundary condition. In this very simple one-dimensional erosional case the advection term competes with the diffusional relaxation term (i.e. heat will diffuse contemporaneously with movement of the crust) so that in the absence of heat production:

$$\mathrm{d}T/\mathrm{d}t = \kappa(\mathrm{d}^2T/\mathrm{d}z^2) + \mu(\mathrm{d}T/\mathrm{d}z) \tag{3.14}$$

If we then add in the heat production term, and note that heat production potentially has the three contributions, we arrive at the complete, though still simplified, one-dimensional description of thermal energy balance in the lithosphere:

$$\mathrm{d}T/\mathrm{d}t = \kappa(\mathrm{d}^2T/\mathrm{d}z^2) + \mu(\mathrm{d}T/\mathrm{d}z) + S/\rho C_p \tag{3.15}$$

Considering the first two terms of the thermal balance equation, it can be seen that the rate of temperature change, given specified heat production, could be dominated by either heat conduction or heat advection depending on the magnitudes of κ and μ and the curvature of the geotherm compared with the gradient of the thermal profile. As the curvature of a thermal profile is strongly dependent on the length scale over which the gradient is required to change, the dominance of conduction/diffusion over advection and vice versa is closely related to the characteristic length scale of the thermal process (e.g. small-scale intrusion versus crustal doubling). This important result is expressed in the Peclet number, P_e:

$$P_e = \mu L/k \tag{3.16}$$

where L is the characteristic length scale, already applied in consideration of the timescales of thermal relaxation in an earlier section. If $P_e << 1$, thermal diffusion dominates and the rocks will approach equilibrium with a steady-state condition over the timescale of the thermal process, whereas in $P_e >> 1$ advection dominates and the rocks remain out of thermal equilibrium with the steady-state lithosphere condition over the timescale of the thermal process (e.g. England & Thompson, 1984).

As emphasised by England and Richardson (1977), the thermal evolution of an orogen is essentially controlled by three processes and their timescales: the duration of thickening; the timescale of thermal equilibration; and the timescale of exhumation. In general, thickening occurs faster than the characteristic timescale for thermal equilibration, with crustal doubling possible in 10 Ma compared with τ of 50 Ma. In contrast, the timescale of exhumation may range from being similar to that of thermal relaxation, to being much shorter. This critically controls whether rocks buried to deep levels heat up significantly, and this depends on how fast they move or advect through the system. Erosion on orogenic scales represents one viable form of heat advection, as it requires crustal material to move through the surface of the orogenic belt. However, exhumation has increasingly been recognised as being a far more rapid and dynamic process in many orogens, with gravitational, buoyancy and tectonic forces acting to drive volumes of hot, low-viscosity deep-level rocks upwards either vertically (e.g. in diapirs) or laterally (e.g. in channels) (e.g. Gerya *et al.*, 2008). The potential evidence for rapid heat advection through fast exhumation of both 'cold' and 'hot' metamorphic rocks and regions in orogens will be explored in Chapter 7.

Further reading

Barber, D.J. and Meredith, P.G. (1990). *Deformation Processes in Minerals, Ceramics and Rocks*. London: Unwin Hyman. See especially the chapter by Knipe on the microstructures in the Assynt region of the Moine thrust, 228–261.

Cowie, P. (1998). Normal fault growth in 3D in continental and oceanic crust. In: *Faulting and Magmatism at Mid-Ocean Ridges*, ed. R. Buck, P. Delaney, J. Karson and Y. Lagabrielle. American Geophysical Union Monograph, **106**, 325–348.

Holness, M.B. (ed.) (1997). *Deformation-enhanced Fluid Transport in the Earth's Crust and Mantle*. The Mineralogical Society Series, **8**. New York: Chapman & Hall.

Hubbert, M.K. and Rubey, W.W. (1959). Role of fluid pressure in mechanics of overthrust faulting. Parts 1 and 2. *Geological Society of America Bulletin*, **70**, 115–205.

Jaeger, J.C. and Cook, N.G.W. (1976). *Fundamentals of Rock Mechanics*. London: Chapman Hall.

Knipe, R.J. (1990). Microstructural analysis and tectonic evolution in thrust systems: examples from the Assynt region of the Moine thrust zone, Scotland. In: *Deformation Processes in Minerals, Ceramics and Rocks*, ed. D.J. Barber and P.G. Meredith. London: Unwin Hyman, 228–261.

McCaffrey, K.J.W., Lonergan, J.J. and Wilkinson, J.J. (2010). *Fractures, Fluid Flow and Mineralization*. Geological Society of London, Special Publication, **155**.

Ramsay, J.G. and Graham, R.H. (1970). Strain variation in shear belts. *Canadian Journal of Earth Science*, **7**, 786–813.

Sibson, R. (1977). Fault rocks and fault mechanism. *Journal of the Geological Society of London*, **133**, 191–213.

Sleep, N.H. and Blanpied, M.L. (1992). Creep, compaction and the weak rheology of major faults. *Nature*, **359**, 687–692.

Spear, F.S. (1993). *Metamorphic Phase Equilibria and Pressure-Temperature Time Paths*. Mineralogical Society of America Monograph, Washington.

Spear, F.S. and Parrish, R.R. 1996. Petrology and cooling rates of the Valhalla complex, British Columbia, Canada. *Journal of Petrology*, **37**, 733–765.

Twiss, R.J. and Moores, E.M. (1992). *Structural Geology*. New York: W.H. Freeman and Company.

White, S.A. and Knipe, R.J. (1978). Transformation- and reaction-enhanced ductility in rocks. *Journal of the Geological Society of London*, **135**, 513–516.

Wibberley, C.A.J., Kurz, W., Imber, J., Holdsworth, R.E. and Collettini, C. (eds.) (2008). *The Internal Structure of Fault Zones: Implications for Mechanical and Fluid-flow Properties*. Geological Society of London, Special Publication, **299**.

4 Large-scale features of orogens: thrusts and folds

Most orogens are found at plate margins, and thus vary substantially in length, width, height and time span of activity. For example, the Hellenides are a relatively small orogen of Cenozoic age where the relation between subducting slab, orogenic wedge and paired HP–HT belt is preserved on a scale of hundreds of kilometres, whereas in South America the Andean orogen is several thousands of kilometres long, and the time span of operation has been much longer, i.e. several hundred Ma. Table 4.1 shows the dimensions of some orogens and their diversity.

Orogens across the Earth are amazingly varied, and there is no 'type' orogen. Their varying widths may be related to different time spans of orogenesis, the type of collision and inherent strength differences between different orogens; for example, an orogen that contains large amounts of basic igneous rocks will be stronger than one that has been thermally weakened. The nature of collision-type orogens is dependent on whether the collision was full frontal or oblique and strike-slip; in other words the docking of two continents can be intense or gentle. There are many examples of strike-slip motions in orogenic belts, a process called transpression. For example, sinistral transpression was an important factor in the collision during the Lower Palaeozoic of Laurentia and Baltica. The rate of convergence of two plates will also be a factor; thus fast-converging margins will accumulate more strain in a given period of time.

Arcuation of orogens and oroclines

Some orogens are notably curvilinear or arcuate, and although this feature of orogens is visible from satellites its origin is controversial: does it result from rotation of a linear orogenic belt, or is it an original feature perhaps reflecting inherited structures such as the shape of a continental margin? Some examples are the Aleutian, Carpathian, Tonga, Betic–Rif, Hellenic, Banda and Scotia arcs. There are different answers from different orogens. In the SW Pacific arcuate ridges and volcanic arcs have been interpreted as oroclines, that is bent orogens. The oroclinal Himalayan belt may reflect the indentation of India into Asia (see later). Similarly the conspicuous oroclines in the Carpathians and Gibraltar probably result from indentation tectonics. On the other hand the orocline of Cenozoic Andes may be an inherited feature. The Banda, Tonga and Aleutian arcs may reflect the fact that the intersection of a subduction zone with the Earth's surface gives rise to a small-circle trace, therefore the arc is primary and geometrical.

Table 4.1 Orogens

Orogen	Width (km)	Length (km)	Dip of S-Z
North American Cordillera m	900	7000	moderate
Sevier orogeny	600	600	
Andes:	–	8000	–
central sector	<800	–	30°
northern and southern sectors	<300	–	2–15°
Western Alps m	200	1100	?Steep
Himalaya m	300	2500	15° now
European Variscan m	600	?	–
Scottish Caledonides, Grampian m	70–120	?	–
Appalachians (Lower Pal.) m	<700	3500	–
Taconic m	? 200	400	–
Urals m	300–400	–	
Scandinavian Caledonides m	200	1400	–
East Greenland m	*c.*200	1400	–
Scottish Caledonides, Scandian Moines	50–80	–	–

m: internal zone with high-grade metamorphism

Weil and Sussman (2004) classified arcuate orogens as follows:

1. Oroclines (the term is from Warren Carey and is derived from the Greek meaning bent mountain) which are caused by the bending of an originally linear belt. The Cantabrian arc in northern Spain (Fig. 4.1) and the Gibraltar arc (see Chapter 5) are examples in which rigid-body, vertical axis rotations of orogens perpendicular to the Earth's surface have taken place. Palaeomagnetism is a valuable tool in demonstrating the origin of curvature.
2. Primary arcs which adopt their curvature early on in the orogeny with no appreciable vertical axis rotation. Arcuation of the Appalachians may be an example of this. Gravitational instability, which may arise during the emplacement of a thrust sheet, results in spreading of the sheet in divergent directions.
3. Progressive arcs which acquire curvature progressively during the deformation history (some of the curvature may be primary, some, due to subsequent deformation).

Every example of arcuation has to be treated on its own merits. Oroclines and primary arcs are end-members, while progressive arcs represent a difficult intermediate group.

In the arcuate Variscan belt of northern Spain, Gutiérrez-Alonso *et al.* (2004) showed by use of palaeomagnetic data that it is a true orocline. The oroclinal bending is associated with tangential longitudinal strain wherein the outer arc is extended and the inner arc compressed (Fig. 4.2). The strain affects the crust and mantle which are thickened in the inner arc and thinned in the outer arc. The thinned mantle of the outer arc was delaminated, and in consequence there was an upwelling of asthenospheric mantle and concomitant melting of the lower crust, giving rise to granitoid intrusions dated at *c.*310–300 Ma.

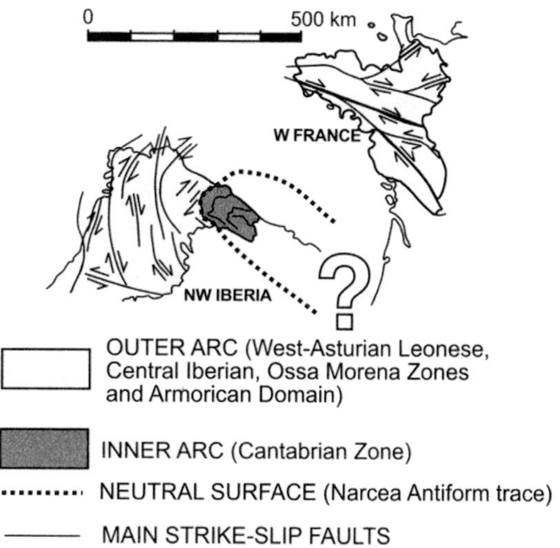

Figure 4.1 Western European Variscan belt in Spain and NW France, showing the arcuate shape of the belt (Cantabrian arc) and the main Variscan faults. From Gutiérrez-Alonso *et al.* (2004), published by permission of the Geological Society of America.

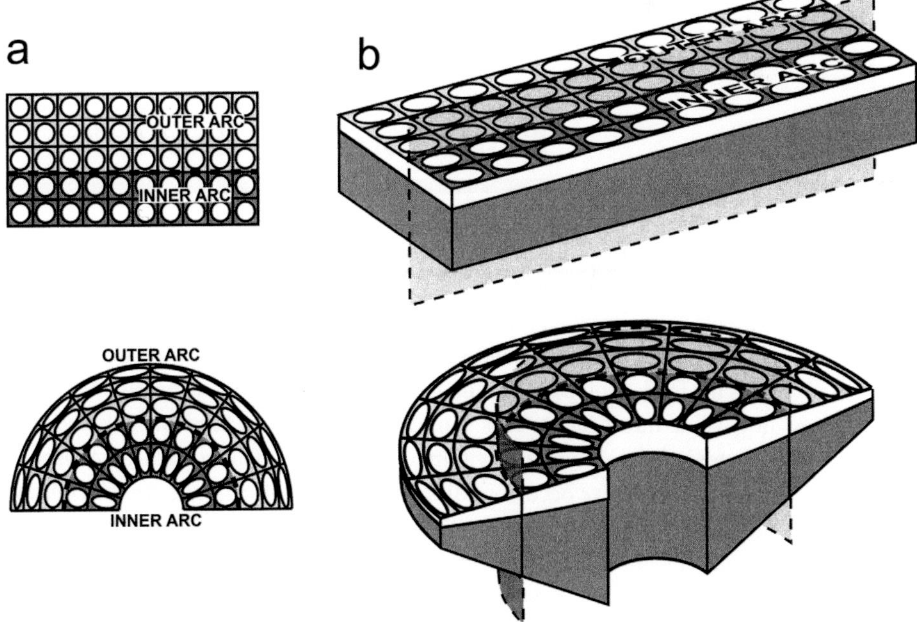

Figure 4.2 Tangential longitudinal strain resulting from lithospheric bending around a vertical axis (a, top view; b, 3-D view). Strain ellipsoids indicate layer-parallel shortening in the inner arc and layer-parallel stretching in the outer arc. The mantle undergoes great thickness changes, being thinned in the outer arc and thickened in the inner arc. From Schellert and Lister (2004), published by permission of the Geological Society of America.

Some examples of the different causes of arcuation are:

1. The shape of sedimentary basins or the future detachment. Thus in collision zones the shapes of coastlines will exert a control on the shape of an orogen, as is suggested in the Appalachians.
2. Buttressing by basement massifs; for example, curvature of the Idaho–Wyoming orogenic belt of the WUSA may be due to the Teton and Gros Ventre basement uplifts. The Uinta Mountain Recess may be caused by the uplift of Precambrian basement.
3. Indentation in which a rigid indenter (a continent or island arc) pushes into weaker material. The classic example is the Himalaya which shows sharp curvature at its western and eastern ends.
4. Wrench faulting motion occurring in basements can exert a control on the configuration of thrust belts developing in the cover. Possible examples are the Ibero-Armorican arc in SW Variscan in Europe and the Andes north of the Scotia arc.
5. Changes in the regional stress field such as divergence in the stress trajectories resulting in divergent flow in a thrust sheet.

The Hellenide arc in the eastern Mediterranean has been explained by McKenzie (1978) by extrusion or escape tectonics. Starting in the late Oligocene or early Miocene (25 Ma), the back-arc basin in the Aegean Sea formed by N–S extension behind the subduction zones of the eastern Mediterranean. At the same time Anatolia moved westward, operating as an indenter extruded by the collision between Arabia and Eurasia. From the late Oligocene until the Pliocene to present, the Hellenic arc bowed to the southwest with increasing concavity to the northeast (Fig. 4.3).

Physical properties of thrust faults

Thrust faults are the dominant features of many orogens and are a major result of crustal shortening during orogenic deformation. One of the great discoveries in tectonics occurred in the 1880s when, almost simultaneously, the Moine and Glarus thrusts were discovered in Scotland and Switzerland respectively (Plate 1). The significance of these discoveries was the realisation that large bodies of rock could be transported over large distances along low-angle thrust faults. This is now recognised as a fundamental mechanism of horizontal shortening and contraction within orogenic belts. Thrust faults are present in many if not all orogens, and they occur on all scales up to major thrusts with displacements of several hundred kilometres. As it turns out, the Moine and Glarus thrusts rank quite low with displacements of less than 100 km, and furthermore, they operated at a relatively modest depth. In contrast, seismic evidence in the Kapuskasing zone in the Canadian shield indicates that thrusts may involve both the entire continental crust and its underlying lithospheric mantle.

The discoveries of thrusts came from quite different lines of evidence. In Scotland Charles Lapworth (1883) postulated the existence of the Moine thrust as a result of finding along the thrust a new rock type which he called mylonite (Greek, milled rock). In other

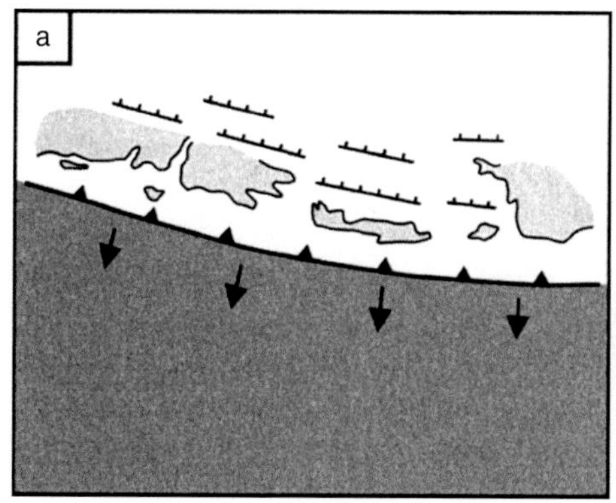

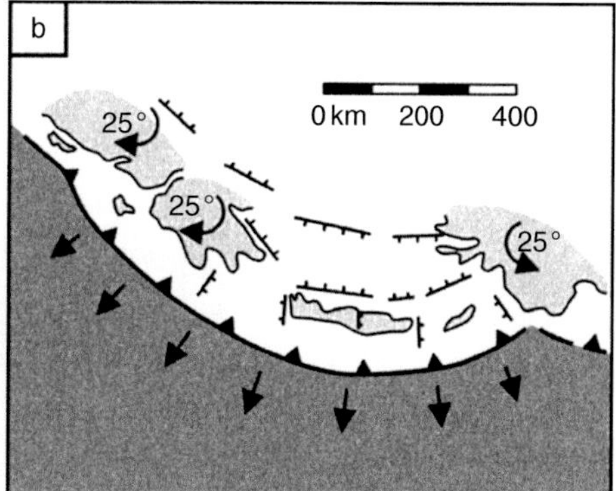

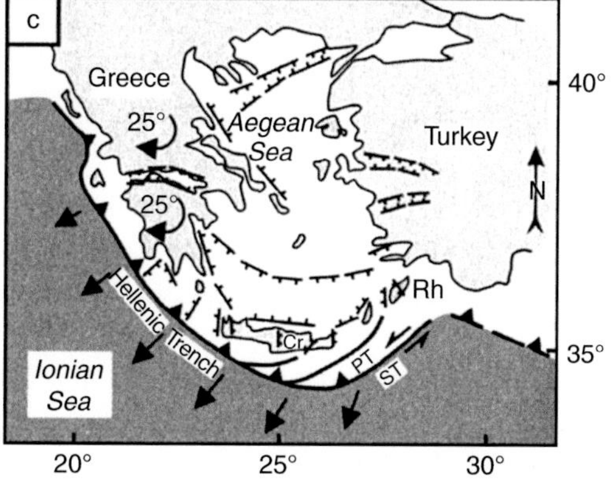

Figure 4.3 Evolution of the Aegean arc from Oligocene to present during subduction of the Ionian Sea beneath the continental lithosphere of the Aegean region. a, Oligocene to early Miocene; b, Middle Miocene to late Miocene; c, Pliocene to present. Curved arrows with numbers indicate sense and amount of rotation. Straight arrows indicate direction of migration of the arc. Note the increase of curvature with time from nearly rectilinear to strongly curved. Cr, Crete; PT, Pliny Trench; Rh, Rhodes; ST, Strabo Trench. From Schellert and Lister (2004), published by permission of the Geological Society of America.

words, the fault was a zone of high strain in which different rock types underwent intense deformation as a result of which their structure and mineralogy converge. In contrast the Glarus thrust in Switzerland was postulated to account for the inversion of the stratigraphical succession, in which older rocks were resting on younger rocks (Fig. 4.4). Marcel Bertrand (1883–4) invoked the Glarus thrust merely by looking at cross-sections of Glarus drawn by Swiss geologists who had recognised the stratigraphical inversion but had drawn back from making the obvious conclusion that a major fault was involved. Instead they envisaged a huge 'double fold', an unlikely geometry but one that minimised the amount of movement demanded by the stratigraphical inversion. By simply connecting up the strata across a valley Bertrand demolished the double fold. A fine piece of armchair geology by Bertrand!

During the nineteenth century, work in Switzerland particularly had demonstrated that rocks could be deformed, a fact that is beautifully displayed on mountain sides in the Alps, but what was envisaged by Bertrand was something of a higher magnitude of deformation, namely that large masses of rock could be transported horizontally for distances of tens of

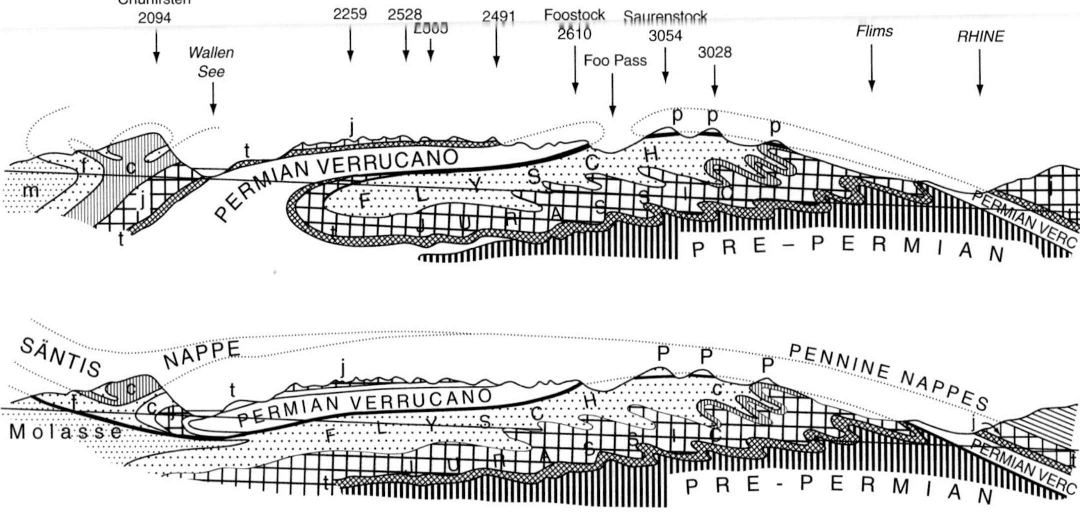

Figure 4.4 North–South cross-section of the Glarus area in Switzerland showing, in the top panel, the so-called double-fold interpretation of the inversion of the stratigraphical succession and below, Bertrand's reinterpretation, disposing of the double fold and replacing it by the Glarus thrust. From Bailey (1935), published by permission of the Oxford University Press.

kilometres. At the time this idea was greeted with general disbelief, but nowadays even the largest thrusts are dwarfed by the motions of huge plates. But the mechanical problems of thrusts still have to be addressed.

In Scotland, Lapworth's work was followed by a thorough mapping of the Moine thrust belt from Cape Wrath in the north to Skye in the south, by a team of British Geological Survey geologists led by Ben Peach and John Horne. In 1907 they published the famous memoir on the Northern Highlands (Peach *et al.*, 1907) which is a classic work on thrust tectonics (Plates 1, 2).

Unless a thrust has been folded by post-thrusting structures (Plate 3) its dip is low on the large scale, and so movement or displacement along thrusts results in a horizontal stacking or piling up of rock strata, generally with older rocks being stacked on top of younger ones. These sliding sheets of rock were called nappes by the early Swiss geologists, and this word (French, sheet, hence napkin) conveys the essential notion. Thus while nappes or thrust sheets vary in length from a few kilometres up to a few thousand kilometres, in thickness they rarely exceed a few tens of kilometres. Thrusts have a leading edge and a trailing edge where they may become detached from their roots. It is easy to imagine this process as something like a landslide, but this analogy breaks down because thrust faults have been identified at all levels in the crust and also in mantle below it. Depending on the depth of the lithosphere involved, thrusts are called thin-skinned or thick-skinned. Thick-skinned thrusts cut down to the mid/lower crust or below and therefore carry basement rocks, whereas thin-skinned thrusts stay mainly in the higher levels of the crust and therefore constitute only a part of the total deformation in the lithosphere.

Thrusts conform to the notion of simple shear because translation of one pile of rocks over another is involved. But pure shear, non-rotational strain is an important process in thrusting as discussed later in this chapter. Models of crustal deformation involve narrow

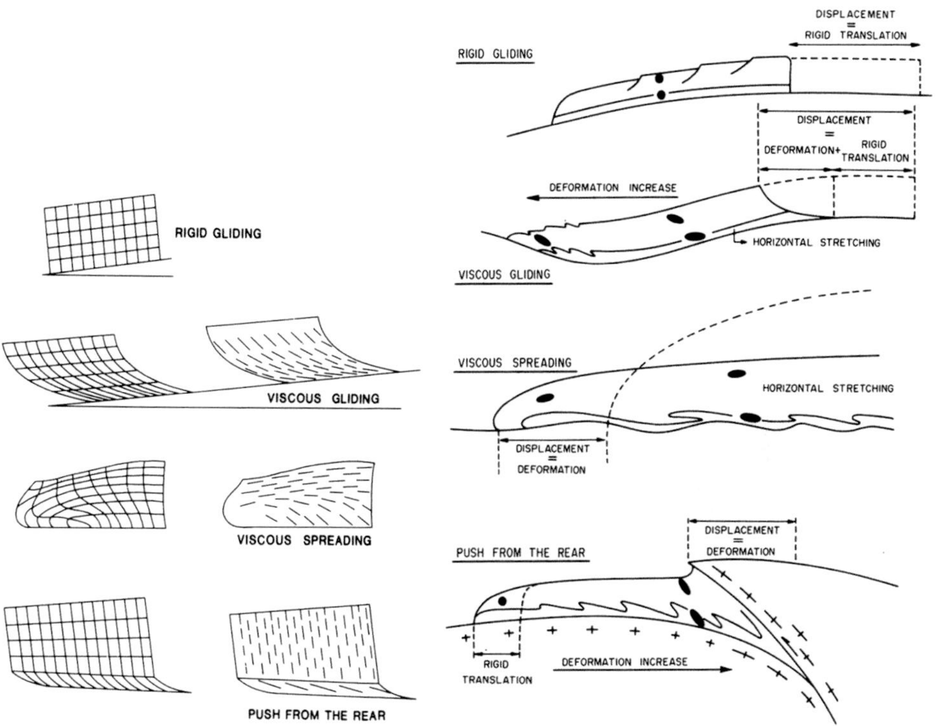

Figure 4.5 Deformation related to four mechanical models for thrusts. Top to bottom: the rigid body gravity gliding; gravity gliding but with viscous flow; gravity (viscous) spreading; push-from-behind. On the left the strain is indicated by an orthogonal grid. On the right the patterns of the trajectories of maximum extension are shown. From Figure 8 of Merle (1986), published by permission of Elsevier.

zones, thrusts, of strongly localised strain along zones of simple shear and a broader area of weaker strain with pure shear (Fig. 4.5). It is manifest that pure shear alone without simple shear is impossible because it causes space problems. This is called strain compatibility. An example of strain compatibility is in Tibet where, according to one of the models (B in Fig. 5.8), the lithosphere seems to have been thickened by vertical stretching in a pure shear regime. On a larger scale than Tibet alone there is compatibility because, according to that model, India indents Asia under a simple shear regime. Because of the thickening the lower crust has become hot and very mobile and has been extruded eastwards from the Tibetan Plateau. The process of extrusion involves simple shear. The lesson is that it is necessary to take account of the strain regimes operating on all scales, in order to understand the processes.

Although orogens are traditionally portrayed as contractional regimes, more recently it has been recognised that normal faults, particularly low-angle normal faults, play an important role in the orogenic process. The extensional regime represented by normal faults may reflect changes in the orogenic wedge as discussed later in this chapter, or they may, as in the Himalaya, be associated with the extrusion tectonics (see Chapter 6). Again low-angled normal faults may be linked to the exhumation of the orogen as an alternative to erosion. This process is called tectonic denudation, during which loss of cover is accomplished by transposition along a low-angle fault.

Doubly vergent orogens

In recent years the development of so-called doubly vergent orogens has been invoked (Willett *et al.*, 1993) (Fig. 4.6). In contrast to the familiar notion that collisional orogens are markedly asymmetric, this concept involves a symmetrical or near-symmetrical orogen. The asymmetry is preserved by the plate convergence but the orogen above the subduction

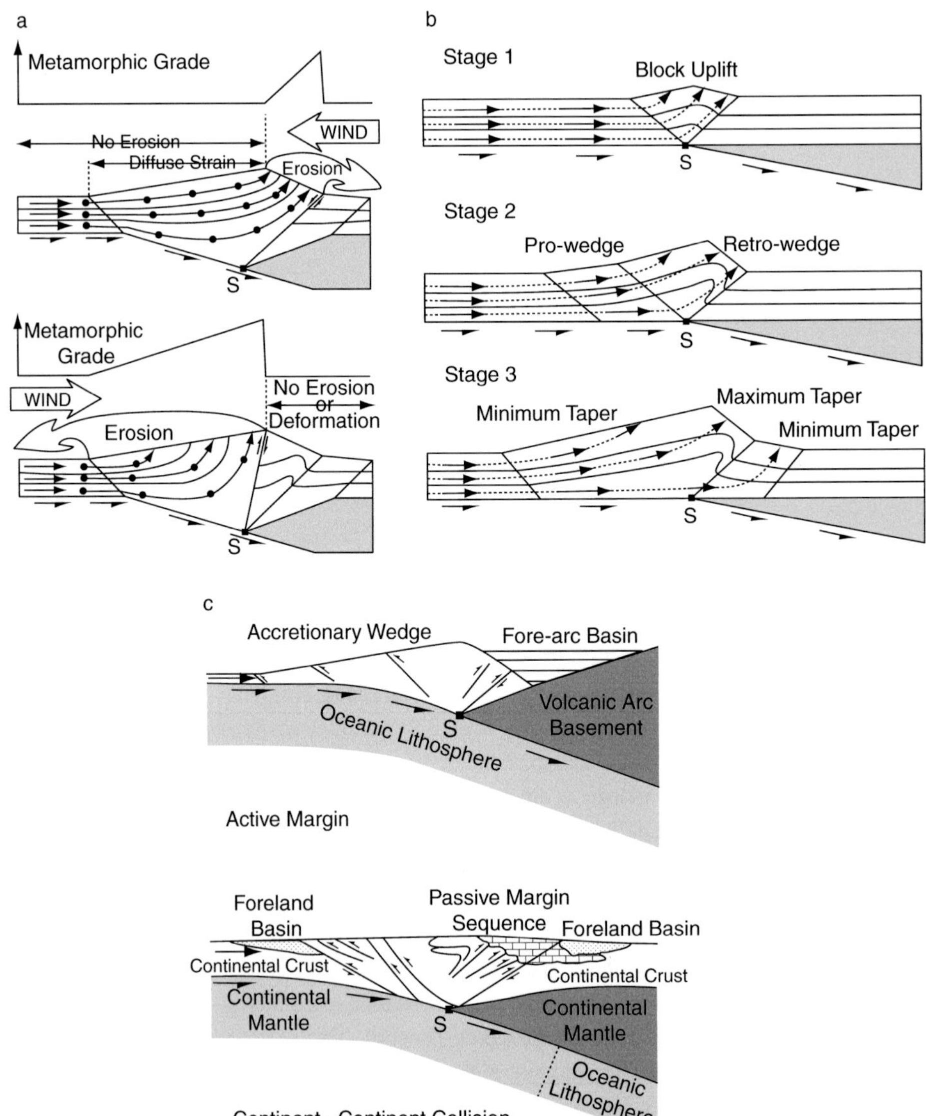

Figure 4.6 Doubly vergent model for orogens showing pro-and retro-wedges, the S-point and growth of the wedge. Modified from Figure 21 of Willett *et al.* (1993), published by permission of the Geological Society of America.

zone shows vergence in opposite directions to form a pro-wedge and a retro-wedge which are separated by the S-point. The S-point (Fig. 4.6) is located at the point where the lower plate subducts beneath the upper or overriding plate. The thrusts are propagated outwards from the S-point at different rates so as to form the pro- and retro-wedges. The Pyrenees show this double vergence. In addition Willett and his coworkers suggested that the Himalaya–Tibetan system is another example of a doubly verging orogen. The Himalaya constitute the pro-wedge and Tibet is the putative retro-wedge. The pro- and retro-wedges do not behave entirely in accordance with the wedge theory set out in this chapter. The pro-wedge may do so but the retro-wedge is steeper and shows the maximum taper angle. However, the doubly vergent orogen model seems to be at odds with the channel flow hypothesis by Beaumont *et al.* (2004), which in the Himalayan case requires that the Himalayan upper crust has been subducted under Tibet (see Chapters 5 and 6). Similarly, the Western Alps, Pennines and Helvetic Alps can be viewed as a pro-wedge with northerly or northwesterly vergence, and the less strongly deformed Southern Alps as a retro-wedge with southerly vergence.

Classification of thrusts

For some years, discussions of thrust mechanisms revolved around the question of whether thrust sheets are pushed from behind by application of compressive stress or whether their motion is controlled by gravitational forces. Perhaps this is not an either/or situation (Fig. 4.5). Some important models which have been put forward to account for thrust tectonics are given below.

(a) Gravity gliding or spreading/flow models

In gravity gliding models the thrust sheet slides down a slope under gravitational forces. Unlike gravity gliding, gravity spreading involves internal deformation of the rocks in the thrust sheet. Gravity spreading deformation may start as a predominantly simple shear phenomenon which then becomes more general (with pure shear) at higher strains owing to late stage gravitational collapse of the orogenic pile. The analogy is with the flow of a glacier. As with glacier flow, motion can be uphill – that is upwards along the thrust plane.

(b) Extrusion of orogen-scale tectonic wedges

An example for this model, favoured by workers such as Kohn (2008), can be found in the Greater Himalaya where the top and the base of an extruded wedge are marked by thrust and normal faults respectively. Searle *et al.* (2003) showed that microstructural data from the Main Central Thrust of Everest region indicate a significant component of plane strain ($k = 1$) pure shear deformation with pure shear components representing 13–53% of the total recorded deformation (see Chapter 3). There is no space problem connected with the pure shear strain path because the general strain results in the transport of the rocks to the Earth's surface. The channel flow model will be discussed in Chapters 6 and 7.

(c) Dislocation model

The name comes from an analogy between the movement of dislocations in crystals and the manner by which thrusts propagate and move by small incremental slips. Thus thrust movement can be likened to the motion of a caterpillar which moves forward owing to increments of small movements along its body. The dislocation model does not specify a particular driving force. The 'dislocation' model is ductile and can be envisaged as differential sub-horizontal displacements within the leading edge of planar mylonite zones with linked transport-parallel compressional and extensional flow in the trailing sections.

Note that (a) and (b) are specifically gravity-driven models.

(d) 'Push-from-behind' model

This is a compressional stress-driven model as the name implies, and it might be called the traditional hypothesis for thrusting. However, the flow mechanisms may be similar to those in some of the other models. The ultimate cause of most thrusts is the compressive stress set up in the lithosphere by convergence of plates, but the secondary consequence of the ensuing lithospheric thickening may be gravity-driven thrusting. In the past gravity-driven and compression-driven thrust models were set up as either/or cases, but in nature the simplicity of cause and effect may be elusive.

In referring to thrust transport it is usual to employ two terms. The *foreland* is the stable region outside the orogen towards which most thrusts have moved. (The term 'vergence' refers to the direction of thrust transport.) The *hinterland* is the region on the other side of the orogen away from which most thrusts have moved. The front of a thrust sheet is called its leading edge, which may be at the Earth's surface, and its rear portion is the trailing edge. Two other terms are useful here: the body of rock above a thrust is known as the hanging wall and that below it is the footwall. A full account of the terminology used in thrust systems is given in Boyer and Elliott (1982) and McClay (1992).

Detachment

Thrusts are detachment surfaces, and obviously they should occur along relatively weak rocks or along the interfaces between weak and strong rocks. There are examples of thrusts running in weak rocks such as evaporites (e.g. the Jura in the Swiss Alps, the Zagros mountains and the Pyrenees). But in the realm of regional metamorphism another agent for softening or weakening rocks comes into play. This is called thermal weakening (see Chapter 3), the implication of which is that thrusts follow horizons where elevated temperatures have softened, perhaps even partially melted, the rocks. There is also reaction softening in which, as a result of deformation and metamorphism, mineral assemblages are altered so as to restore equilibrium. If this process produces flaky micas or other minerals with platy cleavages, e.g. chlorite, then slip is made easier. Reaction softening by hydration of mineral assemblages can be brought about by the ingress of water along the thrust which becomes an easy pathway for fluids. Such a process serves to facilitate sliding. Water, whether it is of meteoric origin or produced by dehydration reactions in rocks, can percolate down into basement rocks and weaken then.

Time spans of thrusts

Measuring the time span of activity on a thrust is difficult. In the Canadian Rocky Mountains the timing of thrust activity can be deduced from the youngest rocks in the thrust sheet and the earliest date for the associated foredeep. More generally, palaeontological and geochronological studies give reasonable though imperfect estimates for slip rates for thrusts with values that usually lie between 5 and 100 mm per year (compare the rates of plate motions); hence the active life for a large thrust sheet may be several million years. That does not mean that a thrust moved at, say, 5 mm a year for every year of its active period. We know that earthquakes are intermittent, so we should regard the stated slip rates as averaged and therefore disguising the intermittency of slip. The dislocation model outlined above suggests that not all parts of a thrust or sliding surface are active at one time; rather, the motion is caterpillar-like. The sliding surface is formed initially by the build-up of stress at a point in the rock, a stress concentration which leads to fracturing giving a micro-crack which then grows or propagates just like the branches of a tree, so that minor faults, imbricates or splays, branch off from major ones. This crack continues to grow longer provided the stress levels are high enough, until we have a fracture along which displacement starts and continues as long as the energy level in the orogen is sufficient. The term detachment (more precisely the thrust pile is detached from its basement) is useful in this context: the French word is *décollement*, or unsticking.

In the Moine thrust zone, geochronology has been used, firstly by dating of the mylonites (*c*.430–408 Ma; Freeman *et al*., 1998) and by dating alkaline igneous intrusions. Many believe that such intrusion was interleaved with slip on the Moine thrust and lower thrusts. Recent zircon U–Pb dating by Goodenough *et al*. (2011) of the Loch Ailsh, Borralan and Canisp alkaline intrusions in the Assynt region suggest that the time span for the main slip on the Moine and Ben More thrusts is no more than a million years at *c*.429–430 Ma. Perhaps that date does not constrain some ductile activity on the Moine thrust, but it is probable that movement on the Moine thrust overlapped intrusion of the early suite of the Borralan intrusion, and movements ceased by the time of intrusion of the late suite at 429 $\pm$ 0.5 Ma. If the slip on the Moine thrust zone is *c*.100 km this gives a slip rate of 5 cm per year. These authors suggested a slip rate for the Glencoul and Ben More thrusts of 2 to 8 cm/a.

Thrust systems

So far we have considered the development of single thrust faults, but orogens are composed of many faults which together form a thrust system reflecting the dynamic growth of a thrust belt over several million years. If we freeze-frame as in a movie, one thrust is active, but move the film on and freeze again and it is seen that the first thrust is now inactive and activity has been transferred to another thrust, and so on to the end of the movie or rather the orogeny! In piles of thrusts the sequence may be top-to-bottom; that is, the younger thrust develops underneath older ones and carries the older ones on its back. This is called 'piggy-back' thrusting as found in the Canadian Rocky Mountains or Helvetic Alps and indeed in many thin-skinned orogens, although often the initial piggy-back sequence is modified as a result of reactivation along the higher thrusts (see below). The first demonstration of a piggy-back sequence was given by Clough, writing in the 1907 *Memoir of the Northern Highlands of Scotland* (Peach *et al.*, 1907). At Dundonnell, the Moine thrust has been folded into a sharp antiform, whereas the lower Sole thrust is planar, indicating a top-to-bottom propagation of thrusts. Leslie (in press) has refuted the existence of the antiform, replacing it by an imbricate stack, but the sequence of thrusting is not in doubt. The recognition of the folding of thrusts during later thrusting has been applied in the Canadian Rocky Mountains (Plate 3) and elsewhere.

The work on the thrusts of the Canadian Rocky Mountains, notably by Price (1981), Dahlstrom (1970) and others, has been very influential in the study of the geometry of thrust systems, particularly because it has provided useful terms to describe the geometries of thrust systems (Boyer & Elliott, 1982). Thrust trajectories may be in smooth upwardly concave ('listric') curves, or, as the thrust sheet moves upwards towards the surface, it may progress in steps called ramps (Fig. 4.7). Ramps may be (a) frontal or longitudinal, parallel to the strike; (b) transverse, nearly perpendicular to the strike; or (c) oblique, at an angle to the strike (Fig. 4.8). Ramps are essential, otherwise thrusts would not place older rocks on younger. Between ramps, thrusts may travel along flats parallel to the bedding or foliation planes (Plate 4).

Imbricates and duplexes

So-called footwall collapse produces a duplex (named after a split level apartment) which consists of a roof thrust and a floor thrust connected by smaller thrusts which develop in sequence. The thrust units in a duplex are bounded on all sides by thrusts and are called horses. Branches coming off a master fault were called imbricates by the original mappers of the Moine thrust zone (Peach and Horne and their associates). But duplex faults are different from imbricates because they curve upwards in smooth curves, asymptotically, and join the master or 'roof' fault (Fig. 4.9). The whole duplex is underlain by a basal detachment. The duplex system marks a sequential development of the thrust system in which active slip is transferred from the master fault to the duplexes and the floor thrust. The emplacement of the duplex causes a footwall thickening and in consequence the roof thrust is deformed by folding. This process may explain culminations or upwarps in thrusts, such as the Assynt culmination in the Moine thrust belt where the Moine thrust is gently upwarped (Fig. 4.10a,b, Fig. 4.11).

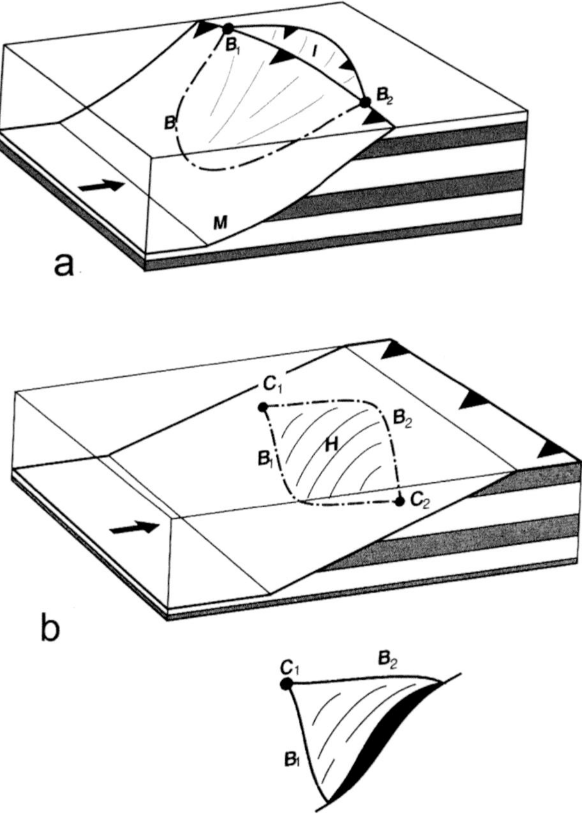

Figure 4.7 Thrust terminology. a, Thrust M with imbricate or splay (I) showing branch line B and tip lines. The branch line intersects the surface at B_1 and B_2. b, A lenticular horse (H) in the footwall of a fault, and below, the horse in 3D. From McClay (1992), Figures 4a and 27, published by kind permission of Springer Science and Business Media.

Modelling

The sequential foreland-directed propagation of thrusts, which is so clearly demonstrated in the Canadian Rocky Mountains, has been simulated in model experiments. The first, in 1989, was by Cadell who was inspired by the discoveries being made on the Moine thrust at that time (Cadell, 1989). In more recent work, the process of foreland-propagating thrust systems has been illustrated by analogue modelling and by sand box experiments (see Stockmal *et al.*, 2007). For example, Lui *et al.* (1992) simulated thrust systems in sand box experiments and demonstrated the progressive foreland-directed spreading of the thrusts. The experiments are clearly relevant to the theory of thrust wedges discussed later in this chapter. The spacing of thrusts was shown to increase with increasing basal friction, and the spacing/thickness ratio of foreland vergent thrusts increases as layer thickness increases. The experiments evaluate factors such as the basal friction of the sliding wedge (see the section on Orogenic wedge theory later in this chapter), and the critical wedge taper in the

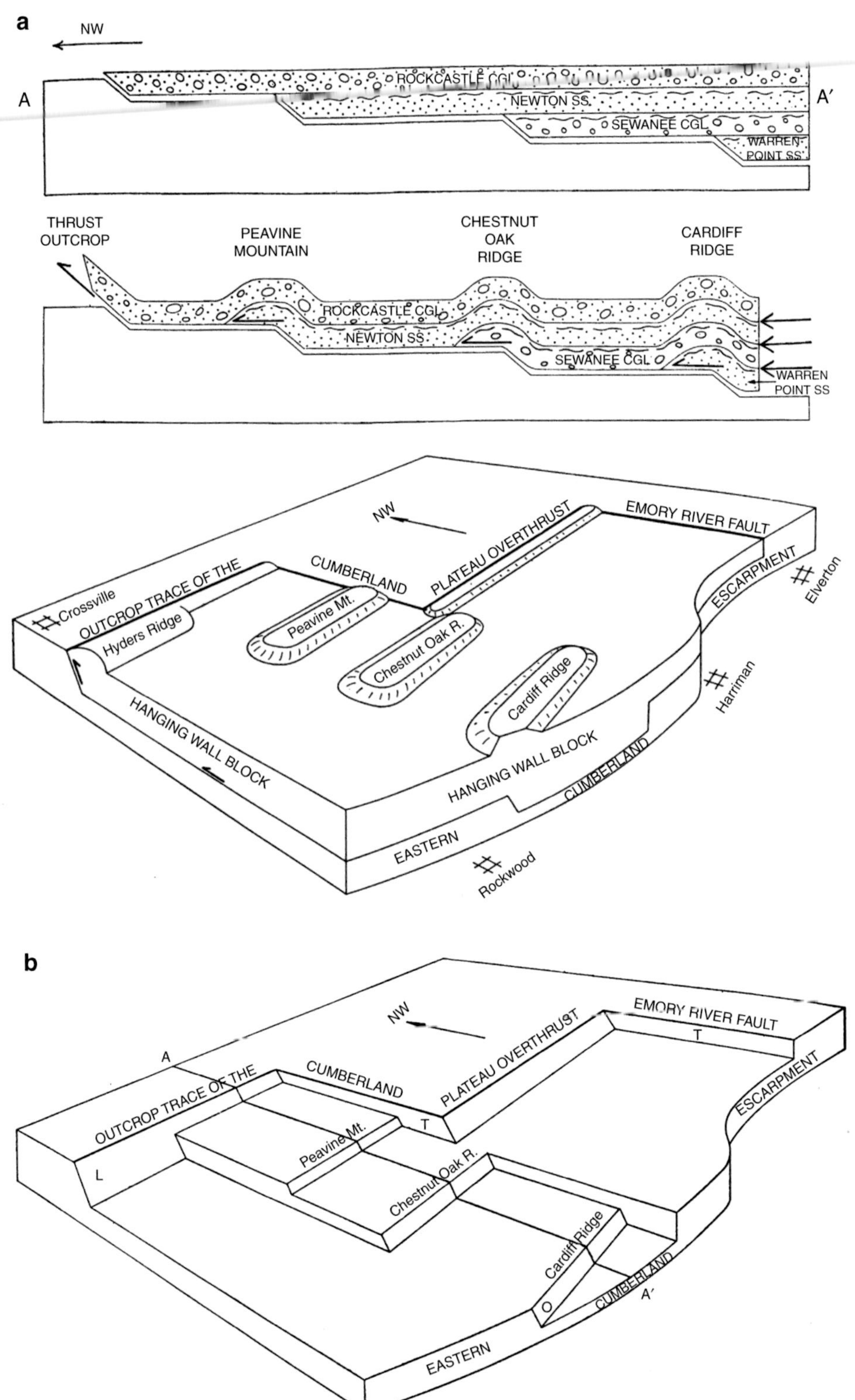

Figure 4.8 The Cumberland thrust in Tennessee. a, The section A–A′ on part b before and after thrusting to show the ramps and ramp-related anticlines. b, Above, the surface geology with anticlines forming ridges and below, the inferred subsurface structure with the hanging wall of the Cumberland thrust removed. L, longitudinal/frontal ramp; T, transverse or lateral ramp; O, oblique ramp. From Wilson and Stearns (1958), published by permission of the Geological Society of America.

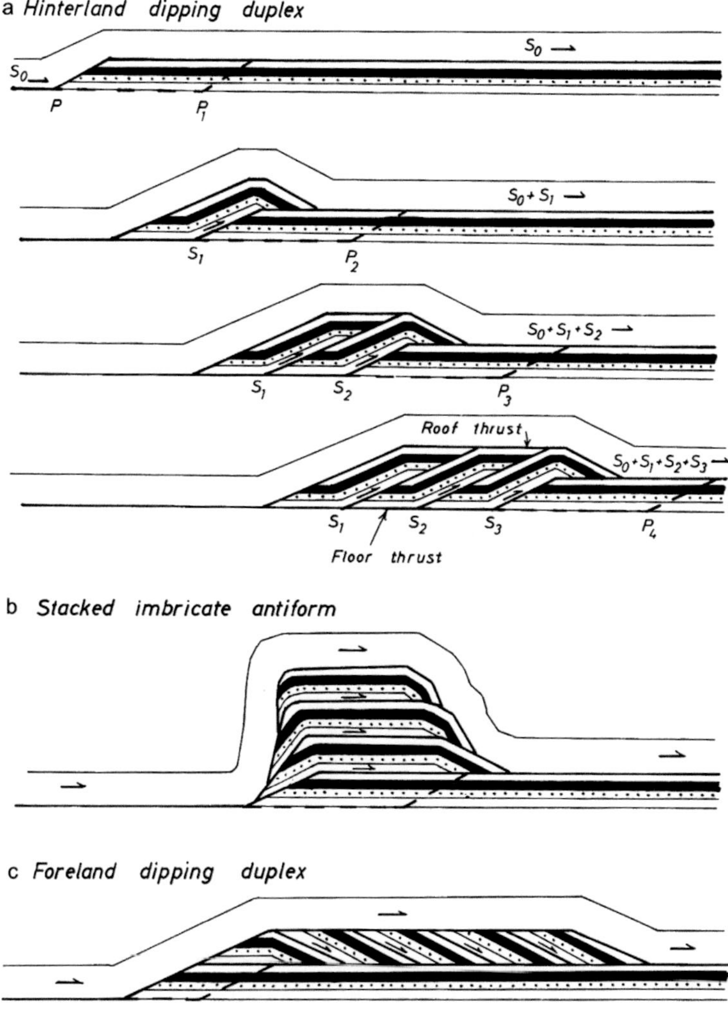

Figure 4.9 Evolution of a duplex by growth of thrusts. a, S_0–S_3, beneath older roof thrust fault. P_1–P_4, trajectories of future faults. b, The stack of duplexes causes folding of the roof thrust. c, A foreland dipping duplex. From Ramsay and Huber (1983, 1987), published by permission of Elsevier.

experiments was close to theoretical predictions using an intermediate value for basal friction of $\gamma_b = 0.47$. For lower basal friction the taper angle is lower. Supra-critical wedges are formed when the basal friction equals or is greater than the coefficient of friction in the wedge and the wedge has a high taper close to the angle of rest for the modelling material.

The contrast between imbricate faults and duplexes emphasises the importance of the places where different thrusts meet, the so-called thrust junctions. A junction may indicate one thrust cutting another or it may mean the thrusts are part of an evolving duplex in which active slip is transferred from one thrust to another (Boyer & Elliot, 1982). It is not easy to decide which. In recent years the simple hinterland-to-foreland, piggy-back, thrust propagation has been criticised, and departure from the rule gives so-called

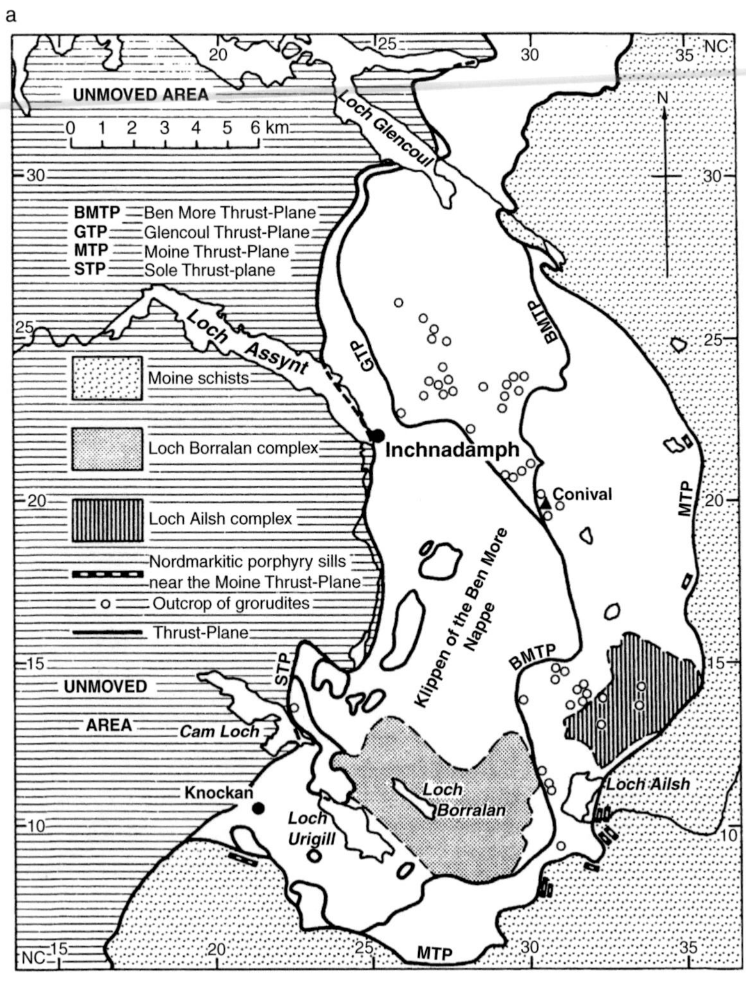

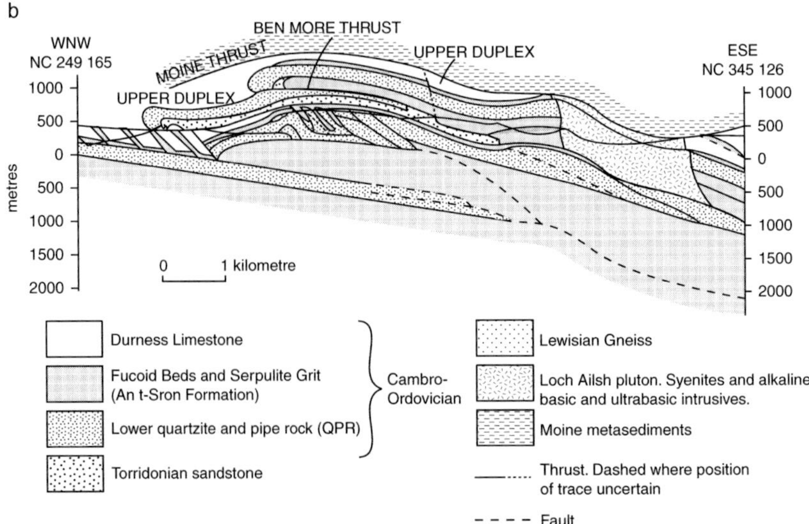

Figure 4.10 Moine thrust belt in the Assynt region, NW Scotland. a, Map to show the Assynt culmination (upwarp of the Moine thrust), and the sub-Moine thrusts. b, Balanced cross-section across the middle of the culmination showing the sub-Moine thrust sheets and horses. From Strachan (2003), published by permission of the Geological Society of London.

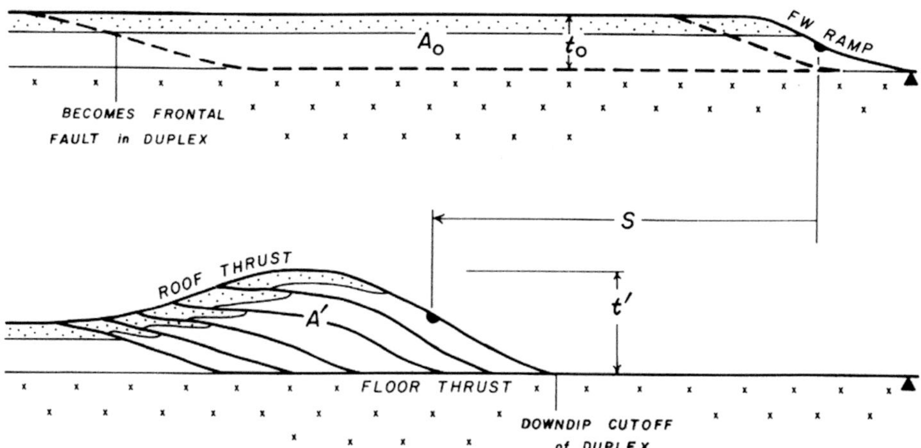

Figure 4.11 Folded roof thrust to a duplex in the Moine thrust zone. From Elliott and Johnson (1980), published by permission of the Royal Society of Edinburgh.

out-of-sequence thrusts which propagate towards the hinterland, in effect reversing the piggy-back sequence. After reviewing these developments, Boyer (1992) suggested that the original idea that a higher thrust became inactive when lower thrusts developed is much too simple because simultaneous thrusting may occur throughout thrust belts. Thus, older thrusts may be reactivated and newer ones may develop in hanging walls. The message is that there are no rigid rules, and each case must be treated on its merits. Tanner (1992) has cited problems with the mechanical evolution of the Boyer and Elliott duplex model and suggested important modifications.

How can time sequences of thrusts be worked out? The Canadian Rocky Mountains provide many answers to that question; for example an older thrust has been steeply folded as a result of a younger thrust moving beneath it (what we might call the 'rabbits running about under a carpet' model), or the thrust sequence may be elucidated using the palaeontological dating of sediments in thrust foredeeps. In addition, truncation of imbricates by a roof thrust may be used, as in the Moine thrust belt.

Relationship between folds and thrusts

The association of folds and thrusts in orogens is a common one but the relationships of folds and thrusts are complex and varied, as follows:

1. Fault bend folds. These are found on ramps and mainly in the hanging wall (Fig. 4.12a). They presumably arise because of shortening strains set up as the rock sequence climbs the ramps.
2. Fault propagation folds. These are associated with, and develop ahead of, a growing thrust, which because it is still actively propagating towards the surface is called a blind thrust (Fig. 4.12b).

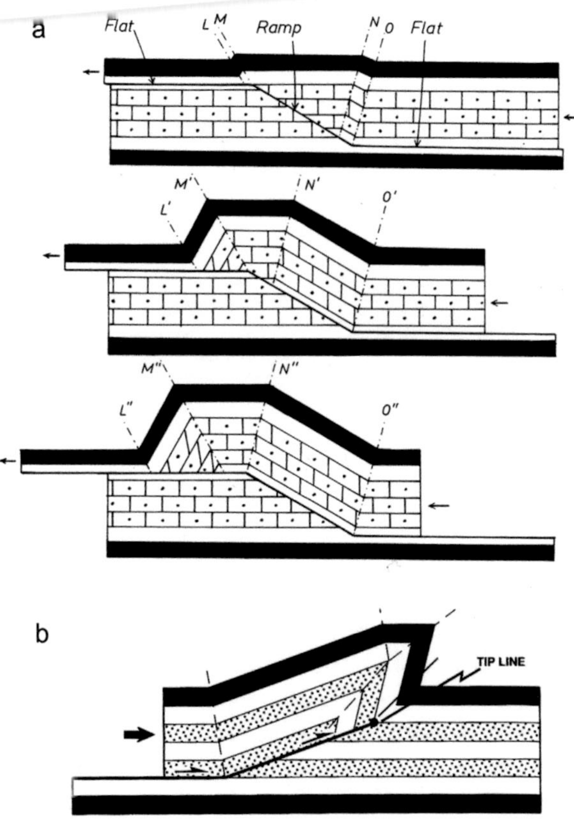

Figure 4.12 Folds related to thrusts. a, Fold developed on a ramp but only in the hanging wall. L, M, N, O show the amplification of the fold. b, Thrust propagation through a fold. The tip line indicates that this is a blind thrust. Panel a from Ramsay and Huber (1983, 1987), published by permission of Elsevier), panel b from McClay (1992), Figure 8a, published by permission of Springer Science and Business Media.

3. Folds in shear zones. This is a wide category encompassing sheath folds, perturbation folds in mylonitic shear zones, and folds developed between two thrusts and synchronous with the thrusting. The Morcles fold in the Helvetic Alps is a good example of the type.
4. Folds which precede thrusts in the manner described by Heim (1922). In these recumbent folds the progressive tightening leads to attenuation and thinning out of the inverted limb, and eventually the whole fold slides along the thrust.
5. Folds which are later than one thrust but earlier than others. Heart Mountain in the foothills of the Canadian Rocky Mountains shows a steeply dipping thrust which has been rotated during movement of later thrusts moving underneath it. The southern Moines of the Scottish Highlands show major folding of the Sgurr Beag thrust which has been dated at *c*.450 Ma; indeed, polyphase folding has affected that thrust. Subsequently the whole thrust stack, including the Sgurr Beag thrust and the polyphase folds, was carried piggy-back along the Moine thrust which is dated at *c*.430 Ma.

Ramsay (1992) has criticised the first two models (Fold types 1 and 2 above), firstly because of the frequent use of kink band geometries which are not common fold styles, and secondly because causal mechanisms are vague and do not exclude the development of fault bend folds in footwalls at ramps. The warning here is against the over-application of a single model. Other models for the Helvetic folds include gravity gliding or spreading, or folding against a buttress formed by a normal fault. In yet another interpretation (see point 3 above and Chapter 5) the Helvetic folds are explained by simple shear and shortening of multilayers between two moving thrusts (see Ramsay *et al.*, 1983 and Fig. 5.21).

Displacement on thrusts

The calculation of displacement on thrusts, especially large ones, is often very difficult. The term displacement is used to describe the change in spatial position of a particle within a body of material. Displacement of a body or thrust sheet leads to three main effects: (1) body translation, (2) body rotation and (3) internal strain (Fig. 4.5).

Failure to recognise these effects may lead to grossly inaccurate estimates of displacement. If a geological feature is present in both the hanging wall and footwall, then the calculation of displacement can be solved easily, especially if no internal strain occurs in the thrust sheet. If we can identify the point where the feature intersects the fault in the hanging wall and the footwall then the displacement can be calculated. Such intersection points are called 'piercing points'. In NW Scotland Clough, in the classic 1907 *Memoir* on that region (Peach *et al.*, 1907), devised a method of calculating displacement on one of the sub-Moine thrust sheets, the Glencoul thrust (Fig. 4.13). This involved matching piercing points (or rather piercing lines) given by the intersection with the Glencoul thrust of the steep contact between two metamorphic complexes in the Lewisian rocks which can be located in the thrust sheet and in the foreland. The result shows that the intersection is a few kilometres further south in the thrust sheet than in the foreland. This might indicate a strike-slip displacement of *c.*10 km, but the kinematic indicators for the thrusting indicate a WNW direction and so a much greater displacement is needed (*c.*24 km) to realign the piercing lines. This example is remarkable because of the parallelism of the mafic dykes (Scourie dykes) that cut the gneisses in the thrust sheet and in the foreland, indicating that the displacement on the thrust involved no rotation of the thrust sheet on a vertical axis.

Another way of calculating displacement is to use klippen, which are outliers of part of the thrust sheet isolated by erosion from the main body of the sheet (Fig. 4.14). The distance between the outer edge of the most external klippe and the front of the main sheet gives a minimum value for the displacement. For example, in the Himalaya the distance from the most external klippen to the main outcrop of the Main Central Thrust in the High Himalaya gives a minimum displacement for the MCT of about 100 km. Intuitively most geologists think it is much more than that. Similarly in the Swiss Alps the Pre-Alpine klippen can be used to give a rough estimate of the displacement on the Pennine nappes.

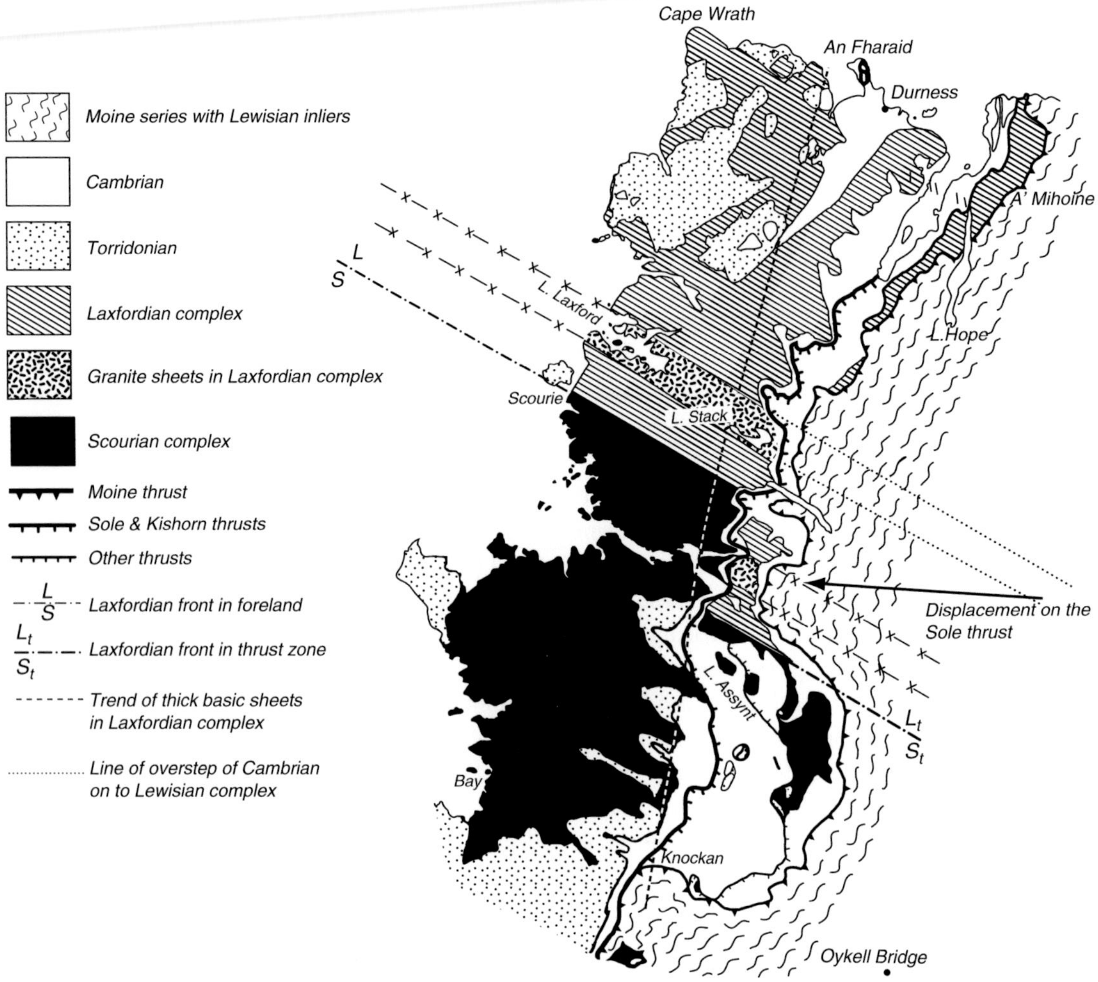

Moine series with Lewisian inliers

Cambrian

Torridonian

Laxfordian complex

Granite sheets in Laxfordian complex

Scourian complex

Moine thrust

Sole & Kishorn thrusts

Other thrusts

$\frac{L}{S}$ Laxfordian front in foreland

$\frac{L_t}{S_t}$ Laxfordian front in thrust zone

Trend of thick basic sheets in Laxfordian complex

Line of overstep of Cambrian on to Lewisian complex

Cape Wrath

An Fharaid

Durness

A' Mihoine

L. Laxford

L. Hope

Scourie

L. Stack

Displacement on the Sole thrust

L. Assynt

Bay

Knockan

Oykell Bridge

Figure 4.13 Map of the northern part of the Moine thrust zone showing how C. T. Clough used offsets of the steep contact of the Scourie–Laxford complexes in order to estimate displacement on the Glencoul thrust. From Ramsay (1969), published by permission of the Geological Society of London.

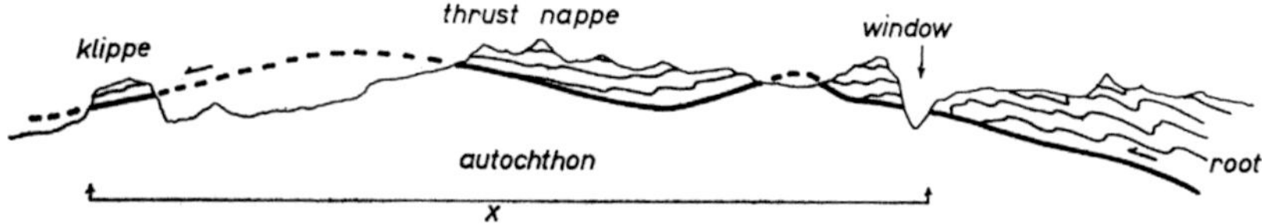

Figure 4.14 A thrust sheet with a klippe and a window. From Ramsay (1969), published by permission of the Geological Society of London.

Unfortunately, these simple techniques may be inaccurate when applied to many thrusts, particularly those sheets in which internal strain occurred during translation (Fig. 4.15). The Glencoul thrust was probably a rigid sheet because the rocks carried in the sheet mainly preserve their pre-thrust features. However, the MCT sheet in the Himalaya was

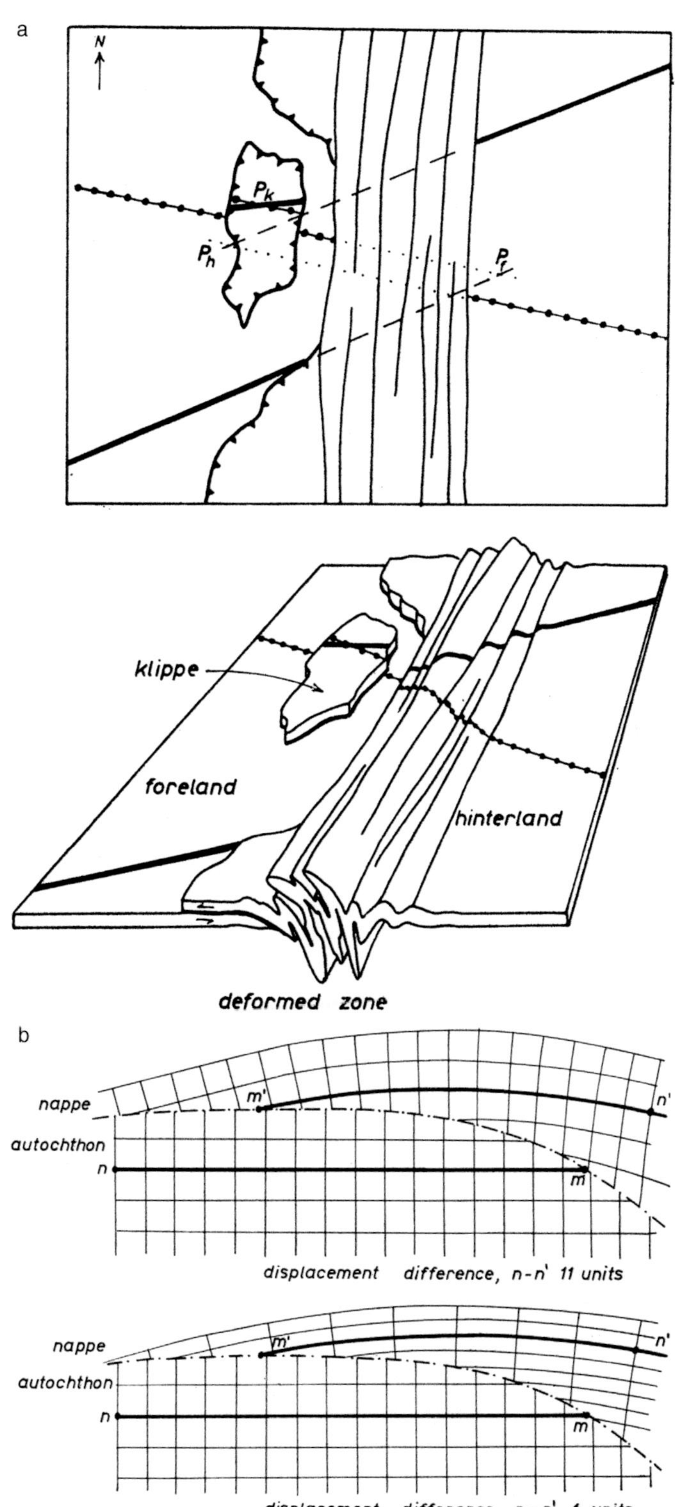

Figure 4.15 a, Offsets of non-parallel linear features (such as dykes or major fold axes) caused by an orogenic belt can be used to calculate shortening or displacement across an orogen. The intersection of these features is determined by

almost certainly translated at elevated temperatures, and the large internal, synthrust, strain has been characterised as general shear, in other words a combination of pure and simple shear. This presents a major problem – not only do we need piercing points, it is also necessary to remove the internal strain. This is possible if there are objects in the rocks that permit the estimation of the internal strain. Deformed pebbles, oolites and fossils and the quartz microstructure have been used for this purpose.

Restored sections

The problem of internal strain in thrust sheets is part of the general difficulty in estimating the amount of shortening in an orogen. In young belts this can be done by analysing the widths and ages of the magnetic anomalies on the ocean floor, in other words using magnetostratigraphy. We have seen before how magnetic anomalies provide a wonderful clock for tracking and timing plate motions. From these data it is possible to calculate the amount of convergence across an orogen. But this is obviously not available in older orogens where the clock – that is, the ocean floor and its stripes – has been subducted. Geometrical methods are available, such as the measurement of the offset of a large-scale structure which occurs on both sides of the orogen. Alternatively, within orogens there is the piercing point technique referred to above. Another way is to use techniques allowing the restoration of the section, that is, by removing the deformation structures. Another term for this is section balancing (Dahlstrom, 1969, 1970, Hossack, 1979) which allows the construction of a restored section, i.e. the section as it was before orogenesis, with the removal of thrusts, folds and other internal strains (Figs. 4.16, 4.17). Balancing means that volume is conserved at all levels in the section; in other words there must be no holes left unfilled. Not surprisingly, results vary in reliability. The technique works quite well in young thin-skinned belts such as the Canadian Rocky Mountains in which internal strain within the thrust sheets and metamorphic changes are not strong. However, in thick-skinned belts where basement rocks are involved, there are many problems in section balancing, and more seriously, if metamorphism was coeval with thrusting then possible volume changes (due to change of volume by diffusion or removal of material out of the section) will complicate the restoration of the undeformed section.

Caption for Figure 4.15 (*cont.*) graphical extension of features P_h in the foreland and P_f in the hinterland. The total displacement and shortening across the orogen can be measured by constructing the vector which joins P_h and P_f. Also shown is an alternative method which can be used where the intersection of the features is found in a klippe (P_k). From Ramsay (1969), published by permission of the Geological Society of London. b, Thrust sheets with rigid body translation (above) and with internal strain (below) are superficially similar. The strain or lack of it is illustrated by a grid. Although the displacement differences shown by m–m′ are identical for the two thrusts, the crustal shortenings measured by n and n′ are quite different. From Ramsay (1969), published by permission of the Geological Society of London.

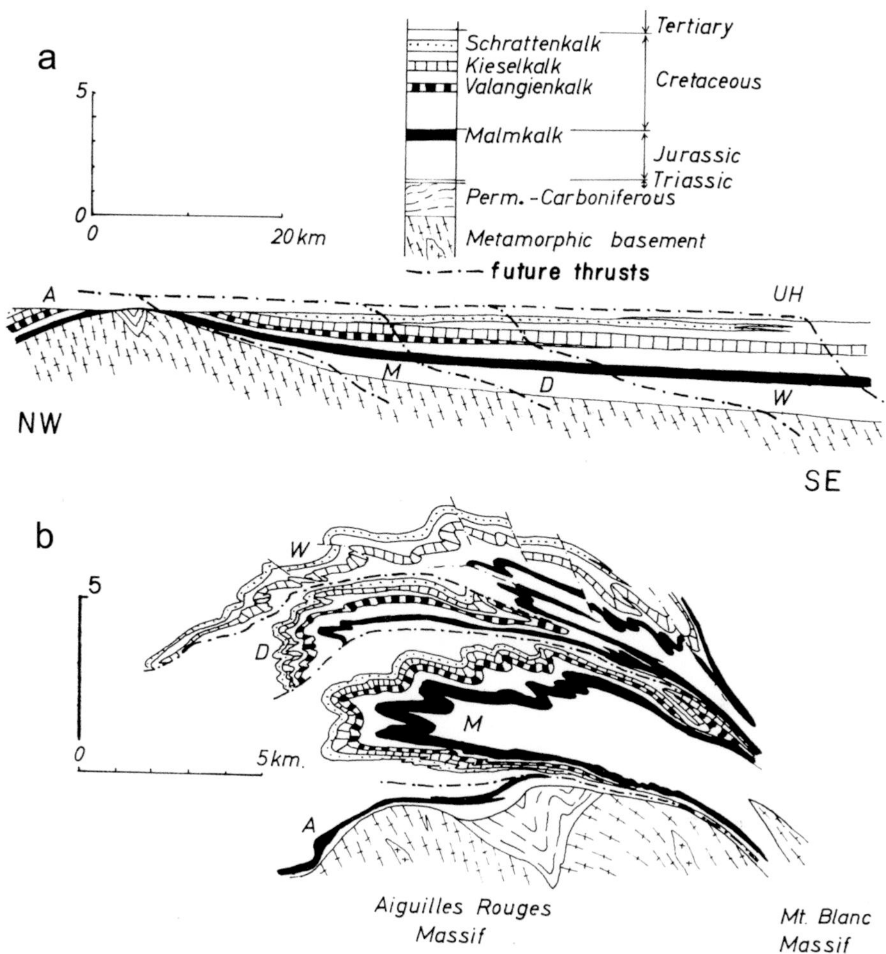

Figure 4.16 Lower nappes in the Helvetic Alps to show the idea of restoration of section. a, Palinspastic section to show Mesozoic shelf sediments above crystalline basement, and b, a profile to show the section after orogenesis. A, 'Autochthon'; M, Morcles Nappe; D, Diablerets Nappe; W, Wildhorn Nappe; UTT, Ultrahelvetide Nappes. b, Basement–cover relations in the Helvetic zone. To construct the restored section requires knowledge of, first, the transport direction and order of formation of the nappes; second, the amount of displacement; and third, the effect of unrolling the internal folds in the nappes. From Ramsay *et al.* (1983), published by permission of the Geological Society of America.

Thrust terminations

So far we have looked at the arrangement of thrusts and thrust systems in two dimensions in a profile section perpendicular to the strike, although time has also been considered. But what do thrusts look like from above, as viewed on a map, for example? Although some thrusts are of very great length along strike, measured in

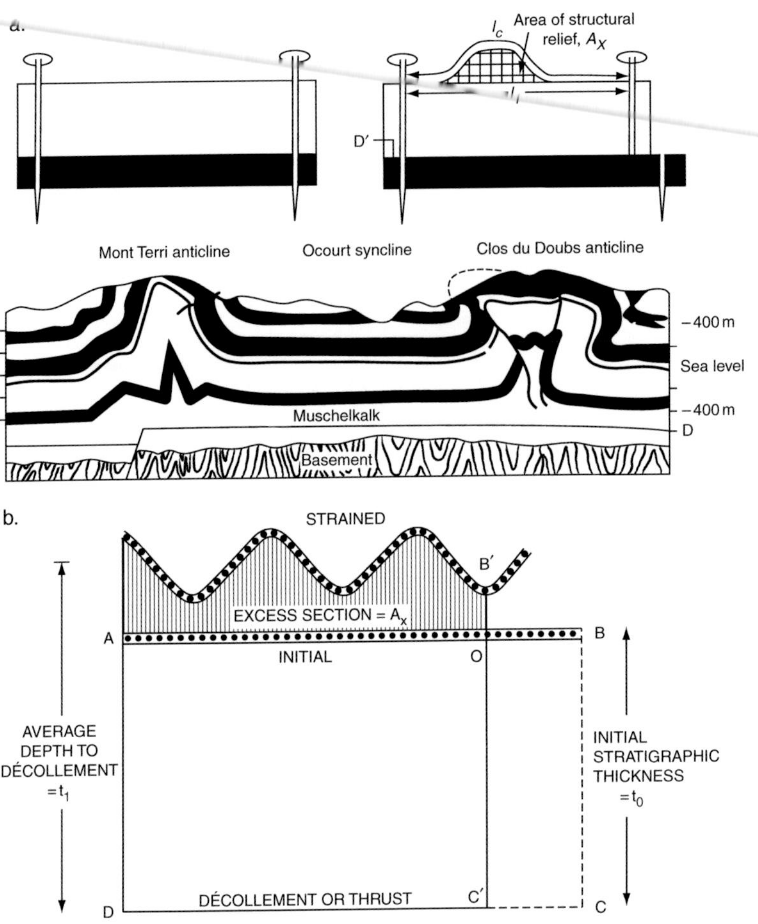

Figure 4.17 a, b, The principles of construction of balanced cross-sections using the Jura with the décollement as an example. Beds of thickness t_0 overlie a thrust or décollement. Plane strain and no loss of volume is assumed in the following. a, Two folds in the Jura section. The décollement separates the folded section from the part below the décollement (for further details on the Jura see Chapter 5). Shortening of a section, the length of which is delimited by pins. Folding above the décollement is shown by the broken pin in the right-hand figure. The part below the décollement is in black. b, Methods of measuring the shortening. Bed AB has a length l_0. After strain the bed is shortened to AO $= l_1$. Shortening $= l_0 - l_1$. After shortening AB is uplifted to produce the Excess Section or Area, A_x. Therefore the initial stratigraphic thickness t_0 is transformed to an average structural thickness, t_1. For the section down to the décollement ABCD = AB′C′D after straining. Excess Area, A_x = OBCC′. Two methods measure the shortening: (1) Bed length method. In a, l_c is the bed length measured around the fold and thus $l_c = l_1$. Shortening is $l_c - l_0$. (2) Volume balancing method. $l_0t_0 = l_1t_1$, therefore $l_0 = l_1t_1/t_0$ (t_0 = depth to décollement, known from tunnels through the Jura). Shortening $= l_0 - l_1$. D, Décollement. (Based on Dahlstrom, 1970, and Hossack, 1979.)

tens, hundreds, even thousands of kilometres, they do not go on forever; in other words they are finite objects which end somewhere at a tip line or termination (see the dislocation model above) (Fig. 4.18a–d). This is an essential property of all faults large and small, and it has some interesting consequences in the understanding of the

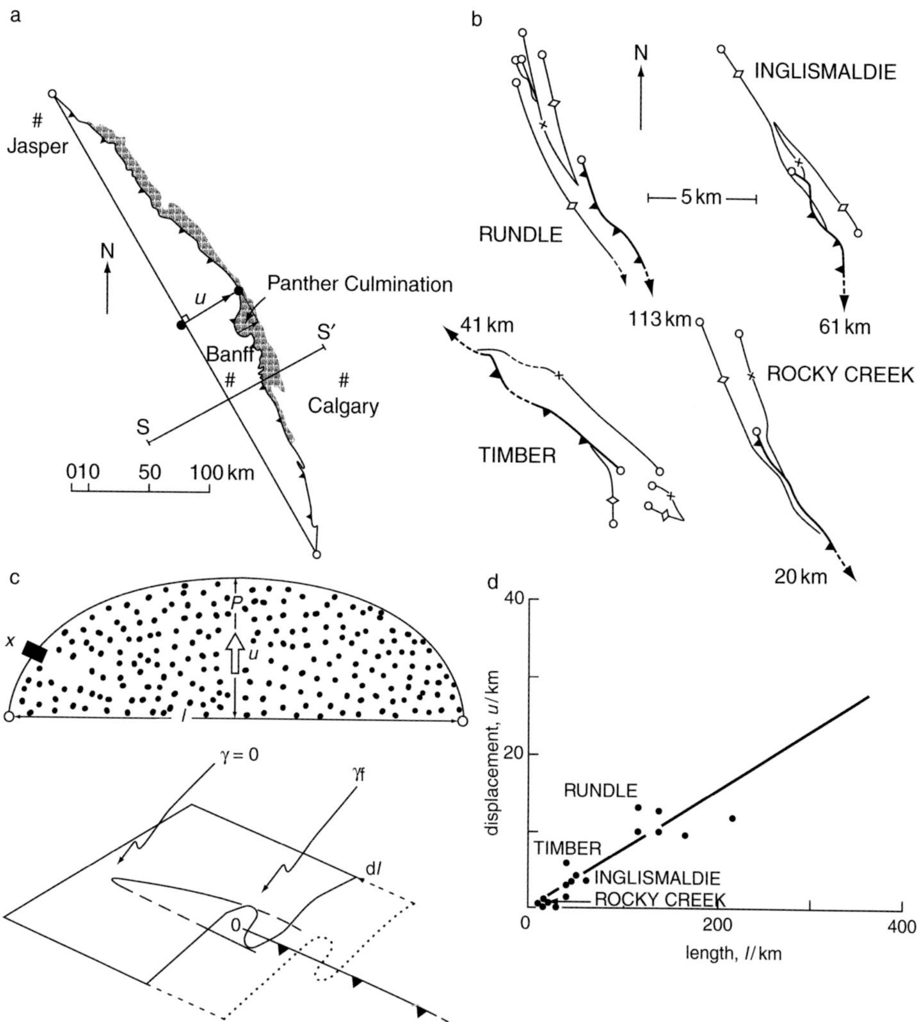

Figure 4.18 Terminations of thrusts in the Canadian Rocky Mountains. a, Map of the terminations of the McConnell thrust, which has a strike length of 410 km. The chord joining the ends of the thrust has a normal bisectrix, *u*, along which maximum displacement has occurred. In the central portion the hanging wall is Mid Cambrian and the footwall is Cretaceous molasse. At the terminations there is Devonian thrust above Permian, therefore the thrust has cut up section in the hanging wall and down section in the footwall. b, Maps of folds at the ends of thrusts in the Foothills and Front Ranges. All the folds start about 0.5 km from the thrust terminations and die out about 8 km from the thrust ends. c, The bow and arrow rule for thrust displacement (see text). Plan view of a thrust which is shaped like a half ellipse with maximum displacement, *u*, in the central region. In the shaded area, *x*, the thrust is about to propagate into the folds. d, Plot of linear relationship of displacement against length of thrusts in the Rockies. From Elliott (1976), published by permission of the Royal Society.

mechanics of faulting. Picture a fault like this: cut a roughly circular disc from a sheet of cardboard. The disc is the fault plane, and displacement takes place over the surface of the disc but is reduced to zero at the edges; that is, the cut represents the tip line of the fault with no displacement. The fault has been produced by propagation from a point source. Outside the cardboard disc there is no fault and no displacement. This is what we mean by a finite fault. Such faults are real, because in coal mines where the horizontal shafts occur at different levels it may be that steeply dipping faults are found at one level but not above or below that level. However, other configurations of finite faults are possible. Terminations of small faults are well known and are characterised by splays off the main fault or horse tail faults; but only in favoured places like Mount Kidd in the Canadian Rocky Mountains is it possible to show the termination of a large thrust sheet, the Lewis thrust (Plate 5). This thrust spans the USA and Canada border. It is *c*.170 km long and has a maximum displacement of 51 km, but towards the terminations the thrust loses displacement. Along strike from the north termination of the Lewis thrust, the thrust is replaced by folds. Similarly in the south as the Lewis thrust approaches the southern termination, displacement is transferred to the folds in the Sawtooth Range and to the Front Ranges (see Boyer, 1992). In plan view, therefore, the Lewis thrust has north and south terminations along the strike, and the displacement increases from the terminations towards the middle of the sheet. A useful way of showing this is the hanging wall sequence diagram (Fig. 4.19) or a stratigraphic separation diagram (Fig. 4.20) which depicts decreasing stratigraphical separation across a thrust towards the tip.

Unfortunately, terminations are only points in space, and the 3D form of the whole fault is unknown. It may be roughly circular like the cardboard disc, but more plausibly it should be pictured as like a bow and arrow (Fig. 4.18), as suggested by Elliott (1976), where the 'arrow' is the displacement vector. Similarly the fault starts from a point source and propagates outwards. Thus the terminations of active thrusts are really points in space and time marking the current position of the leading edge of the thrust. Discontinuous lenticular thrust sheets are called horses, as we have said, and are delimited by a curved branch or tip line (Fig. 4.21). But the branch line may not mark a termination, because the basal thrust of horses may continue beyond the branch line (see Fig. 4.7). A piece of fractured crust with faults large and small should be viewed as a dynamic system in which faults have grown or propagated over time. Mount Kidd, or indeed the cardboard disc, reveals a moment in that propagation.

Some interesting questions follow from the concept of finite faults; for example, how do the faulted parts and the non-faulted parts maintain compatibility in terms of volume? The answer is that volumes are conserved because the amount of displacement decreases towards zero at the termination or tip of the fault (Walsh *et al.*, 2008).

The recognition of structures on different scales raises the question: does the small scale mirror the large scale? For example, is it reasonable to assume that structures in rocks in a roadside outcrop are similar in style and orientation to the really huge structures, those that might fill one or even several mountains? Careful mapping can prove that the assumption is often justified. In the case of faults it follows from the above that there should be a relationship between the size of a fault and the amount of

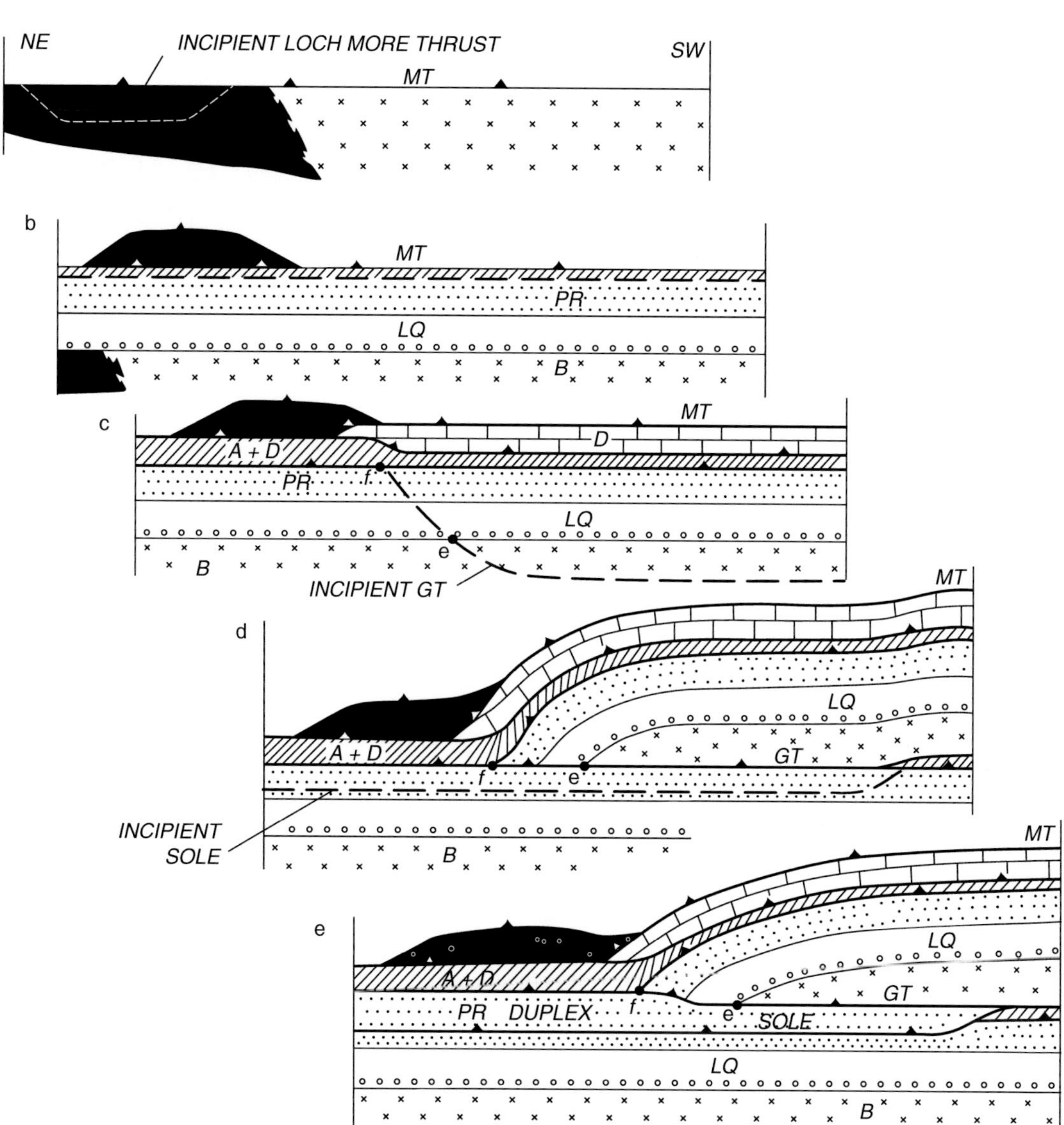

Figure 4.19 Hanging wall sequence diagram for the Moine thrust zone showing the evolution of the hanging wall from stages a to e. Note the successive accretion of horses, including basement horses, onto the Moine thrust sheet. Imagine that you are travelling on the thrust, as in the engine compartment of a train, and you look ahead to see the position of the active thrust at various times. For example the horse which is incipient in c has moved forward onto the active thrust in d. *B*, basement; *LQ*, Lower Quartzite; *P*, Pipe rock; *A+D*, An t-Stron plus Durness formations; *D*, Durness; *GT*, Glencoul thrust; *MT*, Moine thrust. From Elliott and Johnson (1980), published by permission of the Royal Society of Edinburgh.

displacement on it. This is the displacement–length ratio, and it appears that the longer the fault the larger the displacement, but whether the relation is linear is not so clear. The term used here is scale-invariant, and the subject is called fractals. In a piece of fractured crust that exhibits faults large and small, we often find that the relationship is fractal, that

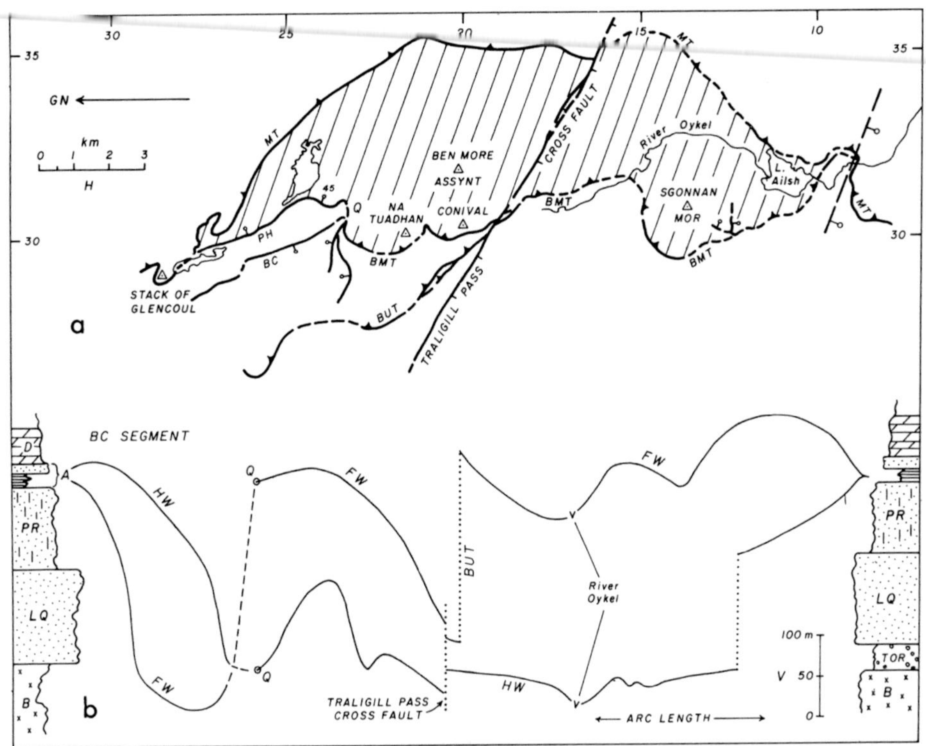

Figure 4.20 a, b, Stratigraphic separation diagram for the Ben More thrust (Moine thrust zone). Note that across the thrust the stratigraphic separation, the difference between the stratigraphic levels in the hanging wall and the footwall, diminishes from the central part towards the northern and southern tips. The course of the Ben More thrust in the northern part follows the older editions of the Assynt map. From Elliott and Johnson (1980), published by permission of the Royal Society of Edinburgh.

is, scale-invariant. This means that there are relatively few very large or very small faults, and most are medium sized.

Transport direction in thrusting

The direction of transport of thrusts is usually towards the foreland and we can speak of foreland-propagation of a thrust system resulting in an asymmetric orogen. There are a variety of small-scale structures in the rocks at or near to a thrust which are used as kinematic indicators to determine slip directions on thrusts. Examples are slickenside lines, stretching lineations and asymmetric folds. These can be called tectonic weather-cocks or, more exactly, kinematic indicators. In addition to foreland-directed thrusts we find some directed towards the hinterland: these are called back-thrusts and they may represent an alternative way of releasing stress in conjugate Coulomb fracture systems.

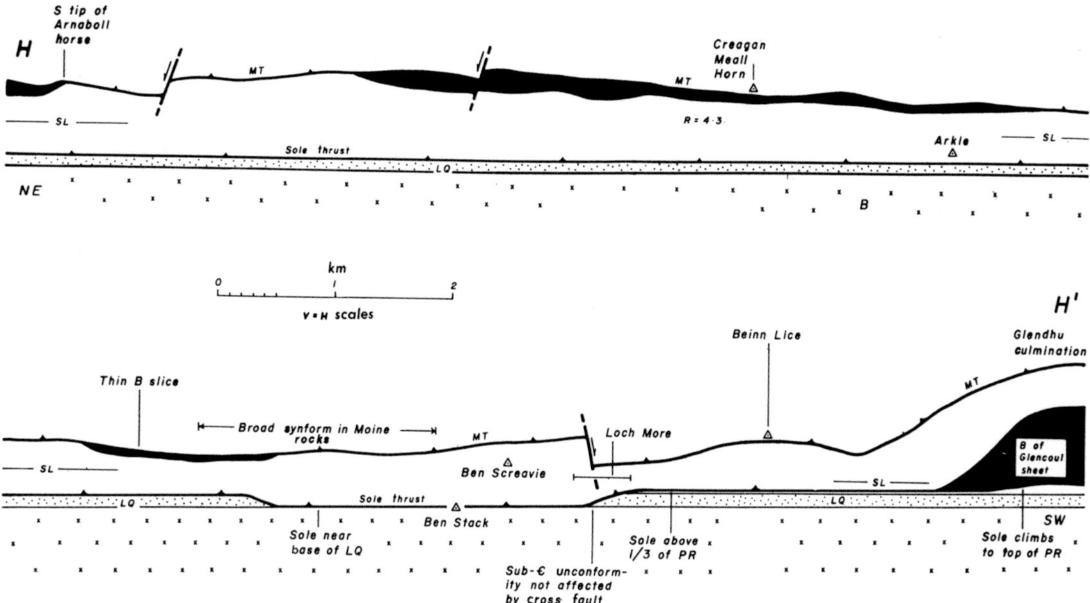

Figure 4.21 Horses of Lewisian basement (in black) which have been dragged along the footwall of the Moine thrust. The view is towards the ENE in the direction of transport. The thrusted section between the Moine and Sole thrusts is mainly Cambro-Ordovician. *B*, basement; *LQ*, Lower quartzite; *MT*, Moine thrust. *R* is the thickening t_1/t_0 where t_1 is the thickness after deformation and t_0 is the original thickness. From Elliott and Johnson (1980), published by permission of the Royal Society of Edinburgh.

Thrust mechanics

Earlier we noted the failure of nineteenth-century geologists to accept readily that thrusts or nappes existed. However, the proposition that large bodies of rocks have been transported horizontally for distances up to several hundred kilometres still needs to be addressed. The early sceptics posed the mechanical problem: how can this transport be attained without reducing rocks to a heap of rubble, in other words without exceeding the strengths of the rocks involved? The mechanical problem is to do with friction and with the stress which the nappe exerts on the basal thrust along which it moves. This is called the normal stress and since thrust planes are usually sub-horizontal in orientation, the normal stress in thrusts will be exerted near-vertically, thereby opposing movement. Imagine a book placed on a glass plate: if the plate is horizontal a considerable force is needed to push the book along, but if the plate is tilted the book can slide quite easily. By tilting the glass the normal stress is reduced and the book moves simply under gravitational forces.

This notion gave rise to the gravity gliding model for thrusts. The trouble with the hypothesis is that thrusts do not usually dip in the direction of sliding; instead, they have moved up-dip. The gravity spreading hypothesis retains the dominant role of gravity but permits thrusts to move up-dip. The prerequisite for spreading is an elevated mass which then spreads horizontally either by viscous flow or by thrusting. Silly putty will do this admirably if enough time is allowed. For a time, many geologists thought gravity spreading could solve the mechanical problem by postulating for thrust sheet emplacement something analogous to the dynamics of glacier flow. Glacier flow is controlled by gravity; they tend to flow downhill but, and this is important, they can flow uphill over irregularities in their substratum. The reason is that the control is not the topography of the base of the glacier but rather the slope of the glacier's surface: provided the surface slope is downhill the glacier will continue to move, even uphill. Is the analogy with thrusting in orogenic belts valid? Although most geologists are now doubtful that it is the whole truth, the idea that gravitational forces contribute to thrust emplacement is very attractive.

Another approach to thrust mechanics sought to address the effect of friction on the thrust, and the normal stress exerted by the rocks in the thrust sheet. Lubricants could facilitate the motion of thrusts, but this dodges the main issue, which is overcoming the normal stress. In 1959 Hubbert and Rubey published an extraordinarily influential paper which addressed this problem (Hubbert & Rubey, 1959). In essence their idea was that thrust sheets or nappes could be regarded as being analogous to a hovercraft which floats across its substratum, water or land, because it is buoyed up by a bag of air: the bag is inflated before flight and the hovercraft is driven forward by a tiny engine. In the case of thrust sheets the buoyancy device comes from fluids in rocks (see Chapter 3). The effect of enhanced fluid pressure is to create a buoyancy or in other words to counteract the normal stress exerted by the rock. The essential requirement is that pore fluid pressures must be high and approaching the value of the normal stress, σ_n, or lithostatic stress. This condition requires an impermeable seal such as shales, to entrap the water in the zone of high fluid pressure. Fortunately this can be tested: oil companies measure fluid pressures in boreholes and confirm high values for fluid pressures. The Hubbert and Rubey (1959) formulation is: $\gamma = \text{Co} + \bar{\sigma}_n \tan \phi$, where γ is the shear stress, Co a cohesive strength term, $\bar{\sigma}_n$ the effective normal stress and ϕ the coefficient of friction. In this equation normal stress is effective stress, which is lithostatic stress minus the fluid pressure. If fluid pressure is high then the right-hand side of the equation is minimised; indeed Hubbert and Rubey argued that the cohesive strength could be minimal, and so the thrust sheet would be buoyant and would float over its substratum with little stress being applied.

The Hubbert and Rubey model calls for the maintenance of high fluid pressures over a few million years, as this seems to be the timescale required for emplacement of large thrusts. At any rate the conclusion must be that the fluid pressure model does not provide a general answer to the mechanical problem of thrusting. A significant development in the search for a comprehensive mechanism for thrust is termed wedge theory, in which fluid pressures are not the only factor involved.

Orogenic wedge theory

In previous chapters we have set out the notion of thrust belts as evolving systems; that is, individual thrusts propagate but on the major scale the succession of thrusts which is

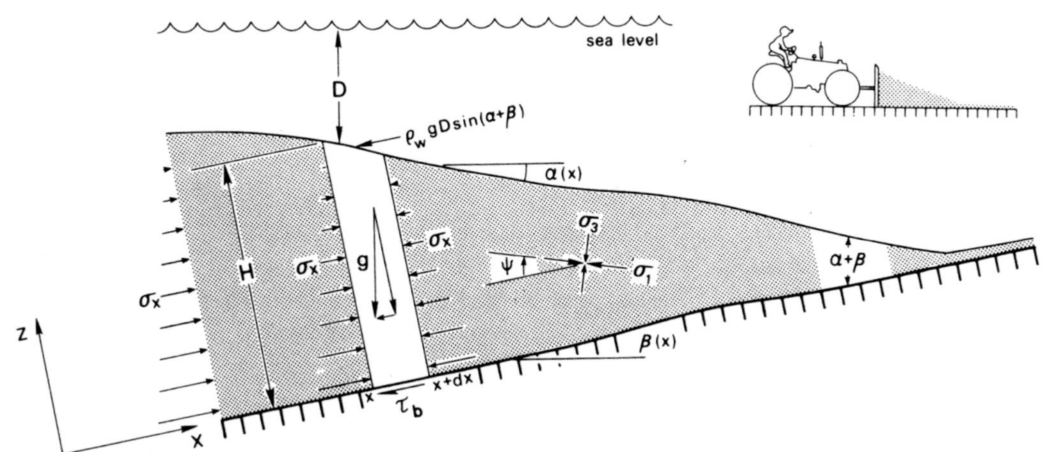

Figure 4.22 Wedge theory to show the snowplough analogy and the terms used in the analysis of wedges of material subjected to horizontal compression and on the verge of Coulomb failure (see text for explanation). *D*, depth below sea level; *H*, wedge thickness; σ_x, stress acting on the back of the wedge; σ_1, σ_2, σ_3, maximum, intermediate and minimum stress acting on an arbitrary point in the wedge; *g*, gravitational acceleration; α, dip of surface slope, β, dip of basal detachment; ρ_w, density of water; ψ, angle between σ_1 and the *x* direction in the wedge; τ_b, basal shear stress. From Davis *et al.* (1983), Figure 1, published with permission.

commonly observed in thrust belts indicates an evolution over considerable time spans. During the 1980s considerable advances were made in the search for a general or widely applicable hypothesis for the mechanics of thrusting and of the evolution of whole thrust belts. The first step was to get the right shape or form for thrust sheets. Hubbert and Rubey adopted a book analogy in which it was assumed that thrust sheets were shaped like boxes, but this is wrong; in reality they are shaped like wedges. The concept of a wedge shape for thrust sheets follows from the trajectory of thrusts which progresses upwards through the crust towards the surface. This shape change is important for the simple reason that in switching from a box to a wedge we have at a stroke taken away half the volume and mass! This might be called Pickford's law (after a certain well-known British removals firm), which states that if you want to move something the easiest way is first to minimise its weight as much as possible.

Wedge theory is a major contribution made by Davis, Suppe and Dahlen (1983) and Dahlen *et al.* (1984). The wedge is divided into a strong part which can withstand the stresses to be applied to it and a weak part at the base of the wedge. Then the theory explores the factors which control the ability of the wedge to slide forward along its weak base (Fig. 4.22).

The wedge model for thrusts suggests an analogy with the action of a snowplough. The plough pushes forward and piles up the snow into a wedge-shaped body which at a certain moment can be pushed forward with ease. The factor which controls when this happens is the shape of the wedge, or, more exactly, the taper, which is the angle between the upper and lower surfaces of the wedge. The crucial idea is that there is a critical taper required for easy sliding. Wedge theory was initially based on accretionary prisms which are composed of material added to a continental margin. The snowplough is replaced by the continent behind the wedge through which stresses are transmitted. In a general case for orogens, exemplified by the Canadian Rocky Mountains, the fold and thrust wedge is driven by the stresses transmitted through the hinterland made of previously deformed and metamorphosed rocks to the west.

The amount of the taper angle is a function of the physical properties of the wedge's strong and weak parts: wedges made of different materials will have different taper angles. Most wedge taper angles are between 1° and 9°. This relationship between taper angle and material properties can be expressed in an equation (see later) in which the angle of taper is related to a term defining the rheological properties of the base (coefficient of friction, fluid pressure) divided by a term which does the same for the rest of the wedge (dip of surface slope of the wedge, coefficient of friction etc.). The wedge theory can be tested provided that enough is known about the thrust sheet and the rocks in it. Previously in this chapter we noted the early recognition of the mechanical problem involved in pushing large sheets of rock for many kilometres without breaking them up. We have already alluded to one solution – the high fluid pressure model. In wedge theory the surface slope of a thrust wedge is an important controlling factor in thrust movements. In the same way as glaciers move up hill as long as there is a gravitational potential defined by the surface slope of the glacier, so relatively thin thrust wedges can be pushed up a basal slope as long as there is sufficient surface slope on the wedge.

The critical taper of the wedge depends on the shear strength of the base. Thrust wedges with low basal strength, for example wedges moving over sediments where the fluid pressure is high, or over rheologically weak salt or gypsum deposits, will require only a low surface slope. However, wedges moving over rocks with high shear strength such as crystalline gneisses or granites will require a steeper surface slope.

Active orogens such as the Himalaya or Taiwan, or the active accretionary prisms situated over subduction zones, are tapered in shape with the thickness decreasing in the direction of thrusting (Figs. 4.23, 4.24). This wedge shape can be analysed as a deforming thrust sheet gliding on a basal décollement. The theory is called Coulomb wedge theory because it is assumed that the Coulomb fracture theory applies to the wedge. The crucial point is that for a given Coulomb strength (resistance to failure by fracture) the wedge has a critical taper defined by $\alpha + \beta$, the slopes of the upper and lower boundaries of the wedge. It is assumed that the basal décollement is a nearly flat plane sloping up towards the foreland at an angle β. The surface of the wedge slopes down towards the foreland at an angle α. A stable taper is required for easy gliding of the wedge and is a function of the physical properties of the wedge and the strength of the base.

If the wedge became thicker because a lower thrust sheet had been added to its volume (remember how the piggy-back thrust system works) then if the base remains the same the wedge must become thinner in order to restore the critical wedge taper. Thinning will be accomplished by normal faulting (extensional faults) similar to landslides. Again if the slope angle becomes too large then the wedge will be extended by a thrust fault in the toe, lengthening the sheet and decreasing the slope angle. Likewise, addition of material to the toe of the wedge, by sedimentation or other processes, causes an adjustment to maintain the critical taper of the wedge. On the other hand if the wedge suffers volume loss during

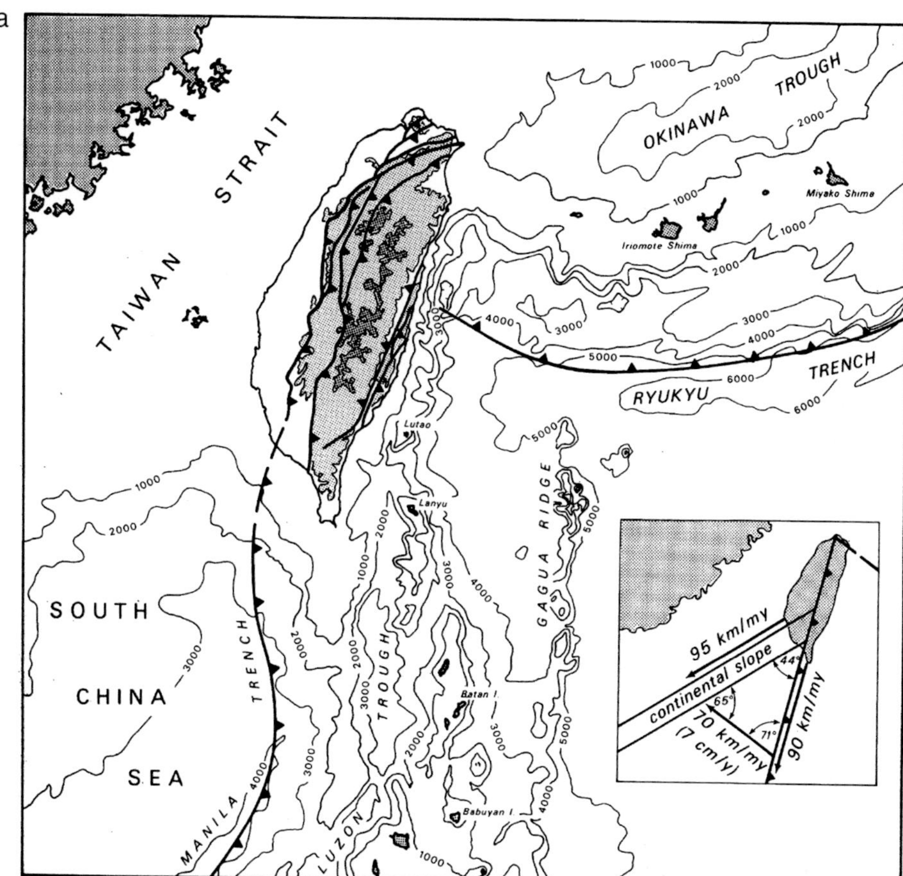

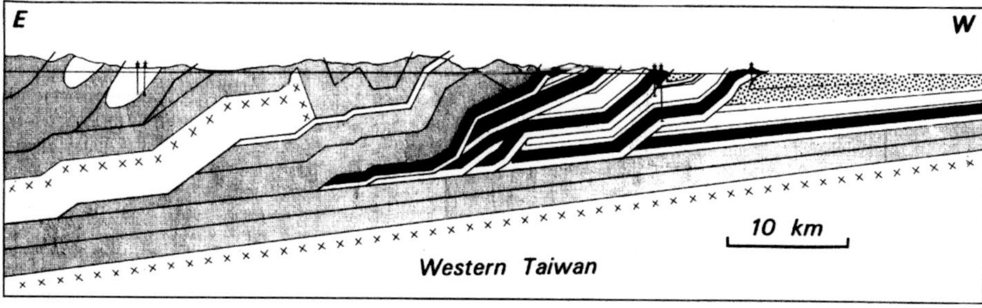

Figure 4.23 a, The thrust belt in Taiwan with plate tectonic setting. The thrust belt formed as a result of oblique arc–continent collision. The arc is in the east of Taiwan. From Davis *et al.* (1983) Figure 9, published with permission. b, Exhumation of the wedge using zircon fission track (ZFT) data. From Kirstein *et al.* (2009), published by permission of Wiley-Blackwell.

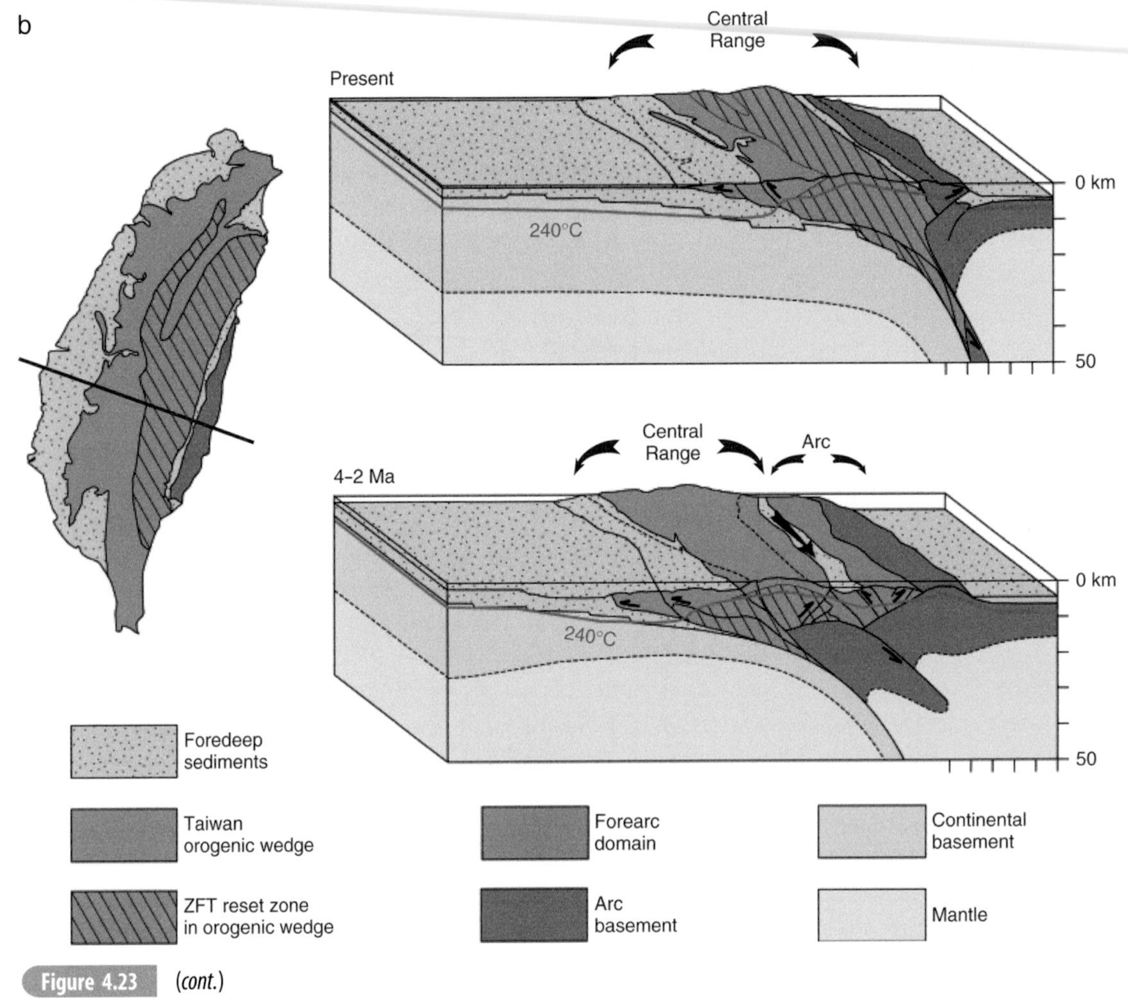

Figure 4.23 (*cont.*)

thrusting, for example by an erosional loss of volume, then critical taper must be restored by thickening of the wedge. Under Coulomb conditions thrust faults (contractional faults) are predicted in this situation. The visible consequences are thrust faults and folds in the wedge, all of which provide a shortening and thickening of the wedge.

Another situation in which the taper becomes unstable is when, as the wedge emerges onto the Earth's surface, it passes over uncompacted water-laden sediments. This then means that the surface slope is too steep.

The dynamics of the wedge are such that the driving forces that push the wedge towards the foreland must be balanced by the resisting forces that tend to prevent motion. The driving forces consist of the traction σ_x which acts on the rear vertical face of the wedge. The Coulomb fracture theory allows us to predict how the strength of the wedge will increase with confining pressure and thus how the normal traction will increase with depth. Furthermore, the strength of the rocks is strongly affected by pore fluid pressure within the

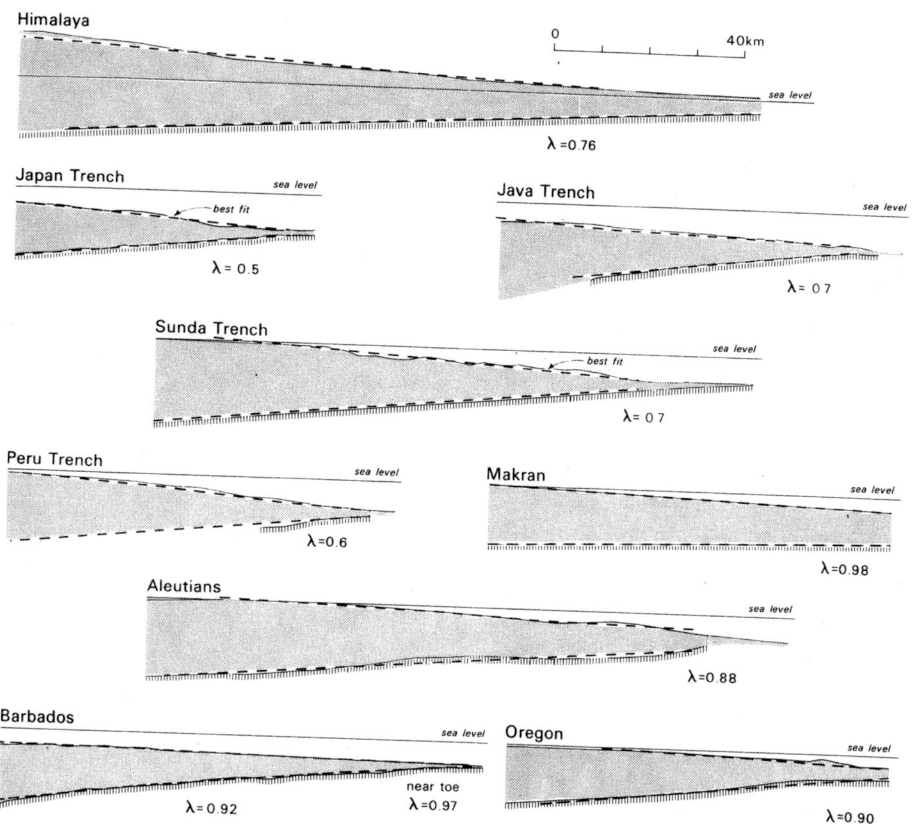

Figure 4.24 Wedge tapers in the Himalaya and several active submarine accretionary wedges. λ, Pore fluid pressure. From Davis *et al.* (1983) Figure 16, published with permission.

thrust wedge. An increase in pore fluid pressure tends to counteract the overburden pressure on the rock and thus tends to weaken the rocks. (Recall the earlier discussion of fluid pressures.)

The surface slope also creates a driving force because at any given level in the thrust wedge the overburden stress (ρgd) at a given point at depth d below the surface is greater than the overburden stress at a point nearer the foreland. Part of the horizontal component of stress, σ_x, is proportional to the overburden, therefore there is a horizontal gradient in σ_x which is equivalent to a horizontal force per unit volume of material. The same condition controls the flow of ice sheets. As already mentioned, another analogy is with dirt or snow wedges that form in front of a bulldozer.

Resistance to thrusting comes from the décollement and is derived partly from the frictional resistance to sliding on the fault given by the average frictional shear stress on the basal décollement, multiplied by the area of the décollement. Again pore pressure on the décollement is important because it determines the effective normal stress on the thrust and thus the frictional resistance to sliding (Fig. 4.25).

Setting up a balance between the driving forces and the forces of resistance gives an equation that accounts for the geometry of the wedge. If the length of the thrust sheet

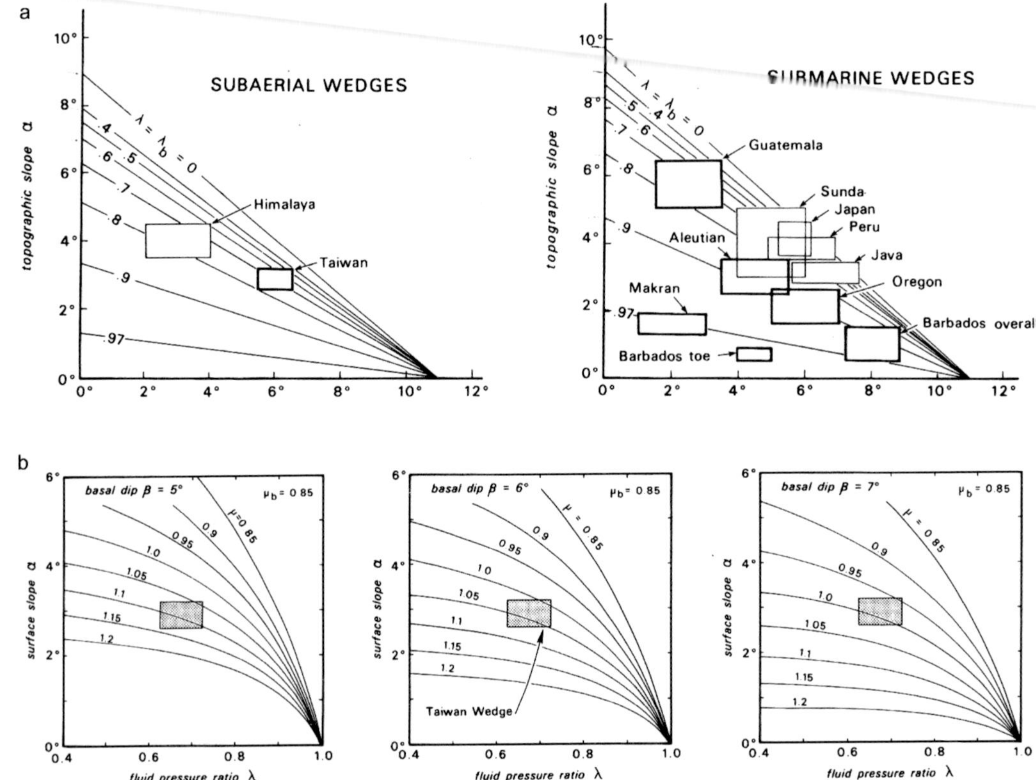

Figure 4.25 a, Plots of surface slope vs. basal dip for various values of λ_b in subaerial and submarine wedges. b, Surface slope vs. fluid pressure ratio λ for three basal dips, $\beta = 6 \pm 1$. By Byerlee's law, $\mu_b = 0.85$ on the basal detachment. Shaded box shows the measured α and λ in the Taiwan wedge. The best fit is for an effective coefficient of internal friction $\mu = 1.03$. λ, fluid pressure; μ_b, coefficient of friction along the basal surface. From Davis *et al.* (1983), published with permission.

increases, the frictional resistance to gliding will increase. To counter this resistance and permit motion of the sheet, the driving force must increase. Since this is not permissible because the rock is not strong enough to withstand the extra stress, the only possible mechanism is to thicken the thrust sheet, thereby increasing the area over which the driving force operates.

The equilibrium condition for wedges is given by the formula:

$$\alpha = \frac{(1 - \lambda_b)\mu_b - (1 - \lambda_i)\kappa\beta}{(1 - \lambda_i)\kappa + 1} \tag{4.1}$$

with:

α – the slope of the décollement;
β – the slope of the surface of the wedge;

λ_b and λ_i – the ratios of pore fluid pressure to overburden pressure along the décollement ($\lambda_b = \rho_b gH / \rho gH$, where H is the average wedge thickness) and also internal to the thrust sheet ($\lambda_i = \rho_i / \rho gt$) respectively;
μ_b – coefficient of friction along the décollement;
κ – a measure of the Coulomb fracture strength of rock in the thrust sheet.

This formula applies to subaerial wedges. For subaqueous wedges, such as accretionary prisms, it must be modified to allow for the pressure of the column of water.

If the décollement is horizontal and the pore fluid pressure is everywhere zero then the surface slope is given by:

$$\alpha = \frac{\mu_b}{(\kappa + 1)} \tag{4.2}$$

Higher coefficients of friction on the décollement result in higher angles for surface slope, reflecting the need to overcome the friction and therefore drive the sheet.

An increase in pore fluid pressure along the décollement will cause the surface slope to decrease. As a result, the resistance to thrusting is reduced and therefore the driving force needed to move the sheet decreases. On the other hand, increases of fluid pressure within the wedge decrease the effective strength of the rock and so reduce the possible horizontal normal stress that can be applied to the rear vertical face of the wedge (Figs. 4.21, 4.22). The wedge described in equation (4.1) is the thinnest body of Coulomb material that can be moved over the décollement. Wedge theory has been exemplified using analogue modelling (Huiqi *et al.*, 1992) and finite element modelling (Stockmal *et al.*, 2007). In the latter modelling, which is based on the Rocky Mountains, the influences of the accretion of a finite thickness layer, flexural isostasy, internal layering and strength contrast, syndeformational erosion and sedimentation are explored. Several models are given, one of which indicates that a very weak basal detachment is required for large thrust sheets to develop. Relative displacements between backstop and the sub-wedge crust are also given; for example, displacement of 240 km is associated with shortening values of *c*.65% for taper angles of around 6°. Huiqi *et al.* (1992) used sand box models to show how the critical taper determines whether or not large thrusts are developed. Thus low taper makes it less likely for thrusts with long aspect ratios to develop, and increase in the taper angle increases basal friction and the likelihood of large thrusts.

The wedge theory has been tested in Taiwan where extensive exploration for oil makes it possible to determine several of the factors in Equation (4.1), for instance pore fluid pressure and the strength of the basal surface. The Taiwan wedge is the result of arc–continent collision during which process oblique collision took place. What information is needed to solve the equation?

Firstly, the angle of taper of the surface slope and the dip of the base can be measured. The surface slope can be determined from a profile of the topography. The dip of the base is more difficult because it is below ground, but this is where geophysics comes to the rescue: if the base is seismically active then the location of earthquake shocks coming from the base fixes its orientation. Fluid pressures in the wedge and along the base are known from borehole data gathered by oil companies. For the other items in the equation – coefficient of friction etc. – it is necessary to make use of laboratory determinations of the values using similar rocks. Obviously this assembly of data cannot be made in many places because there are too many gaps in our knowledge. The discoverers of wedge theory, Davis, Dahlen and Suppe, were therefore able to present a satisfying mixture of theory tested by actual examples. This reminds us that theory is as important in the Earth sciences as in what are regarded as the more mathematically rigorous sciences, but theory must be applied in the field if it is to be convincing. So wedge theory can explain the way actual thrust wedges evolve. This is not to say that it provides the only possible explanation (not everywhere is as convincing a case study as Taiwan).

An illuminating account of wedge evolution in Taiwan is given by Kirstein *et al.* (2009) who described rapid erosion rates in Taiwan using detrital zircon U-Th/He and U/Pb and zircon fission track data (Fig. 4.23b). They dated detrital Plio-Pleistocene sediments in the Coastal Range of western Taiwan in order to document changes in exhumation rate through time. As a result of diachronous arc–continent collision taking place from 6.5 to 4 Ma, wedge thickening occurred in Taiwan and led to the removal of sedimentary cover and the unroofing of metamorphic rocks. There is a time-lag of at least 1.5 Ma between onset of arc–continent collision and the onset of rapid exhumation of the wedge starting at between 2.3 and 3.2 Ma. Exhumation rates are among the highest in the world; for example 3–5 mm/a over the period from 1.5 to 0.4 Ma. These rates continue today. The sediments were deposited in foreland basins in eastern Taiwan.

One slightly surprising result from the discussion of wedges is the prediction that extensional structures can be formed in a compressional regime like an orogenic belt. Until recently geologists imagined that compressional and extensional regimes were mutually exclusive, it must be one or the other. Now we are saying that the two can coexist in one orogen. A remarkable example has been recognised quite recently in the Himalaya. One of the most important consequences of the India /Asia collision is the Main Central thrust or MCT, an enormous thrust wedge which has been transported southwards towards the Indian continent. The high mountain peaks of the High Himalaya are carved out of this sheet. Above the MCT sheet there is a large extensional fault, the South Tibetan Detachment (STD), which can be seen on the upper slopes of Everest. The MCT and the STD can be followed along most of the 2500 km of the Himalaya. The puzzle is that the MCT and the extensional fault were active at roughly the same time, about 20 Ma. Wedge theory provides one explanation: the MCT wedge became too thick and the stability criterion demanded a thinning, hence the normal fault. However, as we will note in Chapter 6, many geologists now prefer to explain the extensional faulting as a result of channel flow; that is, the thrust and normal faulting indicates extrusion of a low-viscosity wedge.

There are particularly instructive examples where the wedge model has been applied to thrust sheets moving over weak rocks such as evaporites. As predicted, such thrust wedges have a very low critical taper, a flat topography and a zone of thrusting that extends far out into the foreland. Good examples are the Himalayan frontal thrusts in Northern Pakistan. In the Salt Ranges and Potwar, thrusts have detached along Cambrian salt horizons and have been transported for *c.*200 km (Fig. 4.26). The salt-based thrust belts only needed to maintain a wedge taper of *c.*1° to continue moving, compared with 8–12° for belts with strong basal detachments. The frontal slopes of the Salt Ranges are *c.*300 km across, with

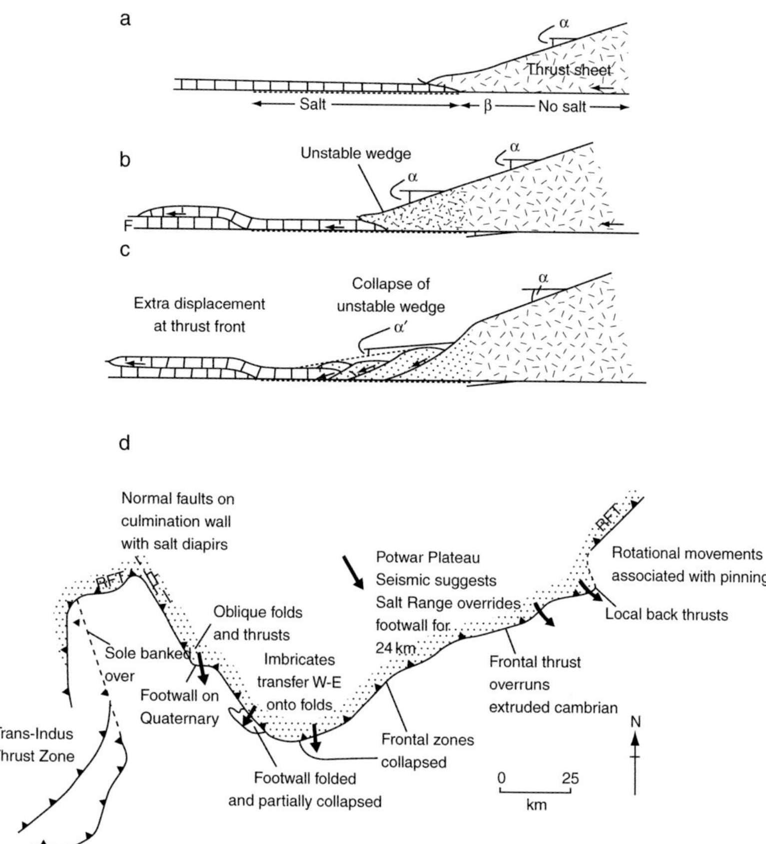

Figure 4.26 Salt Ranges of the frontal ranges of the Himalaya in Pakistan. The Salt Ranges show how the strength of the basal surface controls the taper; in this case it is the presence or absence of salt that determines the wedge geometry. a–c, As the thrust sheet overrides a Cambrian salt horizon the wedge collapses by low-angle normal faults. d, Map of the Salt Ranges showing the variation in slip direction (shown by arrows) because the thrust sheet is pinned at the eastern end of the Salt Ranges where the salt is thin and basal slip is more difficult than in the west. From Coward (1994), published by permission of Elsevier.

a relief <1500 m above the foreland. There is therefore a potential to drive gravitational collapse of the orogen. Their topographic slope is nearly zero, and the fluid pressure ratio is 0.93. And because the evaporites are widespread for most of this region, no internal deformation is needed to increase the taper. In the eastern Salt Ranges, however, the amount of evaporite decreases and the basal slope is *c.*1°. In consequence in the eastern Salt Ranges thrust sheets show internal thrusting, because the diminution or absence of salt makes it necessary to increase the surface slope.

Is it possible to make wedge theory a general theory of thrusting, not only in the upper crust but also at deeper levels where different rheologies apply? Platt (1986) has extended the wedge theory to treat non-Coulomb wedges and has given examples from California, the Alps and elsewhere (Fig. 4.27). This is despite the fact that in metamorphic rocks fracture is less likely because ductile behaviour has taken over; in addition, deeper level

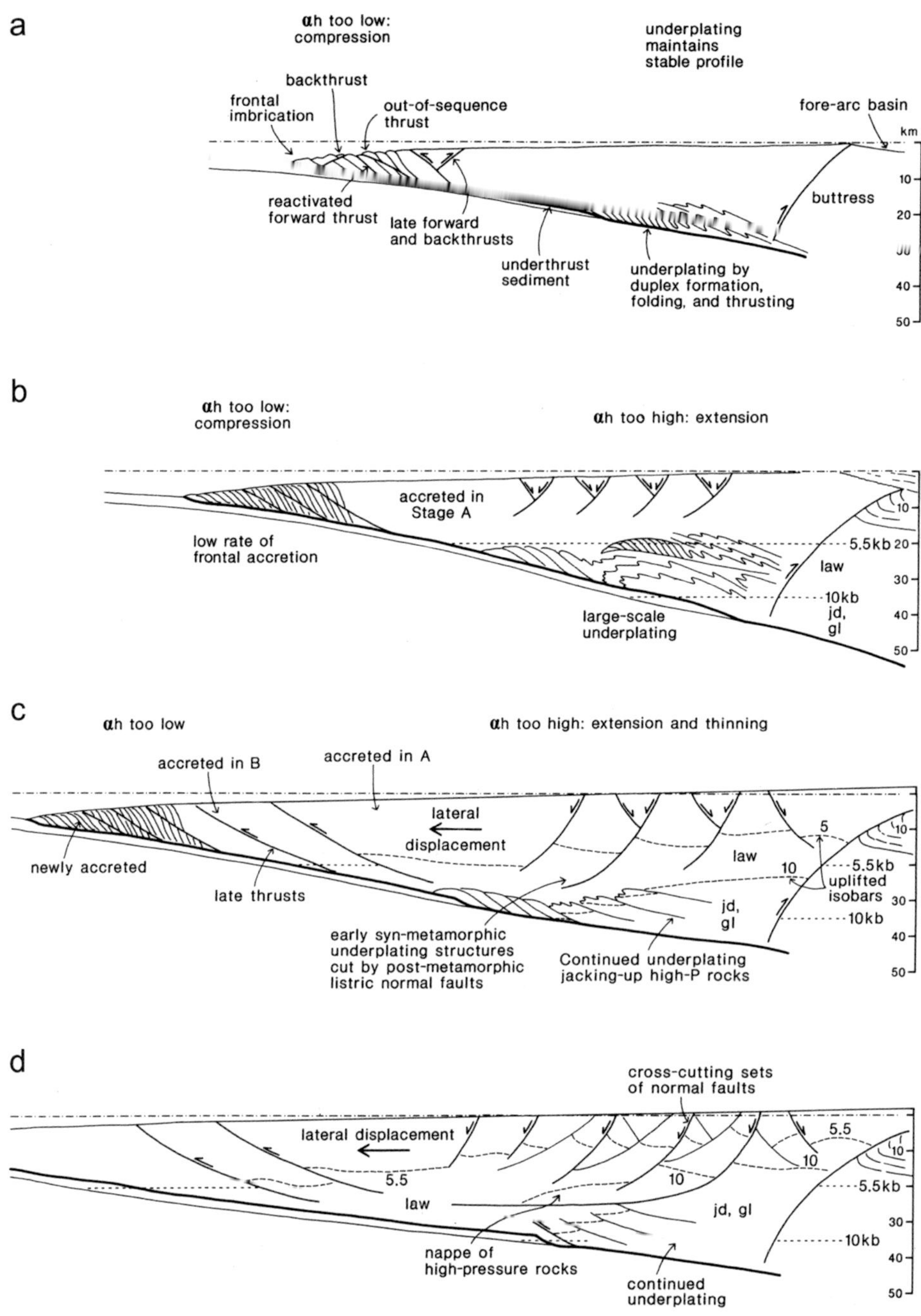

Figure 4.27 Plastic wedge. Cross-sections to show the evolution of a plastic wedge. a, Early stage with frontal accretion dominant; αh (the surface slope times the height of the wedge) is too low in the frontal region which therefore shortens. b, Sediment is underthrust and the rate of frontal accretion is low, therefore underplating is the dominant mode of accretion. αh is therefore too high in the rear of the wedge which extends by normal faulting or ductile flow at depth. The deeper parts of the wedge undergo high *P*/*T* ratio metamorphism. c, Continued underplating and extension have lifted the high-pressure rocks toward the surface. d, In a mature prism, underplating and extension have brought high-pressure rocks to within 15 km of the surface. ge, glaucophane; jd, jadeitic pyroxene; law, lawsonite. From Platt (1986), published by permission of the Geological Society of America.

rocks have low yield strength and have a non-linear viscous rheology. Yet they appear to behave exactly as do Coulomb wedges. What Platt calls plastic wedges will deform internally so as to maintain the wedge shape in which stress exerted by the surface slope will be balanced by the resistance to sliding on the base. The stability criterion for plastic wedges is:

$$\alpha = \frac{\gamma}{\rho g \lambda} \tag{4.3}$$

This is the same as the glacier sliding term, and it indicates a special case where there is no internal deformation and the surface slope is such that gravity-induced slope stress balances the shearing stress at the base.

Plate motions may exert a control on the form of wedges. For example, an increase in the rate of subduction may increase the shear strength at the base of the wedge and this will lead to wedge thickening. Similarly, variations in the rate of plate motion or variations in the partitioning of the plate movement may lead to alternations of contractional and extensional deformation within the wedge. This may account for the polyphase deformation often seen in orogens, in which several apparently distinct episodes of deformation occur during one orogeny. Similarly, when subduction ceases, the orogenic wedge may undergo extensional collapse because the basal drag exerted by the subduction is no longer there. In the deeper parts of the wedge the basal shear strength decreases as the crust heats up, so these parts may collapse by extensional faulting.

Despite the similarities there must also be differences between Coulomb and plastic wedges. For a plastic material the strength is defined by yield strength. In a plastic wedge the strength does not increase with depth and so the surface slope would have to be steeper than for a Coulomb wedge. Sliding of the wedge by plastic deformation which is independent of pressure would be the same under the thicker part of the wedge as it is under thinner parts. Thus plastic resistance to thrust motion would not increase as rapidly with distance from the toe as it does in Coulomb wedges. And the surface slope would be lower. Pore fluid pressures would not have a large effect on the material properties. Predictably, adjustments of the critical taper in plastic wedges are brought about by ductile faulting or folding.

To summarise, Coulomb wedge theory has been influential in the studies of the evolution of orogens around the world. And its application to deeper orogens where Coulomb fracture theory is less applicable has broadened the value of the theory. It provides a ready explanation for the observation that normal faulting, signifying an extensional regime, can be present within a contractional orogen. But we should beware of assuming that wedge theory is the only possible explanation for such a feature.

The discussion of Coulomb and plastic wedges serves as a reminder that rheologies change markedly as depth increases in the lithosphere. Thrusts are not only brittle superficial or thin-skinned features, marked by fault gouge and breccias; some are so-called ductile thrusts formed at deeper levels in the lithosphere. This means that there is no actual break along the thrust and there is no brecciation such as occurs in brittle thrusts. The thrust is a zone of ductile shear marked by mylonite. Mylonite is a highly deformed rock in which brittle deformation is minimal; instead the quartz assemblage is dynamically recrystallised during stress. Crystal plasticity is the dominant deformation mechanism, and quartz grains have deformed in response to stress by gliding on several crystallographically determined planes. Examples of glide planes in quartz are basal planes, prism faces and rhombohedral faces.

Ductile thrusts

The rarity of cataclastic features in mylonites means that many ductile thrusts have been missed. A good example is in the Moines of NW Scotland where several large ductile thrusts were overlooked by the early workers. Their recognition came partly from the discovery of high strain along the thrust but mainly from stratigraphical evidence that the thrusts were superimposing older rocks on the younger, in this case Archaean and early Proterozoic Lewisian upon the late Proterozoic Moines. The simplest explanation of the difference between ductile and brittle rests in the metamorphic temperatures prevailing during the thrust event. However, it is important to bear in mind that variations in strain rate may control whether ductile or brittle behaviour occurs. Ductile thrusts have been described in many orogens, for example the Grenville orogen, the Alps (Plate 6) and of course the Himalaya.

Thus we have hot thrusts and cold ones, with very different characteristics. In the Moine thrust zone, Knipe (1990) has shown that, with time, thrusting changed from ductile to brittle thrusting in the lower thrusts.

Internal strain in thrust sheets

Earlier in this chapter we referred to the contrast between rigid body translation and translation involving internal strain in the thrust sheet. There is a general assumption that strain within thrust sheets is essentially two-dimensional, with no movement of material into or out of the transport plane (i.e. plane oriented perpendicular to the thrust surfaces and containing transport direction). This assumption of deformation with no along-strike extension or contraction greatly simplifies geometric analysis of thrust systems and retro-deformation procedures (Dahlstrom, 1970), which can then be handled as a two-dimensional problem within a constant area transport plane.

Simple shear has been postulated as the principal mechanism for strain accumulation in orogenic belts characterised by gently dipping L–S tectonites (deformed rocks with prominent lineation and foliation) in which the linear element of the grain shape fabric trends parallel to the transport direction. However, plane strain is more the exception than the rule in orogenic belts. Many field studies have clearly demonstrated that plastically deformed rocks making up thrust sheets in orogenic belts are not simple L–S tectonites, but display a wide spectrum of grain shape fabrics ranging from pure S-tectonites (deformed rocks with a dominant planar fabric) through L–S to pure L-tectonites (in which a linear fabric is prominent). Strain data from these studies generally indicate that the maximum principal finite strain direction (x) trends approximately parallel to the inferred transport direction, while the intermediate principal finite strain direction (y) trends parallel to orogenic strike. Most strains plot within the flattening field (S-tectonites) of the Flinn plot, although constrictional strains (L-tectonites) have also been measured (Plate 7).

A contrary view is that observed strain features may be due to the accumulation of different strains giving apparent strains produced by, for example, the superposition of a plane strain tectonic fabric upon an earlier fabric which may be of either sedimentary or tectonic origin. For example, prolate and oblate fabrics may be produced by superposing in varying orientations a tectonic plane strain (either pure shear or simple shear) on an oblate sedimentary fabric. However, it is generally accepted that the observed spectrum of S- through L-tectonites in orogenic belts indicates real 3D strain variation, with obvious implications for both along-strike strains and their attendant space problem. For example, assuming no volume change, which seems to be reasonable in the case of thrust sheet deformation dominated by crystal plasticity and in which evidence for diffusive mass transfer is lacking, along-strike extensions of 11–15% and 20% are indicated by strain integration in the Bygdin (Hossack, 1968) area of the Norwegian Caledonides (Plate 8).

Folds in shear zones

The evidence for along-strike flow leads to a consideration of the evolution of folds in mylonitic shear zones and especially the frequent occurrence of fold axes orientated nearly parallel to the transport direction (Plate 6). Such folds are utterly different from the fault-bend and propagation folds discussed earlier. Their origin has given rise to many controversies, but recent work seems to have solved the problem. Many of these folds are so-called sheath folds (Fig. 4.28) which signify intense flattening during which their axes rotated into near-parallelism with the transport direction. Sheath fold geometry is conical and the fold axis plunges into the outcrop but reappears because of the strongly curvilinear axis. In three dimensions the folds are cone-shaped, like tall dunces' caps. Another name is 'eye-folds' because of their appearance when seen in cross-section.

Another mechanism for producing fold axes parallel to the transport direction is flow perturbations (Alsop & Holdworth, 1997). These are associated with culminations and depressions in thrust sheets, which are caused by segmentation of the hanging wall into surges on culminations, where the forward flow is accelerated, and depressions where flow is reduced. The kinematic picture is akin to that in surge zones evident in landslips and mud flows. The perturbations are bounded by a belt in which strike-slip deformation is needed to accommodate the differential flow in the segments. Both dextral and sinistral strike-slip motions occur on the boundaries of surges and therefore the fold vergences are different. These folds may look like sheath folds, but the origin is quite different and their axes have not rotated. In the Moine thrust zone both types of folds occur but the sheath folds are earlier than the perturbation folds.

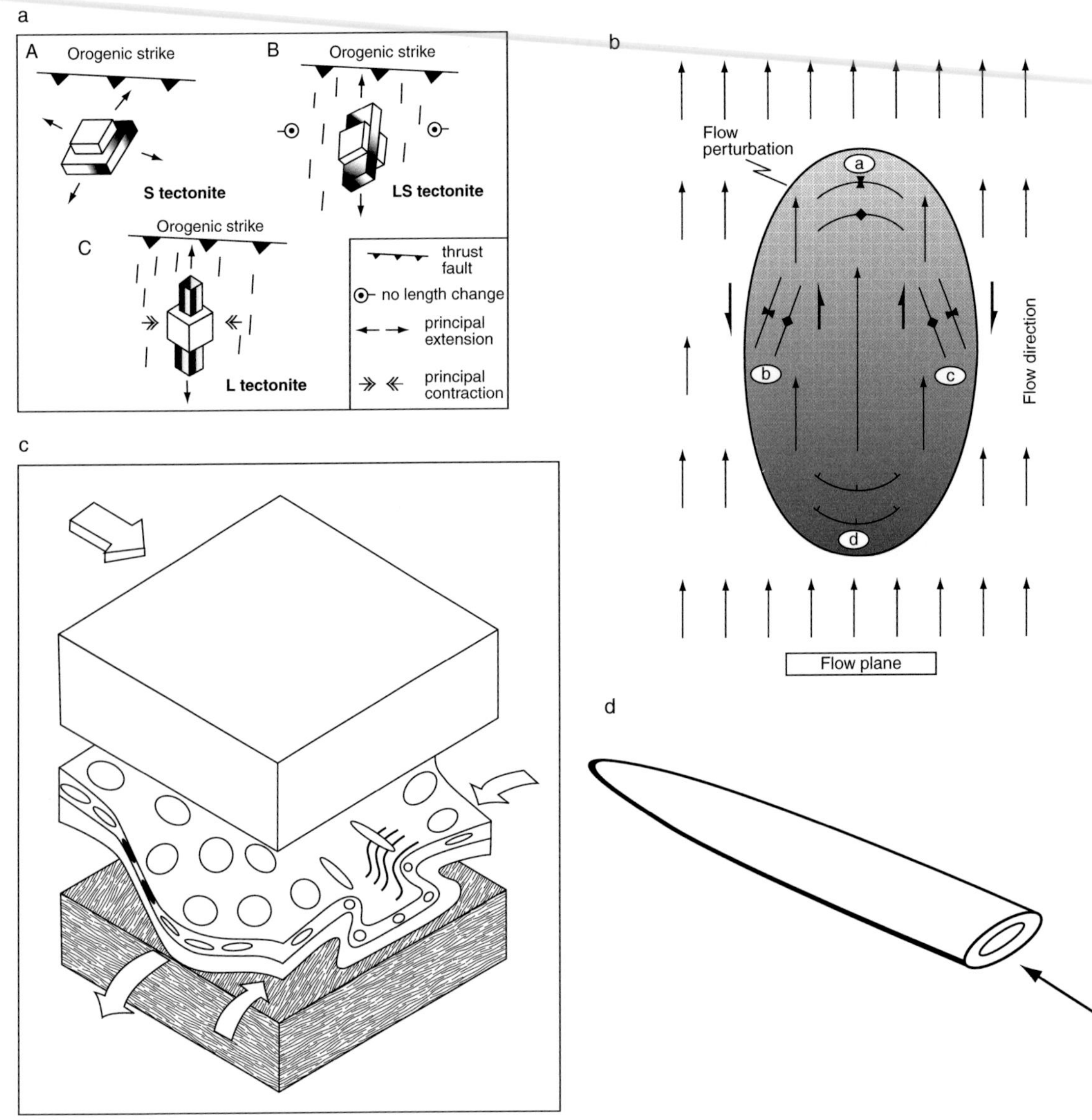

Figure 4.28 Along-strike extension in thrust sheets. a, S-, LS- and L-tectonites (for definitions see text) in thrust sheets shown by the transformation of a cube. Only in S-tectonites is there substantial along-strike extension. Figure by Richard Law, published with his kind permission. b, Plan view of a thrust showing flow perturbation during non-coaxial flow. A local increase in flow velocity is assumed; flow velocities are indicated by the length of the arrows. Predicted deformation structures are shown. Dextral and sinistral shears at the margins of the perturbation develop asymmetric folds, denoted by circled b and c, with opposite senses of vergence. Thus folds with axes roughly parallel to the transport direction are formed on the margins of cells in which accelerated flow occurs. Compressional structures are formed at a and extensional at d. Modified after Alsop and Holdsworth (1997), published by permission of the Geological Society of London. c, Folds with axes near-parallel to the transport direction developed to accommodate along-strike extension in a thrust sheet. From Sylvester and Janecky (1988), published by permission of Norsk Geologisk Tiddscrift. d, Sheath folds formed by the rotation of fold axes to be near-parallel to the transport direction, which is shown by an arrow.

A range of tectonic models have been proposed to explain along-strike extension at various scales. At the relatively small scale, it has been proposed that strike-parallel extensions may be produced by straining of a hanging wall above lateral ramps in thrusts, below which there are progressively developing duplexes. On the other hand, strike-parallel compression may similarly be produced in the hanging wall between adjacent duplexes. At a larger scale, and based largely on scaled analogue experiments, gravity-induced spreading with zones of divergent and convergent flow has been proposed to account for along-strike extension and shortening respectively in penetratively deforming thrust sheets and nappes. At the largest scale, along-strike extension may ultimately be a result of oblique plate convergence.

The potential space problems associated with along-strike strains may be the cause of large-scale features. For example, along-strike extension may produce the arcuate map patterns (salients and recesses) displayed by some orogenic belts. Based on field data from the Caledonides of Norway, Sylvester and Janecky (1988) proposed that along-strike extension associated with zones of oblate fabric development (S-tectonites) may be compensated for by intervening zones of constrictional strain (L-tectonites) in which transport direction-parallel fold hinges develop (Fig. 4.28c), the zones of constriction progressively narrowing along strike as the surrounding flattening zones undergo along-strike extension.

Studies by R. D. Law and others (1986, 1987) showed that the analysis of crystallographic fabrics produced by plastic deformation offers a simple potential method for determining the actual strain path followed. Numerous experimental studies (involving both dislocation creep and dynamic recrystallisation), numerical simulations and analyses of naturally deformed quartz-rich rocks have indicated that in progressive deformation along constant coaxial strain paths there is a simple relationship between strain symmetry and the pattern of quartz *c*- and *a*-axis fabrics that develop; the fabric strengthens but does not change in pattern with increasing strain. Crystal fabrics also offer criteria for assessing shear sense and, in combination with finite strain data, criteria for both quantifying vorticity of deformation and distinguishing between monoclinic and triclinic flow. The assumption is made that the measured fabric reflects an important part of the deformation history rather than just the dying gasps of that history. Also it is assumed that neither continuing dynamic recrystallisation nor thermal annealing have modified or even destroyed the tectonite fabric. At least for coaxial strain paths, analysis of experimentally and naturally deformed quartzites indicates that the skeletal fabric patterns associated with dislocation creep and dynamic recrystallisation remain closely similar in form, and certainly indicate the same strain symmetry.

Non-coaxial superposition of plane strains will result in a highly non-linear progressive development of the finite strain ellipsoid which continuously changes its symmetry (k value). Similarly, numerical modelling of transpressional deformation within a constant kinematic framework indicates that vorticity of flow (defined below) will have an important influence on progressive symmetry (k) change of the evolving finite strain ellipsoid. These models make specific predictions concerning progressive change in shape of the finite strain ellipsoid when the tectonite has not developed along the simple strain path indicated by finite strain symmetry. For example, quartz mylonites on the Alpine Fault

Zone of New Zealand are strongly foliated, but frequently lack a stretching lineation, suggesting flattening strains. These mylonites, however, are characterised by single girdle fabrics indicating plane strain deformation with the principal stretching direction inferred from the crystal fabrics oriented parallel to the local plate motion vector.

Starting in the early 1980s, evidence has been found for strain path partitioning associated with thrust sheet emplacement. Microstructural and petrofabric studies by Law and co-workers in the Moine thrust zone of NW Scotland and the Greater Himalayan Slab of Bhutan (Law, 1986; Law *et al.*, 1987; see Chapters 5 and 6) all indicate strain path partitioning, with the interior of the thrust sheets containing an important component of coaxial vertical thinning and transport parallel extension, while deformation near the thrust sheet margins is dominated by often relatively narrow zones of strongly non-coaxial flow.

Quantitative methods for vorticity analysis have confirmed that at least a component of coaxial (pure shear) straining exists in deformed rocks (Fig. 4.29). The methods depend on the estimation of vorticity in the fabric. Vorticity is a term referring to the different behaviour of the principal strain axes during progressive deformation and is a measure of the average rate of rotation of material lines of all orientations about each coordinate axis of the deformation ellipsoid (see Law, in Searle *et al.*, 2003). Vorticity is zero for non-rotational strains and non-zero for rotational strains. It is assumed that the vorticity vector is perpendicular to the maximum and minimum principal axes of finite strain, and that a non-coaxial flow will lead to formation of fabrics with monoclinic symmetry. However, although the methods do assume a minimum of monoclinic symmetry, they remain valid for general 3D strains and do not assume plane strain ($k = 1$) deformation.

Components of pure shear and simple shear can be quantified in terms of the kinematic vorticity number, Wk, which is defined (for $0 < Wk < 1$) as the non-linear ratio of pure shear ($Wk = 0$) and simple shear ($Wk = 1$) components, assuming a steady-state deformation. Pure and simple shear components make equal contributions to the instantaneous flow at $Wk = 0.71$.

Quantitative vorticity analyses have been carried out in several orogens of various ages, including amongst Cenozoic origins, the Greater Himalayan slab of India and Tibet, the Hellenides of Greece, and the Caledonides of Scandinavia. In addition, other regions of high tectonic strain such as the Alpine Fault Zone of New Zealand have been studied using this approach. In some of these examples it can be demonstrated (e.g. based on strain analysis or crystal fabric patterns) that deformation approximated to a plane strain with principal axes in the transport plane, and in these cases it may be reasonable to assume variable 2D combinations of pure and simple shear in numerical modelling.

Generally, however, the implicit assumption of plane strain deformation remains to be proven (Fig. 4.29). Indeed, in the case of the Moine thrust mylonites, although the principal extension direction indicated by strain and petrofabric analyses does trend parallel to the transport direction, strain symmetry varies from general flattening to general constriction both along strike and within a short distance above or below the thrust planes. For such non-plane strain situations, the analytic solution for calculating transport-parallel stretching must be modified to take into account strain parallel to orogenic strike.

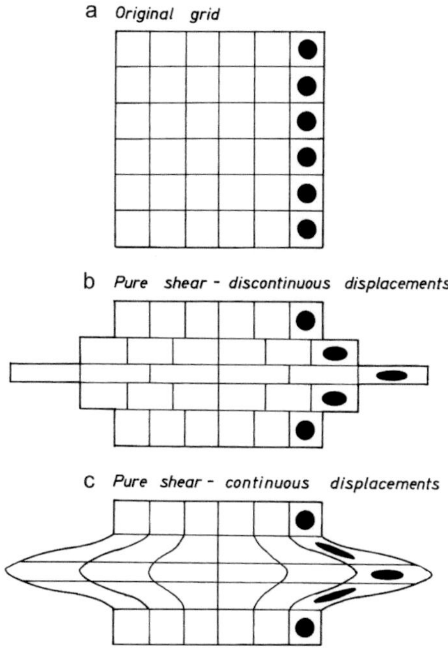

Figure 4.29 Plane strain showing two types of pure shear. From Ramsay and Huber (1987), published by permission of Elsevier.

Rotation of thrust sheets about a vertical axis

Finally, there is the possibility of rotation of thrust sheets about a vertical axis. In the Glencoul sheet in the Lower Palaeozoic Moine thrust zone of NW Scotland, we noted earlier that no such rotation had taken place because Precambrian dykes in the thrust sheet and the foreland were aligned. But elsewhere rotation has been demonstrated by, for example, the use of palaeomagnetism (see Dinarès *et al.*, 1992, for references). Examples of substantial rotations, about 50°, have been described in the Pyrenees. In the Idaho–Wyoming thrust belt rotation of thrust sheets has been attributed to the deflection of transport direction by buttresses of Precambrian massifs. Thrust sheet rotations are discussed again in Chapter 5.

Further reading

Badertscher, N.P., Beaudoin, G. and Therrien, R.M. (2002). Glarus overthrust: A major pathway for the escape of fluids out of the Alpine orogen. *Geology*, **30**, 875–878.

Knipe, R.J. (1990). Microstructural analysis and tectonic evolution in thrust systems: examples from the Assynt region of the Moine thrust zone, Scotland. In: *Deformation Processes in Minerals, Ceramics and Rocks*, ed. D.J. Barber and P.G. Meredith. London: Unwin Hyman, 228–261.

Huiqi, L., McClay, K.R. and Powell, D. (1992). Physical models for thrust wedges. In: *Thrust Tectonics*, ed. K.R. McClay. London and New York: Chapman and Hall, 71–81.

Johnston, S.T. (2004). The New Caledonia–D'Entrecasteaux orocline and its role in clockwise rotation of the Vanuatu–New Hebrides Arc and formation of the North Fiji Basin. In: *Orogenic Curvature: Integrating Palaeomagnetic and Structural Analyses*, ed. A.J. Sussman and A.B. Weil. *Geological Society of America Special Paper*, **383**, 225–236.

Ring, U. and Kassem, M.K. (2007). The nappe rule: why does it work? *Journal of the Geological Society of London*, **164**, 1109–1112.

5 Evolution of orogens

In this chapter we will be looking at the evolution of several types of Phanerozoic orogenic belts. Precambrian orogenesis will be dealt with in Chapter 12. The Himalaya and the Alps are part of a huge belt of Cenozoic age which runs from the Pyrenees through the Balkans into Turkey and on to the Middle East, Pakistan and India into Burma. There is also a leg from the Betic Cordillera to the Rif in North Africa and via Corsica to the Ligurian and Internal Western Alps. These parts were the result of the collision of Gondwanaland (the Late Palaeozoic assemblage of South America, Africa, India and Antarctic) and Eurasia (Europe and Asia). We also consider the Andes and the Caledonides in order to illustrate different types of orogens. For the present, examples are confined to the Cenozoic orogens because, as mentioned above, the younger mountain belts offer a better chance of understanding evolutionary processes in orogenesis than the older deeply eroded belts in which much of the evidence is missing (see Chapter 12).

In the now discarded geosynclinal theory of orogenesis as set out for example in Holmes's *Principles of Physical Geology* (Holmes, 1944), the pre-orogenic phase was a precursor of the orogenesis because the sedimentary and igneous rocks deposited in the geosyncline were already undergoing compression and so were predestined to become involved in orogeny, the point being reinforced by the postulated downward flow of mantle convection cells which led the whole process. Plate tectonics introduced a paradigm shift which included a denial of any link between the events occurring before orogenesis and the orogeny itself; this is well demonstrated by the Swiss Alps which were undergoing extension not compression before orogeny. The attempt to separate temporally extensional and compressional strain events is much too simple. For example, compressional strain in forearc wedges may be synchronous with extensional strain in the back arc, as for example in South America where overall convergence during the Jurassic–Early Cretaceous between the oceanic and continental plates involved synchronous extensional and compressional strains. This is a common feature around the Pacific where back-arc basins are opening during subduction. In addition, as Royden (1993 a,b) has shown, the roll-back and advance of the subduction zone produces alternations of extension and compression of continental margins.

Why should the history prior to orogenesis be relevant in considering orogenesis? Obviously the type of rocks, especially their varying strengths, involved in orogeny will exert a control on the processes. Normal faults formed as a result of crustal thinning during the pre-orogenic phase may in later orogenesis be reactivated as thrusts.

The transition from passive margin to active margin

The switch from passive margin to active margin is marked by important changes. Firstly, the deep sea sediments carried to the Benioff zone are scraped off by the development of small thrusts. Secondly, as oceanic crust is carried down to great depths it may undergo melting because temperatures increase with depth. The magmas so generated will rise and intrude the overriding plate to give a volcanic arc such as we see today in the islands of Japan or Indonesia. Thirdly, sediment deposited in the trench ahead of the overthrust plate, together with oceanic basalts, may accumulate into an accretionary prism near the trench. Fourthly, the loading of the subducted plate by the overriding plate causes it to bend downwards. The structures resulting from the bend are seen as normal faults where the plate is undergoing extension. These faults are responsible for many earthquakes. Fifthly, the subduction of a cold slab of oceanic crust brings density into play, this time to pull the slab down into the weaker hotter asthenosphere. As mentioned previously, this gravitational force acting on the slab may act as a driving force for plate motion. Also there are the 'roll-backs' which are horizontal movements of the hinge of the subduction zone. How long does it take to form an orogenic belt like the Alps or the Himalaya? The obvious answer is that the belt must be younger than the youngest rock deformed in the belt and older than the oldest post-orogenic sediment. If the pre- and post-orogenic sediments contain fossils or can be dated by geochronological methods, then they provide an approximate time span for the orogeny. There is also geochronological evidence available to mark the timing of events during the orogeny. The orogen is usually related to plate motion and therefore orogeny should cease to develop when the subduction ceases, but subduction can start up again. In more recent orogens magnetostratigraphy (see Chapter 1) provides a method of dating the start of collision by indicating a slowing down of plate motion at the time of collision.

It is beyond the scope of this book to give detailed case histories of orogens, the main concern being with processes. Nevertheless it is necessary to include brief accounts of several famous mountain belts, including two continent–continent collision belts, one Cordilleran and one arc–continent collision belt. In the latter example, arc–continent collision was followed by continent collision and transpression. While it is undoubtedly true that there is no such thing as a typical orogen, reflecting the complex interplay of the mechanisms of rock deformation (see Chapter 3), the grand design of orogenic processes is still recognisable despite the different ways in which it operates. This chapter concentrates on the main tectonic events and the peak time of metamorphism in the selected orogens. In Chapters 8 and 9 the cooling histories and exhumations of orogens will be discussed.

Collision belts

The tectonic conditions in continent–continent collision are controlled by the fact that continental crust, unlike the denser oceanic crust, is buoyant and therefore tends to resist subduction. More precisely, the lower lithosphere may be subducted but not, perhaps the less dense mid/upper crust which instead undergoes thickening and isostatic uplift to give mountains. That is why we find mountains at the collision zone. A distinction that reflects the differing densities of oceanic and continental crust can be made between the 'A' subduction which involves subduction of continental lithosphere, and 'B' subduction which involves the deep burial of oceanic crust. B subduction can turn into A type, as for example in the India–Asia collision.

It is important to define what is meant by continent–continent collision, and this is not as easy as might appear at first sight. Thus the India–Asia collision was probably diachronous, and moreover, it is still continuing. Perhaps we should specify the start of collision in a particular place. Most coastlines and continental margins are irregular, giving different collision times for different parts of the coast. It is not easy to determine when the initial contact of two continents has taken place. The northward drift of India started *c.*130 Ma at the time of the break-up of the Gondwana continent. India's drift was fast, and during its journey it passed over a mantle hotspot which produced the Deccan lavas at 65 Ma. The magnetic anomalies of the Indian Ocean basin lithosphere tell us that the Indian continent slowed down greatly from *c.*13 cm per year to *c.*5 cm per year at about 45–50 Ma, and that is assumed to mark the start of collision. Another way of dating collision is to date the sediment that formed near the suture on top of the sediments of the trench and the accretionary prism. These tend to be non-marine, simply because the sea water is no longer there. In addition to post-collisional sediments, in the Himalaya the time of collision is indicated by incoming of Eocene marine sediments (Subathu Formation, see Chapter 9) which is probably due to the downbending of the Indian plate at the collision zone owing to loading by the overriding Asian plate.

Sometimes in collision zones there are places where slabs of oceanic crust have been pushed up onto the continent – the term is obducted. We see this in the Troodos mountains in Cyprus and, in the collisional setting of the Himalaya, the Spongtang ophiolite which is a klippe at the highest structural level in the Himalaya. In the latter this obduction occurs before collision in late Cretaceous times and therefore is no indicator of the date of collision.

In its essentials the collisional process is that one plate overrides another and the lower one is carried down, often to great depths. Therefore the lower plate will be heated and if carried to great depths will develop high-pressure metamorphic mineral assemblages such as jadeite and lawsonite. If the lower plate underplates the upper overriding plate, the latter will be jacked up. This is one way of forming high plateau.

The colliding plates are under stress and therefore break up into several thrusts. These thrusts are not separate as might appear at the surface but join into a detachment surface beneath the orogen which dips at a low angle towards the hinterland. The detachment may be very deep in the lower crust or even in the upper mantle. The positioning of the detachment is determined by rock strength, and it is along the relatively weak rocks that we would expect to find it. The width of the thrust sheets will depend on lateral variations in strength profiles through the crust and upper mantle. Crust that has been stretched before orogenesis starts is thinner than normal and lacks the low-velocity zone near the base of the crust. For these reasons it is relatively strong. Deformation of the crust may be more complex than we have assumed so far. The upper crust may thicken independently of the mid-lower crust. In this case they must be separated by a detachment which allows crustal layers to part company, as it were. For example, the lower crust may not be deformed but just slides passively down the subduction zone while the mid/upper crust undergoes shortening.

Mountain uplift

In terms of elevation of mountains, the continental crust and mantle act in contrary ways. Thus thickening of the continental crust will increase the topographic elevation while thickening of the dense lithospheric mantle will depress the elevation. If the crust thickens but the mantle subducts without thickening then there should be pronounced elevation of the frontal regions of the mountain belt. If the mantle does not subduct but instead underplates a region in the hinterland of a mountain belt, then this region will not rise but will subside. The subsidence leads to the formation of basins of sedimentation on the hinterland.

Symmetry and asymmetry in orogens

The orogens formed by continent–continent collision are frequently asymmetric which means that there is a single dominant direction of transport or vergence, on the thrusts. Broadly the Himalaya are like this. But there are places in the Himalaya where the dominant south vergence is upset by northerly verging thrusts – that is, verging towards the hinterland (so-called back thrusts). These may reflect the influence of inherited structures, the continental crust being re-worked by orogenesis many times and therefore full of weak zones. Another possibility is that the north-verging structures reflect a shear couple operating in the lithosphere; that is, given the near-horizontal maximum compression stress there are two theoretical shear planes orientated at 30–45° to the maximum compression. Cordilleran belts are also asymmetric as shown clearly in the Andes.

In contrast, in some collisional orogens there is approximate symmetry, for example the Scandian orogen in Scandinavia and Scotland formed during the collision of Laurentia and Baltica in the early Palaeozoic. The thrusts verge towards the WNW in Scotland but towards the ESE in Scandinavia, an example of what has been termed a doubly vergent orogen.

Strike-slip orogens

So far we have only considered full frontal collisions. However, it is now apparent that other collisional styles are possible, such as where one continent approaches the other at a low angle. These are called strike-slip dominated orogens or zones of transpression, to indicate a complex type of collision. Such oblique motions occurred during the complicated closures of the Iapetus Ocean in the Early Palaeozoic and are also seen in the Southern Alps of New Zealand.

Uplift and collapse of mountains

The most obvious characteristic of young orogens is that they are mountains in the topographic sense. The implication is that there has been uplift relative to mean sea level as a direct consequence of thickening. There is a maximum thickness sustainable by most segments of crust, and this is governed by the strength of the thickened crust and the support provided by the horizontal stress produced by plate convergence.

Gravitational energy is stored in a mountain belt as a result of the thickening and uplift, and a lessening of the horizontal stress will cause a collapse of mountains. Thus as the crust and/or lithosphere is thickened to produce a mountain belt, an increasing amount of work is needed to counteract this gravitational energy.

When the maximum elevation permitted by gravitational forces is reached in one part of the orogen, crustal shortening spreads or moves to the frontal part of the range. This is called foreland propagation of the mountain belt and will be further discussed in Chapter 6. In the Andes, normal faulting at topographically high levels indicates extension and perhaps the start of mountain collapse. The strengths of rocks play an important role in the spreading or migration of mountain belts. Thus simple migration of thrusting can be prevented if the rock in which the new thrust is trying to form is strong enough to resist the propagation of a thrust. The Tarim basin in China is a good example because the crust there was not thickened during the northward spreading of the deformation zone ahead of the Indian indenter. This is because it is made of strong rocks which transmitted the stress but refused to be deformed by it.

Stages of collision

In collision belts such as the Alps or Himalaya, polyphase deformation is the rule and later deformation episodes will, if sufficiently intense, overprint the structures developed before and during the early stage. How can we imagine the progress of collision tectonics? Searle (2007) suggested that the stages in continent–continent collision can be studied in the Alpine–Himalayan orogen. The stage immediately pre-collision is shown by the Semail ophiolite in Oman (Fig. 5.1). This is the best example of obducted ocean crust in the world. Over a period of *c*.21 Ma in the Late Cretaceous, a slab of ophiolite some 5 km thick and with an areal extent of 550 × 150 km has been emplaced onto a passive continental margin sedimentary succession. The ophiolite contains gabbros, pillow lavas and radiolarian cherts overlying a 12–15 km thick pile of upper mantle peridotites. The ophiolite is situated above a NE-dipping subduction zone.

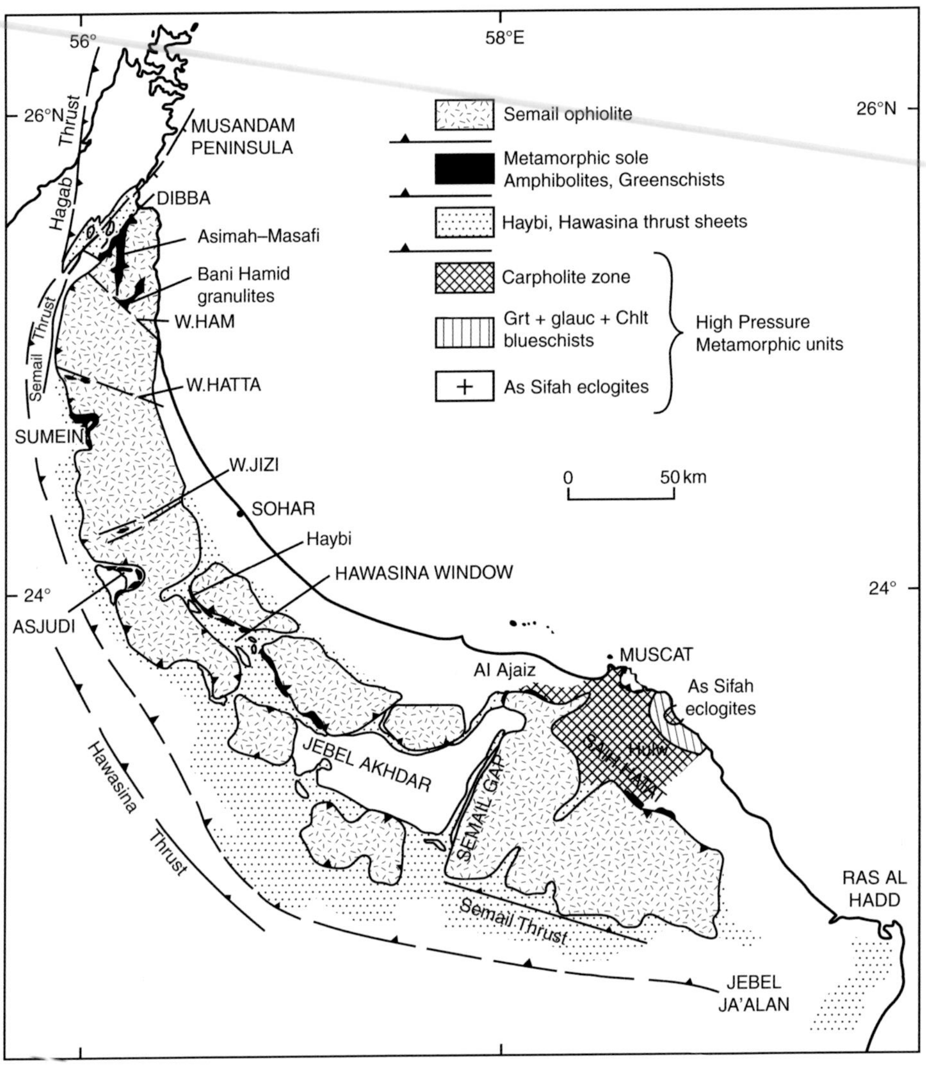

Figure 5.1 Semail ophiolite in the Oman mountains; an obducted ophiolite. From Searle (2007), published by permission of the Geological Society of America.

Along the base of the ophiolite is a 'sole' showing inverted metamorphism, with garnet + clinopyroxene amphibolites formed at 800–840 °C and 10–12 kbar, that is 50–60 km depth. The heat source for the sole was the ophiolite above it, i.e. still hot ocean crust.

The early stages of collision are shown by the Zagros Mountains (Fig. 5.2). These were formed when the NE passive margin of the Arabian plate collided with the central Iranian plate, probably in the Oligocene–Miocene. The Zagros Mountains stretch for *c.*1800 km and are 250–300 km wide.

A good example of the results of the final continent–continent collision process may be seen in the Himalaya. The collisonal process has been intense; moreover, it has lasted about 50 Ma and is still ongoing. So the features developed before and during the collision and

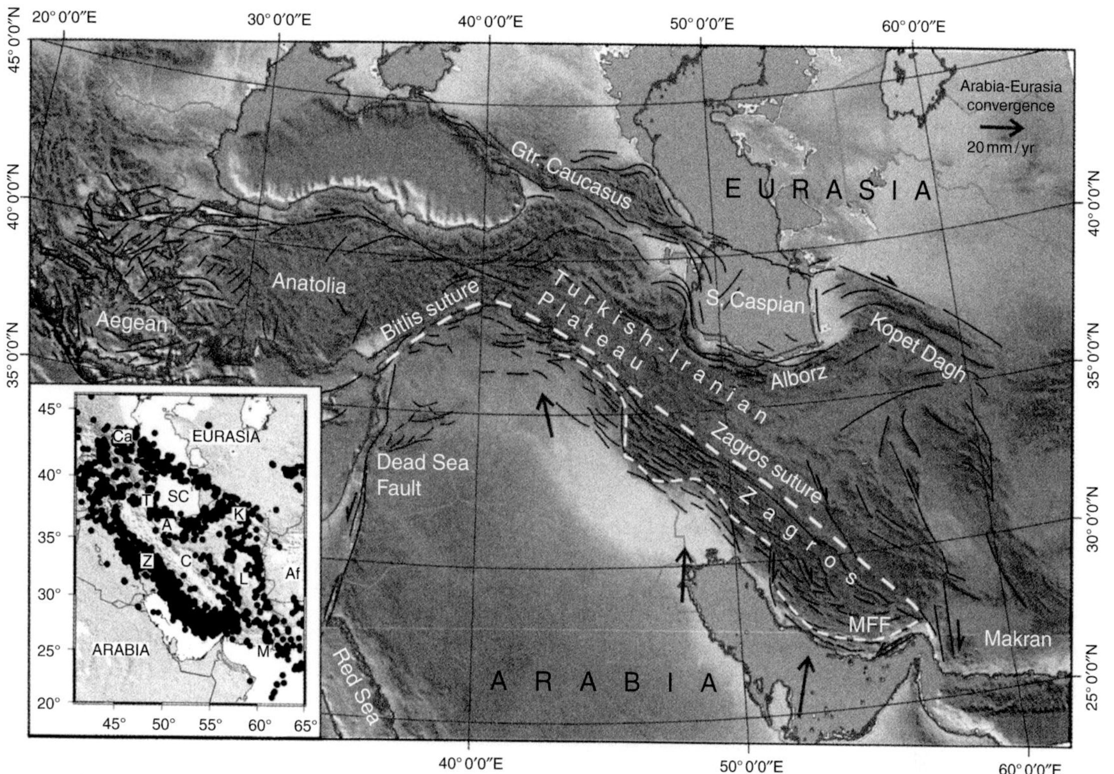

Figure 5.2 The Zagros Mountains to show their situation in relation to the oblique collision of Arabia and the Iranian plate. A, Alborz; Af, Afghanistan; C, Central Iran; MFF, Main Frontal Fault. From Searle (2007), published by permission of the Geological Society of America.

early stages of the collision process have been obscured or destroyed. But something remains. The pre-collisional Spongtang ophiolite in the western Himalaya resembles the Semail ophiolite, and it was obducted onto the north Indian margin in the late Cretaceous. In Oman the leading edge of the continental margin was subducted to depths of *c*.80 km to form eclogite facies metamorphic rocks in the deeper structural levels. However, in the Himalaya the high-pressure rocks including eclogites may belong to the main collision event rather than to the Oman stage. The Zagros stage seems to have been obliterated by the Himalayan orogeny. Earlier it was stated that the full record of orogeny is much more likely to be preserved in young mountains than in old deeply eroded ones. The lesson here is that even in young belts the complete history may be lost.

Collision belts

The history of collision belts can be summarised in the following examples. The emphasis here is on the construction of orogens, exhumation and decay of orogens being dealt with in Chapters 8 and 9.

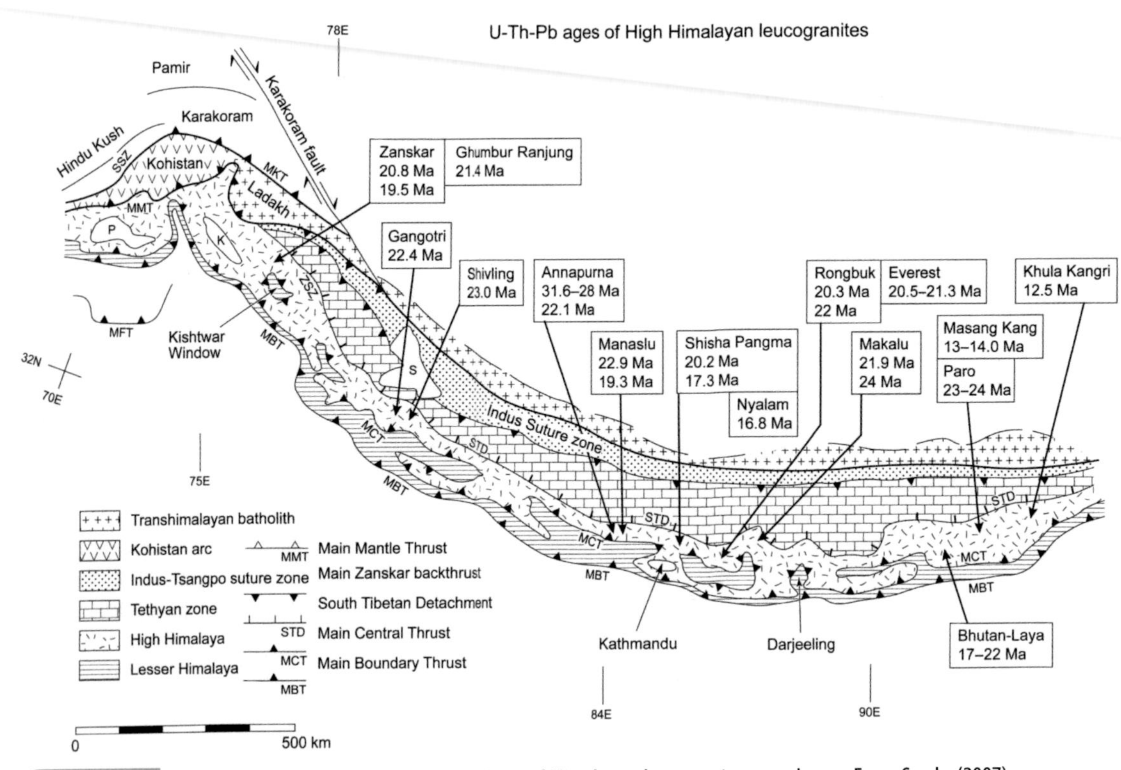

Figure 5.3 Main tectonic zones of the Himalaya. Ages of Himalayan leucogranites are shown. From Searle (2007), published by permission of the Geological Society of America.

The Himalayan Karakoram and Tibetan orogen

This is the largest area of high mountains on Earth, with an average elevation of 4–5 km but rising to a maximum elevation in the peaks of Mount Everest (8848 m) and K2 (8611 m) (Fig. 5.3). It stretches from the Afghan–Pakistan border to Yunan and Burma, over 2500 km. The orogen was formed by the closure of the neo-Tethys ocean following the break-up of the Gondwana supercontinent. India, a part of that continent, started at *c*.135 Ma to drift northwards with Madagascar from East Antarctica, and Madagascar separated from India at *c*.85 Ma. During its journey north, India passed over a mantle plume which caused the eruption of the Deccan plateau lavas in northern India at *c*.65 Ma ago. The present-day hotspot is Reunion. The collision is complex. It started first in the west with a collision between India and the Kohistan island arc, and then came the main collision with the Andean-type continental margin of Asia, marked by calc-alkaline batholiths and volcanics in the Lhasa terrane of Tibet. The Indus–Tsangpo Suture (ITS) is the 'join' between India and Asia, and it is marked by ophiolites, deep-sea sediments deposited on the floor of the neo-Tethys Ocean and Mesozoic island arc volcanics. In Pakistan the Main Mantle Thrust (MMT) between the Kohistan arc and the Indian plate is the likely continuation of the Indus–Tsangpo Suture. The India–Asia collision is not strictly orthogonal. The Indian plate has rotated anticlockwise relative to Asia and therefore there is more shortening in the

Box 5.1 Tectonic and stratigraphic units of the Himalaya

The Himalayan orogen is strongly asymmetric and is made up of several units which are, from south to north:

1. **The foredeep basin of lower Cenozoic to recent sediments**. The basin is being disrupted by the actively propagating Main Frontal Thrust (MFT), the most southern and most recent of the Himalayan thrust systems. Separated from the Lesser Himalaya by the Main Boundary Thrust (MBT) from –
2. **The Lesser Himalayan Sequence**: early to late Proterozoic and early Palaeozoic low-grade sediments, with some Permian and Mesozoic rocks and volcanics. Separated by the Main Central Thrust (MCT) from –
3. **The Greater Himalayan Sequence (GHS)**: mainly late Proterozoic/early Palaeozoic sediments but with Permian or even Mesozoic components. In addition there are some granites of Lower Palaeozoic age, metamorphosed up to high-grade Barrovian metamorphism in the Cenozoic. In the higher levels the Greater Himalayan Sequence is intruded by leucogranites which are early Miocene in age. The sequence is fault-bounded and ranges in thickness from 2–3 km up to *c*.30 km. Separated by the South Tibetan Detachment (STD) from –
4. **The Tethyan Sequence**: early Palaeozoic to late Mesozoic sediments mostly in a low-grade metamorphic state.

If the structural units in the Himalaya are restored to the pre-orogenic section then it is apparent that there is the Proterozoic to late Mesozoic sedimentary sequence and volcanics deposited on the northern margin of the Indian continent (Fig. 5.1.1; Steck, 2003).

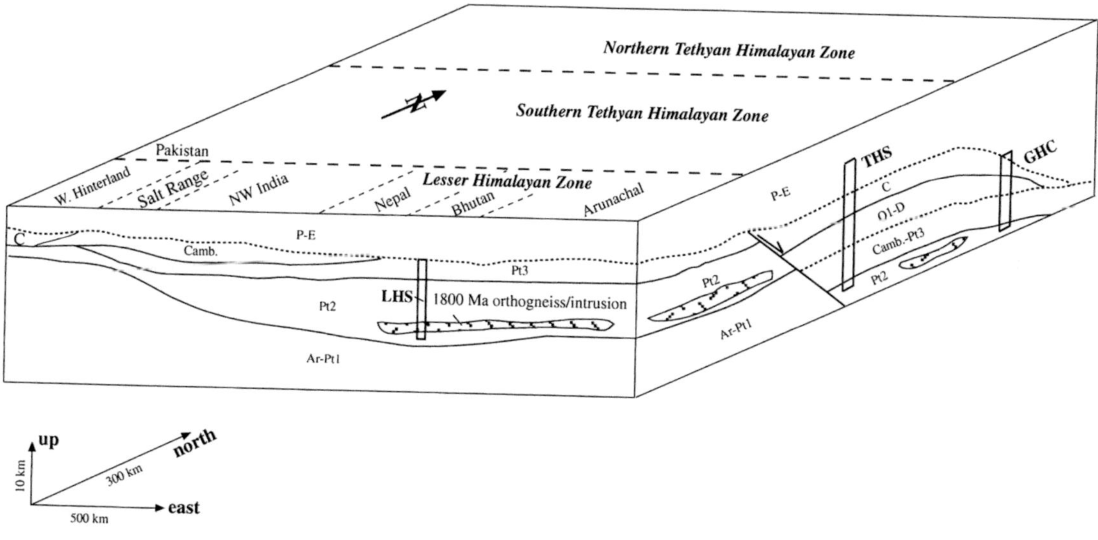

Figure 5.1.1 Passive margin in northern India during the pre-collision stage in the Himalaya. ArPtl, Archaean and early Proterozoic rocks. Pt2, Mesoproterozoic. Camb. Pt3, Cambrian and late Proterozoic rocks. Ol-D, Ordovician and Devonian. C, Carboniferous. P–E, Permian–Eocene. Note that the placing of the Tethyan Himalaya to the south of the Greater Himalaya is based on the down-to-north slip on the STD but ignores previous thrust motion on this fault. (From An Yin, 2006, published by permission of Elsevier).

Box 5.1 *(cont.)*

There has been much controversy over the stratigraphical correlation across the boundary faults of the Himalaya. It is coming to be accepted that the older, early Proterozoic part of the Lesser Himalaya is older than the Greater Himalayan Sequence seen in outcrop. The GHS in part overlaps in age with the Phanerozoic Tethyan Sequence and is equivalent to the younger part (late Proterozoic to Lower Palaeozoic) of the Lesser Himalaya. Low-metamorphic-grade equivalents of the high-grade GHS can be seen in the late Proterozoic to Cambrian Haimanta Group, which is a succession of sandstones and shales occurring in the western Himalaya in faulted contact with the GHS. The conclusion is that the Lesser and Greater Himalayan Sequence and the Tethyan Sequence were all part of a mid/late Proterozoic to Mesozoic passive margin on the northern margin of the Indian continent. The Indian continent migrated northwards during the Mesozoic, eventually colliding with Asia (Fig. 5.1.2, and Plates 10,11,12,13).

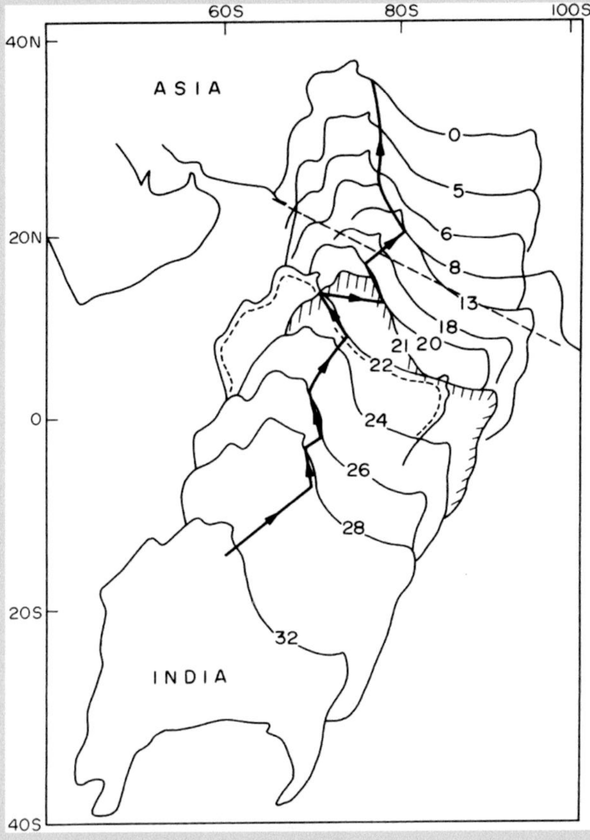

Figure 5.1.2 Progress of India, showing the relative positions of India with respect to Asia which is arbitrarily kept in its present position. Magnetic anomalies are shown by numbers. Anomalies 21 and 22 mark the position of the northern margin of India at the time of collision with Eurasia. The position of the Eurasian margin is indicated by a dashed line. Greater India is shown after the restoration of the thrusts in the Himalayan orogen – the northern margin stretched north from the present margin by 800–1000 km. (From Coward, 1994, published by permission of Elsevier.)

Eastern Himalaya. Collision started at *c*.45–50 Ma, at which stage the northern margin of India lay 800–1000 km north of the present position of the suture.

As we said at the start of the chapter, the Himalaya and the Alps are part of a huge orogenic belt of Cenozoic age which runs from the Pyrenees to Burma, and from the Betic Cordillera to North Africa and the Western Alps. The orogenic belt in SE Asia consists of not only the 2500 km long × 300 km wide Himalaya, but also orogenic belts stretching some 2000 km north of the Indus Tsangpo Suture, for example the high plateau of Tibet, which has an area of at least 1.2×10^6 km^2, more than twice the area of France, and to the north, the Tien Shan and the Altyn Tagh. Thus the Himalaya–Tibet orogen is a remarkable example of diffused deformation, cf. the much smaller width of the Alps.

The Himalayan–Tibetan–Karakoram orogen is the result of the northward motion, at a rate of $<$13 cm/a, of the Indian oceanic lithosphere which carried the Indian continent as a passenger on the Indian plate. The interesting point is that after the initial India–Asia collision at 50–55 Ma, dated by the slow-down at magnetic anomaly 22, the plate motion has continued, with convergence of the Indian and Asian plates but at a slower rate of *c*.5 cm/a. GPS data show that movement continues today at a rate of 4 cm/a. Total shortening across the whole belt varies from *c*.2000 km in the west to 2800 km in the central and east sectors, estimates for total shortening that far outweigh the tectonic importance of the Himalaya in the collisional process. Thus shortening (probably at least 800 km) across the Himalaya only constitutes about 30% of the total lithospheric shortening resulting from the India–Asia collision.

Mechanics of shortening in the Himalaya

The Himalayan orogen provides a remarkable example of the influence of horizontal compressive stress on the one hand and vertical stress (gravitational energy) on the other. The balance of these forces controls the evolution of the orogen from the early Cenozoic until the present day. The Himalaya have undergone polyphase orogeny during the Cenozoic, and orogenic activity has continued up to the present resulting in crustal thickening to 60–70 km. Estimates of total shortening vary but generally fall within the range 800–1000 km. The mechanics of shortening are as follows:

Indentation

In 1924 Argand famously proposed that the Indian continent had underthrust Asia for hundreds or even thousands of kilometres. However the modern interpretation is just as remarkable. The Indian continent has acted as a rigid indenter which pushed into Asia, perhaps reflecting the greater strength of the old Indian lithosphere relative to the weaker Asian lithosphere, which has been weakened by several orogenies during Phanerozoic time. Molnar and Tapponnier (1975) devised an indentation model in which space was created for India by tectonic expulsion of Asian lithosphere along strike-slip faults

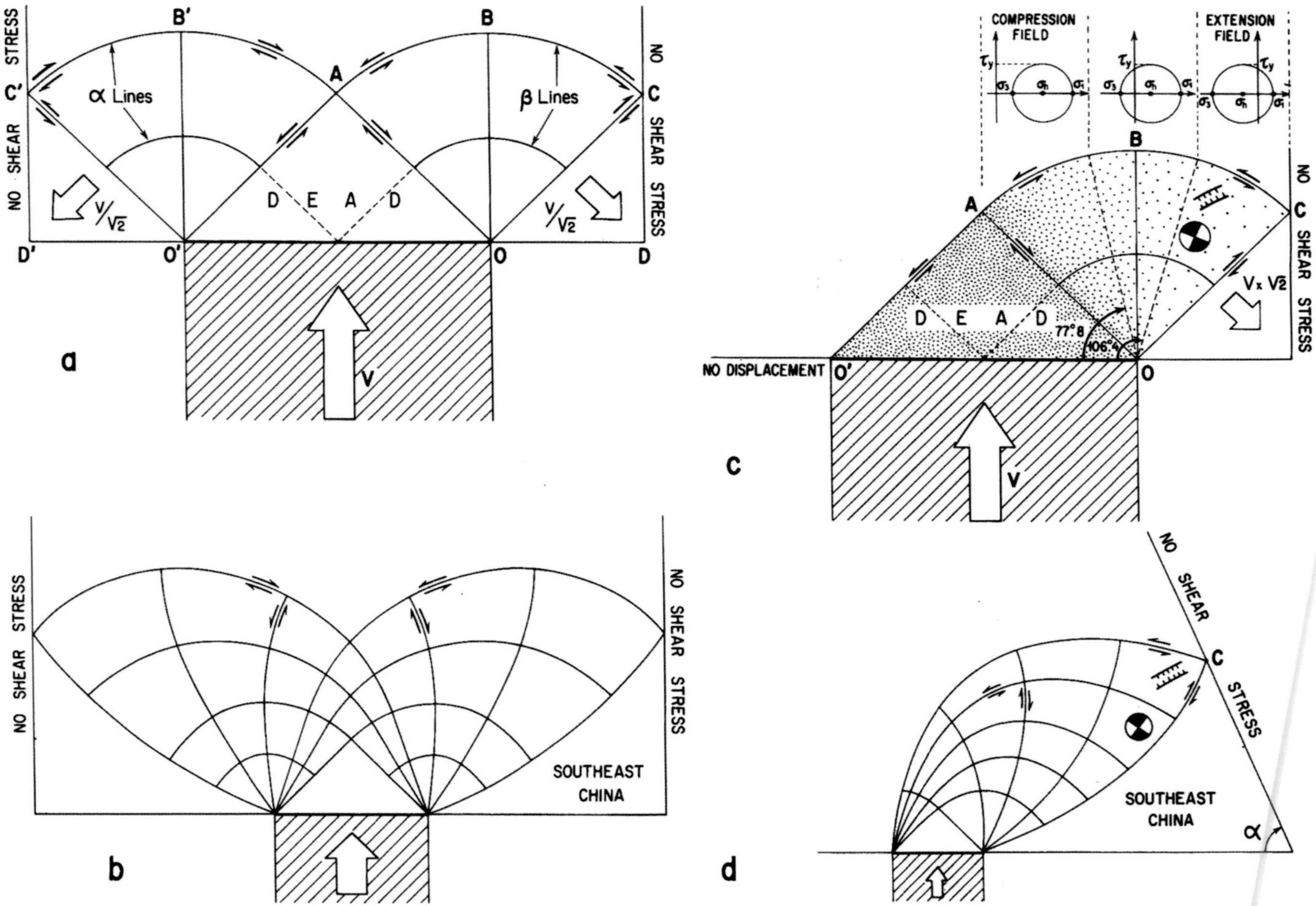

Figure 5.4 Models for a rigid indenter (diagonal ruling) resulting in the extrusion of blocks of continental lithosphere along slip lines. Arrows show the sense of shear on slip lines, large arrows show direction of rigid motion. a–d, The geometry of slip lines with indenters of different sizes relative to the indented block. In c, Mohr circles show states of stress in three regions where σ_1 and σ_3 are compressive, where σ_1 and σ_h are compressive and where σ_h is tensile. The density of dots decreases from areas of high compressive stress and associated thrust faulting and crustal thickening, to areas of tensile stress and associated normal faulting and crustal thinning. From Tapponnier and Molnar (1977), published by permission of the Geological Society of America.

(Figs. 5.4 and 5.5). To picture this we must remember that the northern margin of the Indian continent before collision was <1000 km north of the position of the Indus–Tsangpo Suture. This is because the Indian margin has been shortened, amounting to *c*.70–80% contraction in section across the Himalaya. This estimate translates into a reduction of north India by *c*.500 km up to 1000 km or more, the uncertainty showing the difficulty of restoration, i.e. determining the original undeformed state of the Himalayan region. Bendick and Flesch (2007) suggested that the indenter may have had a complicated shape rather like a prong, and varied in position along strike and with depth.

Although initially India was referred to as a *rigid* indenter analogous with metallurgical working, in fact the structural and metamorphic state of the Himalaya reveals that the indenter was not rigid throughout the Cenozoic. Although thrusting occurred before *c*.35 Ma, after that date the Indian continental lithosphere has been intensely weakened by high-grade metamorphism and it has been disrupted by intense ductile and brittle thrusting. This suggests that there was a change of mechanism, occurring perhaps in the late Eocene or early Oligocene, from rigid indentation to underthrusting of Indian lithosphere. If so then the convergence between the Indian peninsula and the Asian margin over the past 25–30 Ma has been accommodated by shortening of the northern margin of India and probably underthrusting of Indian lower crust beneath Asian crust.

Thrusting

Thrusting in the Himalaya represents the scraping off of Indian mid/upper crust from the lower crust and mantle. After initial collision and prior to the emplacement of the Main Central Thrust, the Tethyan Sequence was deformed by thrust faulting representing decoupling of basement and cover during underthrusting of the Indian plate. Searle *et al.* (2007) in the Western Himalaya show that Tethyan Sequence was thickened by thrust faulting with at least 85 km shortening between the early Eocene and the Oligocene (*c*.50–30 Ma).

One thrust fault described by Vannay and Hodges (1996) in the Annapurna area, termed the EoHimalayan thrust, is of early to mid Eocene age (45–40 Ma) and resulted in major crustal thickening in the Tethyan Sequence. Thus, by mid Eocene to Oligocene, *c*.35 km of cover had been emplaced above the Greater Himalayan Sequence. That means sufficient cover to permit the peak metamorphism in the late Oligocene (*c*.34 Ma).

The MCT is a ductile shear zone some 5–10 km thick, carrying a mid/upper crustal layer in which sillimanite grade metamorphism, migmatisation and partial melting occurred about 21–19 Ma during slip on the MCT (Figs. 5.6a,b and Fig. 5.7; see also Plates 5 and 6). As a consequence of thrusting on the MCT during the early Miocene (23–21 Ma, it appears that the elevation was at least as high as at present resulting in, at least locally, an erosional loss of *c*.20 km of rock in only 2–3 Ma. The low angled MCT was active *c*.25 Ma and carried, for perhaps 200 km, the already thickened Greater Himalayan Sequence over the Lesser Himalaya which may not have been previously deformed in the Cenozoic. Although significant ductile slip on the MCT probably ended about 16 Ma, there is evidence of later brittle slip on the MCT perhaps lasting until late Miocene times

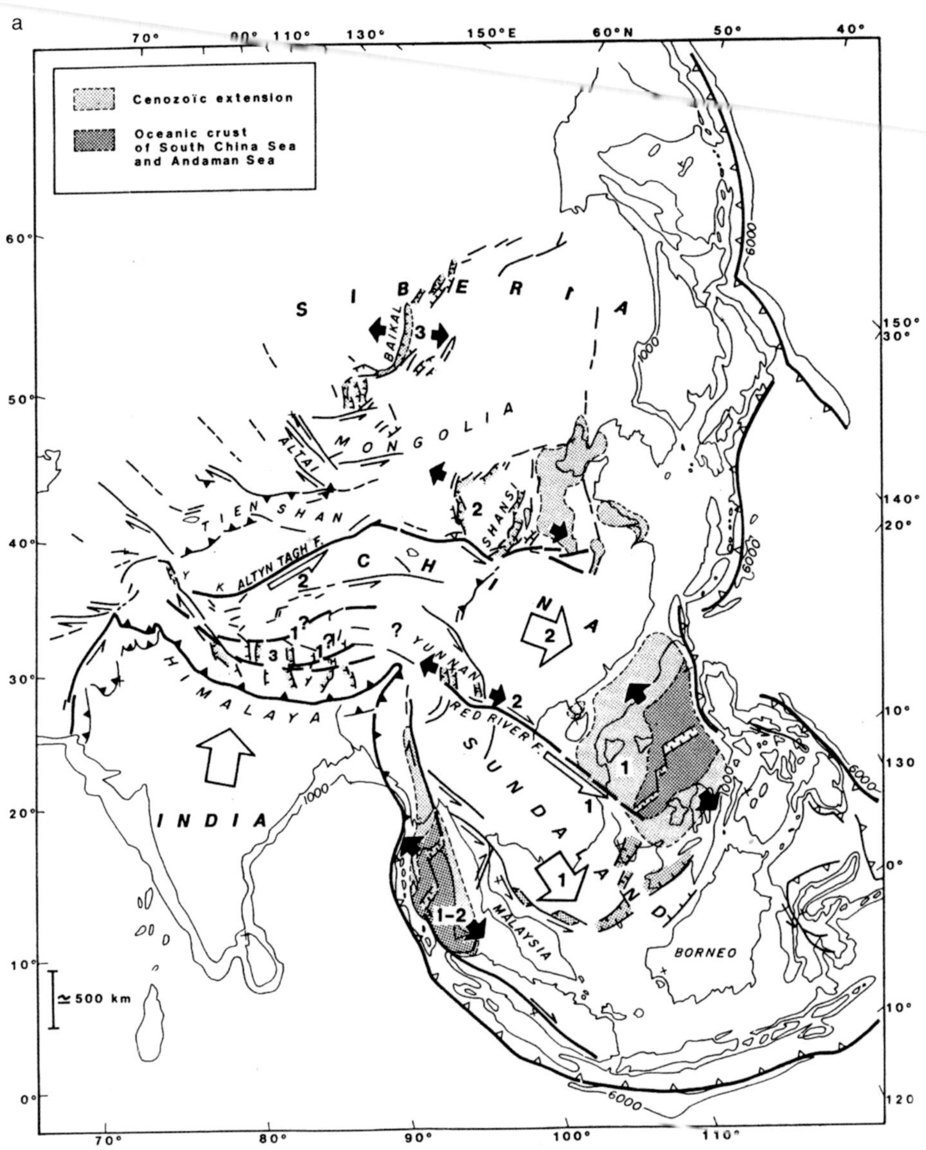

Figure 5.5 a, b, Maps showing extrusion or expulsion along large faults (cf. the slip lines in Fig. 5.4) in East Asia. Heavy lines are major faults or boundaries. Open barbs are present subduction zones of oceanic crust. Solid barbs are thrusts. Large open arrows show major motions with respect to Siberia since the Eocene. Numbers refer to extrusion phases: 50–20 Ma, 20–0 Ma, and 0 to future. From Dewey *et al.* (1988), Fig. 9, published by permission of the Royal Society.

(see Searle *et al.*, 2008). After 17 Ma the isograds in the GHS were folded by open major folds which form conspicuous domes and basins. The thrust systems of the Himalaya propagated down section from the MCT in a piggy-back style; thus the MCT is folded owing to the development of footwall duplexes and the MBT and MFT.

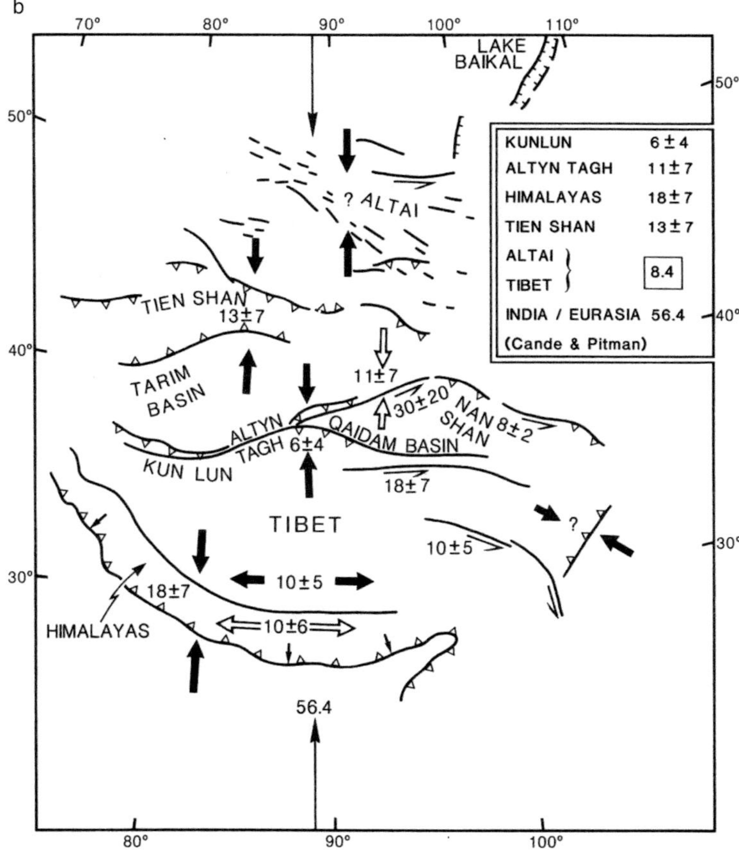

Figure 5.5 (*cont.*)

The MCT is clearly a major intracontinental crust structure, so it is perhaps surprising that there has been a huge controversy over its life span, even location. As Searle *et al.* (2008) have pointed out, it has been positioned along the kyanite isograd, and two faults, MCT1 and MCT2 have been postulated placed above and below the MCT zone; all very confusing. A famous and much debated feature of the High Himalaya is the inverted metamorphism associated with the MCT. Thus the Barrovian sequence from chlorite to sillimanite, which indicates an increase in temperature associated with increasing crustal depth, is found upside-down as we go down section through the MCT zone. More detail is given in Chapter 7. The metamorphic inversion is characterised by telescoped isograds; associated with the MCT zone is clearly a metamorphic break with the highest-grade rocks overlying the low-grade rocks. It is surprising that rarely has strain entered the argument, simply because it would be expected that a major structural feature is marked by a zone of high strain. Partly the problem arises because the stratigraphy is not very clear and so the traditional criterion for a thrust – that it places older rocks on younger – cannot be employed. Perhaps more importantly, the MCT is a ductile fault in which strain is

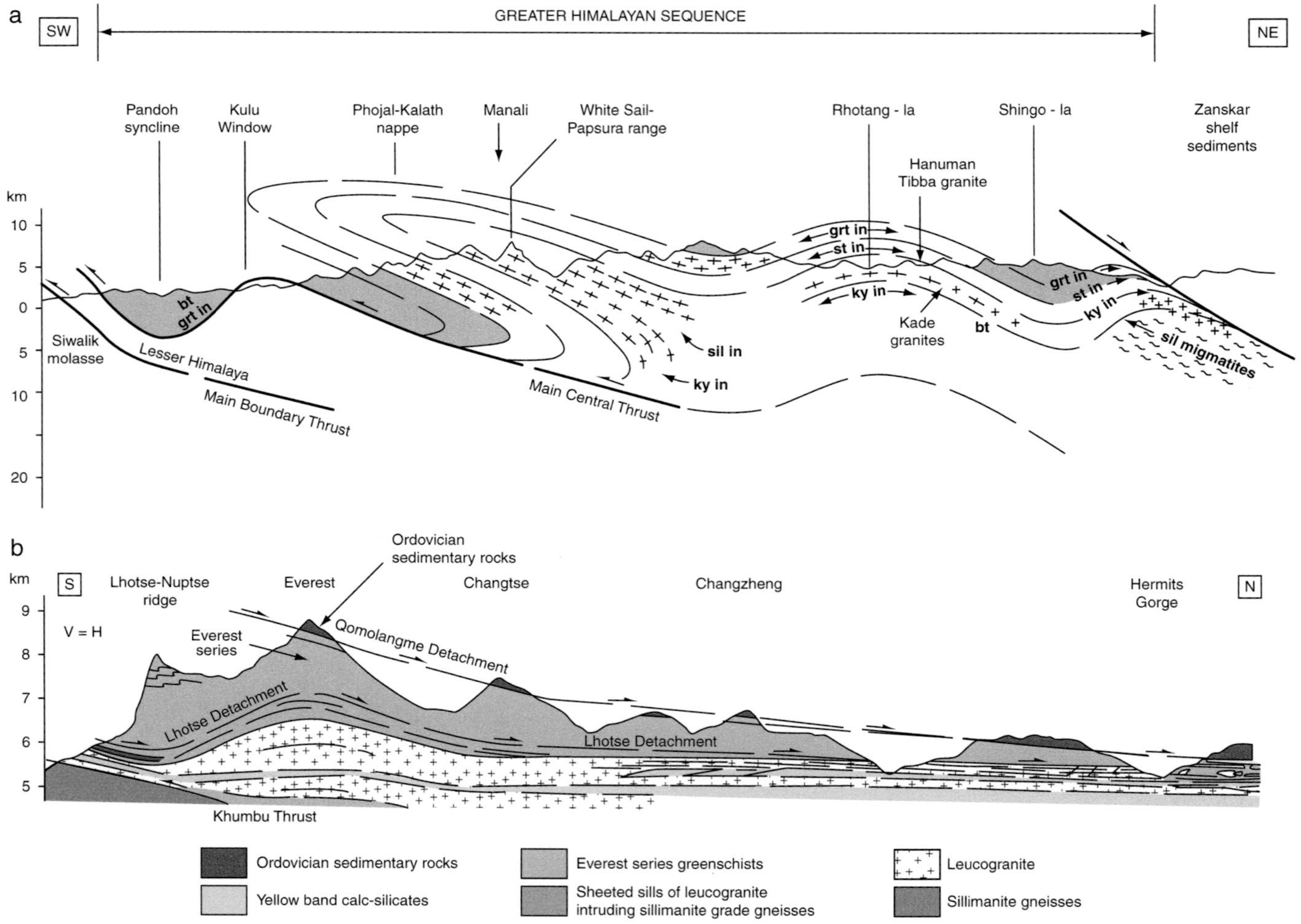

Figure 5.6 Geological cross-sections across (a) the western and (b) the eastern Himalaya. From Searle (2007), published by permission of the Geological Society of America, and Searle *et al*. (2003) Figure 4. bt, biotite; grt, garnet; st, staurolite.

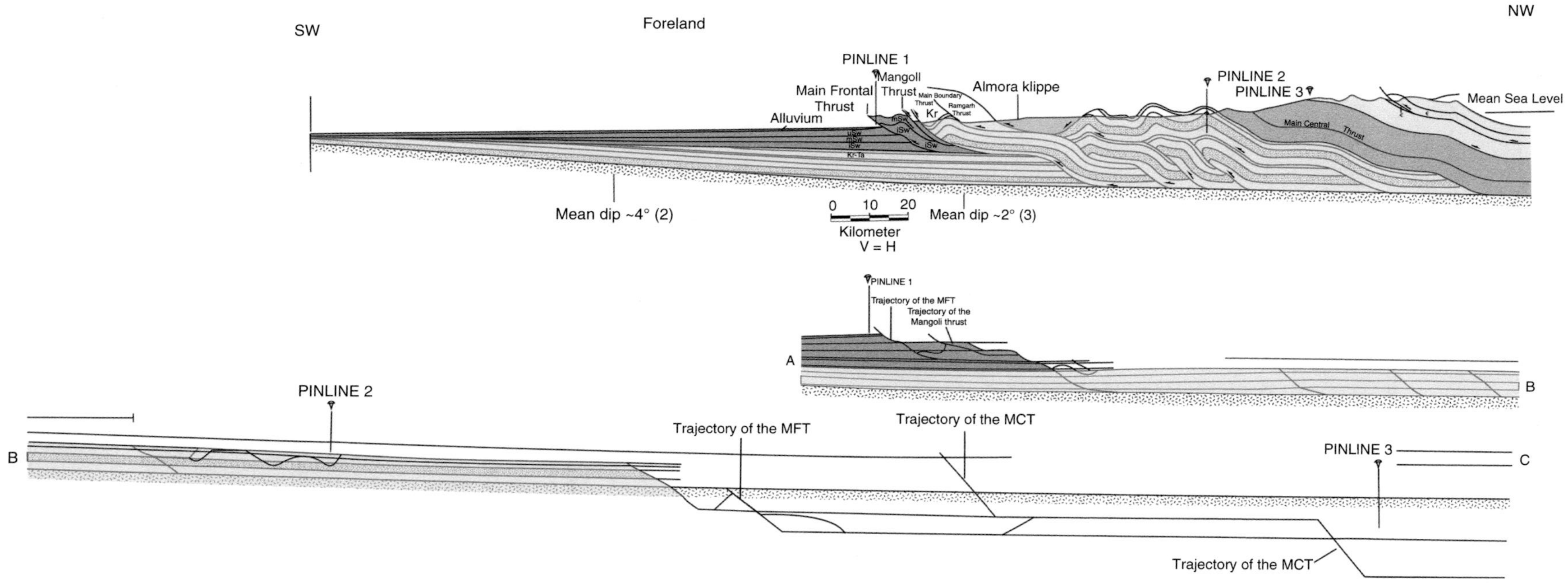

SW

Foreland

NW

PINLINE 1

Mangoli Thrust

Main Frontal Thrust

Alluvium

Main Boundary Thrust

Ramgarh Thrust

Kr

Almora klippe

PINLINE 2

PINLINE 3

Main Central Thrust

Mean Sea Level

Mean dip ~4° (2)

0 10 20

Kilometer

V = H

Mean dip ~2° (3)

PINLINE 1

Trajectory of the MFT

Trajectory of the Mangoli thrust

A

B

PINLINE 2

B

Trajectory of the MFT

Trajectory of the MCT

PINLINE 3

C

Trajectory of the MCT

Figure 5.7 Restored section (ABC) of the Lesser Himalaya and Main Central Thrust sheet in the central Himalaya. From Srivastava and Mitra (1994).

distributed over a zone rather than along a distinct break. Searle's solution is to place the MCT at the base of the high strain zone, the MCT zone.

Goscombe *et al.* (2006), working in Bhutan, supported this approach and regard the MCT as traditionally mapped as an unconformity between the Lesser and Greater Himalayan sequences. In the eastern Himalaya and Bhutan they identify the high strain zone at a much higher structural level within the Greater Himalayan Sequence. This conclusion appears to illustrate Searle's point that too often the MCT has been located at stratigraphical or metamorphic boundaries rather than along high strain zones.

In Chapter 4, we discussed evidence for the rotation of thrust sheets about a vertical axis: further examples are found in the Himalaya (see Treloar *et al.*, 1992). The movement directions of thrust sheets as shown by kinematic indicators, especially stretching lineations, are strongly divergent along the Himalayan chain.

The rotation of thrust sheets took place because of rotation at thrust tips where the displacement is zero. In the Salt Ranges, rotations of *c*.50° are found. According to the interpretation put forward by Treloar *et al.*, the contrast between the south or south–southeast thrust transport directions in Pakistan and the mainly southwest directions of the later thrust systems in India causes convergence and the formation of large-scale antiforms such as the Hazara and Nanga Parbat syntaxes. In the Hazara Syntaxis the MBT is folded over the antiform, while in the Nanga Parbat Syntaxis the MMT is folded and NW thrusting and dextral strike-slip faulting has occurred along the western margin. The rotation of thrust sheets in the Himalaya is due to the pinning of each thrust because of the interference of the Pakistan and Indian and Nepalese thrusts.

The South Tibetan Detachment (STD)

The STD forms the top of the Greater Himalayan Sequence (GHS) and it separates almost unmetamorphosed Tethyan shelf rocks, which were shortened by *c*.80 km in the Oligocene, from metamorphosed GHS rocks that were shortened in late Oligocene and Miocene times (Plate 6). The STD was active from *c*.23 to 16 Ma and is a south-vergent thrust which has been reactivated as a normal fault. Apparently the stress system changed but the later extensional strain exploited earlier zones of weakness. The normal fault is broadly synchronous with the MCT, thus presenting a paradox – what is an extensional fault doing in the compressional stress regime? Like the MCT, the STD shows a polyphase history with late brittle phases.

Underthrusting in the Himalaya

A long-running controversy in studies of the Himalaya concerns underthrusting on the Indian lithosphere beneath Asia (Figs. 5.8, 5.9). No-one continues to believe the suggestion made by Argand that this underthrusting extended for thousands of kilometres north of the suture, but most workers now accept at least 300 km of underthrusting by India. But controversy now centres on the problem of whether all the Indian lithosphere was underthrust or, perhaps more likely, only part of it. For example, DeCelles *et al.* (2002) considered that only Indian lower crust has been underthrust under Tibet, and they propose a balance between the shortening within the mid/upper crust of the Himalaya and the

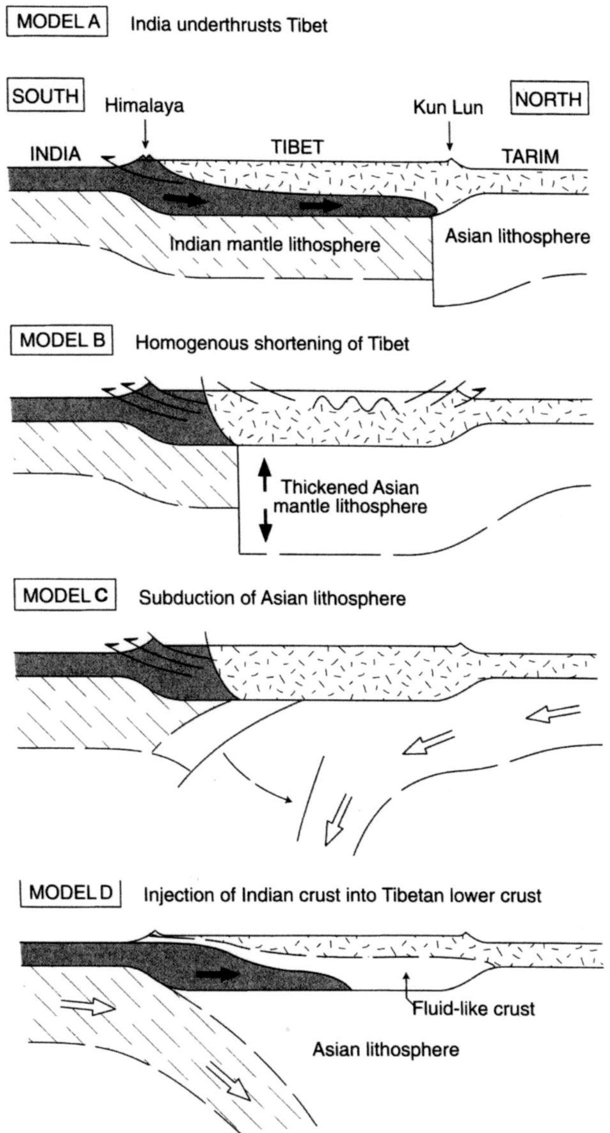

Figure 5.8 Mechanics of lithospheric thickening and thinning and uplift in the Himalaya–Tibetan region. Four models are shown: Model A: Indian lithosphere underthrusts Eurasia following delamination of Eurasian lithospheric mantle along the Moho; Model B: Eurasian lithosphere thickens by vertical plane strain and north–south shortening. A variant of this model (England and Houseman) adds convective removal of the thermal boundary layer of the lithospheric mantle after thickening (see Fig. 5.10); Model C: southwards subduction of Asian lithosphere. Tibetan crust thickens and rises by magmatic underplating; Model D: the injection model of Zhou and Morgan. Indian crust (shaded) is intruded into weakened Asian lithosphere. From Searle (2007), published by permission of the Geological Society of America.

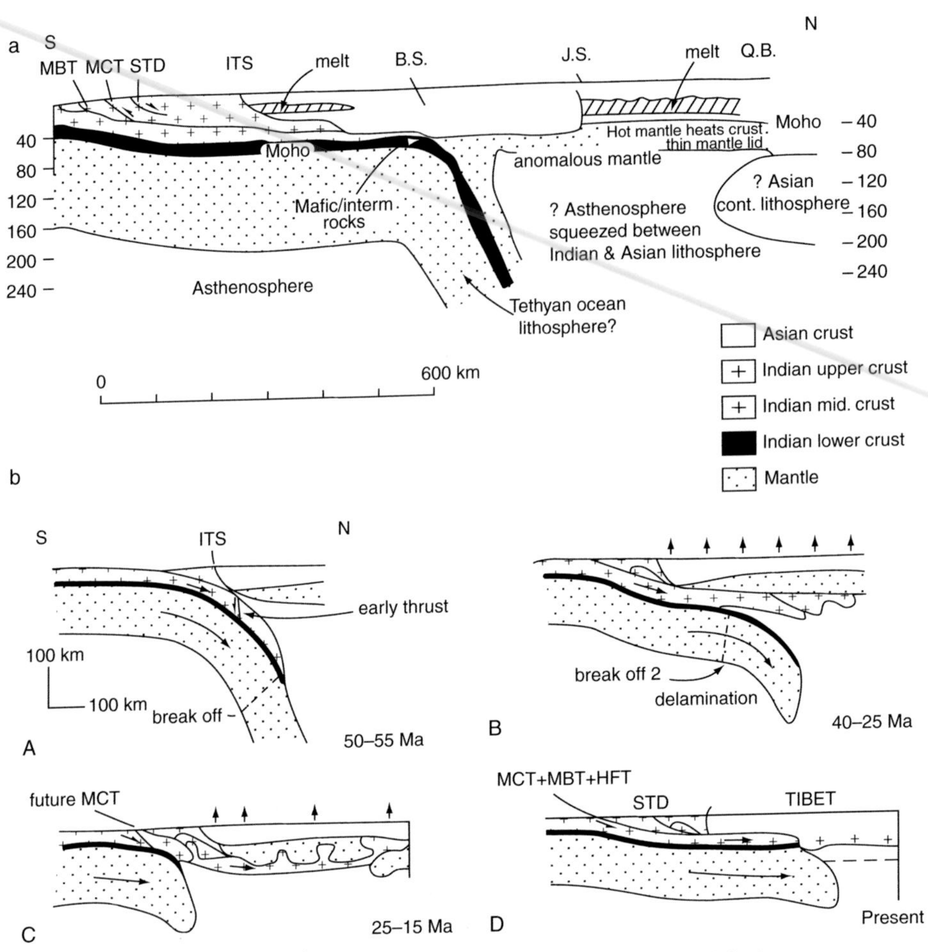

Figure 5.9 a, Speculative Himalayan cross-section and deep structure. This model invokes three subduction zones – one, the familiar north-dipping zone seen at the Indus Tsangpo suture, another beneath the Banggong suture where Indian mantle and lower crust is subducted having travelled for large distances beneath Tibet. The third subduction zone is beneath northern Tibet where Asian continental lithosphere, Indian mantle and lower crust is subducted having travelled for large distances beneath Tibet. Redrawn from Owens and Zandt (1997), Fig. 5, published by permission of *Nature*. b, Sections based on modelling showing a change from steep subduction to shallow subduction of India under Asia. At *c*.40–25 Ma there was slab break-off. Modified from Chemenda *et al.* (2000), published by permission of Elsevier.

amount of underthrusting of Indian lower crust. Thus they suggest that 650–700 km of upper crustal shortening has taken place in the Himalaya balanced by the detachment of Indian lower crust which has travelled for about the same distance beneath Tibet. Substantial underthrusting requires the existence of a flat basal detachment beneath the Himalaya and Tibet. Above the detachment is the Indian upper/mid crust (late Proterozoic/early Palaeozoic) which is involved in the Himalaya thrust sheets. Below the detachment is Indian lower crust, probably granulite facies, and upper mantle, both of which are subducted beneath southern Asia. Decoupling of upper/mid crust from lower crust and upper mantle probably was determined by the contact between the relatively weak upper/mid crust and the strong granulite facies lower crust and the mantle. In Chapter 6 a differing view based on the channel flow hypothesis is discussed. Capitanio *et al.* (2010) invoked the subduction of Indian lower crust, some 5–15 km thick, which having lost its upper crust by the scraping off of the Himalayan thrust sheets is dense enough to undergo subduction under Asia, perhaps far exceeding the commonly quoted 300 km. Furthermore, these authors suggest that the subduction of dense lower crust served as a driving force for the current India–Asia convergence.

Oroclines

A notable feature of the Himalaya and its contiguous orogens is the oroclinal curvature, with conspicuous curves termed syntaxes at the western and eastern ends of the Himalayan chain. The western syntaxis is called the Nanga Parbat Syntaxis and the eastern is the Namche Barwa Syntaxis. Between the syntaxes, the Himalayan belt has an arc-like form which may be inherited from the island arcs which existed on the margin of the Asian continent prior to collision with India.

Whereas the main Himalayan belt results from a head-on continent–continent collision, the associated north–south trending belts seen in eastern Pakistan and in the east in Burma reflect transpressional stresses associated with strike-slip motions as India plunged further and further north into Asia. For example, the active Sulaiman fold belt in the east shows a large-scale Z-shaped bend which is related to sinistral strike slip on the roughly N–S trending Chaman fault, a zone of lateral transpression (Jadoon *et al.*, 1992). In addition these belts may have been rotated as India progressively pushed into Asia.

The metamorphic evolution of the Himalaya

The orogenic evolution of the Himalaya is often divided into two phases: the Eocene to late Oligocene EoHimalayan phase and the early Miocene to recent NeoHimalayan phase. In reality the two phases are continuous. Prior to these events in the Himalaya there was an ultra-high-pressure (UHP) event which was roughly synchronous with collision. The rocks formed in this event are eclogites with the high-pressure polymorph of quartz called coesite. Temperature at the time of formation of the eclogites was 550 °C and pressures were *c.*27 kbar. These pressures indicate that Indian plate rocks had been subducted to depths of 80 to 100 km. The date of the UHP event is *c.*50–47 Ma. It was followed by rapid exhumation up to depths of about 40 km. This decompression is marked by a change in the metamorphic facies to lower-pressure amphibolite facies.

During the EoHimalayan phase (*c*.55–25 Ma), crustal thickening took place probably by means of thrusting, ending with <high-grade Barrovian metamorphism dated by U–Pb on monazite and Sm–Nd on garnet methods at *c*.35–32 Ma. Kyanite–sillimanite assemblages grew indicating temperatures of 550–680 °C and pressures of 10–12 kbar equivalent to a depth of 35–45 km. In the NeoHimalayan phase (30–16 Ma dated by U–Pb monazite) sillimanite-cordierite assemblages grew showing temperatures of 650–770 °C and pressures of 3.7 to 4 or 5 kbar, the equivalent depth being 14–18 km. The NeoHimalayan episode is characterised by partial melting and anatexis and the generation of the Himalayan leucogranites, which are sills intruded from *c*.24 Ma to *c*.17 Ma into high levels in the GHS. The Himalayan leucogranites contain quartz/K feldspar/plagioclase granites with tourmaline, garnet muscovite and biotite. They were derived by decompression dehydration melting of muscovite in the presence of fluids. The youngest high-grade metamorphism is seen in the syntaxes where Plio-Pleistocene sillimanite and cordierite and feldspar assemblages grew associated with partial melting of old crust.

In the Pakistan Himalaya, Treloar (1997) and Treloar and Rex (1990) have shown that regional metamorphism followed quickly after collision which occurred shortly before 50 Ma. Ultra-high-pressure metamorphism was so rapid that Treloar appeals to shear heating associated with Main Mantle thrust which has carried Kohistan arc rocks over the Indian continent. UHP metamorphism was dated at *c*.46 Ma. Also, Treloar *et al.* (2003) regarded the eclogite formation as syncollisional owing to slab pull by the subducted Indian lithosphere. Retrogression of eclogite facies rocks to greenschist facies occurred at 40–42 Ma. Treloar and Rex suggested that peak metamorphism attaining temperatures of 500 °C occurred at *c*.45 Ma, associated with the emplacement of the Kohistan arc onto the Indian plate. Subsequently cooling occurred over 35–23 Ma, on the evidence of hornblende, muscovite and biotite ages.

Karakoram

The last part of the Himalayan–Tibetan orogen to be considered is the Karakoram range on the borders between Pakistan and India with Tajikistan and China. The Karakoram lies to the north of the suture and represents a mainly igneous province on the Andean arc which bordered the Asian continent. The numerous high peaks in the Karakoram include K2, the second highest in the world. Much of the Karakoram is made up of lower crustal high-grade gneisses and granitoid batholiths. Crustal thickness reaches <90 km. The range is seismically active, for example the Hindu Kush seismic zone which is due to subduction of the Indian plate beneath the Tajik basin. Polyphase high-grade metamorphism spans 65 Myr, from pre-collisional (65 Ma) to post-collisional (45, 30, 25, 4 Ma). The large Karakoram batholith is a pre-collisional diorite–granodiorite–tonalite which was later deformed to amphibolite facies gneiss. In addition there are pre-collisional I-type biotite and hornblende diorite–granodiorite of mid Cretaceous age and post-collision S-type monzogranite garnet and two mica leucogranite dated at 20 Ma. The Karakoram is notable for high exhumation and uplift rates.

The Tibetan Plateau

Earlier it was stated that shortening in the Himalaya only accounted for less than half of the total convergence between India and Asia. Does the Tibetan Plateau fill the gap? The Tibetan Plateau is an important part of the orogen in several respects (Figs. 5.10, 5.11, Plate 13). The plateau has an average altitude of 5 km and a low relief (<1–2 km). Together with the Himalaya it constitutes the largest area of high elevation on Earth (Fig. 5.12). It is an area with internal drainage in its central part placed in the rain shadow of the Himalaya. Erosion is low, and hence deep rocks are rarely seen. Structurally Tibet is a collection of several terranes, Kun Lun, Songpan, Ganzi, Qiangtang and Lhasa, which were accreted to the Asian plate in the Mesozoic. The crucial point about Tibet is the crustal thickness which varies between 65 and 80 km; that is well established, but the mechanism involved is very unclear. Further evidence on the Tibetan lithosphere is given in Chapter 10.

The controversies and uncertainties concerning the evolution and the raising of the Tibetan Plateau are as follows. The shortening budget for Tibet set by the estimated post-collisional convergence between India and Asia is *c.*1000 km. The estimated crustal thickness calls for *c.*50% shortening, implying that Tibet was *c.*2000 km across prior to shortening. Yet estimates of shortening across observed thrusts in southern Tibet show only about 300 km shortening during the Cenozoic. This evidence clearly presents a problem for the vertical thickening model (Fig. 5.8) which involves *c.*1000 km of shortening across Tibet. In other words, there is a deficit in the shortening not only of the crust but also of the whole lithosphere which has been thickened (possibly 170–210 km thick). The base of the Indian lithosphere is *c.*170 km deep beneath the Himalaya and *c.*210 km deep beneath the Bangong suture in central Tibet. The small amount of upper crustal shortening of Cenozoic age in Tibet does not exclude the possibility that the lower crust has thickened far more than the upper crust.

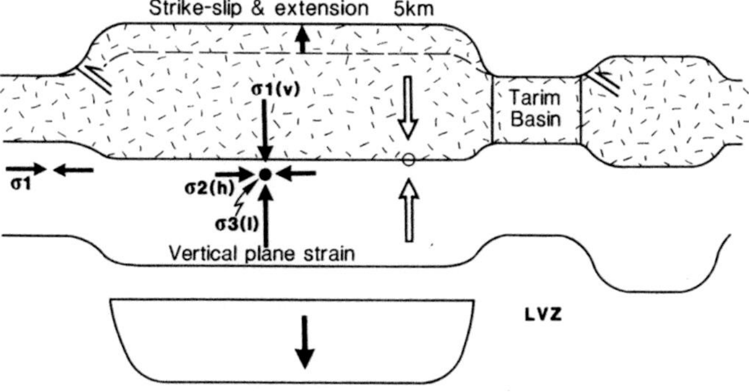

Figure 5.10 To show mantle thinning by convective removal of lower mantle lithosphere (England and Houseman). Uplift and extensional collapse of the thickened Tibetan crust is a consequence of the removal. The stress system in Tibet shows vertical σ_1 in accord with the extensional regime. Crust is stippled, mantle is white. LVZ, low velocity zone. From Dewey *et al.* (1988), published by permission of the Geological Society of London.

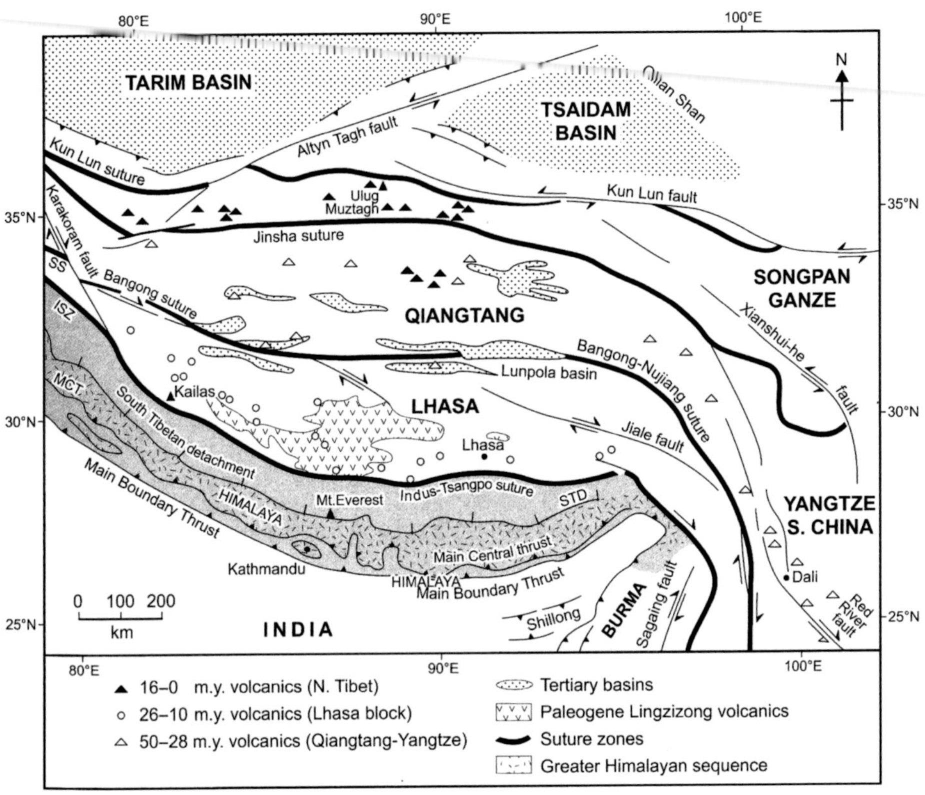

Figure 5.11 Map of Tibet showing the old sutures and the distribution of shoshonitic and ultrapotassic volcanic rocks. From Searle (2007), published by permission of the Geological Society of America.

Pre-Cenozoic thickening in Tibet?

To complicate matters further, Murphy *et al.* (1997) have argued that the crustal thickening seen in Tibet is not all of Cenozoic age, and indeed at least the southern Tibetan Plateau was elevated as a result of late Mesozoic Andean-style orogeny which produced as much as 30–40% of the total shortening. If the late Mesozoic orogeny affected all Tibet, then the deficit of shortening in the Cenozoic rises to 600–700 km. However, it seems likely that the proposed pre-Cenozoic orogeny was restricted to southern Tibet, but still some of the deficit remains. Therefore another mechanism is needed to account for lithospheric thickening of Tibet in the Cenozoic. Underthrusting of Indian lithosphere beneath southern Asia provides one solution for at least the southern part of Tibet (Fig. 5.8). But if the underthrust lithosphere pushed Asian lithosphere out of the way, then there may be no gain in the lithospheric thickness of Tibet. Another solution to the problem of the excess crust/lithosphere in Tibet has been suggested by Zhao and Morgan (1985) who proposed injection of Indian crust into Tibet as a viscous body by lateral extrusion beneath the upper crust of Tibet (Fig. 5.8). The injection was driven by the gravitational head formed by the rise of the Himalaya. It jacked up the Tibetan Plateau.

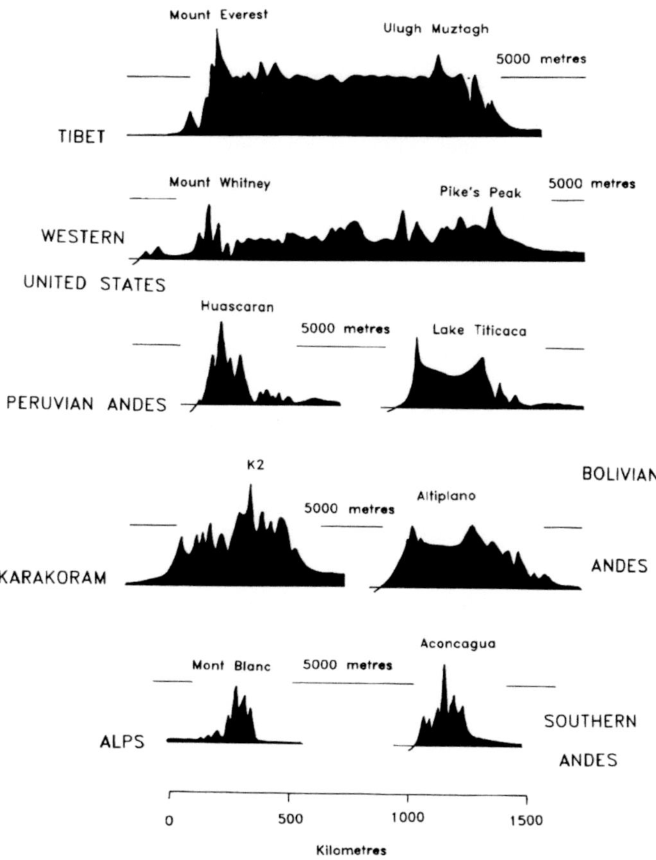

Figure 5.12 Comparisons of heights of mountains around the world to show the unique features of the Himalayan–Tibetan range. From Molnar (1988), published by permission of the Royal Society.

Chung *et al.* (2009) argued that the Tibetan lithosphere was already hot and soft because of the subduction of NeoTethyan Ocean prior to the India–Asia collision. Associated with the subduction there was underplating by basalt of Tibetan lithosphere. The underplating led to the creation of a thermally softened lithosphere in the Lhasa terrane, and the subsequent indentation of India between *c.*45 and 30 Ma resulted in lithospheric thickening and the formation of an orogenic root beneath southern Tibet. Eclogitisation of the root is suggested. The root, composed of high-density lower crust and mantle, foundered during the Oligocene, and an upflow of asthenosphere caused partial melting giving rise to adakitic magmatism (subduction-related andesites and sodic rhyolites) dated at *c.*30 Ma; the result was regional topographic uplift which later spread northwards. The model is consistent with underthrusting at 30–25 Ma of Indian mantle and lower crust under Tibet into the space provided by the foundered Tibetan lithosphere. The underthrusting gave support for the highly elevated southern regions which otherwise might have collapsed after the foundering of the root.

Deep crustal flow

Flow in weak zones in the crust is discussed in Chapter 6. Klemperer (2006) suggested that crustal flow is inevitable under Tibet. The depth of viscosity minima within the crust increases northwards from upper/mid crust beneath southern Tibet to mid/lower crust beneath central and northern Tibet. Klemperer assigned Poiseuille flow to the channel under southern Tibet and Couette flow to that beneath northern Tibet because of the relative weakness of the lithospheric mantle there (see Chapter 6). Present-day flow directions over eastern Tibet broadly confirm Klemperer's kinematic picture (Zhang *et al.*, 2004).

Differences between North and South Tibet

Owens and Zandt (1997) and Klemperer (2006) suggested that the lithospheric mantle under all Tibet is >100 km in thickness (Fig. 5.9a). However, there are notable differences between southern Tibet, which has weak mid crust and strong upper and lower lithosphere, and northern Tibet, which has hot asthenosphere at high levels. Bendick and Flesch (2007) suggested that the viscosity contrast between crust and mantle lithosphere is relatively small beneath northern Tibet. Crustal flow is compatible with crust–mantle coupling under these conditions. Strong seismic anisotropy in Asian lithosphere of central and northern Tibet is characterised by correlation between the fast polarisation direction and the surface direction of maximum shear, requiring mechanical coupling of the lithosphere here. Northern Tibet, beneath a strong lid, hosts Eurasian lithosphere with monotonically increasing seismic velocities and maximum seismogenic thickness of 25 km.

According to Bendick and Flesch, the Moho and the base of the lithosphere are flat under northern Tibet and the lithospheric thickness is 200 km. Therefore there was no loss of mass by convective removal. The surface and the mantle are deforming in a vertically coherent manner. This means that the lithosphere is mechanically coupled and transmits vertical normal stresses associated with gravity collapse into the mantle.

The geophysical work is on the whole consistent with the view that the thickened crust of Tibet in part at least could be due to 200–300 km of underthrusting by lower Indian crust of Archaean granulites and mantle similar to those of the Indian Shield. If so, then it is likely that subduction of this crust to a depth of 60 km or more would mean that the Archaean granulite facies or eclogite facies assemblages are overprinted by Cenozoic metamorphic assemblages. Evidence for the present-day existence of elevated temperatures at depth comes from the INDEPTH profile (Fig. 5.13) which revealed 'bright spots' in the upper/mid crustal layer of Tibet; these spots of mid to high conductivity are thought to be partial melts. Additional evidence of present-day high temperatures and melts at depth under Tibet comes from hot dry xenoliths showing temperatures of 800–900 °C. The volcanic rocks found on the plateau provide important evidence of its thermal evolution. Shoshonitic or ultrapotassic volcanics are widespread and have been dated at 50–30 Ma in central Tibet, 26–18 Ma in the Lhasa terrane and 15–0 Ma in Kun Lun. This time progression from south to north may be due to hot asthenosphere that

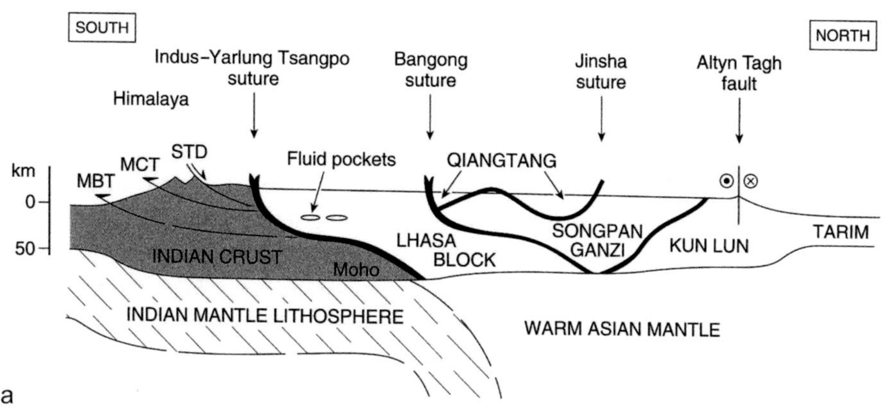

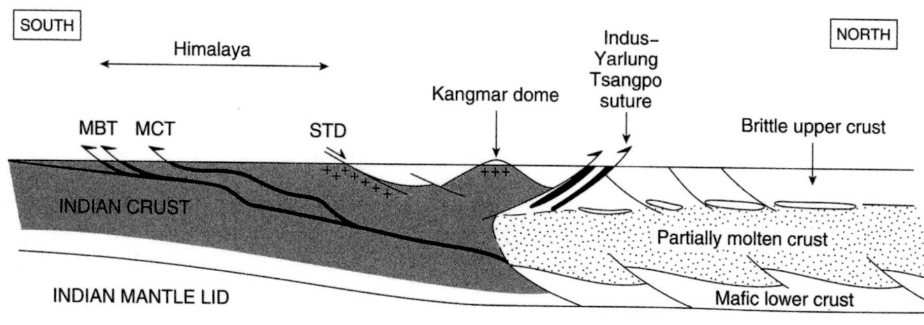

Figure 5.13 Profiles of the Himalayan–Tibetan orogen. a, From Searle (2007), published with permission of the Geological Society of America. b, INDEPTH profile showing partial melting in the mid/lower crust of Tibet. Redrawn from Nelson *et al.* (1996).

was pushed north as Indian lower crust and cold mantle were underthrust from the south (see Owens and Zandt, 1997) (Fig. 5.9a). More discussion on the deep structure of Tibet is given in Chapter 10.

N–S grabens

Another important feature of present-day activity on the Tibetan Plateau is the presence of eight N–S grabens which are interpreted to show a change of stress regime from N–S compression to E–W extension, amounting to about 5–10 km. There are also some strike-slip faults at high levels on the plateau. The late faults have been interpreted as showing a collapse of the plateau. However, at the low topographic levels on the margins of the plateau, a thrust regime exists at low levels, perhaps indicating lateral spreading of the plateau. Only a few of the late normal faults have been dated, for example the Thakkola graben faults dated at *c.*14 Ma by Coleman and Hodges (1995) and faults in central Tibet dated at 13.5 Ma by Blisniuk *et al.* (2001).

Timing of Tibetan uplift

Clearly there is still much uncertainty concerning the history and nature of the Tibetan lithosphere. Even worse, the timing of uplift of the Tibetan Plateau to its present elevation of *c.*5 km is still controversial. For some, e.g. England and Houseman (1986, 1988), major uplift is recent, about 8 Ma, and is a post-thickening response to the convective removal of the thermal boundary layer at the bottom of the lithospheric mantle. The hypothesis is that the mantle became unstable since, as a result of lithospheric thickening, it was surrounded by hot asthenosphere. The protrusion of cold lithospheric mantle into the asthenosphere creates a lateral temperature gradient which drives convection and thereby the removal of the lower lithospheric mantle. Removal led to a rapid increase in potential energy in Tibet and hence rapid uplift.

Some support for the 8 Ma uplift comes from Bull and Scrutton (1992) who have linked long-wavelength folds of the ocean lithosphere south of Sri Lanka, which have been dated at *c.*8 Ma, to a period of maximum stress, and this may be connected with uplift of Tibet at that time. Lease *et al.* (2007) used detrital single zircon evidence that there was marked uplift and erosion of the northeastern and eastern parts (Laji Shan) of the Tibetan Plateau at 8 Ma. In addition there was synchronous uplift of Liupu Shan, 300 km to the west, which suggests that there was uplift at *c.*8 Ma over a broad region. Set against these conclusions are a few (too few!) investigations which use isotopes or palaeobotany to tackle the problem. The model is also refuted by the evidence of a long history of volcanism, and a consensus now is that the major uplift, at least of part of the plateau, is much older than 8 Ma. Let us briefly review the opposition. DeCelles *et al.* (2007) compared oxygen and carbon isotopes in modern and older palaeosols and concluded that the southern part of the plateau, near the Banggong suture, was at its present height at 26 Ma. Other estimates of the timing of uplift of Tibet are from Schoenbolm *et al.* (2006) and Clark *et al.* (2004), working in southeastern Tibet (southeast of the Eastern Syntaxis), who used thermochronology, sedimentology and palaeobotany to postulate a rise of 1.5 km since the Pliocene. They postulated that at *c.*9–13 Ma lower crustal material began to flow into the southeast margin inflating crustal thickness and causing passive surface uplift and river incision. The lower crustal material continued to flow southeast in the Pliocene. This hypothesis presupposes an earlier thickening in the main part of the Tibetan Plateau, probably in mid-Miocene times. Spicer *et al.* (2003), working in southern Tibet, used palaeobotany (the morphology of leaves as an altitude indicator) in the Namling basin as an indicator of increasing elevation; that is, changes of vegetation are expected to occur as the height increases. Moist static energy at that location was determined, and from the study they suggested that the plateau has been at near its present elevation since 15 Ma. Rowley and Currie (2006), working in central Tibet, used analyses of oxygen isotopes in Late Eocene carbonate sediments from soils and lake sediments and proposed that south Tibet was at *c.*4–4.5 km by 35 Ma. Their results were challenged by Molnar, Houseman and England (2006) who pointed out that they come from a small part of the Lhasa block which was thickened during pre-collisional Andean margin development and are therefore irrelevant to the Cenozoic deformational history. Currie *et al.* (2005), investigating oxygen and carbon isotope ($\delta^{18}O$ and $\delta^{13}C$) values in the Oiyug basin in the extreme south of Tibet, showed that the Tibetan Plateau was at 5.2 km high by 15 Ma and that the plateau grew continuously northwards and was uplifted to 4 km by 40–50 Ma. Currie *et al.* estimated the palaeoaltimetry of southern Tibet from the analyses of oxygen isotope concentration of palaeoprecipitation as recorded in Miocene pedogenic calcium carbonate nodules.

These palaeoaltimetry measurements are too few and permit only tentative conclusions, but they appear to indicate that the plateau uplift may be much older in the south of Tibet than in the north. The conclusions are not necessarily in conflict; Lease *et al.* and DeCelles worked in different parts of the huge area that is Tibet. To that extent the data are consistent with a model involving progressive south to north thickening. The story is further muddied by the probability that southern Tibet was already thickened at the time of the India–Asia collision.

Molnar and Stock (2009) suggested that there was a *c.*40% slowing of the India–Asia convergence in the period between 20 and 10 Ma, and at this time the Tibetan Plateau grew outward, especially since 15 Ma. This growth set up radially oriented compressive strain in the regions surrounding Tibet. Increase in deviatoric stress during late Miocene times has been noted in the Indian Ocean. There was an abrupt increase in mean elevation of the plateau by 1–2 km at this time, a statement which is permitted by the palaeoaltimetry measurements in the north of the plateau mentioned earlier. Molnar and Stock attributed the increased elevation to convective removal of the mantle lithosphere.

Turner *et al.* (1993) used geochronological dates on volcanic potassic andesites from northern Tibet. The generation of these rocks is assumed to require thinning of the lithospheric mantle. Argon ($^{40}Ar/^{39}Ar$) dating gives an age of *c.*13 Ma for the beginning of this volcanism and uplift. Dating of the Thakkola graben by Coleman and Hodges (1995) gave 14 Ma as the time of normal faulting there. Neither of these results is consistent with plateau uplift at 8 Ma.

What can be stated with some confidence is that present evidence does not support abrupt uplift of the entire plateau at 8 Ma as proposed by England and Houseman (1986, 1988). Perhaps a gradualist model for the uplift is conceivable, during which the Cenozoic deformation and uplift migrated northwards across Tibet at a rate of 2.2 mm/a. The England and Houseman model has been tested by dating the normal faults that cross the plateau generally with a N–S trend. If these faults are associated with the collapse of the plateau, then they are important evidence. But the story from all too few faults is confusing: some support the 8 Ma date for attainment of greatest altitude but others point to extension starting as early as 19 Ma. As already mentioned, the volcanic rocks provide a test for the magic 8 Ma event and they show ages stretching back to 40 Ma.

In summary, it is clear that there is much work to be done to answer major questions about the Cenozoic history of the Tibetan Plateau, notably its raising and the cause of the crustal thickening.

The Alps

We have considered the Himalayan–Tibet–Karakoram orogen in some detail, but it may be a special case at least among recent orogens if only because of the large

Box 5.2 Mesozoic history of the Alpine region

In the Alpine realm, the Mesozoic history of the European margin was dominated by stretching and rifting on a passive margin, controlled by the sinistral strike-slip motion of Africa and Europe (Fig. 5.2.1). This motion lasted until the early–late Eocene when Africa started to move, first northwards and then after the Miocene, NW–SE. The pre-terminal continent–continent collision history of this region concerned the opening of the NeoTethys ocean which opened because of a phase of spreading which operated from mid Jurassic times in the western Alps to mid/upper Cretaceous times in the east. Part of NeoTethys was oceanic crust, and this is now incorporated into the Alpine nappes.

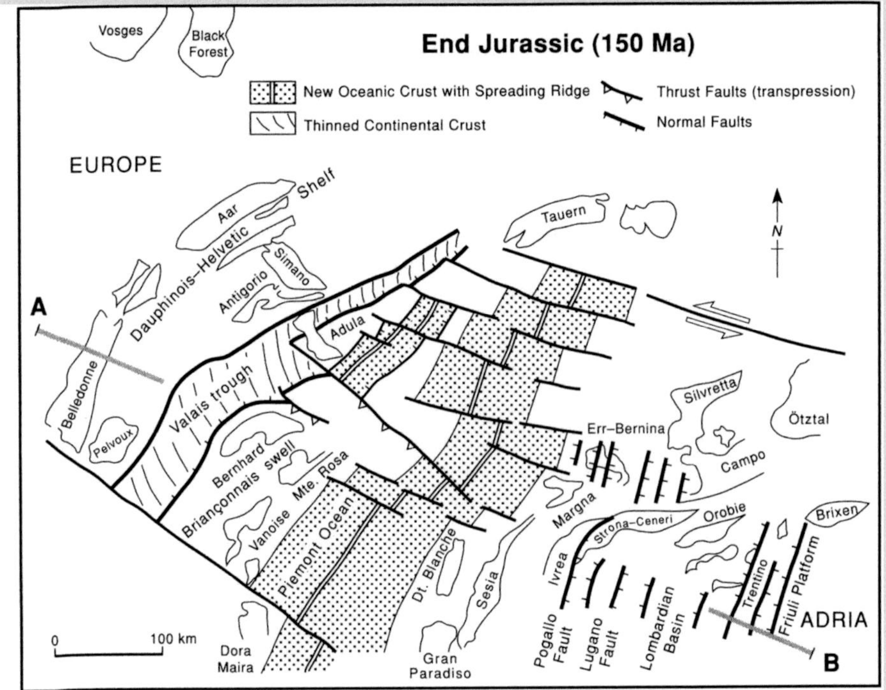

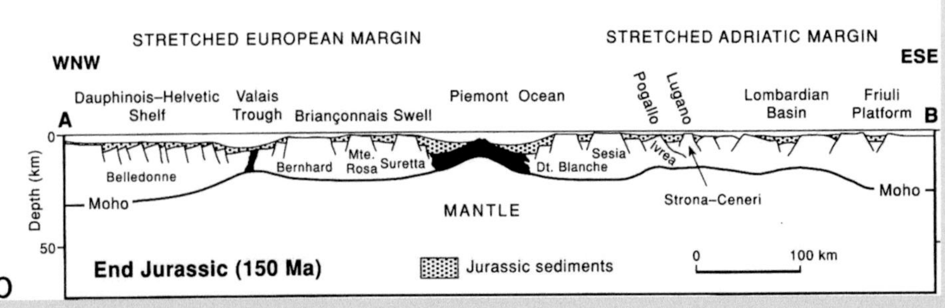

Figure 5.2.1 a, b, Alps in the Mesozoic showing platforms and basins between Europe and Adria. From Pfiffner (1992), published by permission of Cambridge University Press.

Box 5.2 *(cont.)*

The rifts were very active in the Jurassic, which was the time of separation of Gondwana and Laurasia and the opening of the central Atlantic. The tectonic regime in the Alpine area was one of basins and swells, developing in a transtensional regime perhaps related to the W–E strike-slip motions of Europe and Africa. Uplifted platforms or microcontinents such as the Briançonnais platform have a basement of Variscan and older high-grade metamorphic rocks, and a cover of Trias to Mesozoic sediment. The platform was fault-bounded as shown by the breccias and basic volcanism along its margins. Basins like the Valais Trough received clastic sediments and some mantle-derived basalt which was allowed in by the thinned continental crust.

The Piemont Ocean is an important element in the Mesozoic tectonic regime in the Alps. It may have been as wide as the Atlantic Ocean. Eastern NeoTethys terminated south of Adria which remained part of Africa. These motions were synchronous with opening of the central Atlantic. This ocean was short-lived, from the mid Jurassic to mid Cretaceous or Eocene, but it was truly oceanic and contained serpentines, radiolarian cherts and ophiolites.

The southern margin of the Piemont Ocean was Adria, a northern promontory of the African continent. On the Adria margin, crustal thinning under extensional stress produced NNE–SSW fault blocks as in the Lombardy basin, the Triassic Hallstatt basin in Austria and the Monte Generoso basin in Italy. The Sesia zone and the Dent Blanche nappe of the Pennine Alps were derived from the Adria continental margin and the adjacent Piemont oceanic crust. The other Pennine nappes were derived from the platforms and basins of the European margin, e.g. the Great St. Bernard nappe was derived from the Briançonnais platform.

The foregoing interpretation was famously set out by Rudolf Trumpy (1960), and it seemed to sweep aside the geosynclinal model which proposed that the sedimentation phase was merely a forerunner of the orogenesis in the sense that the sediments were destined to become an orogen. However, as is shown in the main text, after Trumpy and until recently it was held that after 130 Ma subduction started and continental collision ensued. Thus the distinctions between the orogenic and sedimentary phases were blurred.

area that has been affected by the collision. Therefore it is appropriate to draw a contrast between this and the style of other continent–continent collision orogens by a brief review of the European Alps. The Swiss Alps are famous as the place where many basic concepts of structural geology were discovered. Indeed, the work in the Alps has contributed many words and terms to the subject, such as basement and cover, nappes, autochthon and allochthon (literally, the same and different country respectively), flysch, molasse, A-type subduction, décollement, klippe, window and traineau écraseur.

The Alps result from the collision between Adria (a promontory of Africa) and Europe, and on a larger scale from the motion of Africa which in late Cretaceous or early Eocene times changed from a westerly direction to a northerly movement towards Europe, first northwards and then, after the Miocene, northwest–southeast. Northward movement of Adria continued until the Pliocene or even later.

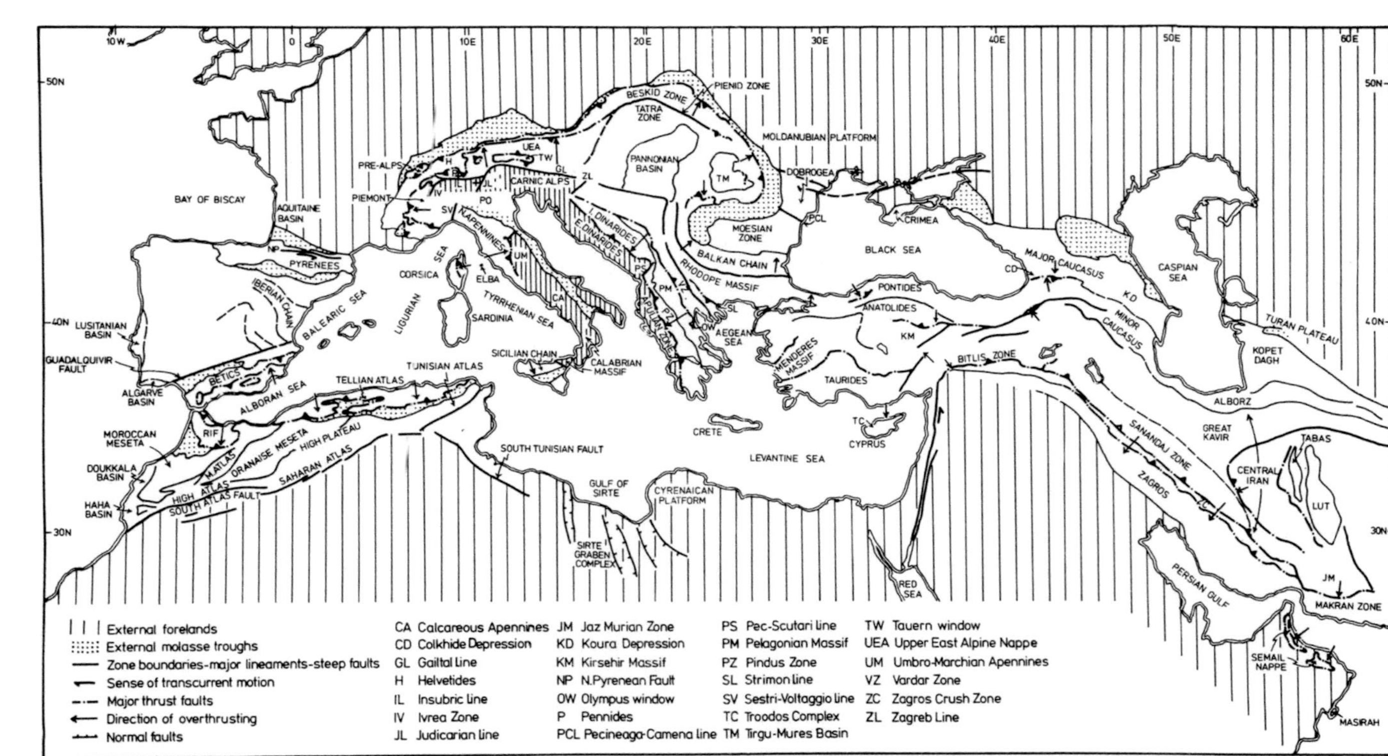

Figure 5.14 Tectonic map of the Alps and contiguous orogens around the Mediterranean. From Dewey *et al.* (1973), published by permission of the Geological Society of America.

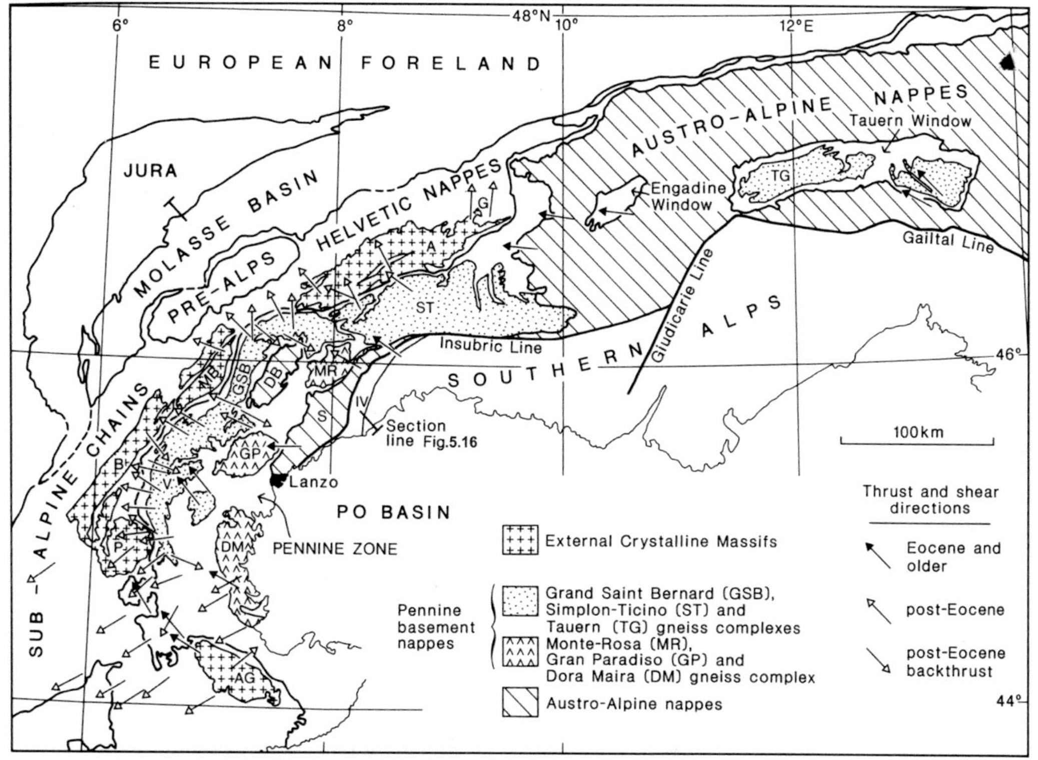

Figure 5.15 The western Alps showing the main tectonic units and thrust vergences around the orocline. A, Aar Massif; AG, Argentera Massif; B, Belledonne Massif; DB, Dent Blanche nappe; DM, Dora Maira Massif; G, Glarus; GP, Gran Paradiso Massif; GSB, Great St Bernard gneiss complex; IV, Ivrea Zone; MB, Mont Blanc massif; P, Pelvoux massif; S, Sesia zone; ST, Simplon–Ticino gneiss complex; TG, Tauern gneisses; V, Vanoise massif. From Platt (1986), Figure 7, published by permission of the Geological Society of America.

Before attempting to unravel the complex history of the Alps the main units of the Alps will be described.

The Alpine chain excluding the Southern Alps is *c*.250 km across (Figs. 5.14, 5.15). Into this rather small space about 800–900 km of rock has been squeezed, according to retrodeformation made by Platt (1986) (Fig. 5.16). Overall the Alpine orogen is asymmetric with northerly vergence towards the European foreland, but there are also southerly directed back-thrusts. It is bounded on the north by the European foreland block and on the south by a hinterland forming the Southern Alps and Po basin. Although the main mechanism of shortening in the Alps was thrusting, there is also folding on a large scale such as can be seen on the spectacular Dent de Morcles nappe in the Helvetic zone or in the Simplon-Ticino and Great St Bernard nappes of the Pennine zone.

There are two main parts to the Western Alps: an internal zone consisting of the Penninic Alps which constitute a metamorphic core to the orogen (Fig. 5.17), and an external zone

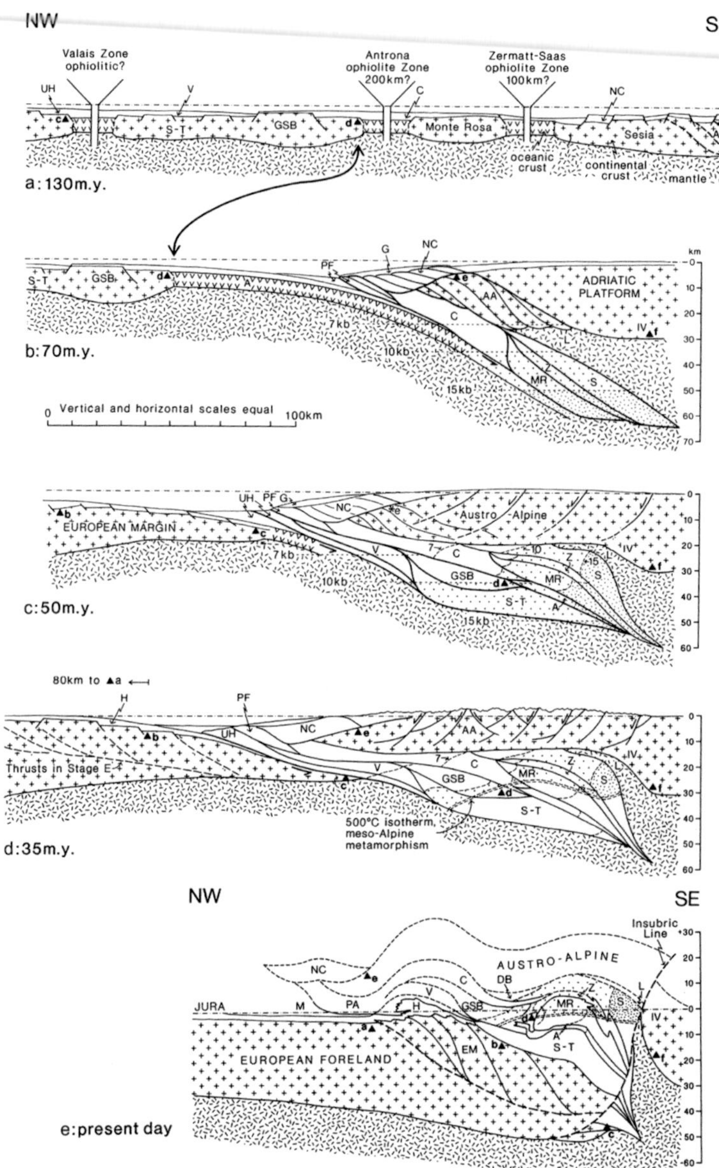

Figure 5.16 Platt's composite restored section of the central Alpine Zone. The whole orogen has been shortened by *c*.800–900 km, the Helvetic region has been shortened by about 120 km. A, Antrona ophiolite; AA, Austro-Alpine basement; C, Combin zone of calc-schists; DB, Dent Blanche nappe; EM, External crystalline massifs; G, Gosau Beds; GSB, Great St Bernard nappe; H, Helvetic cover nappes; IV, Ivrea zone; L, Lanzo and related peridotite massifs; M, Molasse; MR, Monte Rosa nappe; NC, Northern Calcareous Alps; PA, Pre-Alpine nappes; PF, Pennine flysch; S, Sesia zone; ST, Simplon–Ticino Gneiss complex; UH, ultrahelvetic cover and flysch; V, Valais calc-schists; Z, Zermatt-Saas ophiolite and calc-schists. a, Mid Cretaceous – sedimentation in platforms and basins: ocean crust omitted from the basins. b, c, Possible start of subduction. d, Oligocene – main UHP metamorphism in Penninic zone followed by rapid exhumation and

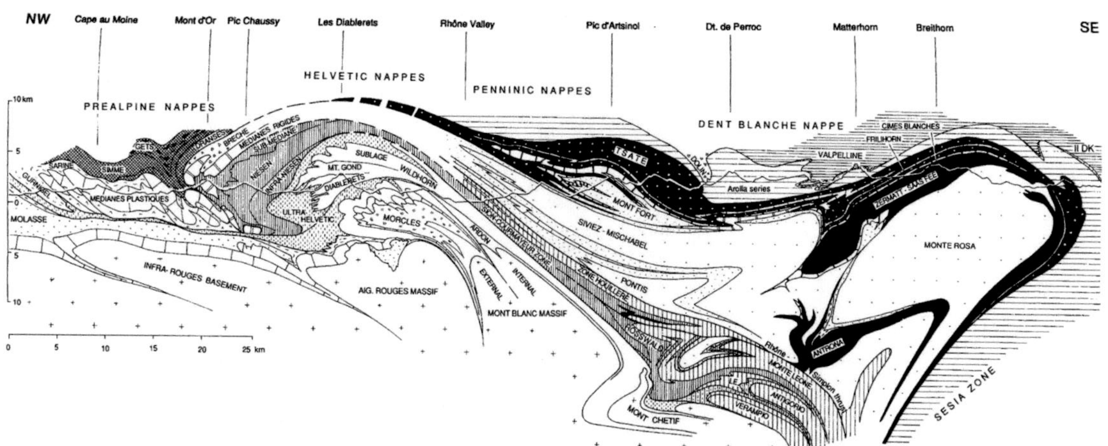

Figure 5.17 Basement and cover relations in the structural domains of the Western Alps. Basement: crosses, dots and vertical lines. In black: obducted ophiolitic assemblages. Most basement-cored nappes include remnants of cover left behind after part removal of cover units (e.g. Pre-Alps). From Esher et al. (1993), published by permission of Elsevier. From Esher *et al.* (1993), published by permission of Elsevier.

which comprises the Helvetic Alps, the Pre-Alps, the Eocene–Miocene North Alpine Foreland Basin and the Jura Mountains. To the south of the internal zone, in the hinterland of the Alps are the Southern Alps.

There are several crystalline massifs in the external and internal zones (Fig. 5.15). In the external zone examples are the Mont Blanc, Aiguilles Rouges, Pelvoux and Mercantour massifs, all of which have been elevated by late Alpine events, while in the internal zone, there are the Dora Maira, Sesia and Gran Paradiso massifs.

These are areas of high-grade metamorphic and granitic rocks of Variscan or older ages which were remetamorphosed in the Alpine orogeny. The massifs in the internal zone were involved in the Alpine ultra-high-pressure (UHP) metamorphism which is also seen in the ophiolitic rocks occurring between the Pennine nappes.

Culminations and depressions

Major feature of the Pennines and Helvetics (Fig. 5.18, Fig. 5.19) are large-scale culminations and depressions of fold axes: an up and down structure along the strike direction.

Caption for Figure 5.16 (*cont.*) Lepontine metamorphism at temperatures of >500 °C which largely obliterates the UHP assemblages. e, Rapid exhumation after the Lepontine metamorphism leads to the present-day backthrusting of the Pennines and uplift along the Insubric Line. **a**, **b**, **c**, **d**, **e**, **f** are reference points. From Platt (1986), published by permission of the Geological Society of America.

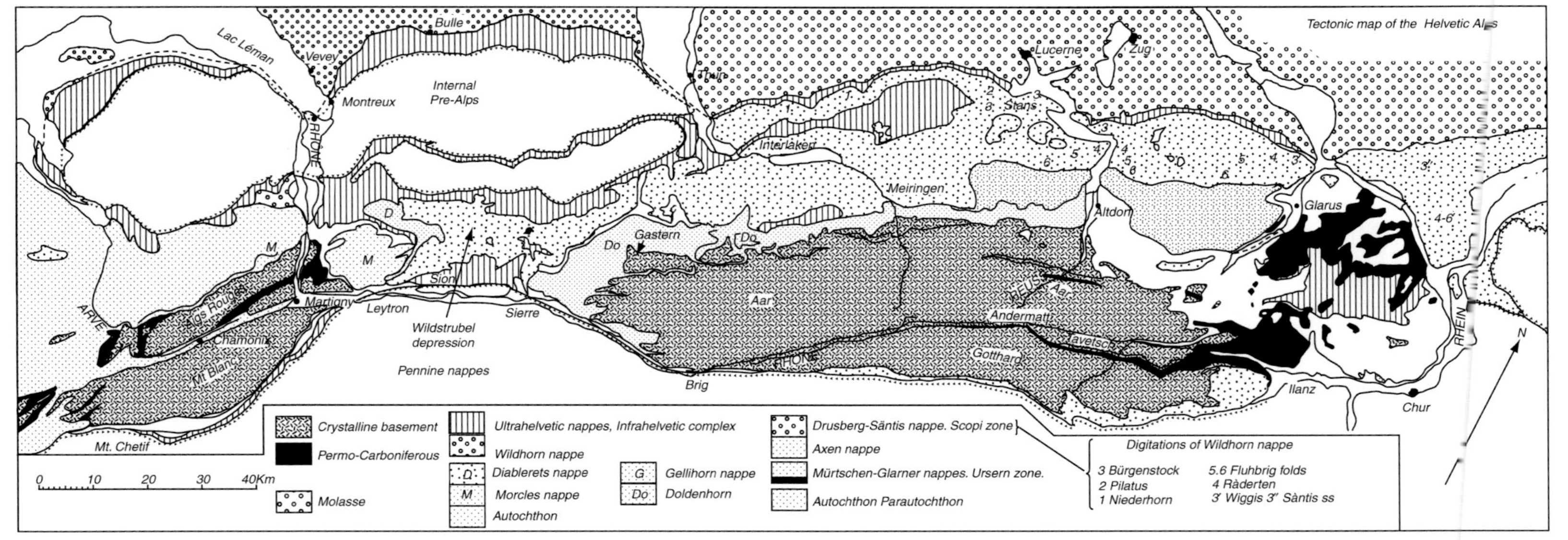

Figure 5.18 Map of the Helvetic zone in the Swiss Alps. From Ramsay and Huber (1983, 1987), published by permission of Elsevier.

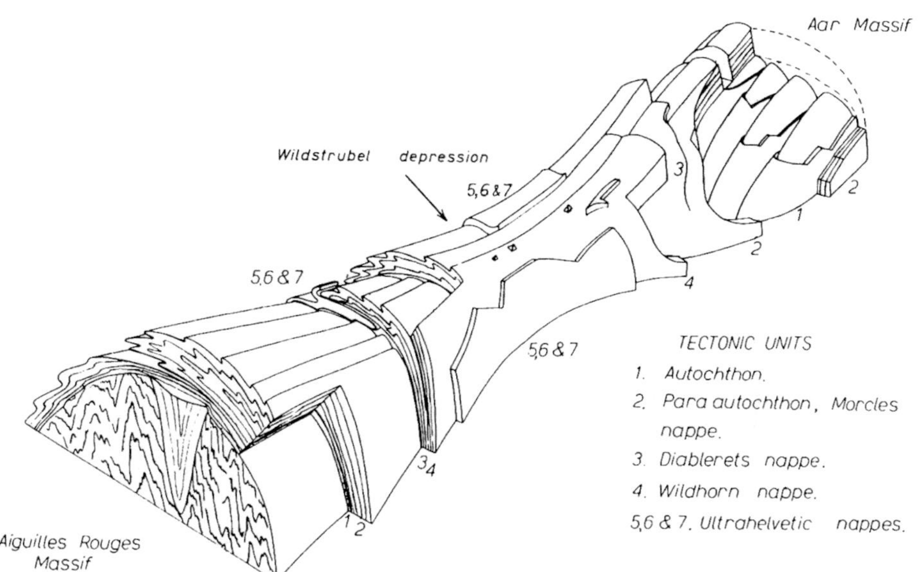

Figure 5.19 Block diagram of the Helvetic zone showing culminations and depressions and basement cover relationships. From Ramsay and Huber (1983, 1987), published by permission of Elsevier.

Because of these culminations and depressions it is possible to examine different structural levels in the orogen simply by travelling along strike from one place to another; thus in the Western Alps around Zermatt, because of a plunge depression the higher Penninic nappes are exposed whereas the plunge culmination in the Simplon–Ticino region means that the lower nappes of the Pennine pile can be examined.

The zones will be described from south to north across the Alps.

Internal zone

The Pennine Alps were formed by crustal thickening in the early Cenozoic (Eocene/Oligocene). Six large thick-skinned nappes were formed each with a cover of Mesozoic and Cenozoic rocks and a basement core of crystalline rocks formed in the Variscan orogeny. The nappe pile may change along the strike, therefore the nappes of eastern Switzerland such as the Suretta, Tambo and Silvretta nappes are not necessarily lateral continuations of the Dent Blanche, Monte Rosa, Great St Bernard and Simplon–Ticino nappes in western Switzerland. Although most are composed of continental crust, basement and cover, some Pennine nappes, notably the Dent Blanche nappe which is the highest nappe in the western Swiss Alps, incorporate sheets of ophiolite, gabbro etc., which show that oceanic crust has been obducted onto the continent. Some of the Penninic nappes may be sheets but others, e.g. the Simplon–Ticino and Monte Rosa nappes, are fold nappes in the Heim (1922) sense.

The deformation style in the basement cored nappes of the Pennines is mostly ductile, in contrast to the later Helvetic nappes of the external zone which show some brittle behaviour.

Metamorphism in the Pennines is of two types: an earlier UHP, and the later Lepontine medium/intermediate-pressure, Barrovian metamorphism which increases in grade down section from greenschist facies in the upper nappes to amphibolite facies with kyanite and sillimanite in the lower nappes as seen in the Simplon–Ticino region within a plunge culmination.

Towards the east, the Penninic Alps plunge beneath the Austro-Alpine nappes which are largely composed of thrust sheets consisting of low-grade carbonates. The evidence for this statement is found in the Tauern and Engadine 'windows' where high grade metamorphic rocks probably belonging to the Penninic Alps, or their lateral equivalents, form an infrastructure (hence the term window) to the Austrian Alps. Possibly the Austro-Alps were formed first, as early as the late Cretaceous, and the Pennine and Helvetic nappes formed beneath them.

External zone

In the Oligocene to Miocene the external zone of the Helvetic Alps was formed by further thickening of the upper crust, forming a pile of six thrust sheets at the base of which is the Morcles nappe of western Switzerland (Fig. 5.19, Plate 14). The Helvetic zone is *c*.300 km long and 40 km wide. The Helvetic nappes developed 'piggy-back' style, in a top to bottom sequence and underneath the pre-existing Pennine nappes. Westwards the Helvetic nappes pass into the much simpler structure of the French Alps, and to the east they disappear in western Austria. This is a good illustration of the fact that thrusts and folds cannot continue along strike indefinitely. Although the dominant structural style in the Helvetics is thin-skinned tectonics, several of the Helvetic thrusts have cut down to the Variscan basement (Fig. 5.20 a,b, Plate 15). Thus the basement and cover are involved in spectacular structures including tight interfolding of basement and cover and thrusting, beautifully shown on mountain sides like the Jungfrau and the Wetterhorn. The structural style of the Helvetics has been referred to by Heim as fold-nappes, the implication being that the folds evolved progressively with inversion and stretching of the lower fold limb leading to thrusting along that limb. Other models have been proposed, such as those based on gravity sliding or spreading; in addition there is the model of Ramsay *et al.* (1983) and Rowan and Kligfield (1992) in which the folds were propagated by layer parallel shortening and rotation during simple shear acting between two thrusts (Fig. 5.21, Plate 14).

One of the spectacular early discoveries in the Alps was the fact that the Helvetic nappes are covered by a thrust pile, some 8 km thick, known as the Pre-Alps and Ultrahelvetide nappes, a thickness which gives us a minimum estimate of the depth of cover to the Helvetics before erosion (Plate 15). The Pre-Alpine and Ultrahelvetide nappes are klippen – that is, far-travelled thrust sheets now detached from their roots by erosion. The problem was, where did they come from? The exotic far-travelled nature of the Pre-Alps was evident early on because the sedimentary facies and fossil assemblages are quite unlike those of their Helvetic footwalls. The difference is not in age but in sedimentary facies, the sediments in the Pre-Alps being composed of deep water deposits in contrast to the shallow water rocks in the Helvetic region. The likely match for the Pre-Alps and Ultrahelvetides is with the rocks of the central and southern part of the Pennine region; that is, they are northern continuations of the Pennine nappes and the internal zone. Esher *et al.* (1993) suggested that the Ultrahelvetide nappes of the external zone may have arisen by décollement and stripping of the cover off the basement nappes of the Pennines. The resolution of the mystery of the Pre-Alpine nappes is a good demonstration of the sort of detective work involving many disciplines that is required in orogen belts.

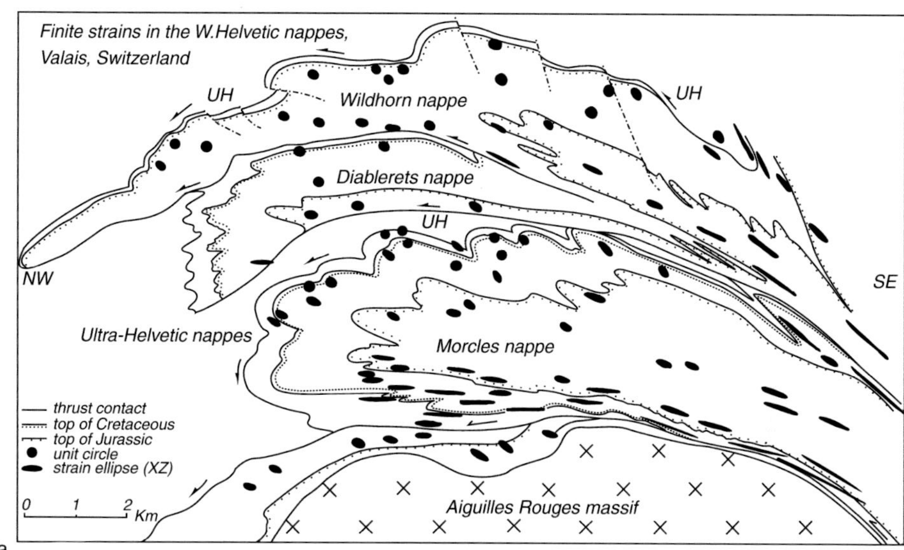

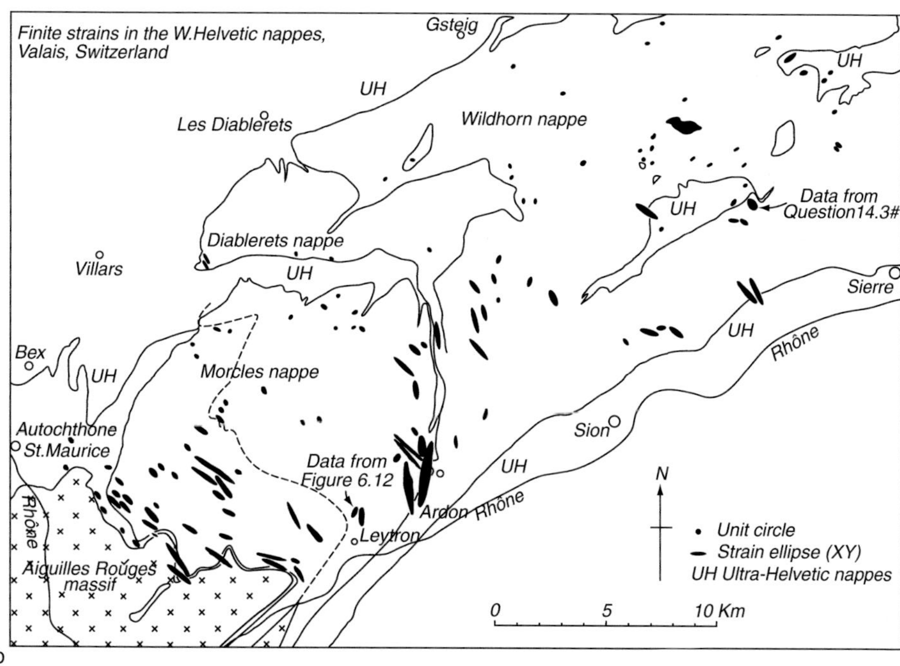

Figure 5.20 a, b, Strain in the lower Helvetic nappes. Plotted on the profiles of the lower Helvetic nappes are strain ellipsoids determined from deformed oolites and other strain markers. Note the variation in the intensity of strain around the folds, with the most intense strain on the inverted limbs. The lower diagram shows the plan view of strain with the principal extension being roughly NW perpendicular to the fold axes (Plate 16). From Ramsay and Huber (1983), published by permission of Elsevier.

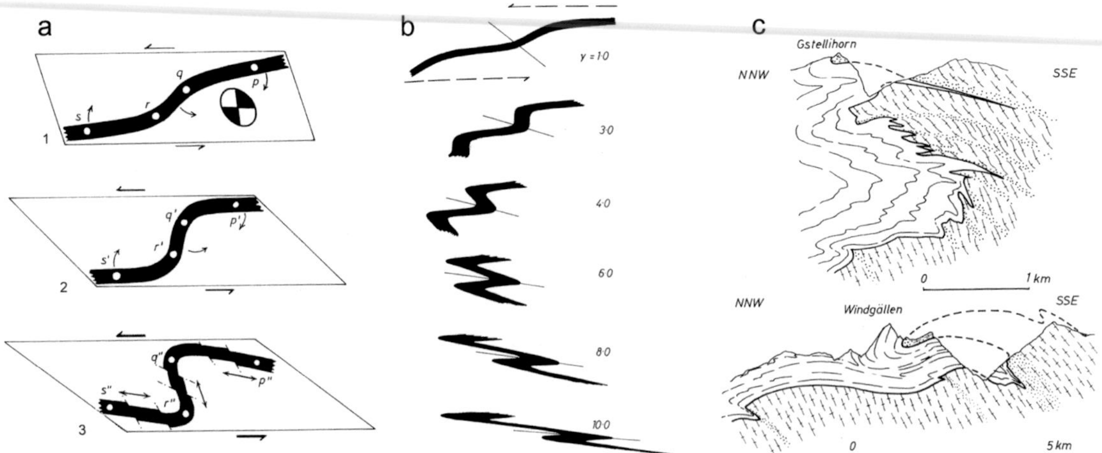

Figure 5.21 Basement-cover relations in the Helvetic zone. a, b, An interpretation of the Helvetic nappes as shear zones. In simple shear, buckling of a layer leads to rotation of fold limbs so that extension of a limb may occur when it rotates into the field of extension in the strain ellipsoid. τ is shear strain. c, Basement involvement in the Helvetic nappes. The dashed lines show shear zones in the crystalline basement indicating that the strong basement reacted differently to the weaker cover which has been folded. After Ramsay *et al.* (1983), published by permission of the Geological Society of America.

The North Alpine Foreland Basin

The North Alpine Foreland Basin contains two parts: a flysch basin of late Cretaceous–Eocene age containing turbidites, and a molasse basin of Late Oligocene–Miocene age containing detritus from the rising Alps. This and other basins associated with the uplift of the Alps will be discussed in Chapters 8 and 9.

Jura mountains

In the Pliocene the Jura mountains were formed as an arcuate belt of open folds above a shallow detachment, situated to the north of the North Alpine Foreland Basin (Fig. 5.22). The Jura belt is composed mainly of Mesozoic limestones and is famous for the décollement ('unsticking') of the folded limestones along a weak zone composed of Permo-Triassic evaporates. Above the décollement is a folded section of open concentric folds; below it there are sediments resting on Variscan basement, all of which shows no such deformation. It is as if the upper section of rocks had slipped on a banana skin. The décollement was recognised by Buxtorf (1907), and its existence presented a problem of volume balancing in the Jura section because the fold deformation clearly did not affect the whole rock section. That problem is now solved: the unsticking surface is a low angle thrust, formed as the Alpine deformation spread northwards.

The Hinterland

The hinterland of the Alps, the Southern Alps, consists of thin-skinned thrusting without large crustal shortening. The Southern Alps are separated from the Pennine zone by a

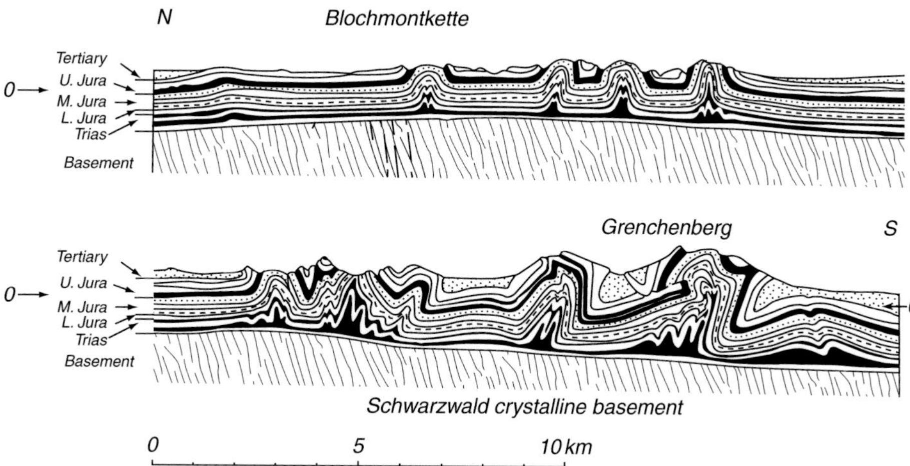

Figure 5.22 Section across the northern Jura. From Ramsay and Huber (1987), vol. 2, published by permission of Elsevier.

steeply dipping fault, the Insubric–Judicaria line, which is a late stage structure associated with the steep zone found in the southern part of the Pennine Alps where the generally gently dipping nappes turn into a steep dip.

Evolution of the Alps

So how did this complicated bit of geology come about? The global context for the Alp region was the break-up of Pangaea and separation of Laurasia and Gondwana which started *c.*180 Ma in the early Jurassic. The opening of the central and later the north Atlantic resulted in an easterly motion of Africa and Europe relative to America, but at times they moved east or west with respect to each other. Later, Africa drifted northwards, and so the Alps are the result of the convergence between Europe and Adria which in Mesozoic times was a promontory of the African continent forming what is now Italy and the Adriatic sea area.

Ultra-high-pressure metamorphism

As already indicated, over 150 years' work on the Alps has given it a crucial role in tectonic studies. It may therefore be surprising to discover that views on Alpine evolution have only recently undergone a complete turn-about. At the centre of this revolution are the UHP metamorphic rocks which until recently were assumed to be of Cretaceous age, the implication being that active subduction and continental collision in the Alpine region started in the Cretaceous. This version was set out by Pfiffner (1986, 1992) and Pfiffner *et al.* (2000) who considered that a rigid terminology like pre- and syn-tectonic breaks down in the Alps because the UHP event, arising from the subduction of European crust, started in the Cretaceous while sedimentation was going on. In brief, in this event there was strike-slip motion, thrusting in the Southern Alps, high-pressure metamorphism and the subduction of ocean fragments in the Eastern Alps.

Pfiffner's proposal is now discarded because more appropriate methods of dating (U–Pb dating of zircon and Sm–Nd dating of garnet) showed that most of the UHP metamorphism occurred in the Cenozoic, and in marked contrast to Pfiffner's model the Mesozoic history of the Alps is portrayed as almost completely anorogenic.

The new Cenozoic dating of the UHP metamorphic event – including eclogites – emerged from the work of many Earth scientists (Amato *et al.*, 1999; Lapen *et al.*, 2003; 2007; Liato and Froitzheim, 2006; Berger and Bousquet, 2008) and much of the new work is discussed in a review by Rosenbaum and Lister (2005). Most of the dates are from the internal crystalline massifs and the Great St Bernard nappe on the European plate, and from the Gran Paradiso, Sesia–Lanzo crystalline massifs of the continental margin of the Adria plate. For example, Amato *et al.* (1999) gave a Sm–Nd and Lu–Hf garnet age of 40.6 ± 2.6 Ma for the UHP event. Rapid exhumation followed this high-pressure event and is reflected in a retrograde change of metamorphic facies to greenschist dated at *c.*38 Ma. This implies an exhumation rate of 10–26 km Ma^{-1}, which is fast. According to Rosenbaum and Lister, Cenozoic ages for the UHP metamorphic rocks prevail (e.g. *c.*45 Ma in the Piemont zone and *c.*35 Ma in the internal crystallines) but in the Sesia–Lanzo zone there may be a late Cretaceous high-pressure event. Resulting from the successive underthrusting of continental crust, older UHP rocks in one slice may overlie younger UHP rocks in a lower slice. The new consensus is not only surprising but cautionary as well, bearing in mind the huge amount of work done on the Alps over a period of about 150 years. Equally interesting is that the present view, that collision is a Cenozoic occurrence, is actually a return to the traditional view!

The depth of subduction

The depth of subduction of continental and oceanic crust (A- and B-type subduction) is shown by the geobarometric estimates from the metamorphic assemblages in the subducted continental and oceanic rocks of the Pennine nappes of the internal zone of the Alps, e.g. Zermatt ophiolites (remnants of the NeoTethys ocean now in the Dent Blanche nappe), 10–12 kbar; Monte Rosa nappe, 16 kbar; Great St Bernard nappe, <10 kbar. The cover of the St Bernard nappe must have been detached from its basement, because it shows no evidence of high-pressure metamorphism. In the Dora Maira massif, now dated at 38–35 Ma, the presence of the mineral coesite indicates that pressures of 35 kbar were attained during the event, which suggests a subduction to depths of 100 km.

These pressures mean that European crust has been subducted to depths of *c.*35–56 km. At the lowest exposed structural levels in the Penninic Alps, in the Simplon–Ticino nappes found in a culmination, eclogites indicate comparable amounts of subduction. The crustal subduction and UHP metamorphism represent the early Alpine phase in the Alps.

Underplating and the orogenic wedge

As we have seen, the large high-pressure terranes in the Alps indicate massive underplating of both ocean crust and continental crust onto the Adria plate. In Chapter 4 it was shown that such underplating results in instability in the orogenic wedge. In consequence the wedge must have extended and thinned at its rear end allowing the uplift of the high-pressure terranes, a further consequence being that many of the tectonic contacts in the Alps are low-angle normal faults rather than thrusts, although they may have formed along older thrusts. The lateral spreading in the hindward part of the wedge caused the emplacement of older high-pressure rocks over lower-pressure rocks; for example, the Sesia–Dent Blanche nappe overlies the Combin and Zermatt units and the Great St Bernard nappe.

The current view on evolution

The Cenozoic evolution of the Alps involves rapid subducton of crust, followed by rapid exhumation with Barrovian metamorphism, followed quickly by uplift. The main collision of Europe and Adria started in the Eocene and is marked by the beginning of subduction of the southern part of the European margin or rather the Piemont Ocean. Schmidt and Kissling (2000) have placed the full collision as starting at about 35–40 Ma, but this does not seem to leave enough time after collision for the crust to thicken and heat up in order to give peak metamorphism at *c*.35 Ma, the traditional age for it. The flysch of the lower part of the North Alpine Foreland basin appears to register the start of collision at or before 45 Ma. Some would start the collision in the Maastrichtian (latest Cretaceous) so perhaps the time of collision was diachronous along the Alpine belt. Allen *et al.* (1991) and Sinclair and Allen (1992) interpreted the flysch as being formed as a result of flexural bending of the European plate due to the loading by thrust sheets which advanced at a rate of 3–4 mm/a in the central Alps but at 10–14 mm/a in western Switzerland. The inference is that collision and subduction of European crust started at least 5 Ma before the UHP metamorphism, which obviously only formed after subduction of crust by 50–100 km.

After the subduction and emplacement of the Pennine nappes and consequent crustal thickening, there was a so-called Mesoalpine or Lepontine metamorphism, a medium–high temperature Barrovian event dated at *c*.35 Ma (but see below). The medium pressures associated with the Lepontine metamorphism indicate that exhumation had started soon after the high-pressure event. The Lepontine metamorphism increases in grade downwards through the nappe pile, i.e. the highest grade, kyanite and sillimanite is seen in the Simplon–Ticino nappes at the base of the nappe pile.

The recent datings of the high-pressure metamorphism more or less coincide with the traditional age for Barrovian metamorphism. For example revised ages for, first, Barrovian metamorphism and second, in bold, the high-pressure metamorphism are: 38 Ma and **38 Ma** (Amato *et al.*, 1999), 31.5 Ma (Janots *et al.*, 2007), 32 Ma and **37–40** Ma (Brouwer, 2000), 34 Ma and **38/43 Ma**, 34–36 Ma and **43 Ma** (Meffan-Main *et al.*, 2004). Lapen *et al.* (2003) described a range of Lu–Hf garnet ages of 50–38 Ma for the Zermatt ophiolite. They interpret 40.6 Ma as the date of peak UHP conditions, and cooling was taking place at **40 Ma**. They deduce burial rates of 0.4 to 1.4 cm/a, a rate which is comparable to the shortening rate of the converging European and Adria plates. In the Dora Maira massif (Rubatto and Hermann 2001), high-pressure metamorphism occurred at 35 Ma followed by rapid exhumation leading to greenschist facies conditions by 32 Ma. This implies a rate of exhumation of *c*.50 km per Ma. The closeness of the ages for the high-pressure and later Barrovian metamorphisms indicates rapid exhumation of the

European crust, e.g. 7 mm/a or 3.3 mm/a, rates which are fast enough to explain the different depths involved in the two metamorphisms. Rubatto and Hermann described a high-pressure metamorphism at 35 Ma and exhumation rates of 3.4 to 1.6 cm/a. Such rates cannot be explained entirely by erosion and are more likely to be due to buoyant rebound, slab break-off and/or tectonic denudation.

Continent–continent convergence after final closure of the ocean: the late movements in the Alps

From the mid Oligocene onwards, following nappe emplacement, UHP and Lepontine metamorphism, the Alps underwent complex structural events, which are associated with or coincide with the rise of the Alps and formation of the molasse in the North Alpine Foreland Basin and the Po basin. Although northerly continental convergence continued, extensional normal faulting occurred, indicating E–W orogen-parallel stretching in the early Miocene. Likewise, extensional faulting in the Austro-Alps is linked to the unroofing of the Tauern window. Another late feature is the steepening and back folding of the Pennine nappes north of the Insubric Line. The steep zone has been intruded by granites of mantle origin which may have been channelled along the Alpine suture.

The late structures may be related to indentation (see below) by the Adria microplate which split the European crust apart (see below) and produced in the Internal zone a large flower structure or pop-up, bounded on the north and south by north- and south-directed thrusts of post-Oligocene age (Fig. 5.23). Vertical displacements of *c*.20 km are envisaged along the Insubric Line and along back thrusts in the Pennines.

Dextral strike-slip tectonics

Convergence of Europe and Adria continued after collision, as is shown by deformation of Late Oligocene and Miocene age in the Helvetic zone, the thrusting on the Insubric Line, and the thrusting in the Southern Alps. Convergence continued into Pliocene times in the Jura, and it may still be operating.

Motions along the Insubric Line were probably complex, including not only dip slip mentioned above but also dextral slip which occurred in the mid Oligocene and Miocene, more or less coinciding with the peak of exhumation. Laubscher (1992) suggested a dextral regime in the Alps in the late Oligocene and early Miocene involving dextral strike-slip of *c*.150 km on the Insubric Line. He envisaged 'peels' of Alpine rocks all involved in the transpressive regime. Note that this regime was roughly synchronous with the northerly thrusting on the Helvetic nappes. Later dextral motion on the Insubric Line may be due to the lateral escape of the European block to the east (Fig. 5.24). To the north in the external zone, late deformations include upwarping of the Morcles nappes and the Helvetic nappes by the rise of basement blocks such as the Monte Rosa and the Aiguilles Rouges.

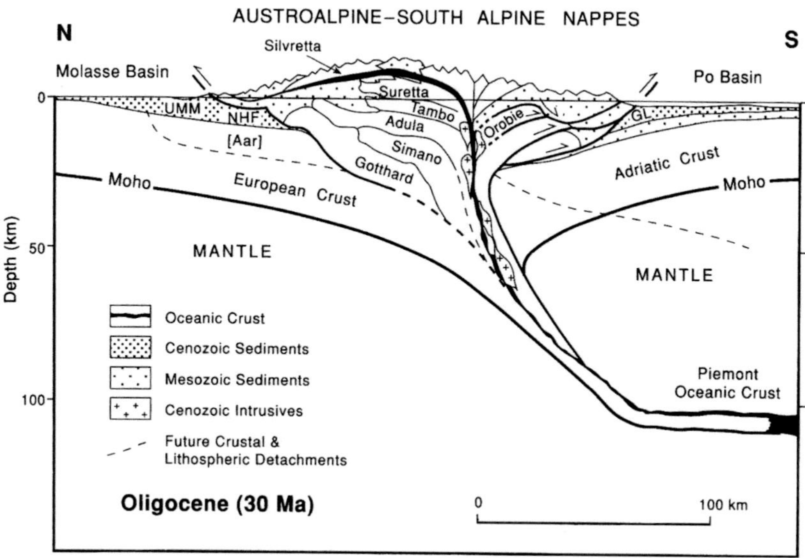

Figure 5.23 Section across eastern Switzerland shows Adria (Adriatic crust and mantle) as a late stage indenter which was forced into the thickened European Plate, splitting it apart. As a result the central Penninic nappes were uplifted and eroded with backfolding near the steeply dipping Insubric Line. The South Alpine nappes show southerly vergence. From Pfiffner (1992), published by permission of Cambridge University Press.

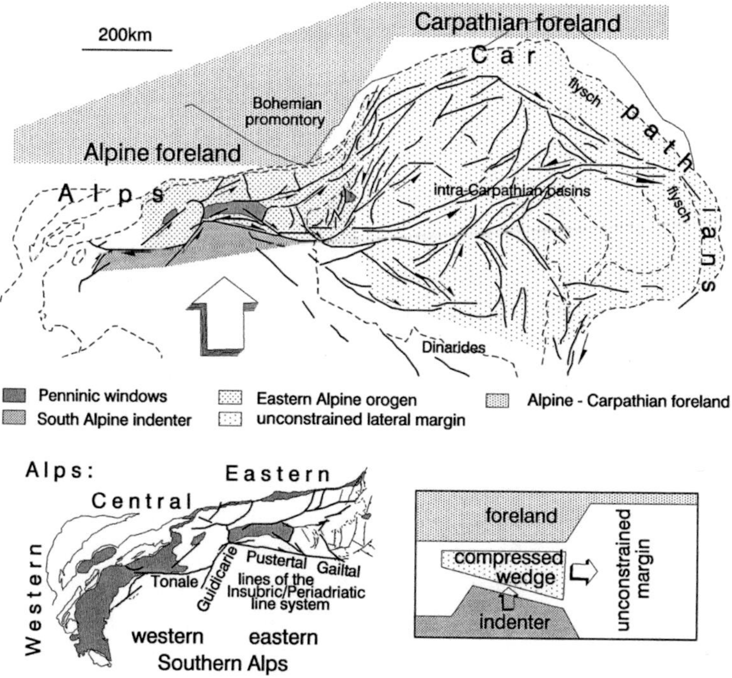

Figure 5.24 Indentation tectonics in the Alps and the Carpathians showing the main faults and windows. Strike-slip faults are shown by an arrow. From Ratschbacher *et al.* (1991a, b), Figures 1 and 2, published with permission.

Indentation tectonics?

Several writers have followed the Himalayan example and invoked indentation tectonics in the Alps (Fig. 5.23). The role of rigid indenter was played by Adria. Unlike the Himalayan indenter, Adria was the upper overriding plate and therefore its action as indenter dates from the exhumation phase following deep subduction of parts of the European plate (Fig. 5.24). Another difference may be that the indentation in the Alps was late stage rather than being initiated at the start of collision. Thinned continental crust was delaminated, the upper crust being peeled off from its basement and thrust north. As in the Himalaya the supposed indenter was deformed and Adria crust was shortened by imbrication of the upper crust, the youngest imbrications being seen in the Southern Alps which were coeval with the Helvetics.

Jiménez-Munt *et al.* (2005) have used analogue modelling with a thin viscous plate model to depict the behaviour of the Adria plate after collision from 35 to 0 Ma. This follows from Platt (1986) who showed radial transport vectors in the Alps around the western orocline (Fig. 5.15). In the western Alps, west-directed thrusting was related to rotations of the Adria indenter, and it has been suggested that Adria acted as a rigid cold indenter so that the Briançonnais zone, now transformed into the Great St Bernard nappe, was thrust to the west ahead of the indenter. Thus the Adria indenter produced the oroclinal bend in the western Alps. The orogen was a weak zone squeezed between the Adrian indenter and the strong European foreland. Between 35 and 6 Ma about 150 km of WNW displacement of Adria occurred, followed by 100 km between 6 and 0 Ma.

Another supposed example of indentation tectonics is the formation of the Carpathian orocline by the eastward lateral escape of lithosphere away from the Alpine region (Fig. 5.23). The strongly arcuate Carpathian belt is bordered on the west by the extensional Pannonian basin, and the whole system can be viewed as a roll-back of the subduction zone in the Carpathians.

A conspicuous large-scale feature of the Alpine belts around the Mediterranean region is the marked curvature (Plate 17). Rosenbaum and Lister (2004) suggested this is due to oroclinal bending of an originally linear fold and thrust belt (Fig. 5.25). The bending occurred from *c*.25 Ma up to the present and involved the anticlockwise rotation of Corsica, Sardinia and Italy (Fig. 5.26). The oroclinal development is caused by a roll-back and rotation of a subduction zone from an originally northerly dipping to roughly WSW dip at present. The Apennines developed by the docking of Corsica and Sardinia with the palaeomargin of Adria. This motion initiated extension, operating synchronously with the slow convergence of Europe and Africa, across the western part of the Mediterranean. Marine basins were formed, some like the Ligurian and Tyrrhennian seas with attenuated continental crust and also oceanic crust. These back-arc basins were due to the roll-back. Seismicity and magmatism in the southeastern part of the Tyrrhenian Sea marks a Benioff zone which dips steeply to the northwest. The slab beneath Calabria is being subducted down to *c*.670 km.

Another part of the oroclinal system is seen in the Rif–Betic region, where the Rif mountains of northern Morocco and the Rif mountains of southern Spain form a sharp arc

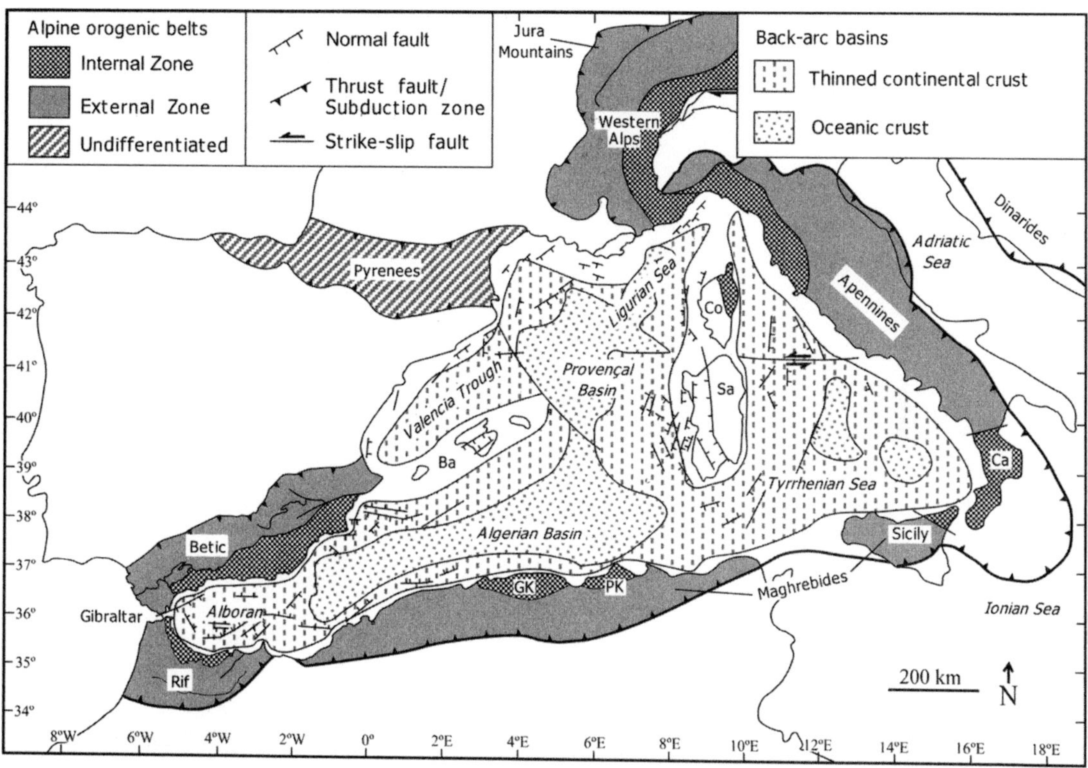

Figure 5.25 Tectonic map of the Alpine orogen in the western Mediterranean. Ba, Balearic Islands; Ca, Calabria; Co, Corsica; GK, Petit Kabylie; Sa, Sardinia. From Rosenbaum and Lister (2004), published by permission of the Geological Society of America.

(Fig. 5.27, Plate 17). The arc is shown to be oroclinal in nature by palaeomagnetic evidence of block rotations of Miocene age around vertical axes. The mountains are separated by the Alboran Sea, the floor of which is made up of the internal parts of an orogen showing high-grade metamorphism which underwent intense, roughly N–S extension operating along low-angled detachments in the early Miocene. The Alboran Domain is probably part of the orogenic belt including the western Alps, Corsica and Calabria, and it included the UHP rocks of the Alps, Apennines and Calabria. The Alboran Sea has been interpreted as a mountain chain which collapsed during post-orogenic extension. The evolution of the Alboran Sea region has been interpreted as resulting from (a) back-arc extension due to roll-back of a subduction zone or (b) extension induced by the break-off of a subducted lithospheric slab or (c) convective removal of the lower lithosphere. Rosenbaum and Lister (2005) preferred explanation (a), in which an easterly dipping subduction zone progressively became more arcuate during roll-back.

Platt *et al.* (2003) and Crespo-Blanc (2007) interpreted the Gibraltar Arc as an orocline resulting from indentation by the Alboran Domain. The plate driving force was the convergence of Africa and Iberia. The Pre-betic and Sub-betic zones in southern Spain are thin-skinned thrust wedges composed of Mesozoic to Cenozoic rocks which have been

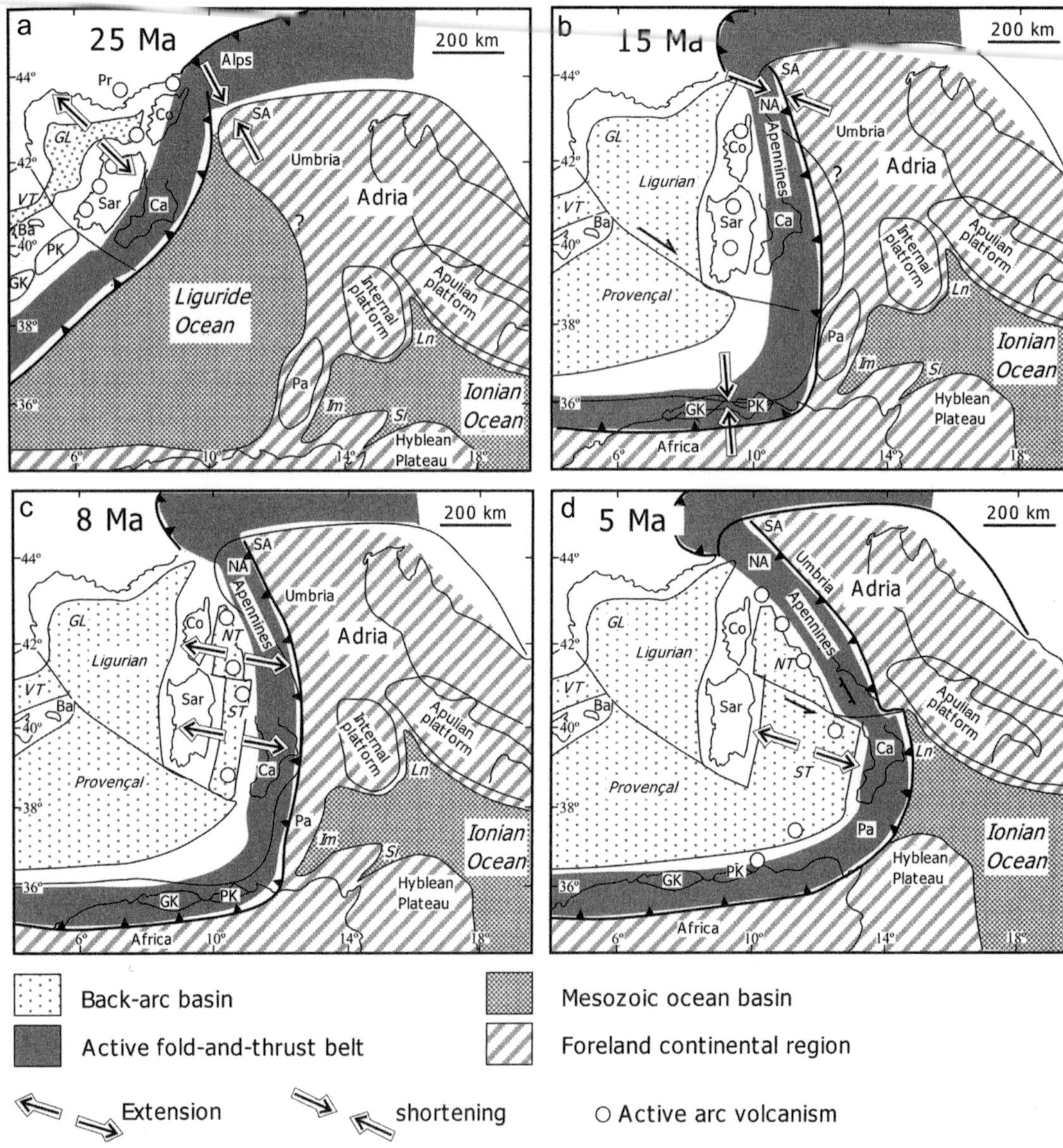

Figure 5.26 Reconstruction of the tectonic evolution of the Western Alps, Apennines and other belts. This involves an anti-clockwise rotation of the Apennines between 25 and 5 Ma. a, Early Miocene. b, Middle Miocene. c, Late Miocene. d, Pliocene. Key as for 5.25 plus: GL, Gulf of Lion; Im, Imerese; Ln, Lagonegro; NA, Northern Apennines; NT, Northern Tyrrhenian; Pa, Panormide platform; PK, Petit Kabylie; Pr, Provence; SA, Southern Alps; Sar, Sardinia; Si, Sicanian; ST, Southern Tyrrhenian; VT, Valencia Trough. From Rosenbaum and Lister (2004), as in Fig. 5.25.

detached from a Variscan basement. Near Gibraltar the strike swings to roughly north–south, and then in the external Rif belt in Morocco, to roughly east–west. The orocline is caused by the westward migration in Oligocene to Miocene time, of the supposed indenter, the Alboran Domain which is a western continuation of the Alpine orogen and is made up

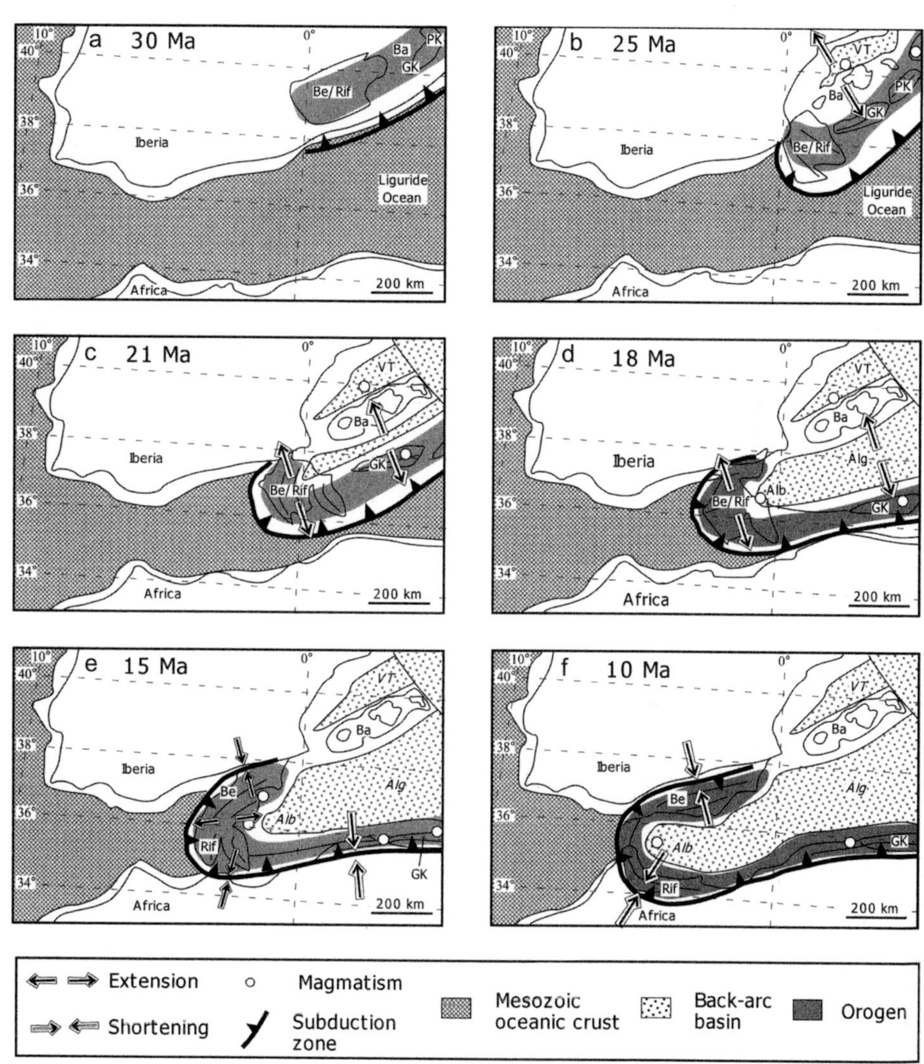

Figure 5.27 a–f, Reconstruction of the evolution of the Rif–Betic orogen between 35 and 10 Ma, showing the evolution of the orocline. Alb, Alboran Sea; Alg, Algerian Basin; Ba, Balearic Islands; Be, Betic; GK, Grand Kabylie; PK, Petit Kabylie; VT, Valencia Trough. From Rosenbaum and Lister, as in Fig. 5.25.

of thrust sheets of polyphase metamorphic rocks of Palaeocene age. The extensional collapse of the Alboran Domain during the Miocene led to the opening of the Tyrrhenian and Ligurian seas and, most relevant to the orocline development, the westward motion of the Alboran Domain and the consequent indentation tectonics. The oroclinal bending of the orogen resulted in it being stretched by strike-slip faulting.

The northern foreland of the Alps underwent deformation in mid Cenozoic times. Evidence of this deformation comes from the Paris and Aquitaine basin, the North Sea graben, most of these being mid Cretaceous in age, and the Rhine Graben, which is mid Cenozoic. These may be related to the Alpine orogeny or the opening of the Atlantic.

Comparisons between the Alps and the Himalaya

In the Alpine chain, the early Eocene/Oligocene date for the continent–continent collision suggests a similar time span to that of the Himalaya. Differences are notable, however: for example, the Alpine belt does not show the great width as in the Himalayan–Tibetan orogen. Excluding the Jura, the width of the Alps is *c.*200–250 km, slightly less than the Himalaya. However estimated shortening is about the same in the two belts. If as suggested, the collision may even be as late as mid Oligocene, then it is likely that the strain rate in the Alps was higher than in the Himalaya. In these belts regional metamorphism followed and was consequent upon crustal thickening. But an interesting difference emerges: in the Himalaya peak metamorphism followed collision by *c.*20 Ma, but in the Alps a much shorter time may have elapsed, perhaps as little as a few Ma. Another interesting point concerns the high-pressure metamorphism in the two belts. In the Himalaya and the Alps, deep subduction associated with eclogites was followed by rapid exhumation. Whereas in the Himalaya peak high-grade Barrovian metamorphism followed this rapid exhumation after *c.*20 Ma, in the Alps this sequence was much faster. In the Alps some workers have invoked an indentation model, the indenter being the Adria lithosphere which may have been a northern promontory of the African continent.

Lister *et al.* (2001) have reviewed evidence in the 15,000 km of the Alpine–Himalayan orogen for episodicity, marked, for example, by changes in time from active compression to extension, what they call tectonic switches. It is clear that orogens involve polyphase events with refolding of earlier by late folds as well as successive episodes of metamorphism. Many years ago, Stille postulated that Earth's history had been divided between periods of active orogenesis and periods of quiescence. This is clearly wrong because globally something is always happening, but what Lister *et al.* address is the possibility that events within the Cenozoic orogen are synchronous. They cite the Eocene start of collision in the Alps and Himalaya, the near synchroneity of peak metamorphism (*c.*35 Ma) in the Himalaya and deep subduction of continental crust with eclogitisation in the western Alps, and a tectonic switch from compressional tectonics and extensional tectonics in different parts of the Alpine–Himalayan orogen. Various possible causes for the episodicity are (a) lithospheric break-off, (b) slab retreat or advance, (c) orogenic surges, (d) migration of thrust sheets and (e) accretion. One problem with this approach is that although events may be roughly synchronous they are dissimilar in nature; for example the early Miocene extension in the Aegean, and the end of contraction in the Pyrenees are both linked to the channel flow event in the Himalaya.

Cordilleran belts: the North and South American Cordillera

The Andean Cordillera in South America is *c.*8000 km long and <800 km wide, and is the classic Andean-style continental margin orogenic belt as defined by Dewey and Bird (1970), situated at the margin between oceanic and continental lithosphere, but with no continent–continent collision (Fig. 5.28 a,b). However, the long history of subduction along the western margin of South America raises questions about the status of the Andes as a classic Andean belt. The Cordillera is caused by the eastward subduction of the Pacific Ocean lithosphere at the off-shore Peru–Chile trench and subduction zone. At first glance the geology of the Andes is simple, but there are hidden complexities and uncertainties. The orogen consists of a western Cordillera, in places a central zone, and an eastern Cordillera sometimes called the Sub-Andean belt. The Andean margin has been a subducting Cordilleran margin for about 500 Ma; furthermore it has been a collision zone with the arrival of the Precordillera (Cuyania terrane) from Laurentia in the mid to late Ordovician and the Cilenia terrane in the Devonian. There was also arc–continent collision involving the inversion of the Rocas Verdas basin in the Southern Andes during the Mesozoic to Cenozoic. The main intrusion of the bulk of the granitic plutons, which are a conspicuous feature of the South American Cordillera, took place during neutral to extensional stress states along the South American margin. For example, the Peruvian batholiths were intruded between 188 and 37 Ma. Thus the recognition of exotic terranes, if correct, means that the Andes orogen is actually a collage of far-travelled terranes which are microcontinental blocks that have been transported along the margin of the Americas. Although these terranes are restricted in space and time, they do indicate collisional tectonics. But the important point is that the terranes were accreted onto the Cordillera in Palaeozoic and Mesozoic times; accordingly we should restrict the Dewey and Bird (1970) definition of an Andean-type orogen to the late Mesozoic and Cenozoic history of the Andes. All this emphasises the dangers in trying to set up a 'type' or 'classic' orogen, given the great variety that they display.

Subduction of the Nazca plate

The Nazca plate is segmented by roughly ENE-trending transform faults which are aseismic outside the spreading rise segments. The descending plate beneath the Andes can be delineated by seismicity down to *c*.600 km. The rate of subduction varies from place to place and through time, 2–10 mm/a according to Wdowinski and O'Connell (1989) or 100 mm/a according to Dewey and Lamb (1992).

There is along-strike variation in the dip of the subduction zone from flat to moderately steep. The seismic energy associated with the flat parts is several times greater than the adjacent steeper parts. It is important to realise that the present dip of the subduction zone may not have applied during the entire compressional phase.

Volcanism

There is a notable link between volcanism and the orientation of the subducted plate. Present volcanic-free sectors of the Andes in the north and south of South America show a rather flat dipping subduction zone whereas the sectors with active volcanoes, notably the Central sector, overlie a subduction zone dipping at least 30° east (Fig. 5.28b). Barazangi and Isacks (1976) have shown that the Nazca plate is divided into five

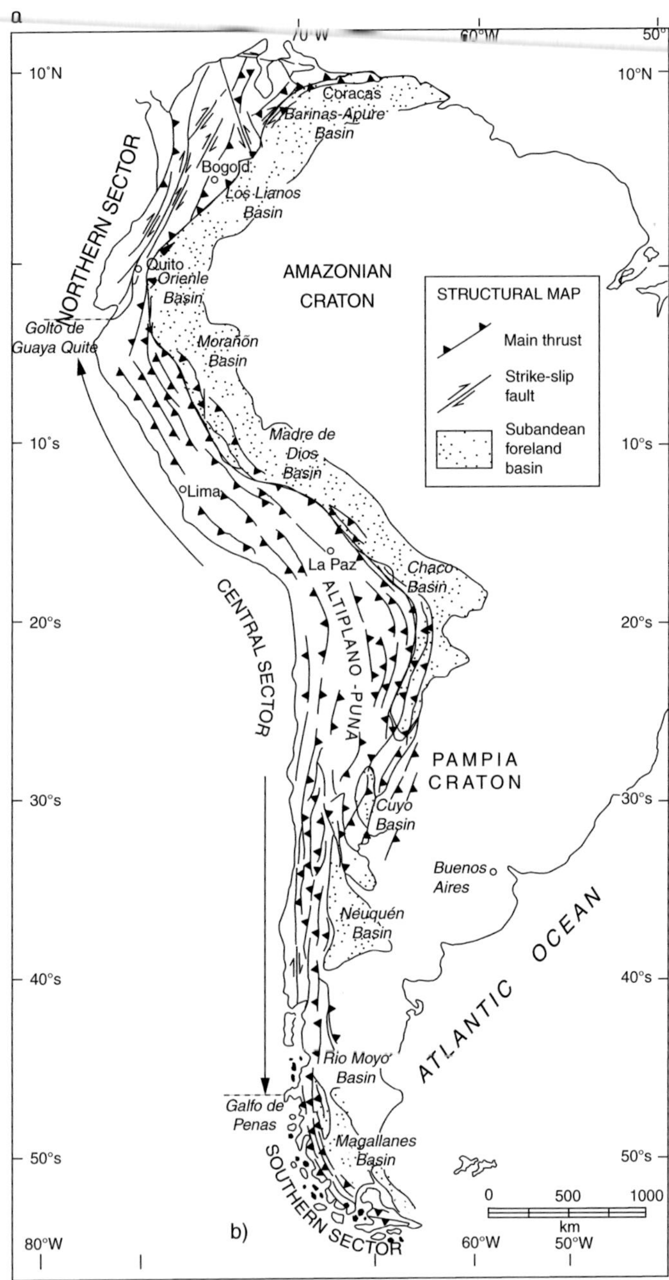

Figure 5.28 a, The sectors of the Andean orogen with trends of thrusts. b, Volcanic zones, contours show the geometry of the subducted Nazca plate beneath the Andes. GPS strain velocity data shown by arrows. From Figure 1 of Ramoz (1999), Fig.1.

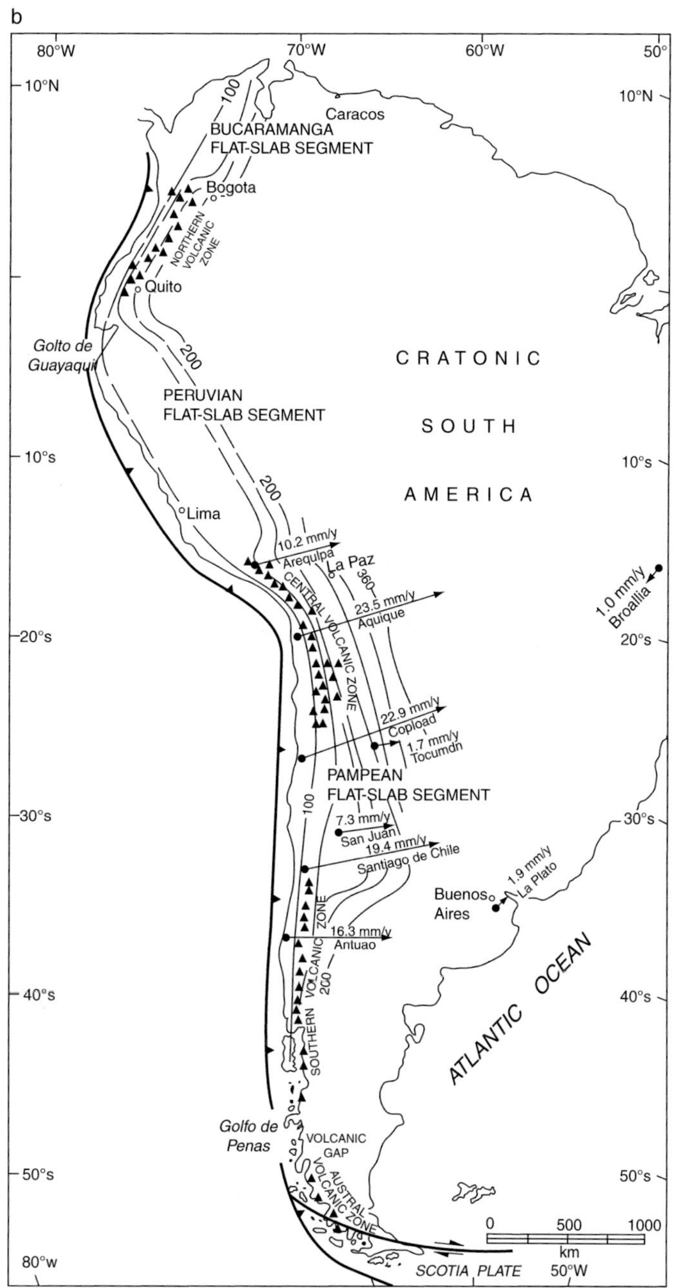

Figure 5.28 (cont.)

segments as it descends beneath South America, and at least two of the faults bounding these segments roughly coincide with the contact between sectors of the Andes. For example, the Nazca Ridge roughly coincides with the contact between the Northern and Central sectors; similarly the Challenger ridge lines up with a break in the line of volcanoes. In the Northern Andes (Colombia) the subduction zone is steep and there are several active volcanoes of basaltic andesite to andesite composition. The Central volcanic zone between Arequipa and north Chile contains hundreds of volcanoes of andesitic and dacitic composition which are mantle-derived magmas modified as they ascended through thick crust. The Southern volcanic zone 33° 30′ S – 46° 30′ S in the southern part of the Central Andes has late Cenozoic and active volcanoes in a 1000 km long volcanic chain. The lavas are basalts and basaltic andesites. In the Southern sector there are few volcanoes.

In the Central sector the coastal Cordillera is composed of Jurassic to mid Cretaceous arc sequences which are cut by the Atacama strike-slip fault. The Cordillera Oriental is composed of deformed Palaeozoic sediments and Cenozoic magmatic rocks, and the Sub-Andean belt is largely made up of Palaeozoic rocks. Together the Cordillera Oriental and the Sub-Andean zone form a fold and thrust belt with mainly easterly vergence.

Sectors of the Andes

Ramoz (1999) described the way that the Andean Cordillera is divided into Northern, Central and Southern sectors (Fig. 5.28a). Another important major feature of part of the Northern and Southern sectors is that the subduction zone is mainly flat dipping, *c.*10°, but in the Central sector the subduction zone mostly dips at *c.*30°. In the Northern and Southern sectors the Benioff zone is flat and the plate descends without asthenosphere intervening between it and the South American plate. In the Central sector the dip is mostly 25° to 30°, and asthenosphere does intervene between the descending plate and the South American lithosphere; it is here that active volcanoes occur. We may summarise briefly the characteristics of the sectors.

The Northern Andes show accretionary events involving ocean crust in Jurassic, late Cretaceous and Palaeogene times. In the Central Andes which extends from 40° S to 46°–30° S, there is a complex tectonic history driven by subduction. The northern part of the Central sector shows extensional tectonics and subduction during the early Mesozoic and layer compression and magmatism with foreland-directed thrusting. The central part of the Central sector from 14° to 27° S is characterised by moderately steep subduction and volcanic activity during Miocene time. Steeper geothermal gradients here are associated with crustal thickening. The Central sector is very variable, with a flat subduction sector 27°–33° 30′ S showing intense deformation and little or no magmatism. South of the flat subduction sector as well as thrusting there is significant strike-slip motion. The Southern Andes, 46° 30′-52° S, shows strike-slip motion and a fold and thrust belt.

In central north Peru there are no volcanics now but there were in the Miocene–Pliocene, suggesting that the Benioff zone was steeper then. At the present day, the topology of the subducting slab is interesting. The dead transform faults such as the Nazca Ridge may have acted as zones of weakness which ruptured during subduction and controlled the topology. However, Cahill and Isacks (1992) used seismicity to determine the geometry of the Nazca plate beneath the central Andes and suggest that the plate is warped, not fractured, with a transition from gentle to steep dips. Tinker *et al.* (1996) show that the warping reflects the dynamic evolution of the South America–Nazca plate system, that is, it is actively bending at depth along an axis parallel to the dip direction of the plate. But before subduction it was an evenly dipping slab. The interesting point here is that the warp in the Nazca plate appears to coincide with the major embayment of the coastline of South America, the Arica embayment. The curved coastline in turn is associated with the Bolivian orocline in which the strike of the rocks swings from NW to roughly N–S. The origin of the orocline is unknown but perhaps it reflects older structures in the South American crust (pers. com. I. W. D. Dalziel). Others suggest a bending of the Andean orogen, a remnant of a hotspot-generated triple-junction rift system or left lateral strike-slip. Whatever the origin, the trench in the embayment has undergone convex seaward bending and this trench geometry possibly imparted a warp to the surface of the Nazca plate.

Control over the upper slab deformation by the convergence direction of South America and Nazca

Where the convergence is perpendicular to the trench the slab has a uniform dip. Away from this region, upper slab flattening occurs. The convergence vector favoured an orthogonal convergence in the Peruvian Andes but on the Chilean margin oblique convergence produced a partitioning of strain with strong strike-slip motion. The northern Andes underwent partitioning in the Palaeogene and Neogene due to the northeast trend of the continental margin which was not orthogonal to the convergence vector. As a result, dextral strike-slip faulting occurred. Overall there was an intermittent obliquity of convergence which changes along the whole Andean belt, with dextral clockwise vorticity in the south and sinistral anticlockwise vorticity in the Central sector. The result is a partitioning of deformation into strike-parallel (strike-slip faults) and strike-normal (thrust faults) displacements.

Structural units and shortening in the Andean belt

On a large scale there are several notable features about the Andes. Firstly, there are no exposed Andean-age high-grade metamorphic rocks except in the south where the Cordillera trends west–east along the North Scotia Ridge transform fault. Presumably the high-grade metamorphic rocks and basement are present sub-surface. Secondly, all parts display an eastern Cordillera, Sub-Andean, fold and thrust belt in which the vergence is eastwards towards the continental margin. Thirdly, evidence of collided ophiolite terranes is confined to the north and south Andes. As well as those mentioned above there are marginal basins and back-arc basins which opened and closed in the Jurassic. Orogeny is diachronous along the Andean belt and the tectonic style is mainly thin-skinned thrusting in the Northern and Southern sectors but thick-skinned thrusting in the wide Central sector where maximum crust shortening took place, namely 320 km of crustal shortening, during the Cenozoic. Away from this sector crustal shortening diminishes, from 160–140 km at 30° 32′ S to only 44–20 km at 37°–39° S. In the Central sector a western Cordillera Occidental is separated by the elevated Altiplano Plateau from an eastern Cordillera Oriental which is made up of an east-verging fold and thrust belt. In addition the lateral growth of the orogen is marked by a succession of foreland basins, particularly well seen in the wide Central sector.

Thick-skinned or thin-skinned thrusts?

Although the exposed Mesozoic–Cenozoic Andean orogen apparently involves thin-skinned thrust belts, deep thrust structures are likely, in common with some continent–continent collision belts (Fig. 5.29). The published cross-sections across the Andes depict the thrusts as flattening downward into a shallow décollement. The thrust vergence is generally eastward towards the continental interior, but back thrusting is common especially in parts of the western Cordillera where the pattern of thrusts indicates divergent vergence west and east. Thin-skinned thrusting is probably true in the Sub-Andean belt but it fails to explain the crustal thickening in the west which may reflect substantial underthrusting of cratonic Shield beneath part of the Andean belt. In addition deep crustal flow, even perhaps channel flow (see Chapter 7), has been proposed. The conclusion is that although it may be correct to depict the thrusts as mainly supracrustal (thin-skinned), some Precambrian basement is involved in the contraction. Apart from the Cordilleran core complex in the extreme south there is no evidence of metamorphism. If the crust is in places twice the normal thickness then metamorphism is predictable at depth. High temperatures at depth are indicated by conspicuous S-type granites and andesites showing melting conditions in the lower crust and mantle.

Timing of crustal thickening

The first crustal thickening in the Mesozoic–Cenozoic Andes took place in the mid Cretaceous. Further crustal thickening occurred from the late Cretaceous and Cenozoic to the present day. In the Central Andes and under the Altiplano Plateau, principal shortening took place between the Eocene and the mid Miocene and important uplift occurred in the late Miocene. In the eastern Sub-Andean foreland, shortening was late Miocene to Holocene. Lopez (2006) suggested that the Altiplano plateau began rising slowly at 30 Ma and that rapid uplift in the foreland fold belt started at 10 Ma. The orogen

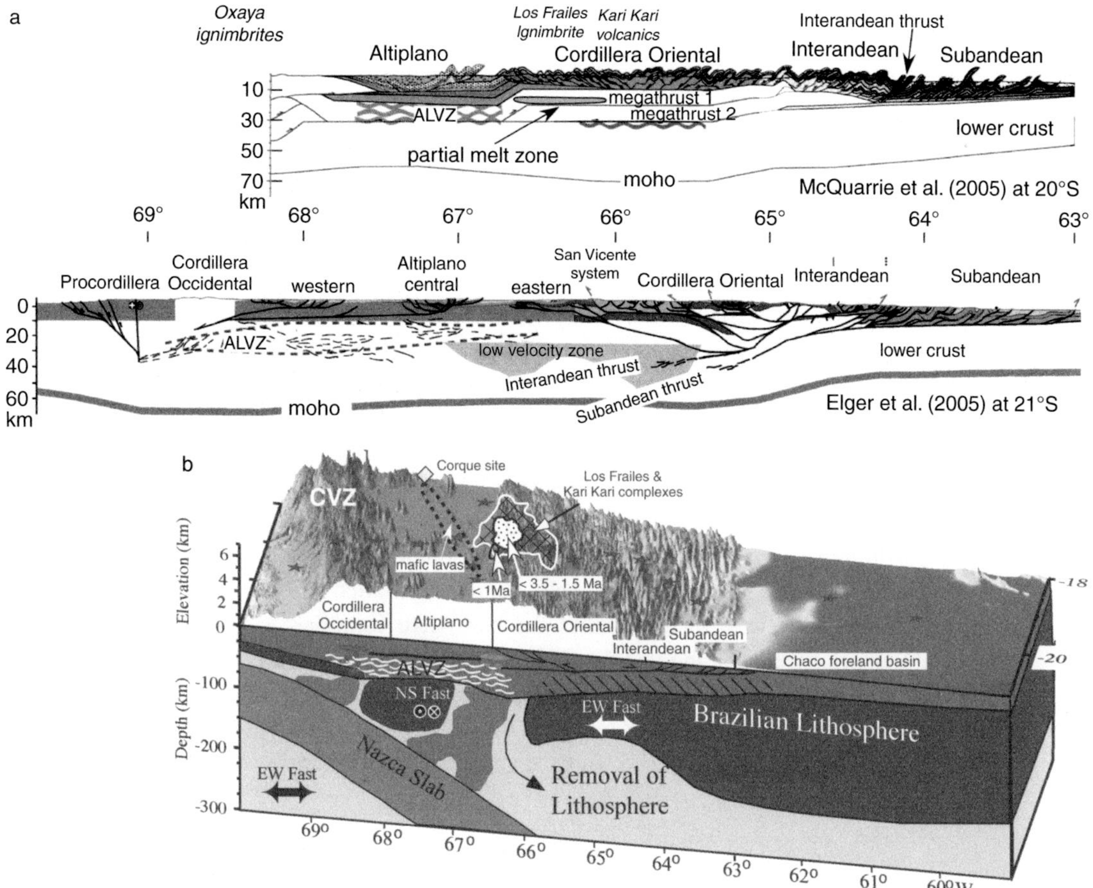

Figure 5.29 a, Sections across the central Andes and b, block diagram to show proposed underthrusting of the Brazilian Shield. The sub-lithospheric lithosphere is delaminated under the Altiplano–East Cordilleran border but not completely removed except under the central Altiplano. Dark grey shows fast upper mantle P waves, white and shaded grey shows slow P waves. ALVZ, Altiplano Low Velocity Zone; CVZ, Central Volcanic Zone; NS fast, EW east, seismic wave velocities. From Macquarrie *et al.* (2005), published by permission of Elsevier; Elger *et al.* (2005) and Beck and Zandt (2002), Fig. 8.

therefore may represent a zone of continuum deformation with periods of quiescence. Two major periods of continuum tectonics have been recognised, 80–55 Ma and 27–8 Ma, and appear to be related to changes in the convergence rate of the plates. Kendrick *et al.* (1999) and Lopez (2006) consider that Nazca–South America convergence has slowed over the past 25 Ma, despite the fact that much of the Andes orogen formed at that time.

In the Central sector, deformation within the Cordillera Occidental, the Altiplano and the Cordillera Oriental started between 48 and 38 Ma with the development of both east and west verging thrusts. Deformation rates reached a maximum of *c*.8 mm/a in the late Oligocene. Then deformation migrated to the Sub-Andean belt at *c*.8–7 Ma where deformation rates are 8–16 mm/a.

Shortening across the Andes is variable and rather model-dependent, but it is likely that in the eastern fold-thrust belts 250 km of shortening occurred.

Neotectonics

The Andes shows significant Neotectonics with active faults (Plate 18). GPS geodetic surveys reveal a remarkable picture. The GPS strain velocities decrease smoothly across the continent, indicating broad distributive shortening across the whole of the Andes. In contrast, geological and seismic data show concentrated shortening in the Eastern Cordillera and the foreland thrust belt. Therefore it is thought that the GPS velocities are partitioned into transient elastic strain in the west but permanent strain in the eastern foreland thrust belts. Kendrick *et al.* (1999) adopted a viscoelastic model for the Andean crust, elastic over short periods, viscous over geological timescales and plastic when stresses exceed the yield strength of the trench fault. Leffler *et al.* (2009) showed at Arequipa a shortening rate of 13±3 mm/a to the northeast relative to the stable continental interior and suggested for that region a part locking of the Nazca–South America subduction zone and small shortening in the Eastern Andes.

Lui *et al.* (2000) found two differing shortening rates from the GPS geodetic work: 30–40 mm/a across the Andes but only <15 mm/a from geologically determined rates in the Sub-Andean fold and thrust belt. Shortening started *c*.30 Ma. The strain velocities decrease from west to east across the belt. Like Kendrick *et al.*, they explained the difference by concluding that the GPS rates refer to elastic and permanent strain whereas the geological rates refer to permanent strain only. These strain velocities on the continent may be compared with 68–78 mm/a for the Nazca plate relative to the stable part of South America. The amounts of shortening vary from *c*.2 to 19 mm/a.

High plateau

The existence of high plateaux in the Andes, the Altiplano Plateau in Chile and Peru and the Puna Plateau in Argentina, invites Himalayan comparisons. The Altiplano Plateau in the Central sector is the larger: with an area of 600,000 km^2, a width of 100–200 km across and an average elevation of 3650 m, this is considerably smaller than Tibet. The Altiplano Plateau is composed mainly of late Precambrian and Palaeozoic deformed basement.

Notable features of the Altiplano are:

1. 10 Ma erosion surfaces which post-date the last compressive phase in the western Cordillera. At 10 Ma the plateau was at no more than half the present elevation. Uplift rates have roughly doubled in the past 5 Ma. In the Puna Plateau of Argentina there was major uplift beginning 10–5 Ma ago and culminating at *c*.2 Ma. Raymo, Ruddiman and Froelich (1988) cited this and the Himalayan–Tibetan example as examples of significant increases in mountain uplift in the late Miocene which resulted in increased rates of chemical weathering (see Chapter 11).
2. Quaternary high-potash volcanism.
3. The presence of some lithospheric mantle, but this may be absent in places.
4. Crustal thickness of 60–80 km. Surface elevation is <4 km.

5. The Altiplano Plateau has been an intermontane basin with extensional and contractional deformation throughout most of the Cenozoic, as witnessed by the existence on the plateau of Cenozoic continental deposits which are 4–10 km in thickness.

Deep structure of the Andes

There are many questions about the nature of the deep structure of the Andes, even though there is a considerable body of geophysical evidence available. The mechanism for crustal thickening in the Andes is still uncertain: was it whole lithospheric thickening by either homogeneous strain or thrusting, magmatic underplating, asthenospheric wedging, crustal wedging or underthrusting of South American continental crust? Perhaps the problem of the mechanism of crustal thickening is most serious in the Central sector in which crustal thickness is <70 km in the western Cordillera, 60–65 km under the Altiplano plateau and *c*.70 km in the eastern Cordillera. In general lithospheric thickness is estimated at 80 km (Wdowinski and O'Connell, 1989), but under the Altiplano the lithosphere is <120 km thick. In the Altiplano there are north to south variations in crustal thickness. The *c*.60 km thick crust could have resulted from thickening of the mid and upper crust, but as we see below there are other interpretations.

We may consider some of the models which have been presented for the deep structure. Isacks (1988) suggested that the Andean upper and lower crust was decoupled from the lithospheric mantle (cf. the lower crustal injection model of Zhao and Morgan in Tibet). Decoupling occurs on a weakened zone reflecting the presence beneath it of a hot asthenospheric wedge located between the subducted and overriding plates (Fig. 5.29). Thus the hot wedge has weakened the upper lithosphere, and its presence is reflected by the fact that magmatism is present in the Central sector but absent in sectors where the subduction dips at a gentle angle. The asthenospheric wedge is there because the dip of 20°–30° of the subduction zone provided the space necessary for it. Therefore the wedge is missing in the northern and southern segments of the Andes where the subduction zone is flatter, the belt is narrower and the shortening is less than in the Central sector. Almendinger and Gubbels (1996) invoked pure and simple shear models to explain the crustal thickening and uplift in the Altiplano Plateau, these models operating from *c*.30 Ma until *c*.10 Ma. After 10 Ma, deformation shifted into the Sub-Andean foreland where thrusts propagated eastward. As an expression of simple shear they propose underthrusting of the Brazilian Shield, resulting in lithospheric thickening. The flanks of the plateau are characterised by thin-skinned thrusting, and on the Altiplano Plateau itself there is a lack of much internal relief and active neotectonic phenomena. The pure shear mode is characterised by more rugged internal relief and thick-skinned deformation to depths of at least 30 km. Furthermore they speculate that some delamination of the lower lithosphere has occurred beneath the Altiplano, the effect being to decrease lithospheric thickness. In the Puna Plateau, however, pure shear was much more important than underthrusting of Shield.

Beck and Zandt (2002) used combined receiver functions and surface wave dispersion data in order to view the central Andes as an active continental margin belt developed by the squeezing of weak western South American lithosphere between the Nazca plate and the underthrusting Brazilian Shield (Fig. 5.29). The latter has been underthrust as far west as 65.5° W, a distance of 100–150 km, but it does not underthrust the whole Altiplano. The Altiplano crust consists of a brittle upper crust decoupled from a ductile, very weak lower crust in which flow occurred. A low-velocity zone under the Cordillera Oriental and the Altiplano probably indicates a zone of partial melting. In contrast, in the Eastern Cordillera the upper crust is still coupled across the basal thrust of the fold thrust belt to the underthrusting Brazilian Shield. However, this model cannot account for all the lithospheric thickness beneath the Altiplano because, as Almendinger and Gubbels (1996) suggested, to the west of the shield thrust wedge there is a partial loss of mantle lithosphere beneath the Altiplano plateau. Thus delamination of lower lithosphere may have taken place in the southern Altiplano where the lithosphere is thinner than normal, partial melts are present and the elevation is higher because of lithospheric thinning. Finally, if the Altiplano was only 1–2 km high at 10 Ma ago (see item 1 of the list of notable features of the Altiplano above), then at least 2 km of uplift must have occurred since then. Perhaps this uplift was due to phase changes at depth in the lower crust from eclogite to less dense granulite facies (see Chapter 10). Yet another mechanism of crustal flow in the Andes is mentioned in Chapter 7: that is, channel flow involving flow of crust along the strike between the two plateaux, the driving force being the excess gravity potential of the Altiplano and Puna plateaux.

Kay and Coire (2009) have presented an integrated structural and magmatic history for the Central sector involving varying dip of the subduction zone, westward drift of South America, subduction and southward drift of the Juan Fernandez Ridge under South America, and slab roll-back. The main magmatism was andesitic-dacitic but there were also episodes of basaltic and shoshonitic magmatism. Kay and Coire (2009) described a delamination of eclogitic crust during steepening of the subducted slab. They suggest that the south Altiplano was uplifted in the late Miocene because of lower crustal flow, lithospheric thickening and some delamination. They question the wholesale delamination of the lower crust and lithospheric mantle proposed by Molnar and Garzione (2007). Oxygen isotopes in carbonates, which are well dated by magnetostratigraphy and $^{40}Ar/^{39}Ar$ methods, indicate a rapid uplift in the late Miocene. At 12–10 Ma the Altiplano was at 500 m, at 706 Ma it was at 2000 m and at 6–8 Ma it was at 4000 m, giving an uplift rate of *c*.1, mm/a.

At the beginning of this section on the Andes we noted the apparent simplicity, but clearly this is wrong as shown by recent work. Much focus is on the deep structure as revealed by geophysical techniques which have revealed the complexities beneath the Altiplano Plateau and other parts of the Central sector. Even a feature as simple as a plateau is in fact extremely complicated. The magmatism seen at the surface provides many clues about the evolution of the deep structure and the behaviour of the subducting slab at depth.

The North American Cordillera

The North American Cordillera extends from Alaska to Mexico and perhaps is continuous with orogens in east Siberia. The Cordillera varies in width from *c*.100–200 km in

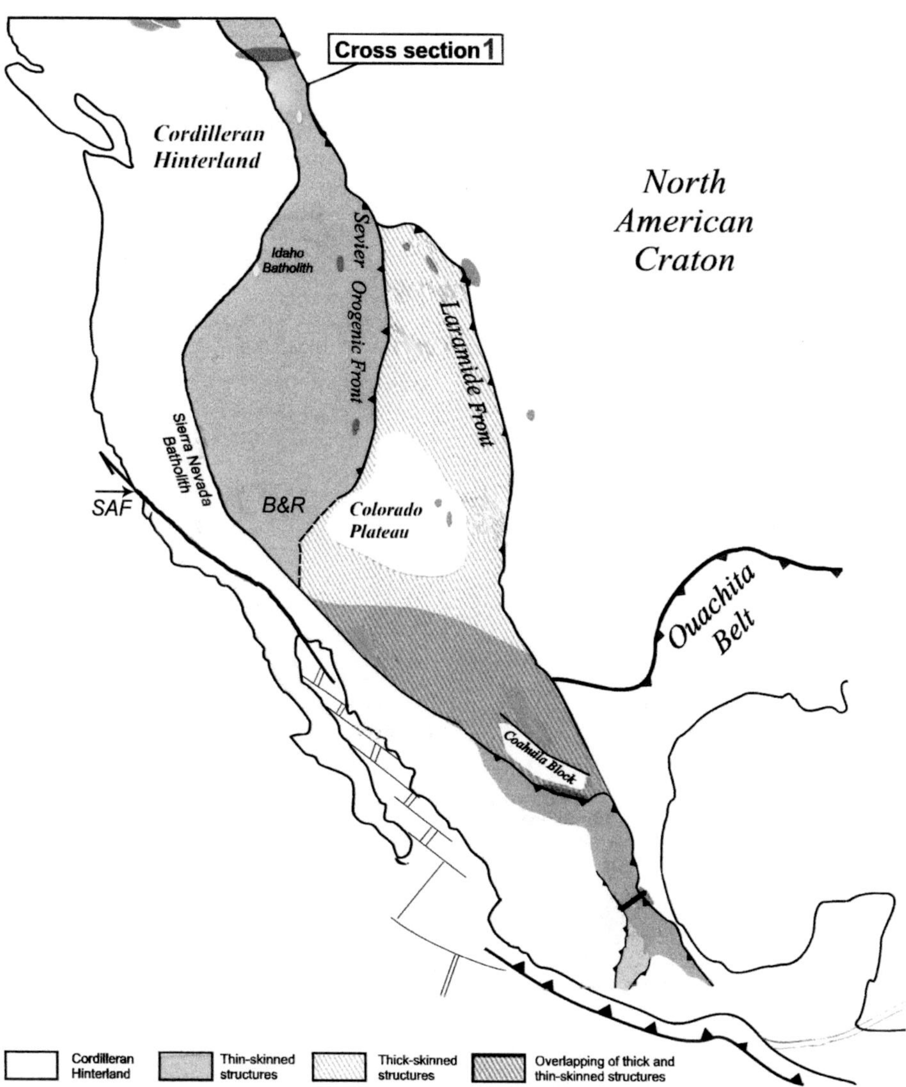

Figure 5.30 Map of the southern part of the North American Cordillera showing major tectonic features and igneous rocks, and the distribution of the Laramide and Sevier orogens, both of which are mainly thin-skinned thrust belts. However, Precambrian basement is involved in the Laramide belt. The Cordilleran hinterland is composed of batholiths and up to high-grade metamorphic rocks, including some blueschist assemblages showing Mesozoic and Cenozoic ages. The Idaho batholith is dated at 105–90 Ma, the Sierra Nevada batholith is dated as having important intrusions at 160–150 and 100–85 Ma. B and R, Basin and Range extensional province; SAF, San Andreas transform fault. Cross-section is Fig. 5.33.

Canada to *c.*1000 km in the USA (Figs. 5.30, 5.31, 5.32). It formed on an active plate margin, and its complex late Precambrian to recent history includes not only compressional orogens but major strike-slip faulting along the San Andreas fault and post-orogenesis extension tectonics seen in the Basin and Range. The San Andreas fault is a transform fault connecting

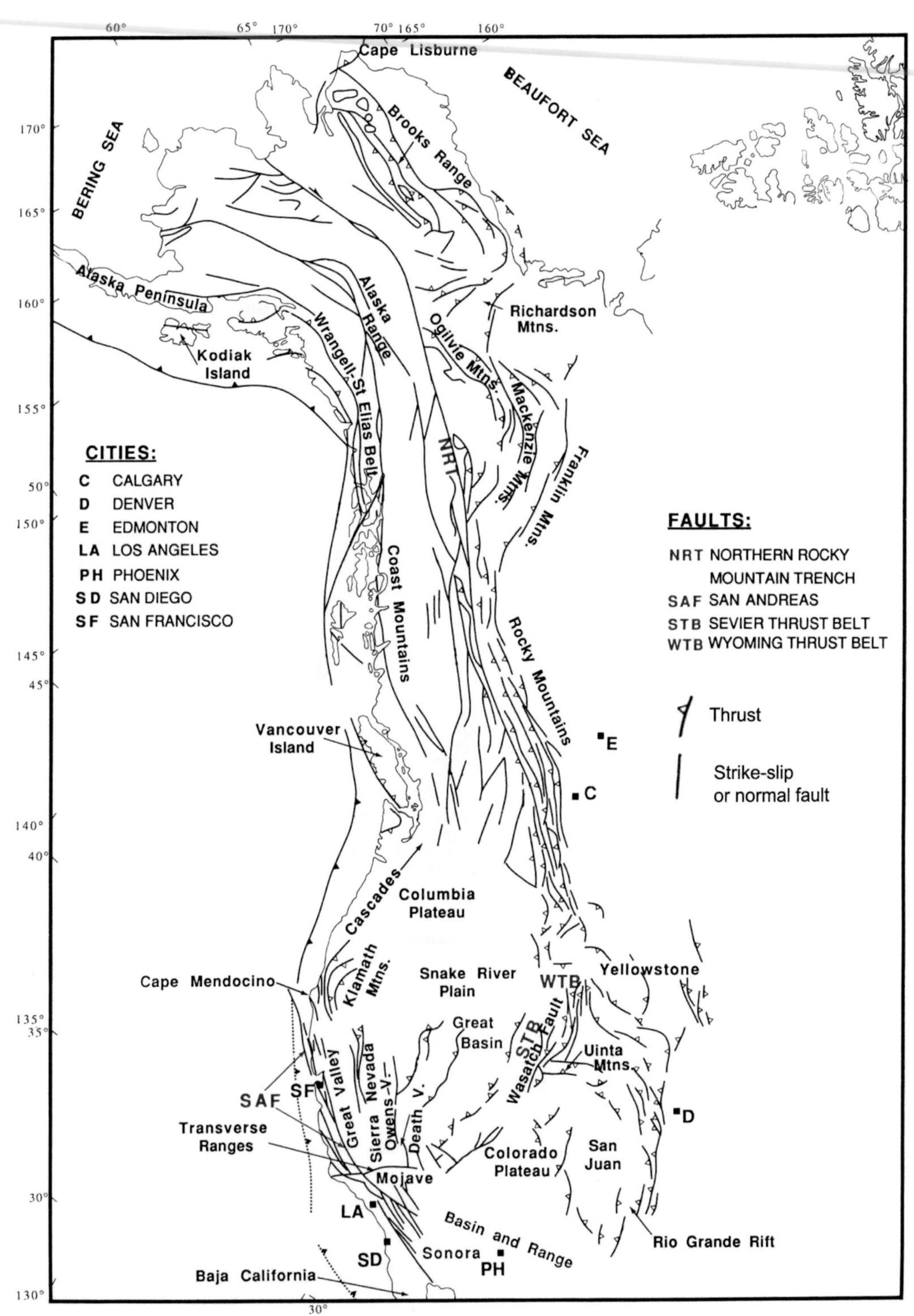

Figure 5.31 The major faults in the North American Cordillera. Modified from Oldow *et al.* (1989), published by permission of the Geological Society of America.

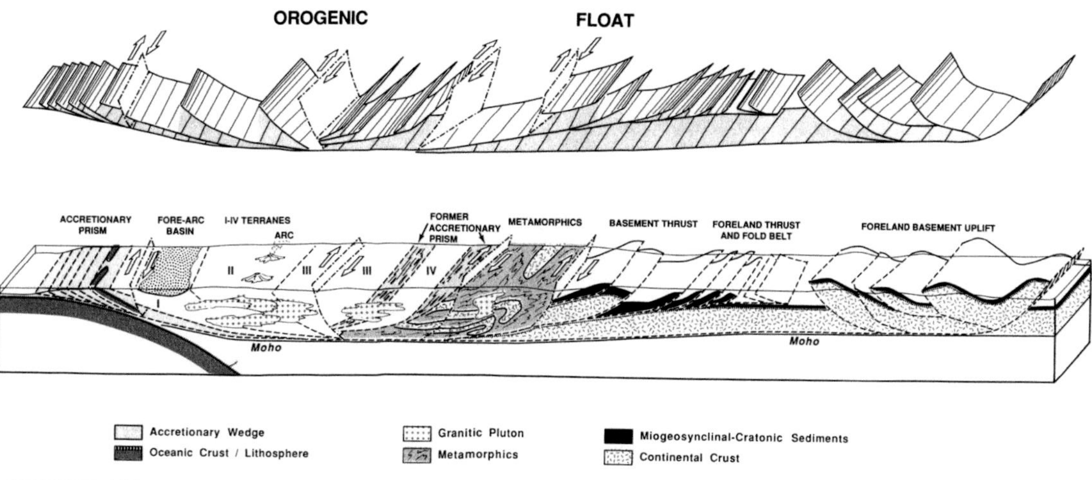

Figure 5.32 Idealised section from the Pacific margin to the continental interior across the North American Cordillera in the western USA. 'Orogenic float' conveys the idea of a basal detachment. Terranes I–IV are shown. From Oldow *et al.* (1989) published by permission of the Geological Society of America.

the Mendocino triple junction to the Gulf of California. In addition, other conspicuous strike-slip faults are the Walker Lane dextral fault of Mesozoic to recent age and the sinistral Mojave–Sonora fault of Jurassic age. Broadly, the orogen has a crude bilateral symmetry with west vergent structures on the west, and east vergent on the east. In the west, ocean crust has been underplated onto the continental crust of North America. The Pacific foredeep and accretionary prism show Cretaceous thrusting and folding.

Collision tectonics?

There is still controversy over whether the Cordillera is collision-related or not. In the west, active folds and thrusts occur in the Coast Ranges of California and in the accretionary prism near the Pacific coast. The Franciscan of western California consists of volcanic, volcanogenic and deep water sediments with ophiolites and ophiolitic melanges. This complex dates from Palaeozoic to Cretaceous, even Cenozoic times. Volcanic rocks were formed up to recent times as in the arc volcanics of the Aleutian islands and continental USA. Calc-alkaline rocks were formed in SW USA and Mexico.

Compressional tectonics are prevelant in Mesozoic times. The eastern part of the Cordillera is a foreland fold and thrust belt formed in the mid/late Mesozoic. In Canada, high-grade metamorphics in the Shuswap and Purcell belts occur in the western Cordillera and the younger Canadian Rockies fold and thrust belt in the east. In the WUSA the Sevier orogen of eastern California and Nevada is a thin-skinned foreland fold and thrust belt developed in Albian–Maastrichtian times with a vergence of thrusting towards the continental interior (Plates 19, 20). To the east of the Sevier belt is the Campanian–Eocene Laramide orogen in which basement uplifts are conspicuous (Fig. 5.30).

The non-collisional models for the North American Cordillera suggest either low-angle subduction or variations in the subduction rate, in order to account for the large width of the belt. Collisional models involve arc–continent collisions dating from the Lower Palaeozoic with consequent orogeny, e.g. the Antler orogeny age. But there is a more complex model called the exotic terrane model which proposes long-distance travel of blocks along the western coast of America and collision with the American continent in early to mid Jurassic times. Wrangellia and Stikinia are names given to these microcontinents, both of which contain a large amount of arc-related rocks such as deep-sea sediments and ophiolitic rocks.

Extensional tectonics

The phase of compression gave way to extension during the later Cenozoic and the formation of the Basin and Range. Some authors, Platt and England (1993) for example, have drawn comparison with Tibet where extensional tectonics and mountain collapse followed compression and crustal thickening. The Cordillera is associated with large batholiths such as the Sierra Nevada, mostly of late Mesozoic or Cenozoic age but some as old as Palaeozoic and early Mesozoic. High-grade metamorphic rocks are rare. They occur in the metamorphic core complexes (see Chapter 7) which formed by Cenozoic extension in the USA, Canada and northern Mexico. However, regionally metamorphic rocks in the Cordillera are generally low grade. They include blueschist facies seen in the Franciscan of California and also Mexico and Alaska.

The Canadian Cordillera

In Canada the Cordillera comprises a western belt of high-grade metamorphic Main Ranges and the Frontal Ranges and an eastern belt consisting of the low-grade Foothills Rocky Mountain foreland fold and thrust belt (Fig. 5.33). Together these belts show generally eastward-verging thrust systems. The foreland fold and thrust belt is over 100 km in width, and consists of low-grade metamorphic platform carbonate sediment rocks of Palaeozoic to Late Cretaceous age.

The whole Cordilleran belt of Alberta and British Columbia was formed in response to subduction along the western margin of North America. Between the end of the Jurassic and the late Cretaceous about ten major thrust sheets were formed in the Foothills foreland fold and thrust belt, in a period of about 100 Ma (Fig. 5.33). The dating of the thrusting is facilitated by the excellent fossil dating so that the life span of an individual sheet is indicated by fossil dating of the synthrust sediments which were eroded off the sheet. The style is thin-skinned and the thrust pile slid on a basal detachment situated above undeformed rocks. This thrust-dominated regime is characteristic of the southern Foothills belt, but in the northern Foothills folds are the dominant mechanism of crustal shortening.

The so-called piggy-back model of thrust propagation, in which the high level thrust formed first and then late thrusts were propagated beneath it, is convincingly demonstrated in the Rockies. The duplex model for thrust propagation was developed in this orogen.

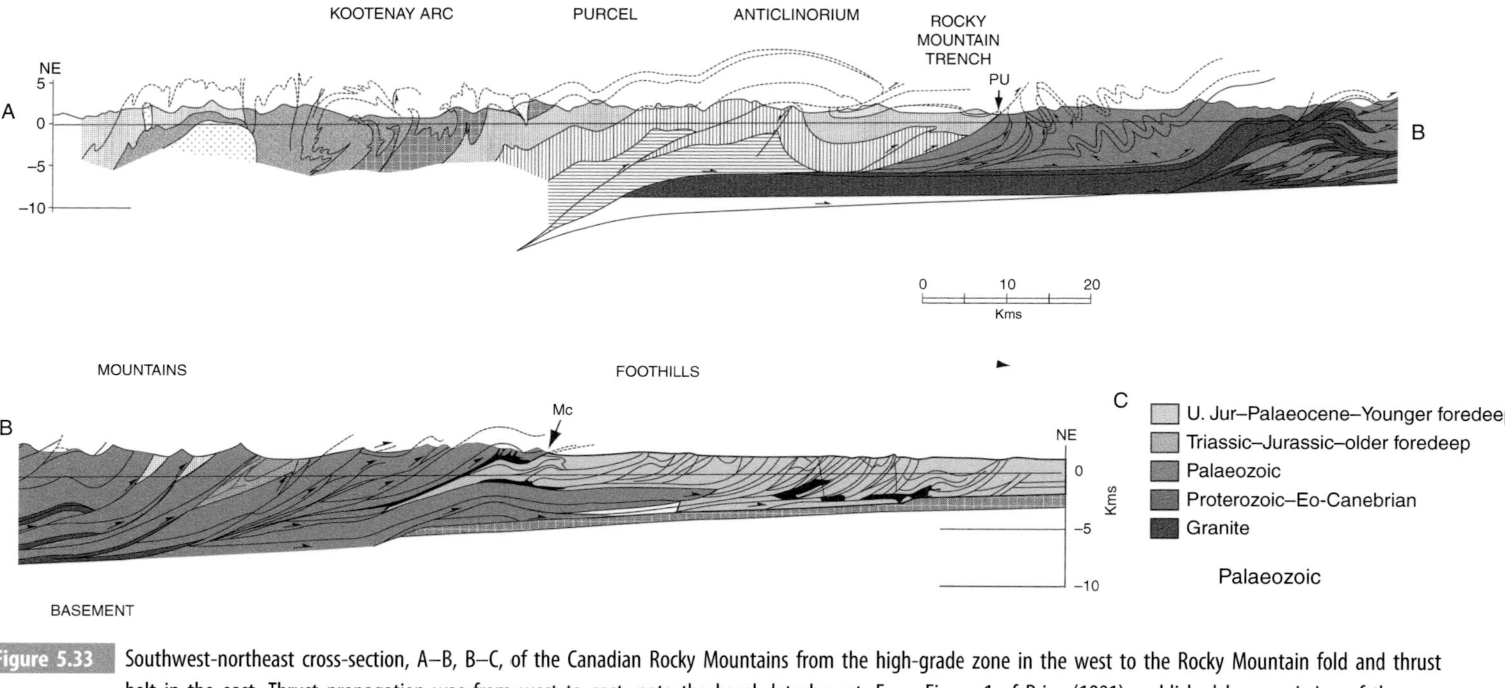

Figure 5.33 Southwest-northeast cross-section, A–B, B–C, of the Canadian Rocky Mountains from the high-grade zone in the west to the Rocky Mountain fold and thrust belt in the east. Thrust propagation was from west to east; note the basal detachment. From Figure 1 of Price (1981), published by permission of the Geological Society of London.

Timing in orogeny

The Himalaya, Alps and Andes are examples of long-lasting orogenies. But other orogens appear to have been formed rapidly. For example, the Grampian orogeny (see below) in Scotland shows a polyphase deformation history with major folds and up to high-grade metamorphism. Although the orogens quoted above show phases, the crucial point is that the Grampian orogeny exhibits similar characteristics but the orogeny all occurred in about 10 Ma, as deduced from stratigraphy and geochronology. This raises interesting questions about strain rate and the rate of thermal diffusion through the crust required to produce the metamorphism quickly.

Oblique collision belts

Strike-slip motions and oblique collision are important processes in orogeny, and the main example is the Lower Palaeozoic Caledonian orogeny. So far we have mainly considered the younger orogens of the Earth, especially those formed in the Cenozoic; Cenozoic orogens including active orogens like the Himalaya are great natural field laboratories in the attempt to understand the processes in orogenesis. But we have to remind ourselves that the Cenozoic constitutes less than 1.5% of the entire history of the Earth. So what of the rest? But as we go back in time, more and more of the evidence for ancient oceans and continental drift is lost.

The traditional view of continent–continent collision was that it was full-frontal (that is, the two continents collided in a direction which was roughly perpendicular to the subduction zone) as in the Himalaya. That is hard collision. However, in recent years the importance of oblique collision has been recognised, in which substantial strike-slip, orogen-parallel motion takes place. Gentle docking of two continents can be called soft collision. The strike-slip regime is termed transpressional to indicate the association of strike-parallel and strike-perpendicular motion.

A good example is the collision of Arabia and Asia which resulted in the Zagros mountains (see later). In that case the angle between the plate vector and the strike of the Zagros belt is about 40° or less. Other examples are the late stage strike-slip that occurred in the Southern Alps of Europe, and the collision of Baltica, Laurentia and Avalonia during the closure of the Iapetus Ocean in the Early Palaeozoic Caledonian orogen.

An extreme case of strike-slip collision involves the assembly of blocks (terranes) of continental crust which were once remote from another and therefore showed distinct geological histories. This is the collage or 'exotic terrane' model and it has been applied to the western Cordillera of the Americas, the proposal being that offshore blocks or terranes have moved along the western coast of the Pacific ocean for huge distances, hence the name 'exotic terrane'.

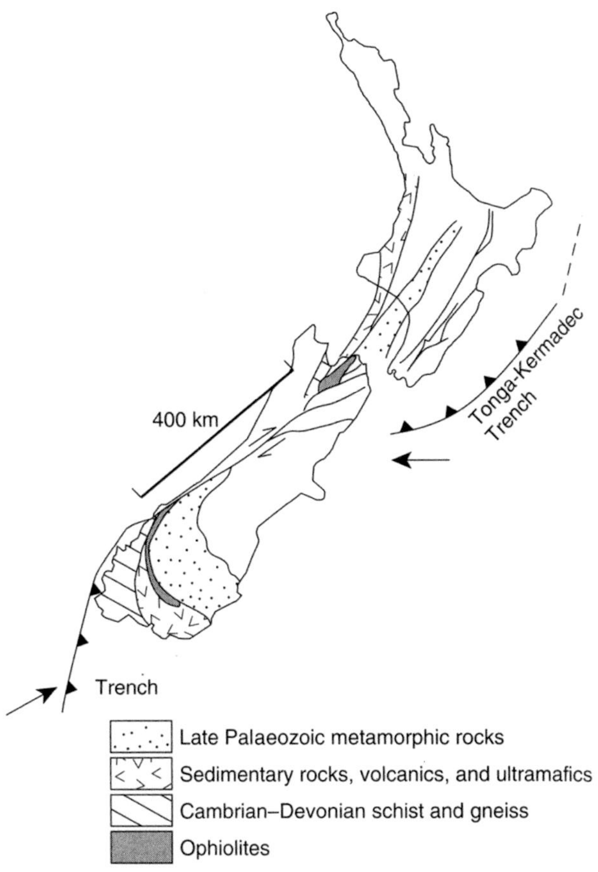

Figure 5.34 New Zealand, with the Alpine dextral transform fault shown by arrows. Note the splays at the eastern end of the Alpine fault in South Island.

The three most famous strike-slip faults are the San Andreas in California, the Great Glen fault in Scotland and the Alpine fault in New Zealand. The Southern Alps of New Zealand (Fig. 5.34) are related to the active Alpine fault which formed as a result of the relative motions of the Pacific and Australian plates: the latter is moving ENE relative to the former. The fault and the associated structural features adjacent to it form a transpressional orogen. It is uncertain when the fault was initiated, possibly in the Cretaceous but mainly in Miocene–Recent times.

The strikingly linear Alpine fault is a *c*.500 km long transform fault which links the Puysegur trench in the south with the Tonga–Kermadec trench system southeast of North Island. Near the termination at the Tonga–Kermadec trench the Alpine fault shows splays. About 480 km of dextral slip has taken place along the Alpine fault which generally dips at *c*.50° SE. En echelon fold axes are found in a 25 km zone adjacent to the fault: they are consistent with dextral shear on it. In addition, vertical motion on the Alpine fault is shown by exhumed Mesozoic and Cenozoic greenschist to amphibolite grade metamorphic rocks (the Rangitata and Kaikoura orogenies, the latter being initiated *c*.10–25 Ma ago), by tilted lake strand lines and by study of sediment records. Norris and Cooper (2001) focused on the late Quaternary displacement on the Alpine fault and showed a partitioning along the fault from oblique slip in some parts to a dominance of more strike-slip or dip-slip components in others. Adams (1981) recorded a slip rate from 5–10 Ma as 12–24 mm/a and an uplift of 15–20 km. Overall the strike-slip rate is *c*.37 mm/a but the recent rate varies along the fault from 10 to 29 mm/a with an average of 27 ± 5 mm/a. Dip-slip varies from 0 to 8–12 mm/a with an average of 10 mm/a, the highest reading being in the Mount Cook area (3764 m).

Caledonides of the North Atlantic

The term Caledonian is really a sack name for orogenies taking place during Lower Palaeozoic times. This extremely complex belt comprises early arc–continent and later continent–continent collisions as well as strike-slip motions (Fig. 5.35). The continents involved are Baltica (Scandinavia, parts of northern Russia and northern Germany), Avalonia (a long peninsula to Baltica which contained England and Wales, and southeast Newfoundland), and Laurentia (the North American continent including NW Newfoundland, Scotland, northern Ireland and East Greenland). The Cambrian–early Ordovician margin of Laurentia against Iapetus is now seen in NW Scotland, in East Greenland and the eastern part of North America. Other oceans coexisting with Iapetus included the Mirovia and Rheic oceans.

Caledonides in North America

The Silurian–Devonian Acadian orogenic belt in New England marks the Laurentia–Avalonia collision zone. Further south into the Appalachians the Caledonian belt or Piedmont belt shows the collision between Laurentia and Carolinia with Africa, which was part of Gondwana. Incidentally the western part of the Appalachian belt is called the Alleghenian and is of Variscan age. Long ago E. B. Bailey spotted that whereas in Europe the Variscan belt lies south of the Caledonides, in western America it lies to the west. In other words the Variscan belt has ‘stepped over’ the Caledonian belt. Bailey cited this as evidence for continental drift and the opening of the Atlantic.

The British Caledonides

In the Caledonides of the British Isles (Fig. 5.36) there are the following blocks or terranes, from northwest to southeast: the foreland composed of Archaean–Early Palaeozoic rocks, the Moine Supergroup (North Highland Terrane) of late Proterozoic age which underwent a

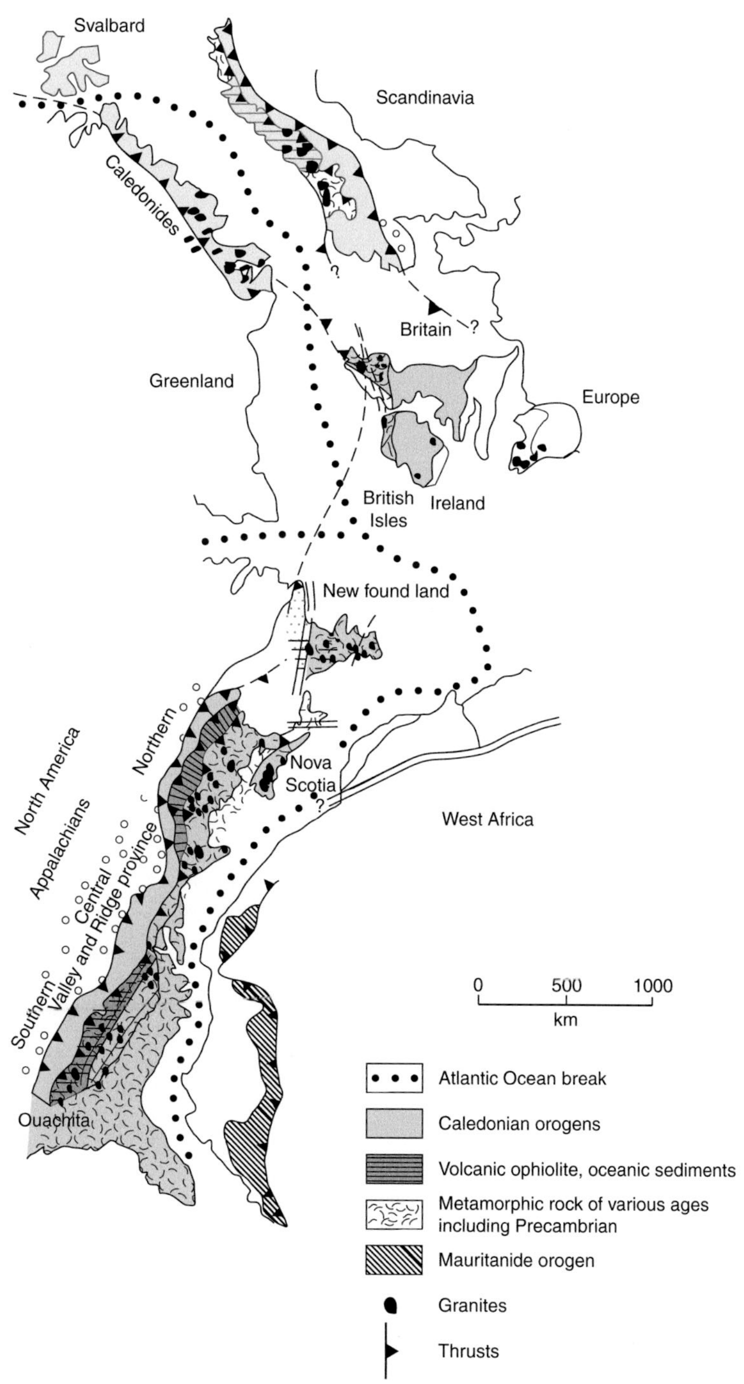

Figure 5.35 Caledonides of the Atlantic region, before the opening of the Atlantic. The 'Caledonian' belt is polyphasal, with orogenic events ranging in age from the Late Cambrian to the Devonian (see text).

Box 5.3 Plate motions, palaeogeography and stratigraphy of Neoproterozoic to Early Palaeozoic rocks in the Northern Caledonides

The late Neoproterozoic to early Cambrian history of the Earth includes several puzzling features. Thus there were one or more global glaciations; there was the explosion of the Ediacaran fauna (630–600 Ma) and Cambrian faunas; the opening of the Iapetus ocean and the Tornquist Sea; and the assembly of the Gondwana supercontinent by the closing of the Mozambique, Adamastor and Braziliana oceans. Gondwana is clearly important, and this supercontinent resulted from the assemblage of Africa, South America, Australia, Antarctica and India (Fig. 5.3.1). At the beginning of the Neoproterozoic, Laurentia, Baltica and Amazonia formed a single continental block representing the last fragment of the Rodinia, which was the supercontinent assembled *c.*1.0 Ga ago during the Grenville orogeny (Fig. 5.3.2). Rodinia was surrounded by the Mirovoi Ocean, and it started to disperse at 800–750 Ma. The placing of continents depends on palaeomagnetism and palaeogeography, but unfortunately there is still a great deal of debate about the evidence, which is far from perfect (see Pisarevsky *et al.*, 2008). The opening and closure of the Iapetus ocean are the key events in Early Palaeozoic times, and there is agreement about its rift-drift stage and that the assembly of Gondwana was more or less synchronous with Iapetus. Thus the final assemblage of Gondwana occurred as late as 530 or 520 Ma, while the opening of the Iapetus Ocean was clearly under way by 540 Ma.

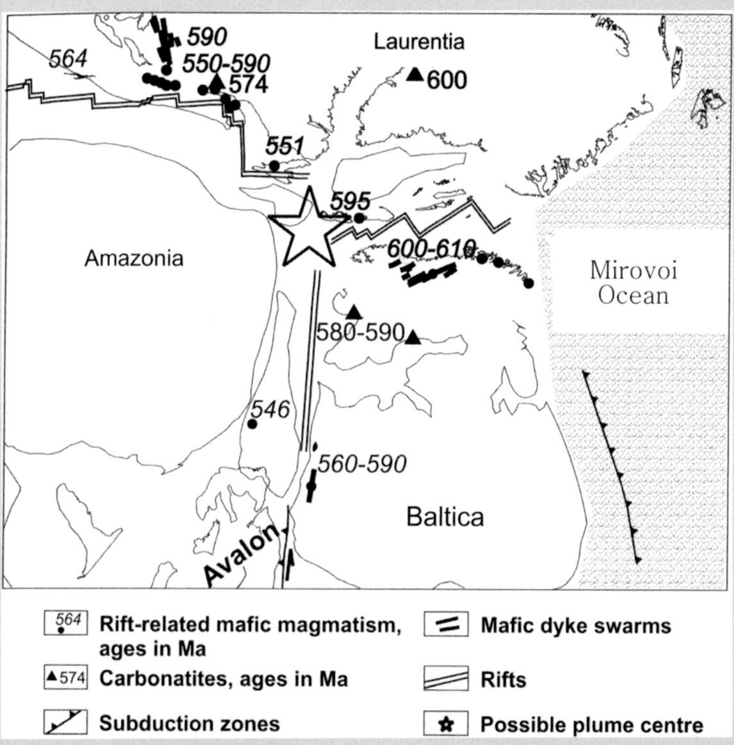

Figure 5.3.1 The assembly of Baltica, Amazonia and Laurentia immediately prior to the opening of Iapetus. From Pisarevsky (2008), Figs.1, 2 and 16, published by permission of the Geological Society of London.

Box 5.3 *(cont.)*

The opening of the early Palaeozoic Iapetus Ocean can be recognised by several events. During the latest Proterozoic and Cambrian times, rift-related mafic magmatism occurred along the eastern margin of Laurentia, the Scandinavian margin of Baltica and the SW margin of Baltica. Dyke swarms formed in three branches of supposed plume-related rifting. Two branches coincide with the strike of the Grenville and

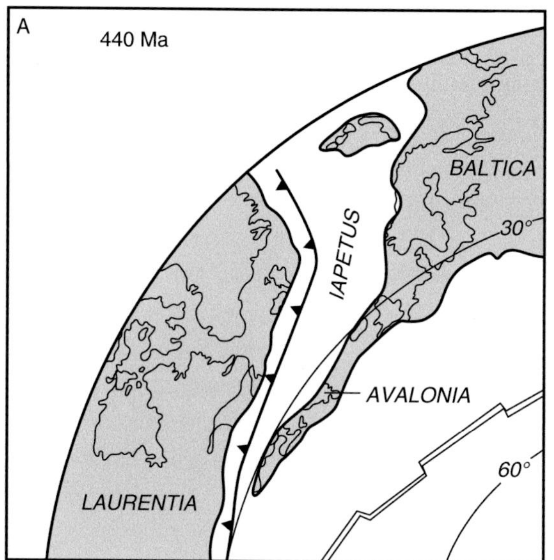

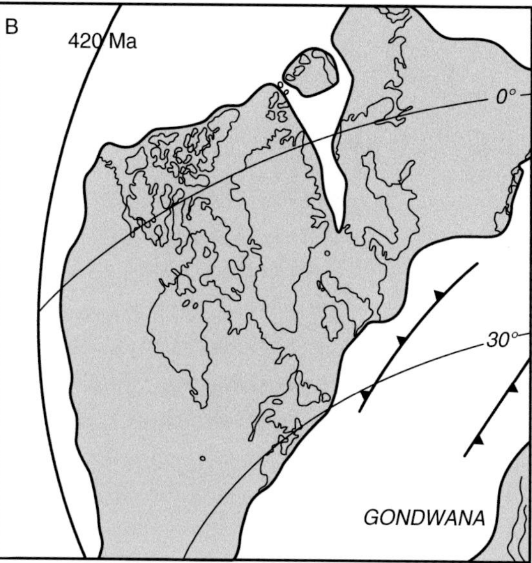

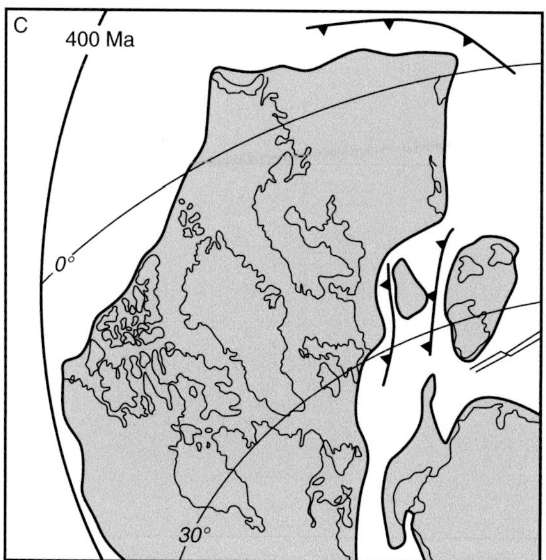

Figure 5.3.2 a–c, Caledonides: palaeomagnetic plots to show the Iapetus ocean from Tremadoc–Arenig to Caradoc–Ashgill times. From Dewey and Strachan (2003), published by permission of the Geological Society of London.

Box 5.3 *(cont.)*

Svecofennian orogens, but one cuts across the strike. In the Scottish Dalradian, basaltic magmatism indicating continental thinning leading to small ocean basins occurred at *c*.590 Ma. The rifting included some initial failures which were successful at later attempts. East Iapetus opened first, then west Iapetus formed when Amazonia and Laurentia broke apart. Baltica and Laurentia were well separated by 550 Ma. After 600–580 Ma, passive margins with sediment deposition formed around Iapetus.

The microcontinents called Avalonia, Carolinia and Cadomia are remnants of Amazonia and northern Gondwana (Nance *et al.*, 2008, Fig. 5.3.2). They are called the peri-Gondwanan terranes which were first attached to Gondwana between 665 and 650 Ma and were detached from it at *c*.540–515 Ma. Avalonia, which was accreted to Laurentia at *c*.400–415 Ma, is seen in New England, SE Newfoundland, and South Britain. Cadomia is seen in NW France and the Iberian peninsula, and bits in Germany and the Czech Republic. At 635 Ma there was Andean arc-related, calc-alkaline and felsic magmatism along the Amazonian and African margins.

The Scottish Caledonides comprises the late Proterozoic Moine Supergroup (which underwent a tectonothermal event in late Proterozoic times), the late Proterozoic to early Palaeozoic Dalradian metasedimentary rocks as well as the Cambro-Ordovician rocks found in NW Scotland. These rocks together with similar aged successions in the Appalachians are passive margin sequences of Laurentia that were later involved in Caledonian tectonometamorphic events. In the Caledonides around the north Atlantic, the late Proterozoic to Lower Palaeozoic sequences show remarkable similarities. Most notably, the Cambro-Ordovician succession in Greenland, Scotland and Newfoundland shows that the passive margin of Laurentia against the Iapetus ocean is recognisable despite the disruption of this margin by the opening of the Atlantic (Fig. 5.3.2).

Following early Ordovician arc–continent collision, subduction commenced on the northern border of the Iapetus Ocean together with the development of a southward-growing subduction-related accretionary prism. This can be seen in the Southern Uplands of Scotland where turbidites range in age from Llandeilo to late Wenlock (*c*.424 Ma). The prism shows a progressively later southward transition from the oceanic Moffat Shale facies to ocean trench turbidites. The turbidite facies becomes younger southwards marking the growth of the prism.

Initiation of ocean closure is indicated by the first arrival in the English Lake District of Iapetus detritus from the Southern Uplands in late Wenlock time (*c*.424 Ma), although the Lake District had earlier become a foreland basin subsiding beneath the load of the Southern Uplands prism in the late Llandovery deepening episode in the Stockdale Group. In Ireland the first arrival of detritus from north of the suture into basins south of the Iapetus suture was in the Wenlock (430–424 Ma), therefore there may have been a diachronous collision between the Scottish and Irish sectors because of irregular margins.

tectonothermal event in late Proterozoic times, the Dalradian Supergroup (Grampian Terrane) of later Proterozoic to early Palaeozoic age, and finally the Midland Valley with Devonian and Carboniferous strata and the Southern Uplands (Central Terrane) with Ordovician and Silurian sediments bounded on the south by the Iapetus suture. The Dalradian Supergroup in SE

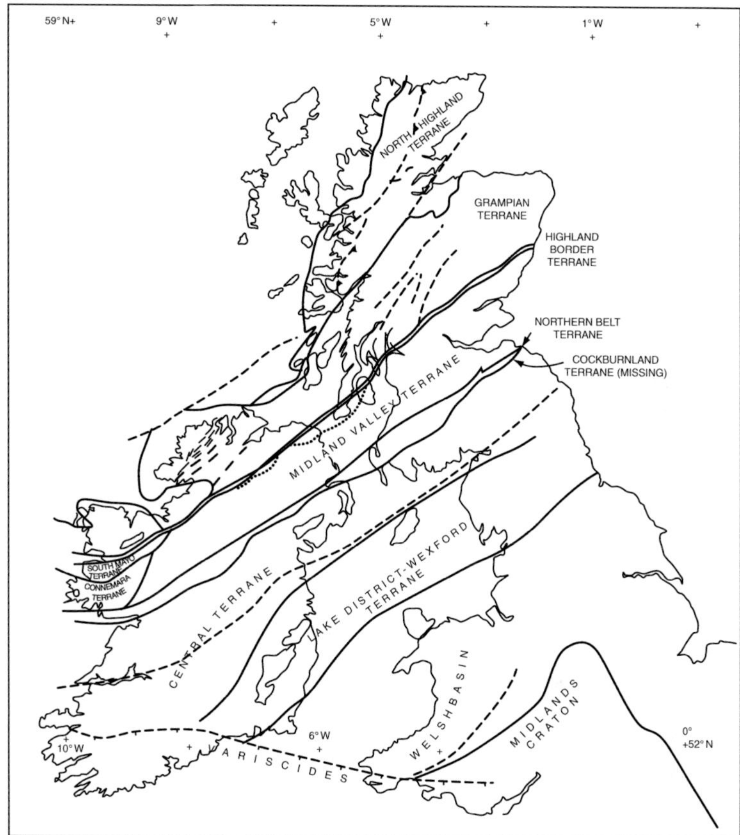

Figure 5.36 Caledonides in Britain, to show the main structural units and important faults.

Scotland is a late Proterozoic metasedimentary and meta-igneous succession deposited on the Laurentian margin. The Dalradian succession changes upwards from a shallow water quartzite–carbonate–shale assemblage to a deeper water turbidite greywacke assemblage. Although at one time these blocks were regarded as having been always together as we see them today, it is now realised that they were assembled by major strike-slip motions.

The Palaeozoic history of the North Atlantic region includes two collisions. The first is an arc–continent collision of early/mid Ordovician age called the Grampian orogeny in Scotland and the Taconic orogeny in the USA, and the second, the Scandian orogeny of Silurian age, during which Iapetus closed by the collision of Baltica and Laurentia. The Ordovician event will be discussed first (Fig. 5.37).

During the earliest Palaeozoic the Midland Valley of Scotland formed an island arc positioned at the western margin of the Iapetus ocean and separated from the Laurentian continental margin by a small back-arc ocean basin. The arc is now hidden by a cover by Devonian, Carboniferous and younger rocks, and its existence can only be deduced from the igneous fragments of it found in the later rocks. The arc can be traced westwards through northern Ireland into Newfoundland and then into New England where the arc–continent collision is called the Taconic orogeny of mid Ordovician age. In the early

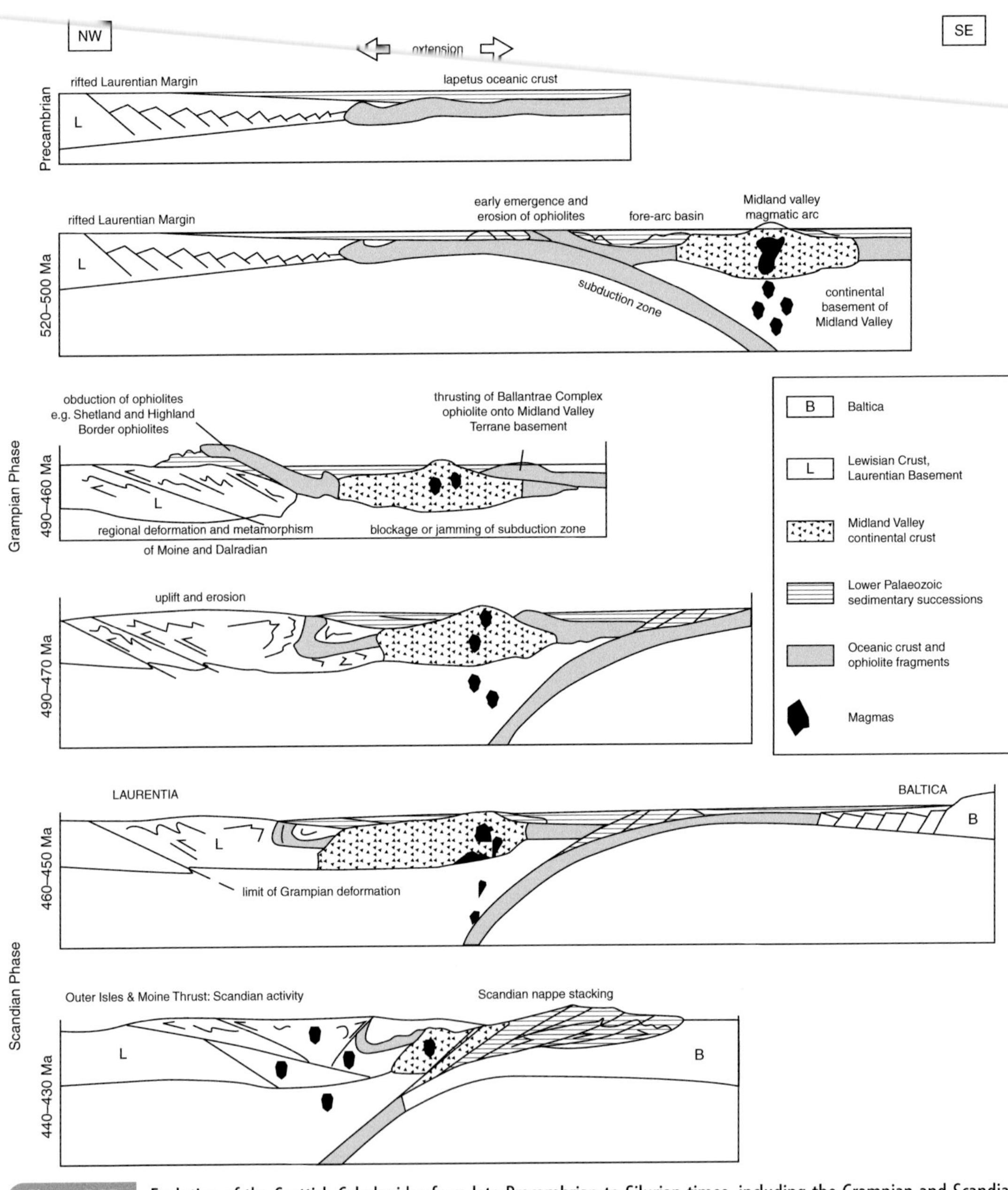

Figure 5.37 Evolution of the Scottish Caledonides from late Precambrian to Silurian times, including the Grampian and Scandian orogenies. From Searle (2007), published by permission of the Geological Society of America.

Ordovician the arc collided with the Laurentian margin, including the Dalradian terrane as a result of SE subduction of Laurentia beneath the arc. The arc–continent collision resulted in the Grampian orogeny in the Dalradian and Moine terranes of Scotland at 470–460 Ma.

The Grampian orogeny

The early Ordovician Grampian orogeny involved crustal thickening and high-temperature/intermediate-pressure Barrovian and low-pressure/high-temperature Buchan metamorphism. The orogeny is polyphasal with the formation in the earlier deformation episodes of the southeast-verging recumbent Tay Nappe. Later deformation episodes are post peak metamorphism and formed upright folds including the major monoclinal Highland Border Downbend which developed a steep zone near the Highland Border.

Obducted ophiolites

If Dewey and Shackleton (1984) are correct, crustal thickening was increased by obduction of a thick slab of oceanic crust emplaced onto the margin of the Laurentian continent and on top of the Dalradian rocks of the southern Highlands of Scotland and in Ireland. Ophiolites obducted onto the Laurentian margin are seen in Unst in Shetland and the Bay of Islands in Newfoundland. In Ireland, the Tyrone Igneous Complex is interpreted as an ophiolite which was obducted onto the Laurentian margin during the Grampian orogeny and prior to *c*.470 Ma. In Scotland, ophiolite is found at Ballantrae on the northern margin of the Southern Uplands on the Ayrshire coast. In the Grampian Highlands the proposed obducted ophiolite has been completely eroded away.

Another important ophiolite obducted occurrence is the Highland Border Series along the Highland Boundary fault in Scotland and Ireland. This series occurs in a narrow, <1 km wide belt sandwiched between the Dalradian and the Devonian and Carboniferous rocks of the Midland Valley. A part of the Highland Border Series resembles an ophiolite with highly altered ultramafic rocks, basaltic pillow lavas, black graphitic mudstones, limestone and cherts. In addition to this assemblage there are arenites which may be assigned to the Dalradian. The Highland Border Series or Complex has been variously interpreted as ocean floor material either in situ or transported laterally to dock with the Dalradian or as an obducted ophiolite. Recent work by Tanner reported by Leslie (2009) puts forward another view – that the ophiolites represent a 'Ligurian' or 'Iberian' margin now recognised in modern settings, which means an ocean-continent transition setting for the Highland Border Series where serpentinised peridotite derived from the subcontinental mantle was exhumed during continental break-up.

The Scandian orogeny: closure of Iapetus

At 440–430 Ma, final closure of the Iapetus ocean started and led to the collision between Laurentia, Baltica and Africa. Whereas in Scotland and Greenland the thrust vergence was to the NW and WNW, in Scandinavia the slightly later thrusts (*c*.420 Ma) verge to the ESE or SE.

The deformation and metamorphism assigned to the Silurian Scandian orogeny did not affect the Grampian belt (Fig. 5.37). However, in NW Scotland the Neoproterozoic Moine Supergroup, a largely metasedimentary succession, has been disrupted by foreland-propagating thrusts of Scandian age (*c*.450 to 425 Ma). The Sgurr Beag thrust, along with other ductile thrusts (Naver, Swordly thrusts) within the Moine outcrop, certainly pre-dates the Moine thrust but the date of *c*.450 Ma for the age of the Sgurr Beag thrust suggests they are either late Grampian or early Scandian features.

The western margin of this part of the Caledonides is formed by the Moine thrust zone which probably developed between 430 and 425 Ma. Activity on the Moine thrust was roughly synchronous with, or slightly later than, high-grade (sillimanite grade) regional Barrovian metamorphism in the Moines.

Sinistral transpression during the Scandian orogeny

During sinistral transpression from about 435 to 425 Ma, Baltica collided with Laurentia to form the Scandian orogen in which strain was partitioned into belt-orthogonal thrusting and belt-parallel strike-slip faulting. Avalonia collided with Laurentia by means of a highly oblique sinistral transpression. This was a soft collision in contrast to a hard collision in Ireland.

According to Dewey and Strachan (2003) the progress of the Baltica–Laurentia collision is marked by a change in the structural regime within the accretionary prism of the Southern Uplands, involving a switch from orthogonal shortening to sinistral transpression in the late Llandovery at or just before 430 Ma. In western Ireland in the South Mayo Trough (Fig. 5.36), a complex sinistral transpression affecting the Ordovician and Silurian rocks is dated at post-Wenlock and pre-Lower Devonian.

Kinematic indicators of strike-slip movements

Dewey and Strachan have used stretching lineations to detect the transport directions for many parts of the Scandian orogenic belt. Scandian transport vectors show an arcuate curve across the Moine outcrop (Figs. 5.38–5.40). Thus transport along the early Scandian thrusting in the internal part of the belt, along the Sgurr Beag/Naver thrusts, was directed to the NNW or NW, but later, during the slip on the Moine thrust, the transport direction rotated anticlockwise to a WNW direction, orthogonal to the orogen.

In Scandinavia early in the Scandian orogeny there was eastward obduction of an ophiolite obducted complex, followed by the emplacement of thrust sheets onto Baltica. The lower to intermediate thrust sheets comprise basement complexes and sedimentary cover rocks derived mostly from the margin of Baltica. The higher thrust sheets contain sheets of oceanic crust and continental basement possibly derived from Laurentia. The Scandian orogeny in Scandinavia probably developed from 430 to 410 Ma and involved deep burial (80–100 km) of the Baltica basement rocks, as indicated by coesite-bearing eclogite. The overall transport direction is SSE–ESE but in detail it varies considerably, including transport parallel or slightly oblique to the trend of the orogen. In general,

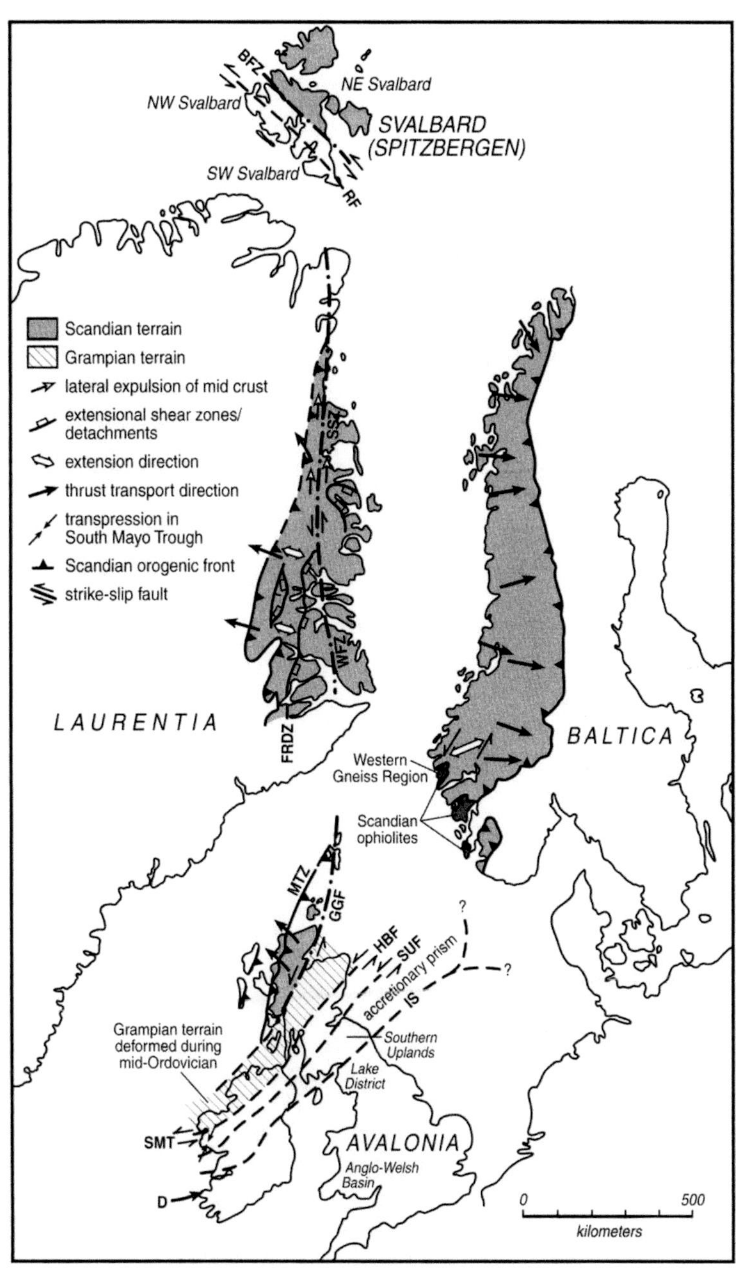

Figure 5.38 Tectonic map of the northern Caledonides showing the Grampian and Scandian terranes. BFZ, Billefjorden Fault Zone; D, Dingle; FRDZ, Fjord Region Detachment Zone; GGF, Great Glen Fault; HBF, Highland Boundary Fault; IS, Iapetus suture; MT, MTZ, Moine Thrust Zone; SUF, Southern Upland Fault; SMT, South Mayo Trough; WFZ, Western Fault Zone. From Dewey and Strachan (2003), published by permission of the Geological Society of London.

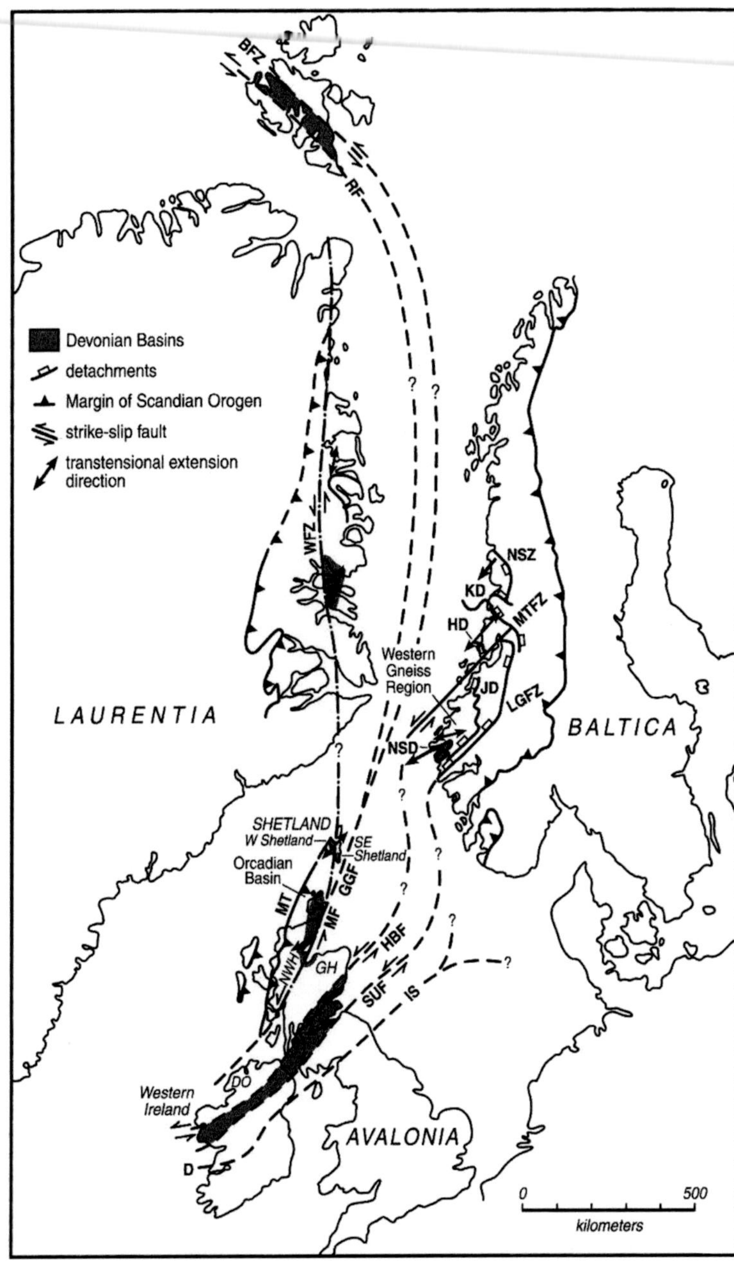

Figure 5.39 Sinistral transpression in the Caledonides of North Atlantic. Key as in Fig. 5.37 plus DO, Donegal; GH, Grampian Highlands; NWH, North West Highlands; HD, JD, KD, LGFZ, MTFZ, NSZ, extensional detachments and faults in the Norwegian Caledonides; SMT, South Mayo Trough. From Dewey and Strachan (2003), published by permission of the Geological Society of London.

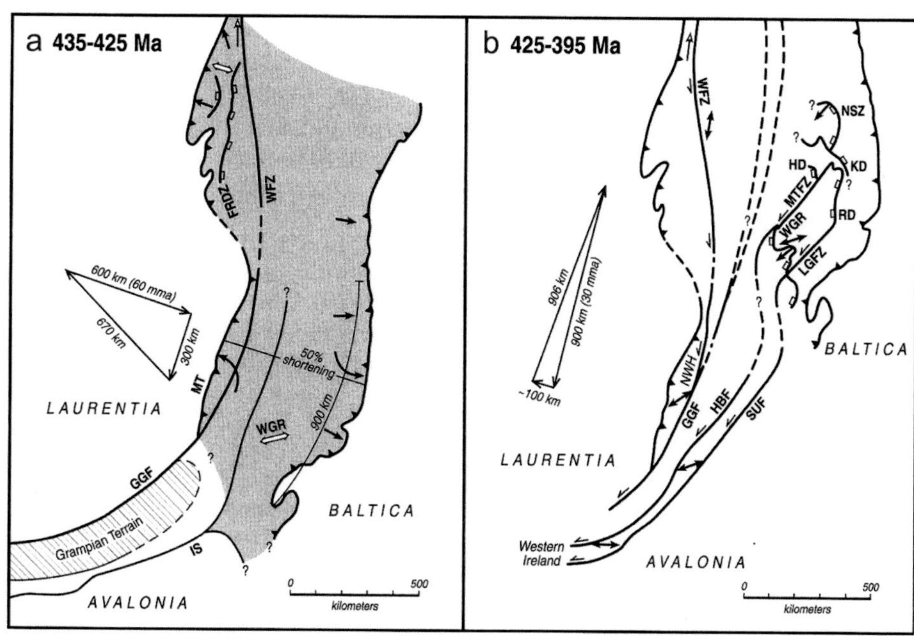

Figure 5.40 a, Silurian reconstruction of the Caledonides with 900 km of sinistral motion on the Great Glen Fault removed, showing the principal Scandian transpressional structures, giving orthogonal convergence between Baltica and Laurentia at a rate of 60 mm a^{-1} (see triangle). Lower estimate for plate motion is 30 mm a^{-1}, the maximum is 67 mm a^{-1}. b, Late Silurian–Devonian reconstruction of the Caledonides showing the principal transtensional structures. WGR, Western Gneiss Region in Norway; for others see Figs. 5.37 and 38. Triangle shows 900 km of sinistral slip. From Dewey and Strachan (2003), published by permission of the Geological Society of London.

transport directions rotated anticlockwise to become nearly orthogonal to the margin as the thrust sheets overrode the Baltic foreland.

In East Greenland, Neoproterozoic rocks of the Eleonore Bay Supergroup rest on an Archaean–Palaeozoic basement. From Silurian to early Devonian times these rocks were involved in oblique plate convergence which gave a complex orogenic history. The kinematics of this region are complicated, with sometimes orogen-parallel movement and north-directed flow, as in the late Silurian. Crustal-scale partitioning of strain involved orogen-parallel flow in the mid crust but, at the same time, east–west shortening by west-directed thrusting at higher structural levels.

In NE Svalbard, the Heckla Hoek sequence of Neoproterozoic age is in part equivalent to the Eleonore Bay Supergroup in East Greenland. Different histories are found in NW and SW Svalbard reflecting the fact that these regions have been juxtaposed by major strike-slip in the late Silurian–Devonian. The Heckla Hoek sequence underwent a tectonometamorphic event in the Silurian, comprising early recumbent folds with west vergence and later intense sinistral transpression. Early amphibolite facies metamorphism was overprinted by greenschist facies. There was sinistral translation of NE Svalbard relative to East Greenland.

In summary, Scandinavia, East Greenland and NW Scotland show sinistral transpression: all have foreland transport propagating deformation sequences with anticlockwise rotation of transport from orogen-normal to orogen-parallel. The whole kinematic framework was dominated by sinistral transpression which became progressively partitioned into orogen-normal and orogen-parallel sinistral slip. Whereas the Laurentia–Baltica collision produced a Himalayan type belt with 50% shortening, the Laurentia–Avalonia collision produced only weak deformation as in the Southern Uplands, or low-grade polyphase clockwise intersecting cleavages.

Switch to orogen-parallel motion

At *c*.425 Ma, relative motion between Laurentia and Avalonia–Baltica became orogen-parallel until 410 Ma, and the main expression of the regime was the Great Glen Fault and its possible northern continuations in Shetland (the Walls Boundary fault) and East Greenland (the Western fault zone).

The Great Glen fault involved orogen-parallel transport and was later than the Moine thrust movements, being mainly active from 425 to 410 Ma. Although the Great Glen Fault must have had at least 700 km slip along it, this is only part of the total sinistral strike-slip taking place from 435 to 395 Ma between Laurentia and Baltica. Thus the Grampian terrane in the Southern Highlands of Scotland was some 700 km southwest of its present position where it escaped the Scandian orogeny. Likewise, restoration of the late Silurian strike-slip brings southern Scandinavia against NW Scotland and northern Scandinavia against East Greenland. Magmatism associated with transtension is shown by the Newer Granites of Scotland which are calc-alkaline I-type granites emplaced in pull aparts – that is, extensional areas in the post-425 Ma sinistral shear regime.

Additional strike-slip motion was supplied by the Highland Boundary Fault and the Southern Uplands fault. Including these faults increases the total sinistral slip across Scotland to about 1200 km, which within the given time span indicates a rate of plate motion of 40 mm/a. However, a more conservative estimate of 900 km total sinistral slip (adding the slip on the Great Glen, Highland Border and Southern Upland faults) gives a rate of plate motion of 30 mm/a, and this amount of slip puts Scotland against SW Norway during the Scandian orogeny. Assuming that 50% shortening occurred across the orogen, then the orthogonal convergence between Laurentia and Baltica was 600 km at a rate of 60 mm/a. Combining the orogen-parallel and orogen-normal components of relative motion of Laurentia relative to Baltica gives a rate of 67 mm/a.

Sinistral transtensional regime

From 410 to 395 Ma, continuing transcurrent motion, plus some orogen orthogonal extension, produced a transtensional regime associated with deposition of the Old Red Sandstone in Scotland. A thrust and strike-slip fault regime accompanied the deposition of the Old Red Sandstone which was laid down in continental basins such as the Orcadian basin. Large open folds, for example those seen in the Midland Valley, indicate that transtension was accompanied by weak orthogonal compression across the basins. In

SW Norway the exhumation of coesite-bearing eclogites took place by sinistral transtension, with strains of 430% E–W stretching, 75% vertical shortening and 25% N–S shortening. On a larger scale, in the Caledonian orogen there was an orogen-wide transition at *c.*400 Ma in mid Emsian times, from sinistral transpression to sinistral transtension. This may have been due to a change in plate motion between Baltica and Laurentia from oblique sinistral convergence through orogen-parallel motion to oblique sinistral divergence, caused by the Scandian orogeny.

Oblique collision in the Zagros

The Zagros mountains stretch for *c.*1800 km and are 250–300 km wide. They form a link between the strike-slip plate margin of the Dead Sea transform fault along the northwestern margin of the Arabian Plate and the remnant oceanic crust in the Gulf of Oman to the east. The suture zone in the Zagros consists of dismembered ophiolites, deep-sea sediments and molassic sediment. The structure of the Zagros is quite simple, major upright concentric and periclinal NW–SE trending folds that affect a *c.*10 km thick Phanerozoic sedimentary sequence. Apart from in the suture zone, thrusting is negligible. Crustal shortening is small, about 50–70 km. Below the folded pile is a detachment zone along a weak zone formed by Precambrian salt. Salt diapirs are common in the Zagros, and the salt domes have risen up through the 10 km thick folded succession. It is assumed that the Precambrian basement below the salt horizon has been subducted to depths of 45 to 50 km beneath the Iranian plateau.

The NW–SE thrusts in the Zagros mountains are cut by transverse strike-slip faults but the main concern here is with the evidence for orogen-parallel strike-slip motion. Khalil and Azizi (2008) have described minor structure and microstructural kinematic indicators which indicate a component of dextral strike-slip. This is predictable given that the plate vectors in the Arabian plate are oblique to the trend of the Zagros.

Intraplate tectonics

The orogens so far discussed are the result of the interaction of quasi-rigid lithospheric plates moving across the Earth's surface and the tectonic forces at plate margins. However, intraplate orogens, such as the Tien Shan, Mongolian Alti, and the Atlas mountains in North Africa occur far from plate boundaries but can all be linked to the Cenozoic collisions. In central Australia there are several orogens formed in the Proterozoic and Palaeozoic which are clearly not on plate margins and not related to collision. Thus there are no ophiolites or other indicators that ocean closure or collision was involved in their formation. The Alice Springs orogen is one of them. It is *c.*100 km across and developed in the Devonian and Carboniferous over a period of *c.*100 Ma. It is marked by a strong gravity anomaly. Others are the Peterman Mountains and the Arunta

Block. Metamorphic grade is greenschist to amphibolite facies. The Alice Springs orogen contains thrusts but only *c*.50–100 km of N–S shortening and is the most recent orogeny on the Australian continent. The zones formed by the orogens have been foci of tectonic activity since the early Proterozoic. They represent weak zones often formed along pre-existing zones of intense deformation, therefore showing continental rejuvenation. Significantly, the Canning Basin to the west of the Alice Springs orogen is more or less synchronous with the orogen but is an extensional basin.

The intraplate orogens are weak zones surrounded by strong lithosphere of the Australian Shield. Furthermore, the weakness has persisted since the early Proterozoic. Roberts and Houseman (2001) interpreted the cause of these orogens as a clockwise rotation of the plate which caused the compression in the orogens and extension in the Canning Basin. The rotation may be due to E–W traction or clockwise movements associated with rifting and collisions on the northern part of Gondwanaland.

Most of the interpretations of intraplate orogens recognise them as weak zones, but whereas some favour as a cause remote far-field stresses involving the transmission of horizontal stresses from a subduction zone, others favour vertical forces in the mantle and delamination or gravitational instabilities at or near the base of the thermomechanical lithosphere. Upwelling of asthenosphere increases the geothermal gradient, so the Moho temperature is *c*.520 °C, 10% higher than the surrounding Shield. Another possibility is an elongate mantle plume.

Sandiford *et al.* (2001) and Braun and Shaw (2001) employed a thin viscous plate model and numerical modelling to explain the features of intraplate orogens. They investigated the role of tectonic processes in modifying heat source distribution in intraplate settings and the thermal-mechanical consequences of changes in heat-producing elements in the upper crust induced by earlier tectonic activity.

Cooling is associated with the exhumation of heat-producing deep crustal rocks. Redistribution of material changes the chemistry and thermomechanical structure. A so-called 'thermal lock' means that the cooling associated with exhumation of deep crustal rocks comprising low heat-producing mafic granulites led to lithospheric strengthening and therefore the end of orogeny.

Sandiford *et al.* suggested that the original crustal section in the intraplate orogens consists of an uppermost crust of unmetamorphosed sediments above a mid–upper crust of granites and metasediments and a mid–lower mafic to felsic crust. Geochemical structuring occurred by either crustal differentiation with large-scale melt transfer or a redistribution of material during deformation, sedimentation and erosion. A combination of these processes markedly affects length scales associated with compositional variation in the crust. Mid crustal granites have incompatible elements concentrated at higher crustal levels. If these granites are exhumed, there is loss of incompatible elements from the exhumed region and into intraplate basins.

Further reading

Allmendinger, R.W., Jordan, T.E., Kay, S.M. and Isacks, B.L. (1997). The evolution of the Altiplano–Puna plateau of the central Andes. *Annual Review of Earth and Planetary Science*, **25**, 139–174.

Allmendinger, R.W., Figuerea, D., Snyder, D. *et al.* (1990). Foreland shortening and crustal balancing in the Andes at 30° S latitude. *Tectonics*, **9**, 4, 789–809.

Berger, A. and Bousquet, R. (2008). *Subduction Related Metamorphism in the Alps: Review of Isotope Ages Based on Petrology and their Dynamic Consequences. Geological Society of London, Special Publication*, **298**, 117–144.

Bernet, M., van der Beek, P., Pik, R. *et al.* (2006). Miocene to Recent exhumation of the central Himalaya determined from combined detrital zircon fission track and U/Pb analysis of Siwalik sediments, western Nepal. *Basin Research*, **18**, 393–412.

Bousquet, R., Goffée B., Henry, P., Le Pichon, X. and Chopin, C. (1997). Kinematic and petrological model of the Central Alps: Lepontine metamorphism in upper crust and the lower crust. *Tectonophysics*, **273**, 105–127.

Catlos, E.J., Harrison, T.M., Kohn, M.J. *et al.* (2001). Geochronologic and thermobarometric constraints on the evolution of the Main central thrust, central Nepal Himalaya. *Journal of Geophysical Research*, **106**, B8, 16177–16204.

Chemenda, A., Burg, J.-F. and Mattauer, M. (2000). Evolutionary model of the Himalaya–Tibetan system geopoem based on new modelling, geological and geophysical data. *Earth and Planetary Science Letters*, **174**, 397–409.

Clark, M.K. and Royden, L.H. (2000). Topographic ooze: Building the eastern margin of Tibet by lower crustal flow. *Geology*, **28**, 703–706.

Cobbold, P.R., Rossello, E., Roperch, P. *et al.* (2007). In: *Deformation of Continental Crust: The Legacy of Mike Coward*, ed. A.C. Ries, R.W.H. Butler and R.H. Graham. *Geological Society of London, Special Publication*, **272**, 321–343.

Coward, M. (1994). Collision tectonics. In: *Continental Deformation*, ed. P. L. Hancock. Pergamon Press, 264–288.

DeCelles, P.G., Robinson, D.M., Quade, J. *et al.* (2001). Stratigraphy, structure, and tectonic evolution of the Himalayan fold-thrust belt in western Nepal. *Tectonics*, **20**, 487–509.

Dewey, J.F., Cande, S. and Pitman, W.C. (1989). Tectonic evolution of the India-Asia collision. *Eclogae Geologica Helvetiae*, **82**, 717–734.

Dickinson, W.R. (2009). Anatomy and global context of the North American Cordillera. In: *Backbone of the Americas: Shallow Subduction, Plateau, and Ridge and Terrane Collision*, ed. S.M. Kay, V.A. Ramos and W.R. Dickinson. *Geological Society of America Memoir*, **204**, 1–29.

Ding, L., Kapp, P., Yue, Y. and Lai, Q. (2007). Postcollisional calc-alkaline lavas and xenoliths from the southern Qiangtang terrane, central Tibet. *Earth and Planetary Science Letters*, **254**, 28–38.

Elliott, D. (1976). The energy balance and deformation mechanisms of thrust sheets. *Philosophical Transactions of the Royal Society, London*, **A283**, 289–312.

Elliott, D. and Johnson, M.R.W. (1980). Structural evolution of the northern part of the Moine thrust belt, NW Scotland. *Transactions of the Royal Society of Edinburgh: Earth Sciences*, **71**, 69–96.

Fielding, E., Isacks, B., Barazangi, M. and Duncan, P.C. (1994). How flat is Tibet? *Geology*, **22**, 163–167.

France-Lanord, C., Derry, L. and Michard, A. (1993). Evolution of the Himalaya since Miocene time: isotopic and sedimentological evidence from the Bengal Fan. In: *Himalayan Tectonics*, ed. P.J. Treloar and M.P. Searle. *Geological Society of London, Special Publication*, **74**, 605–621.

Gapais, D., Cagnard, F., Gueydan, F., Barbey, P. and Ballevre, M. (2009). Mountain building and exhumation processes through time: inferences from nature and models. *Terra Nova*, **21**, 188–194.

Gerbault, M., Martinod, J. and Hérail, G. (2005). Possible orogeny-parallel lower crustal flow and thickening in the central Andes. *Tectonophysics*, **399**, 59–72.

Godin, L., Grujic, D., Law, R.D. and Searle, M.P. (2006). Channel flow, ductile extrusion and exhumation in continental collision zones. In: *Channel Flow, Ductile Extrusion in Continental Collision Zones*, ed. R.D. Law, M.P. Searle and L. Godin. *Geological Society of London, Special Publication*, **268**, 1–23.

Hall, R. (2002). Cenozoic geological and plate tectonic evolution of SE Asia and the SW Pacific: computer-based reconstructions, model and animations. *Journal of Asian Earth Sciences*, **20**, 353–431.

Harrison, T.M., An Yin and Ryerson, F.J. (1998). In: *Boundary Conditions for Climatic Changes*, ed. T. Crowley. Oxford University Press, 38–72.

Henry, P., Le Pichon, X. and Goffée, B. (1997). Kinematics, thermal and petrological model of the Himalaya: constraints related to metamorphism within the underthrust crust and topographic elevation. *Tectonophysics*, **273**, 31–56.

Hetényi, G., Cattin, R., Vergne, J. and Nábělek, J.L. (2006). The effective elastic thickness of India plate from receiver functions, gravity anomalies and thermomechanical modelling. *Geophysics Journal International*, **167** (3), 1106–1118.

Hodges, K.V. (2000). Tectonics of the Himalaya and southern Tibet from two perspectives. *Geological Society of America Bulletin*, **112**, 324–350.

Holdsworth, R.E., Stewart, M., Imber, J. and Strachan, R.A. (2001). The structure and rheological evolution of reactivated continental fault zones: a review and case study. In: *Continental Reactivation and Reworking*, ed. J.A. Miller, R.E. Holdsworth, I.S. Buick and M. Hand. *Geological Society of London, Special Publication*, **184**, 115–137.

Huerta, A.D., Royden, L.R. and Hodges, K.P. (1998). The thermal structure of collisional orogens as a response to accretion, erosion, and radiogenic heating. *Journal of Geophysical Research*, **103**, 15287–15302.

Hurford, A. (1986). Cooling and uplift patterns in the Lepontine Alps, South-Central Switzerland and age of vertical movement on the Insubric Line. *Contributions to Mineralogy and Petrology*, **92**, 413–427.

Isacks, B.L. (1988). Uplift of the Central Andean Plateau and bending of the Bolivian orocline. *Journal of Geophysical Research*, **93**, No. 84, 3211–3231.

Johnson, M.R.W. (2002). Shortening budgets and the role of continental subduction during the India–Asia collision. *Earth Science Reviews*, **59**, 101–123.

Khan, S.D., Walker, D.J., Hall, S.A. *et al.* (2009). Did the Kohistan–Ladakh island arc collide first with India? *Bulletin of the Geological Society of America*, **121**, 366–384.

Knipe, R.J. (1990). Microstructural analysis and tectonic evolution in thrust systems: examples from the Assynt region of the Moine thrust zone, Scotland. In: *Deformation Processes in Minerals, Ceramics and Rocks*, ed. D.J. Barber and P.G. Meredith. Unwin Hyman, 228–261.

Kohn, M.J., Catlos, E.J., Ryerson, F.J. and Harrison, T.M. (2001). Pressure–temperature–time path discontinuity in the Main Central thrust zone, Central Nepal. *Geology*, **29**, 571–574.

Law, R.D., Searle, M.P. and Godin, L. (eds.) (2006). *Channel Flow, Ductile Extrusion and Exhumation in Continental Collision Zones. Geological Society of London, Special Publication*, **268**. An excellent up-to-date source for work on channel flow not only in the Himalaya but many other parts of the world.

Le Pichon, X., Henry, P. and Goffée, B. (1999). Uplift of Tibet from eclogites to granulites-implications for the Andean Plateau and the Variscan belt. *Tectonophysics*, **273**, 57–76.

Lister, G.A., Forster, M.A. and Rawling, T.J. (2001). Episodicity in orogenesis. In: *Continental Reactivation and Reworking*, ed. J.A. Miller, R.E. Holdsworth, I.S. Buick, and M. Hand. *Geological Society of London, Special Publication*, **184**, 89–113.

Lui, H., McClay, K.R. and Powell, D. (1992). Physical modelling of thrust wedges. In: *Thrust Tectonics*, ed. K.R. McClay. Chapman and Hall, 71–81.

MacQuarrie, N. (2002). Initial plate geometry, shortening variations and evolution of the Bolivian Orocline. *Geology*, **30**, 867–870.

McCaffry, R. and Nabalek, J. (1998). Role of oblique convergence in the active deformation of the Himalayas and southern Tibet. *Geology*, **26**, 691–694.

Morag, N., Avigad, D., Harlavan, Y., McWilliams, M.O. and Michard, A. (2008). Rapid exhumation and mountain building in the Western Alps: petrology and $^{40}Ar/^{39}Ar$ geochronology of detritus from Tertiary basins of southeastern France. *Tectonics*, TC 2004, doi:1029/2007.

Oncken, O., Chong, G., Franz, G. *et al.* (2006). *The Andes – Active Subducting Orogen*. Springer.

Parrish, R.R., Gough, S.J., Searle, M.P. and Waters, D.J.W. (2006). Plate velocity exhumation of ultra high pressure eclogites in Pakistan. *Geology*, **34**, 989–992.

Ramsay, J.G. (1981). Tectonics of the Helvetic nappes. In: *Thrust and Nappe Tectonics*, ed. M. McClay and N.J. Price. Geological Society of London, Blackwell Scientific Publications, 293–309.

Robertson, A.H.F. (2000). Formation of mélanges in the Indus Suture Zone, Ladakh Himalaya by successive subduction-related, collisional and post collisional processes during Late Mesozoic–Late Tertiary time. In: *Tectonics of the Nanga Parbat Syntaxis and the Western Himalaya*, ed. M.A. Khan, P.J. Treloar, M.P. Searle and M.Q. Jan. *Geological Society of London, Special Publication*, **170**, 333–374.

Robertson, A.H.F. (2007). Overview of tectonic settings related to rifting and opening of Mesozoic ocean basins in the eastern Tethys: Oman, Himalayas and Eastern Mediterranean regions. In: *Imaging, Mapping and Modelling Continental Lithosphere Extension and Breakup*, ed. G.D. Karner, G. Manatschal and L.M. Pinheiro. *Geological Society of London, Special Publication*, **282**, 325–388.

Searle, M.P., Law, R.D. and Jessup, M.J. (2006). Crustal structure, restoration and evolution of the Greater Himalaya in Nepal–south Tibet: implications for channel flow and ductile extrusion of the middle crust. In: *Channel Flow, Ductile Extrusion and*

Exhumation in Continental Collision Zones, ed. R.D. Law, M.P. Searle and L. Godin. *Geological Society of London, Special Publication*, **268**, 355–378.

Searle, M.P., Simpson, R.L., Law, R.D., Parrish, R.R. and Waters, D.J. (2003). The structural geometry, metamorphic and magmatic evolution of the Everest massif, High Himalaya on Nepal–south Tibet. *Journal of the Geological Society of London*, **160**, 345–366.

Sinclair, H.D. and Allen, P.A. (1992). Vertical versus horizontal motions in the Alpine orogenic wedge: stratigraphic response in the foreland basin. *Basin Research*, **4**, 215–232.

Tapponnier, P. and Molnar, P. (1976). Slip-line field theory and large scale continental tectonics. *Nature*, **264**, 319–324.

Tapponnier, P. and Molnar, P. (1977). The relation of the tectonics of eastern Asia to the India–Asia collision: an application of slip line field theory to large scale continental tectonics. *Geology*, **5**, 212–216.

Tapponnier, P. and Molnar, P. (1979). Active faulting and Cenozoic tectonics of the Tien Shan, Mongolia and Bayka regions. *Journal of Geophysical Research*, **84**, 3425–3459.

Trench, A. and Torsvik, T.H. (1992). The closure of the Iapetus ocean and Tornquist sea: new palaeomagnetic constraints. *Journal of the Geological Society of London*, **149**, 867–870.

Vannay, J.-C. and Grasemann, B. (2001). Himalayan inverted metamorphism and syn-convergence extension as a consequence of a general shear extrusion. *Geological Magazine*, **138**, 253–276.

Vernon, A.J., van der Beck, P.A., Persaw, C., Foeken, J. and Stuart, F.M. (2009). Variable late Neogene exhumation of the central European Alps: Low-temperature thermochronology from the Aar Massif, Switzerland and the Lepontine Dome, Italy. *Tectonics*, **28**, 5004, doi:10.1029/2008T Coo 2387.2009.

Williams, H. (1984). Miogeoclines and suspect terranes of the Caledonian–Appalachian orogen: Tectonic patterns in the North Atlantic region. *Canadian Journal of Earth Sciences*, **21**, 887–901.

Windley, B.F. (1995). *The Evolving Continents*, 3rd edition. John Wiley and sons.

6 Lateral spreading of orogens: foreland propagation, channel flow and weak zones in the crust

The plate tectonic model satisfactorily explains the location of orogens but it is less successful in explaining the complex evolution of these belts. For example, it is abundantly clear that with time most orogens grow laterally or horizontally by the foreland-directed propagation of thrust sheets. As a result the locus of thickening of the lithosphere migrates horizontally across the continent. However, plate tectonics does not tell us why. In this chapter we try to give reasons and discuss spreading involving different mechanisms, in all of which gravity forces play an important role.

Plate convergence responsible for orogeny can operate for tens of millions of years, as seen in the Alps or Himalaya, but this is not always the case. Quick orogeny, lasting for as little as 10 Ma, is possible, as is illustrated by the Grampian orogeny that was discussed in the previous chapter. If convergence continues for long periods then the crust may either continue thickening at the existing site of the orogen or spread laterally away from convergent plate boundaries. Presumably, in view of the frequency of the latter, it may be concluded that spreading laterally involves less work than a continued thickening of the initial site of the orogen, with the implication that hitherto undeformed regions are deformed in the later stages of the orogeny. In this concept, already thickened regions do not continue thickening, or more precisely, while they may undergo refolding and consequent thickening, this is not adequate to accommodate the required total contractional strain over the whole orogenic phase.

Cause of orogenic spreading

The usual interpretation of orogenic spreading is that it has to do with the elevation of the mountains. If they grow too high, work must be done against gravity, and the necessary energy provided by horizontal stresses arising from plate convergence may not be capable of supporting the mountains. If so, then further increase in elevation is prevented and the orogen spreads laterally. However, this problem with energy balance can be avoided if erosion keeps pace with the increases in elevation due to horizontal shortening. Under such conditions the height of the mountains is stable and there is no gain in gravitational energy. In recent orogens there are now considerable data on erosion rates and strain rates, perhaps enough to attempt an answer to the point. For example, in the Himalaya a shortening rate of *c.* 1.5 cm/a (less than the rate of plate convergence) has operated for at least the past 20 Ma. However, it would seem that erosion rates reached this level only for short time spans, so that in general, strain rate greatly exceeded the erosion rate, which is probably generally no higher than 1 mm/a (see Chapter 8). Unless other processes operated, therefore, the conclusion is that in general it is unlikely that erosion can maintain a given surface elevation, that is one that satisfies the force balance within an evolving orogen. By 'other processes' we are referring to tectonic denudation: that is, removal of cover by low-angle normal faulting. Some examples of orogenic spreading follow.

The Mesozoic Canadian Rocky Mountains show that after the initiation of plate convergence the strain spread eastwards across the North American continent in Jurassic and Cretaceous times, first forming the western high-grade orogenic belt in British Columbia and then migrating eastwards into the previously undeformed so-called Rocky Mountains foreland thrust belt. The thrusting in the foreland thrust belt is clearly sequential; therefore during the later stages the more westerly parts transmitted horizontal compressive stress but did not undergo large strain. In other words, the thrusts in the west were not reactivated and out-of-sequence thrusting is not recorded. In the Himalaya the MCT sheet is likely to be at least 300 km wide, and its leading edge and external klippen almost reach the present mountain front. Post-MCT events consist of accretion of crustal slices onto the thrust, thus thickening the orogenic wedge, perhaps as a response to volume loss by erosion of the MCT sheet. An Yin (2006) suggests that the Lesser Himalaya underwent deformation prior to the emplacement of the MCT sheet, but undeniably the main thrust events in the Lesser Himalaya are post-MCT.

A rather different type of orogenic spreading is illustrated by the Moines in NW Scotland where two orogenies are involved: a mid Ordovician Grampian orogeny (arc–continent collision) followed a few Ma later by the Scandian orogeny (continent–continent collision) during which first the Naver thrust formed, and then, by foreland-directed thrust propagation, the Moine thrust. The Scandian orogeny extended further west into the Laurentian foreland than did the Grampian orogeny, because the Moines in the belt between the Naver thrust and the Moine thrust seem to have escaped deformation during the Grampian event. If so, the Naver thrust must have detached near to the western front of the Grampian orogeny. In the Alps, lateral spreading is shown by the late stage, Pliocene-aged, Jura mountains which form the most northern part of the Alps. The Miocene-aged Helvetic Alps developed under a cover or carapace of the thin northern part of the Penninic Alps and may not have represented a major spreading event. In the Andes, lateral spreading is indicated by the succession of foreland basins in the central segment.

Orogenic collapse

The examples of the spreading of orogens so far discussed have focussed on contractional tectonics, but spreading by extension should also be mentioned. One aspect of this feature is orogenic collapse in which a thickened orogen returns to normal thickness by extension and the operation of normal faulting. Examples of collapse are Tibet and the Basin and Range in the North American Cordillera, both of which show extensional tectonics, Basin and Range type and metamorphic core complexes, synchronous with or succeeding a compressional tectonic regime. The Basin and Range is characterised by thin crust

< 25 km thick, high heat flow and a high Bouguer gravity anomaly and the Basin and Range topography. The overall extension of the upper crust is 300–400% occurring in the past 16 Ma. Extension rates vary from 5 to 13 mm/a. In the North American Cordillera generally the start of extension is diachronous and varies from 50–55 Ma in the Omineca belt including the Shuswap Complex, one of the metamorphic core complexes described below, situated to the west of the Canadian Rockies foreland fold and thrust belt, to 16 Ma in the Basin and Range. The extension was associated with phases of andesite, rhyolite and basaltic magmatism.

A comprehensive explanation for the extensional regime in the western USA is not yet available. Many believe that the extension followed contractional tectonics. As mentioned in Chapter 2, Platt and England (1993) invoked the Houseman–England model of convective removal of the lower mantle lithosphere following thickening. Others suggest that the crustal extension and gravitational collapse was driven by buoyancy forces during continued convergence and compression across the orogen including in the foreland fold and thrust belt. Similarly it is not clear how the extension was controlled by plate motions in the western Pacific. The North American continent overrode the Pacific–Farallon plate junction in early to middle Cenozoic times. The change from shallow-dipping to steep subduction of the Pacific Plate took place between 70 and 40 Ma, and may have led to thermal weakening of the thickened lithosphere and extensional stresses.

Metamorphic core complexes

Another example of spreading comes from the metamorphic core complexes of which some thirty were first recognised in the western USA, Canada and Mexico located between the belt of granite batholiths to the west and the foreland thrust belt to the east. The largest core complex is the Shuswap Complex in the Omineca belt of British Columbia. Crustal shortening occurred in the Jurassic and Cretaceous after which extension occurred in the Early Cenozoic.

The core complexes provide evidence of extension of the lithosphere of the western USA in late Mesozoic and early Cenozoic times. The ages of the core complexes vary from place to place between 48 and 14 Ma, the age range reflecting the migration of magmatic activity. High-grade metamorphic rocks and plutons have been exposed during extension and tectonic denudation of their cover. The Cordilleran core complexes display Mesozoic and Cenozoic ages, therefore the metamorphic rocks do not constitute a simple basement: they have probably been reactivated during Cenozoic extension. The metamorphic core is separated from the unmetamorphosed or low-grade cover by a low-angle detachment which shows the characteristics of a normal fault. The faulting is polyphasal involving earlier ductile slip along mylonite zones and later brittle deformation. Controversy has centred on the nature of the detachment which is seen by some as a rheological boundary between ductile deformation below and generally brittle deformation above the detachment which may have great slip displacement. For others (e.g Wernicke and Burchfiel, (1983), the detachment is a major zone of simple shear along which displacements of many tens of kilometres have occurred. During simple shear, steep faults may have rotated into a low dip.

The core complexes in the western USA are attributed by Dickinson (2009) to roll-back of the subducted Farallon oceanic plate, while later (post-17 Ma) Basin and Range block faulting may have reflected torsion in the continental crust derived from transform slip on the San Andreas and related faults. There is further discussion of core complexes in Chapter 7.

Channel flow

A new concept of orogenic spreading is called 'channel flow', which in the Himalaya refers to the nature of the flow occurring during the emplacement of the Main Central Thrust sheet. Rather different proposals of mid or lower crustal flow were made for the Basin and Range where a low viscosity channel has flowed in response to extensional tectonics. In the Himalaya, Zhao and Morgan (1985) were the first to suggest that subducted Indian crust had been injected into the weak lower crust of Tibet, the injection causing elevation of the Tibetan Plateau by hydraulic pressure. All this work is based on the notion that the lithosphere has not behaved as a uniform body; instead it contains certain layers which undergo viscous flow. Beaumont and Jamieson and others (2004) introduced the concept of channel flow in the Himalaya and triggered an outpouring of papers admirably reviewed by Nigel Harris (2007) and by Godin *et al.* (2006).

The word 'channel' conveys the idea of a flowing body of rock which is confined by rigid walls. In the Himalaya it means a viscous, fluid-filled channel lying between two rigid sheets. The upper bounding surface for the channel is the South Tibetan Detachment, a reactivated south-verging thrust, later a normal fault with a down-to-the-north sense of slip. The lower bounding surface is the MCT which has a top-to-south sense of slip. The mechanics of channel flow requires the apparent near-synchronicity of the activity on the MCT and the STD, and a wealth of data confirms this. Both these faults have been traced for about 2000 km along strike which means that the supposed channel is enormous (Figs. 6.1, 6.2).

The date of initiation of the MCT is unknown but is generally thought to be late Oligocene to early Miocene, with perhaps the main slip along it lasting until *c.*16 Ma. The ductile motion on the STD normal or stretching fault is dated between *c.*23 and 15/16 Ma in the west, central and east Himalaya, but *c.*14–10/11 Ma in Bhutan. Later brittle slip on the STD is dated at 15–13 Ma. These faults post-date Eohimalayan crustal thickening and regional high-grade Barrovian metamorphism but are in part coincident with peak sillimanite grade metamorphism and partial melting of the lower part of the MCT sheet. Estimates of the amount of slip on the STD vary from 180 km in the Everest area to 60–40 km in the western Himalaya. The estimates of slip on the MCT range from a minimum of 100 km to 200 km. The along-strike ends of the channel are uncertain but probably occur near to the western and eastern syntaxes. In the west there is a junction between the MCT and STD and this may mark the western limit of the channel. In addition the thickness of the channel probably changed over time, as evidenced by the upward propagation of the normal faulting and a downward propagation of thrusting in the MCT

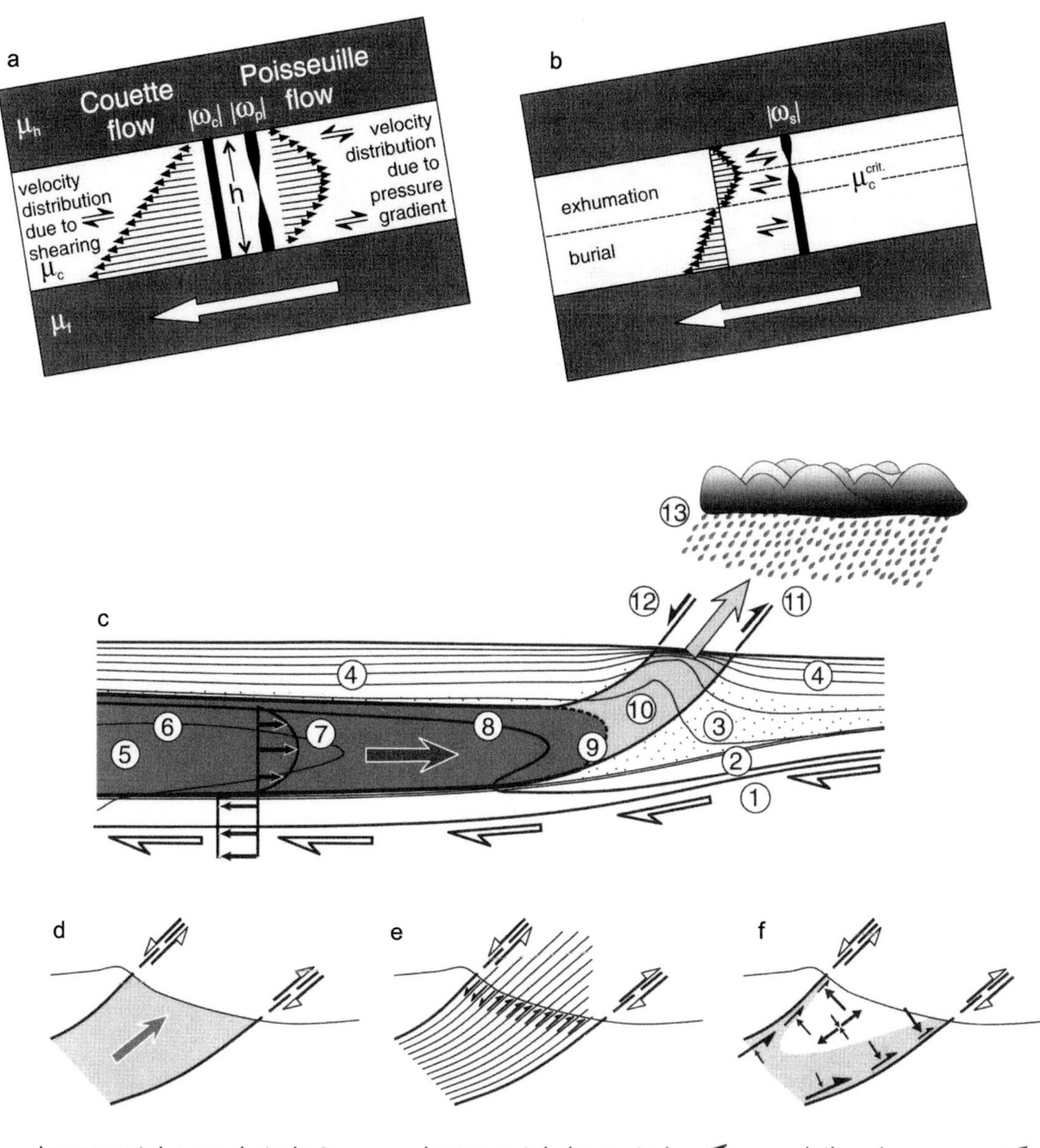

Figure 6.1 Flow patterns in a channel of width *h*. a, Viscosity is shown by μ_c and is lower than in the walls to the channel. The end-members of flow are (left) Couette flow with velocity caused by shearing and (right) Poiseuille flow with velocity caused by pressure gradient within the channel. ω_c is the velocity in pure Couette flow, ω_p is the velocity in pure Poiseuille flow. b, For a given velocity of the subducting plate and channel width there is a critical velocity of the channel material which the Poiseuille flow will counteract and cause return flow and therefore exhumation of the channel. The part of the channel which remains dominated by the induced shear will continue being underplated. ω_s is the vorticity in a hybrid channel flow. c, Application of the model to the Greater Himalayan Sequence to show the effect of climate and erosion on the extrusion and exhumation of the channel. 1, Lithospheric

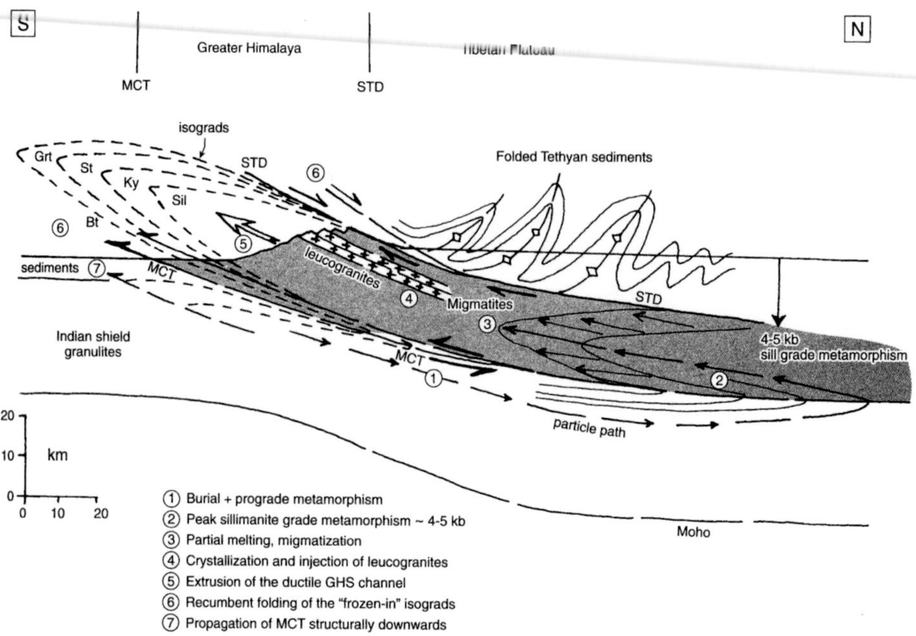

Figure 6.2 Channel flow in the Himalaya. From Searle (2007), published by permission of the Geological Society of America.

zone. The postulated extrusion of the MCT sheet between two faults showing contrary motions is supported by Vannay and Grasemann (2001) who interpreted the strain in the MCT sheet as general shear – that is, combined pure shear and simple shear. Quartz fabric work by R. Law (see Searle *et al.*, 2003 and 2006) on rocks from the MCT sheet confirmed the general shear: the sheet was thinned substantially by vertical shortening as it moved, and the pure shear component may be as much as 40%.

Extrusion and channel flow

Two terms are used with reference to the Himalaya: extrusion and channel flow. They are not synonymous. Extrusion can refer to the movement of rigid blocks such as has been suggested by Molnar and Tapponnier (1975), Tapponnier and Molnar (1979) and Tapponnier *et al.* (1986) with reference to the extrusion and expulsion of rigid blocks of Asian

Caption for Figure 6.1. (*cont.*) mantle; 2, lower crust; 3, mid-crust; 4, upper crust; 5, weak crustal rock; 6, isotherms; 7, schematic velocity profile during return channel flow; 8, 750 °C isotherm structurally below which partial melting starts; 9, rheological tip of the channel; 10, extruding crustal block; 11, lower shear zone of the extruding block; 12, upper shear zone of the extruding block; 13, focused surface denudation. d–f, Possible strain distribution in an extruding block. On right is general shear with pure and simple shear components. From Godin *et al.* (2006), published by permission of the Geological Society of London.

crust along strike-slip faults in response to the indentation process. Extrusion of the channel occurs when it meets the current mountain front where erosion controls the rate of flow and exhumation of the channel. However, channel flow involves a markedly different ductile rheology with viscous flow within a fluid-filled channel. The weak layer flows laterally as a result of high temperatures in the Greater Himalayan slab and partial melting in the lower part of the slab which led to the generation of the Himalayan leucogranites. According to Harris (2007), a melt fraction, F, of $c.0.4$ reduces viscosity by $c.50\%$ and lower melt fractions, e.g. $F = 0.07$, give a marked loss of strength. Using the velocity structure in the low-velocity zone beneath southern Tibet gives $F = 0.07$–012 and using magnetotelluric data gives 0.05–0.14. These factors have caused a reduction in the viscosity of the Greater Himalayan slab to an estimated value of $c.10^{19}$ Pa s. However, the channel is not composed of completely fluid or molten rock as seen in lavas which have much lower viscosities.

Couette and Poiseuille flow

The subject of fluid dynamics provides a theoretical basis for channel flow. Two types of channel flow are recognised: *Couette flow*, in which the induced simple shear across the channel produces a uniform vorticity across the channel, and *Poiseuille flow*, or 'pipe flow', which occurs between stationary plates and the highest velocities produced by the pressure gradient are in the centre of the channel. The shear is in opposite senses for the top and bottom of the channel. Maurice Couette and Jean Poiseuille were French scientists working in the nineteenth century. Poiseuille was interested in the flow of blood through narrow channels, leading him to devise a law for the flow of liquids in pipes – he also gave part of his name (Poise) to a unit of viscosity.

Driving force for channel flow

According to the channel flow model, gravity was the driving force for the movement on the MCT and the Oligocene–Miocene (NeoHimalayan) phase of crustal thickening in the Himalaya. Gravity flow is directed away from the Tibetan Plateau and towards the foreland, that is, contrary to the north-directed India–Asia plate convergence which continued during the south-directed gravity-driven flow. The south-directed flow in the channel is due to the horizontal gradient in lithostatic pressure produced by the high elevation of the Tibetan Plateau in relation to the Himalayan foreland. As shown in Chapter 5, a high elevation for at least part of the plateau is likely to have existed by early late Oligocene or Miocene times, which is what the gravity-driven model requires. Finite displacement in the channel depends on its geometry, on the viscosity, and on the displacement rate of the bounding plates. Channels can tunnel their way through the crust without breaking surface because they find their way along zones of lowest strength (Fig. 6.3).

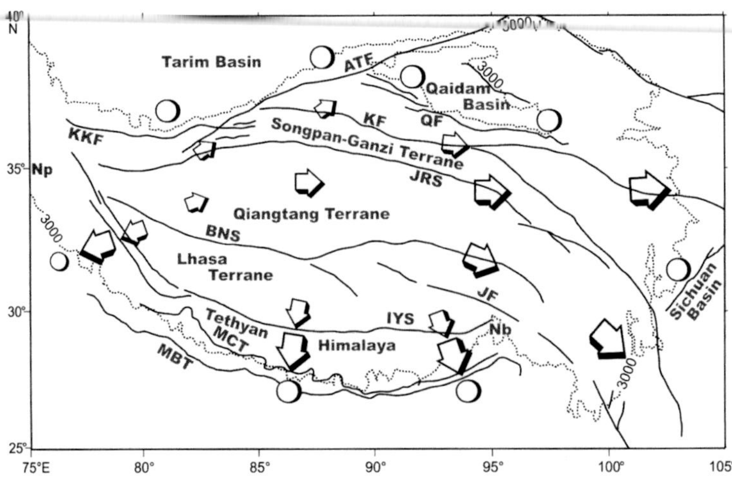

Figure 6.3 Directions of middle/lower crust channel flow in Tibet (open arrows) and regions of no flow (open circles). The gravitational potential energy drives the Poiseuille channel flow between subducting Indian lower lithosphere and the brittle upper crust of the Tethyan and south Lhasa terrane. The southern boundary channel flow is the Himalayan topographic front. The northern boundary may be the Karakoram–Jiali fault system or Banggong–Nujiang suture in the west and the Jiali fault system in the east. North–south compression and east–west extension drives Poiseuille/Couette flow eastwards beneath the Qiangtang terrane. Flow bifurcates north and south of the rigid Sechuan terrane. ATF, Altyn Tagh fault; BNS, Banggong suture; JRS, Jinsha river suture; IYS, Indus–Yarlung suture; JF, Jiali fault; KKF, Karakoram fault; KF, Kun Lun fault; Np, Nanga Parbat; QF, Qaidam fault; Nb, Namche Barwa. From Klemperer, in Godin *et al.* (2006), Figure 6, published by permission of the Geological Society of London.

On the other hand exhumation of the channel will take place if erosion is focussed on the front of the channel, in which case there is a balance between the rate of extrusion and orographically controlled high rainfall and erosion, such as occurs today on the southern slopes of the High Himalaya. Note that those slopes did not exist during channel flow, but no doubt there was a comparable mountain front during that flow.

Channel flow and erosion

An interesting feature of the channel flow hypothesis is the link with erosion and precipitation. The extrusion of the Greater Himalayan Sequence was vigorous in the early Miocene, implying mountain uplift, and the prediction is that erosion would be enhanced then. The focussed precipitation would serve as a driving force for the exhumation of the low-velocity channel. The summer monsoon is a particular example of focussed precipitation at the present topographic front of the Greater Himalaya. The Tibetan Plateau is on the leeward side of the Himalaya and therefore escapes most of the monsoon rain. In consequence, erosion from the Tibetan plateau is low and most of the drainage is internal. The maximum rainfall occurs in a belt stretching about 30 km south from the Main Central

Thrust, as pointed out by Wobus *et al.* (2003). In other areas, too, the maximum rainfall occurs immediately south of the MCT in a belt bounded on the south by a recent thrust (see also Thiede *et al.*, 2004). Apatite fission ages indicate very recent exhumation, and these ages can be correlated with rainfall patterns. Another point which supports the linkage of climate and tectonic activity is that the monsoon increases in intensity from west to east. Rainfall increases from *c.*5 cm/a in the eastern Himalaya to *c.*450 cm/a in Bhutan. Those who argue that precipitation drives channel flow would regard the decrease in Ar^{40}/Ar^{39} mica ages from west to east, from 22–18 Ma in the west to 13–11 Ma in the east in Bhutan, as evidence that this reflects an increase in exhumation rates over a 10–20 Ma period (see Harris, 2007). The inference is that exhumation of the channel has lasted longer in the eastern Himalaya.

If the ages of the Himalayan leucogranites mark the time of active flow, then according to this inference it would be expected that younger granites occur in the east. The data support this because whereas the leucogranites of central Himalaya range in age from 23–19 Ma those in the extreme east are as young as 12 Ma.

Low-viscosity layers in the crust

Although this anticipates parts of Chapter 10, it is appropriate to discuss here something about the strength of the lithosphere that is implied by the channel flow hypothesis. The model calls for a weak low-viscosity layer in the mid–lower crust, in accordance with the increase of temperature with depth. This means a layered crust and the possibility that there can be mechanical detachment of the ductile weak layer. This is opposed to the England and Houseman model (Chapter 5) which proposed a thin viscous sheet in which continuous deformation occurs, for example homogeneous thickening of the whole lithosphere. Alternatively there is a model in which thickening only takes place in the crust while the mantle is not thickened – a non-continuum model. The channel model calls for a strong mantle and a strong upper crust. This is the so-called 'jelly sandwich' model for lithospheric rheology.

Tests for the channel flow model

Although this does not prove it to be correct, the channel flow model is consistent with the available structural, metamorphic and geochronological data of the Himalaya. For example, the MCT has long been regarded as a sheet which moved while still 'hot', although the precise temperatures prevailing during slip are not entirely clear. Hubbard (1989) has described syn thrust hornblende suggesting that temperatures during slip were over 500 °C, and as mentioned above sillimanite metamorphism was occurring during slip on the MCT. The Miocene leucogranites, which comprise only a few per cent of the total volume of the Himalaya, in the central sector have a range of ages from *c.*24 Ma to about

17 Ma, that is roughly syn MCT slip. The leucogranites occur at high levels in the MCT sheet and have formed from melts which rose through several kilometres of the Greater Himalayan slab. This is essential evidence supporting the presumed low viscosity of the slab at the time of channel flow. Perhaps the model falls down when we consider the geology of the external klippen of the MCT sheet: for example, at Kathmandu neither the STD or leucogranites are found. This probably means that the STD has cut up section to the south and therefore overlies the rocks in the external zone. Perhaps the external zone allows observation of a part of the MCT sheet which was pushed ahead of the main channel which carries the melts and leucogranites.

The channel flow model as set out by Beaumont *et al.* (2004) implies major subduction of Indian upper/middle crust beneath Tibet which was then heated prior to its return journey as a channel to the Himalaya. The date of leucogranites seen in the Greater Himalaya marks the beginning of that journey. The 'bright spots' of the Tibetan crust may represent melting and possibly a new channel in the incubation stage. This view has been contested by DeCelles *et al.* (2002) who maintained that the Indian upper/middle crust did not migrate across or become subducted, north of the Suture, rather it has been scraped off and incorporated in the thrusts of the Greater Himalaya.

Clearly the channel hypothesis requires the existence, at least since the early Miocene, of a continuous low-angle detachment under the Himalaya and southern Tibet. This is confirmed by the INDEPTH seismic profile which suggested that at present the basal detachment of the Himalaya continues as a gently dipping surface north of the Indus suture and beneath the Tibetan plateau for about 300 km. However, Klemperer (2006) has disputed the view that subduction was flat throughout the whole of the Cenozoic, and his view is supported by the existence of early Cenozoic eclogites in the Indian plate which suggest steep subduction of Indian crust at an early stage in the collisional process.

Klemperer suggested an early phase of steep subduction which was ended by slab break-off, after which flat subduction took over. This is not a problem for the model outlined above, provided that the change from steep to flat subduction occurred in late Oligocene or early Miocene times when the supposed channel started to move south towards the Himalayan region.

Whether or not the impressive evidence of high temperatures at depth in Tibet means there is a new channel in the waiting, there is no doubt that there is something of importance there. Evidence for the elevated temperatures of the Tibetan mid crust comes from heat flow measurements at the surface which are well above the normal for continental crust. What is not clear is whether the heat flow data is mostly confined to the graben and normal faults which cut across the Tibetan Plateau, or whether the evidence can be taken to show that all the Tibetan mid/lower crust is hot. The evidence of a hot mid/lower crust comes from the line of the INDEPTH profile which may not be representative of the whole of Tibet. Differences in the thermal state of the Tibetan lithosphere are well known: for example North Tibet, that is the part north of the Banggong suture, has a warm mantle whereas south Tibet has a cooler mantle, the temperature difference being 200–300 °C. Evidence of elevated palaeotemperatures under Tibet is given by Ding *et al.* (2007) who reported xenolith evidence of high temperatures in the buried lower crust of northern Tibet and the southern Qiangtang terrane, the heating being dated at or by 28 Ma.

They attribute the heating to the Oligocene underthrusting of Indian lower crust and lithospheric mantle which had reached some 200 km north of the Indus Tsangpo Suture by the Oligocene. This underthrusting and consequent crustal thickening was responsible for the calc-alkaline magmatism in southern Tibet. A different cause is given for the ?Oligocene–Miocene (ultra) potassic magmatism in northern Tibet, namely northward subduction of Asian mantle lithosphere beneath the Banggong suture (reactivated in the Cenozoic as a thrust system) and southward subduction of Asian mantle along the Jinsha suture, the Fenghuo–Ganzi thrust belt (Fig. 5.9a). Between the subduction zones there appears to be an upwelling of hot asthenosphere which resulted in melting of the lithosphere and volcanism.

Deep crustal flow in Tibet

This has been proposed for other parts of the Himalayan–Tibet orogen, for example beneath eastern Tibet, Clark and Royden (2000) (see also Royden *et al.*, 1997; Klemperer, 2006) invoked a lower crustal channel some 15 km thick, involving flow of a Newtonian fluid (Poiseuille flow) (Fig. 6.3). This corresponds to the jelly sandwich model for the strength of the lithosphere described in Chapter 10. As before the channel flow is in response to a lateral pressure gradient between the high Tibetan Plateau and the surrounding lower ground. It is assumed that the plateau has been raised in the past 20 Ma. The presence of a weak crustal zone under Tibet is shown by the zone of low velocity and high electrical conductivity identified in the INDEPTH profile (Chapter 5).

Two types of plateau margin are recognised: one with a low topographic gradient which occurs when the lower crust beneath the margin is weak (viscosity is 10^{18} Pa s) and the other with steep topographic gradient when the lower crust is strong and lateral flow is inhibited (viscosity is 10^{21} Pa s). In the low gradient margin, the lower crust escapes from beneath the plateau and may flow for as much as 1000–2000 km until it encounters areas of strong crust such as the Tarim Basin where it flows around them. Clark and Royden attribute changes in drainage patterns in southeastern Tibet to Neogene plateau uplift. In contrast to the channel flow in the Himalaya, direction of flow of the lower crust is generally away from the central plateau to the northwest, northeast, east, southeast and south.

There are some questions about this model. Firstly, does it conflict with the channel flow model of Beaumont and others in which channel flow is entirely southwestward towards the Himalayan foreland? Secondly, does the Beaumont channel flow correspond to the ‘steep margin’ case outlined by Clark and Royden (2000)? If so, the viscosity would therefore be much higher than envisaged in the analysis by Beaumont. Thirdly, the Beaumont model involves channel flow in the mid–upper crust, not in the lower crust as suggested by Clark and Royden. Their channel model corresponds to the tunnelling channel (subsurface) described by Beaumont, but in the Himalaya the channel has reached the surface; thus we can stand on the channel, and it is fairly well dated by stratigraphy. Another possible objection to the Royden model is that it is now recognised that differing structural regimes occur at different levels in the Tibetan Plateau. At high levels, on the plateau, extensional normal faulting prevails in the recent extensional tectonism, but at low levels as seen at the eastern margins of the Plateau thrusting occurs, perhaps indicating a lateral flow. The conflict between the channel flow models for eastern Tibet and for the Himalaya may be more apparent than real. Perhaps Tibet is like a large misshapen bun with jam filling, from the middle of which material is extruded in all directions when the bun is squashed. However, the space problem remains: did the channel make space for itself like an intrusive sill or did it push material away ahead of it? Unless there is a loss of material there would be a thickening of the lithosphere for a distance of 1000–2000 km outside the plateau. Klemperer, using seismic (seismic reflection, receiver functions, S- and P-wave analyses) and magneto-telluric experiments in Tibet along several profiles, has resolved some of these problems but more importantly confirms the existence of a fluid-controlled weak zone in the Tibet mid–lower crust. First and foremost, Klemperer's contribution is an advance because it draws on evidence from across the plateau rather than one profile, but these techniques have also thrown new light on the composition, temperature and fluid content of the Tibetan lithosphere (Fig. 6.2). The geophysical data strongly support the presence of melts in the Tibetan mid/lower crust. Using Poiseuille and Couette flow models both separately and combined, he is able to present a picture of deep channel flow over a large part of the Tibetan Plateau.

The flow is driven by gravitational pressure gradients already outlined (so this is the squashed bun model). Klemperer pointed out that the differentials of viscosities invoked by Clark and Royden are more important than the actual values given. The movement vectors shown in Figure 6.2 have been partly derived from GPS surveys as well as the well understood point that while N–S compression dominates along the southern fringe of the plateau and within the Himalaya, across the plateau the current stress system is E–W extension, hence the easterly flow shown on the figure.

Is channel flow a special case confined to the Himalaya?

In order to answer the question above, the following assumptions must be made for channel flow in an orogen. They are (a) a high plateau forming a hinterland which gives rise to lateral gravity gradients; (b) motion on a lower thrust and a higher normal fault, as such a mechanical system allows for the flow and extrusion of the channel; (c) weakened, low-viscosity, hot rocks; (d) a jelly sandwich model for continental lithosphere (see Chapter 10), that is one allowing viscous flow between strong upper crust and mantle. Finally, although this is perhaps not essential to the model, it is presumed that Indian mid/upper crust has underplated Tibetan crust and was then heated up to partial melting conditions. Then it was returned to the Indian sector during channel flow along with the melts which are seen as the early Miocene leucogranites of the Greater Himalaya. Can these conditions be met anywhere else in the world?

Brown and Gibson (2006) argued that the Canadian foreland thrust belt was driven by channel flow occurring in the high-grade western part of the Cordilleran orogen in British Columbia. They cite the high pressures detected by geobarometric studies in the high-grade rocks in British Columbia and western Alberta as evidence of substantial crustal thickening and high elevation, which promoted gravity flow to the east driving the foreland thrusting. Williams and Jiang (2005) proposed channel flow in the high-grade part of the Canadian Cordillera. They dispute the common assumption that the thrust-dominated regime of the Rockies foothills can be applied to the high-grade metamorphic area (the Purcell and Monashee mountains etc.) that exists to the west of the foreland thrust belt. Instead they invoke ductile deformation during which foliations and fold axial planes are rotated into a near-horizontal attitude. Thus the crustal section exposed in the west is a huge ductile shear zone, several kilometres thick, in which although discontinuities occur they cannot be compared to the discrete thrusts seen in the foothills belt. They considered channel flow to be probable in the high-grade belt but it is not clear that it passes the tests set out above for channel flow. With more certainty it is obvious that the above stated criteria for channel flow are absent in the Alpine orogen. There is no STD equivalent coupled to the thrusting as in the Himalaya, and no large high plateau in the hinterland comparable to Tibet. The hinterland is found in the Southern Alps where the crustal thickening is mostly thin-skinned.

Channel flow in the Andean orogen has been invoked by Gerbault *et al.* (2005), with transfer of weak lower crustal material from the Puna Plateau in the south to the Altiplano in the north. The gravity-based driving force arises from the fact that the Puna Plateau is *c.*800 m higher than the Altiplano. With a minimum viscosity of 10^{19} Pa s, this flow is calculated to have reached 1 cm/a, and it resulted in the transfer of lower crust, with a thickness of more than 50 km into the Altiplano crust, in 5 Ma. From seismic attenuation evidence, it is suggested that whereas in the Puna Plateau the crust was underlain mostly by asthenosphere, in the Altiplano the crust is mostly underlain by thick lithospheric mantle. On the other hand, McQuarrie (2002) has proposed flow of lower crust from the Altiplano to the Puna Plateau. She did this to explain the high plateau of the Puna and the 50 km thick crust there. These models are reminiscent of proposals of the lower crustal extrusion eastwards from beneath the Tibetan Plateau (Clark and Royden etc.), but in the Andes there is no evidence of a gradient in the gravity potential comparable with that generated by the huge high plateau of Tibet, such as is needed to drive the longitudinal channel flow. In older orogens it is even more difficult to test the channel flow model because too much evidence has been lost. Merschat *et al.* (2005) proposed channel flow in the Piedmont zone of the Appalachian orogen, the northwesterly channel flow being resisted by the rigid block of the Brevard zone. The criticism of this suggestion must be that there is no evidence of a normal fault above the Piedmont zone which is comparable to the South Tibetan Detachment, nor is there evidence of a high plateau to provide the gravity-controlled flow. The plateau would have existed on the western coast of Africa but there is no sign of large crustal thickening in the Mauritanides, which orogen was adjacent to the Appalachians prior to the opening of the Atlantic. The conclusion must be that the jury is still out on the question of whether the channel flow model is widely applicable to orogenesis through time. In the next chapter there is further discussion of the channel flow hypothesis.

Further reading

Cobbold, P.R., Rossello, E., Roperch, P. *et al.* (2007). In: *Collision Zones*, ed. A.C. Ries, R.W.H. Butler and R.H. Graham. *Geological Society of London, Special Publication*, **272**, 321–343.

Dewey, J.F. and Lamb, S.H. (1992). Active tectonics of the Andes. *Tectonophysics*, **205**, 79–89.

Hodges, K.V., Ruhl, K., Schildgen, T. and Whipple, K. (2004). Quaternary deformation, river steepening, and heavy precipitation at the front of the Himalayan ranges. *Earth and Planetary Science Letters*, **220**, 379–389.

Jamieson, R.A., Beaumont, C., Medvedev, S. and Nguyen, M.H. (2004). Crustal channel flow: 2. Numerical models with implications in the Himalayan-Tibetan orogen. *Journal of Geophysical Research*, **109**, BO6407, 1–24.

Royden, L.R. (1993). The tectonic expression slab pull at continental convergent boundaries. *Tectonics*, **12**, 303–325.

Trewin, N.H. (ed.) (2008). *Geology of Scotland* 4th edition. Geological Society of London.

Wdowinski, S. and O'Connell, R.J. (1989). A continuum model of continental deformation above subduction zones: applications to the Andes and the Aegean. *Journal of Geophysical Research*, **94**, 10331–10346.

7 Metamorphism in orogeny

Introduction

Metamorphism is a fundamental process affecting the crust and lithosphere (Miyashiro, 1961, 1973; Ernst, 1975; Brown, 2009). Its significance for the crust is clearly demonstrated by any geological map of the continents, vast areas of which are underlain by metamorphic rocks of various types, from the slates and schists of such regions as the Alps and Appalachians to the grey gneisses of Scandinavia, Canada and Antarctica. Metamorphic rocks are those rocks produced as a result of changes in the physical conditions affecting pre-existing, or precursor, rocks at some stage of their residence in the deep Earth. Crustal protoliths, such as sediments, volcanics, intrusive igneous rocks and pre-existing metamorphic rocks, will undergo metamorphism through the operation of several processes acting on the rocks as they are subjected to changes in pressure and temperature over time, and a variety of products will result. Similarly, precursor mantle-derived ultramafic rocks will be metamorphosed in response to changes in physical conditions that accompany their incorporation in orogenic systems. The metamorphic processes affecting protoliths include mineral recrystallisation, reactions involving minerals (solid–solid reactions), reactions involving minerals and fluids (e.g. dehydration reactions) and at high temperatures reactions involving the production of partial melts and related mineral–melt interactions resulting from melt migration. Mineral dissolution and precipitation may also occur in response to the access of fluids, and those crystal-plastic processes activated by and facilitating deformation may also proceed contemporaneously with metamorphic reactions. Provided the timescale of metamorphism is long enough for equilibrium to be attained or closely approached, the mineralogies of the products will depend on the physical conditions (*P*, *T*), the presence and nature of any fluid phase, and the compositions of the protoliths. The textural features of the products will depend in a complex way on the physical conditions, mechanisms of mineral reactions, and presence and nature of fluids and melts, as well as on the timing and amount of applied stress.

Understanding this complexity and utilising this to invert the metamorphic observations that can be made on individual rocks – including the mineralogy and mineral chemistry, microtextures, age and isotopic information – as well as such observations extended to whole regions of metamorphism, is the principal challenge of metamorphic geology. As shown in Figure 7.1, the realm of pressure–temperature (*P–T*) conditions now recognised as being embraced by metamorphism associated with orogeny is very large: from 4 to nearly 60 kbar, or about 10 to nearly 200 kilometres in depth, and from 250 °C to

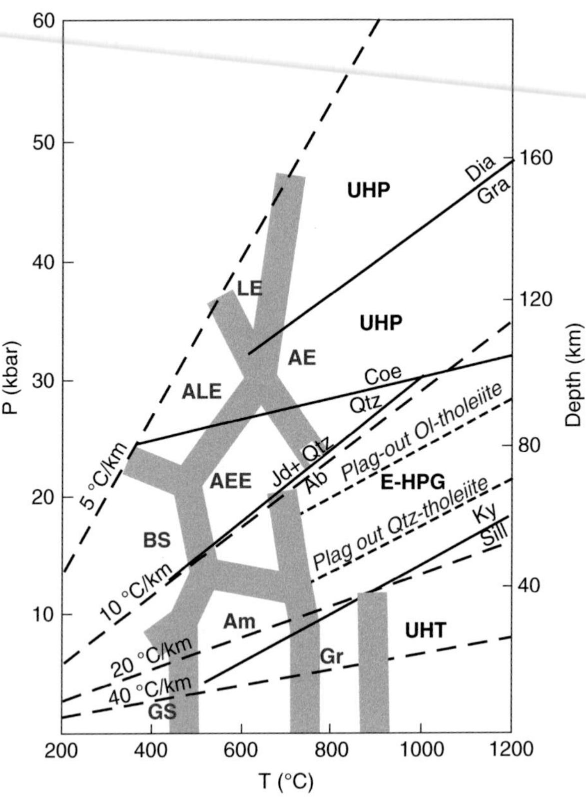

Figure 7.1 The realm of metamorphism depicted in terms of pressure (*P*, kilobars) and temperature (*T*, in degrees Celsius), or *P–T* space. Facies and facies boundaries (grey lines) and positions of the principal mineral reactions are based on Brown (2007). Plagioclase-out lines for mafic rocks are based on Green and Ringwood (1967). Facies labels are as follows: AE, amphibole eclogite; AEE, amphibole–epidote eclogite; ALE, amphibole–lawsonite eclogite; Am, amphibolite; BS, blueschist; Gr, granulite; GS, greenschist. The broad division of metamorphism according to generalised *T*-depth gradients (dashed lines) is based on Brown (2007), Harley (1998a) and Chopin (2003). UHP, ultrahigh-pressure metamorphism (high-*T* end of the high *P*/*T* facies series); E-HPG, eclogite–high-pressure granulite metamorphism (high-*T* end of the medium *P*/*T* facies series); UHT, ultrahigh-temperature metamorphism (high-*T* end of the low *P*/*T* facies series).

over 1100 °C. These *P–T* conditions, mapped out on the basis of analysis of hundreds of metamorphic belts or regions, form arrays between d*T*/d*z* gradients of 5 °C/km and 50 °C/km, and hence must reflect significant differences in the thermal structures and tectonic evolutions of the orogenic systems or stages to which they relate. Prior to embarking on a description and synthesis of current concepts on metamorphism and orogeny, we will consider the historical development of the subject and its methodologies in order to provide a context for the present developments.

The history of the study of metamorphism may be split into four broad phases, the earlier of which continue as underpinning for the later ones. The first phase involved exhaustive documentation in order to establish consistencies and ascertain what observable features – in this case textures and mineralogy – could be employed to introduce a systematic approach. The second phase utilised theoretical developments, particularly in thermodynamics and chemical equilibria, coupled with experimentation to allow the observations to be placed into a context involving the key variables – in this case pressure and temperature – but without similarly detailed consideration of the temporal evolution of those conditions and the implications of such evolutions. The third phase improved the approaches to defining the pressure–temperature domain of metamorphism, integrating the roles of other variables and processes (e.g. fluid flow, melting), and coupled this with the gradual introduction of approaches to address time-dependent processes, elucidate *P–T*–time histories, and allow consideration of metamorphic records in terms of the temporal evolution of orogenic systems. The current phase involves a reconsideration of the metamorphism in more holistic terms and its integration into a dynamic frame of reference in which its systematics are visualised as resulting from the interconnected and time-dependent relationships between tectonics and physical-chemical processes. The construction of increasingly sophisticated thermo-mechanical models to describe crust and lithosphere behaviour during convergence and collision, and use of metamorphic records as tests of the models, have been central to development of current views of how metamorphism is linked to orogeny and, conversely, how the evidence available in metamorphic rocks can be used to constrain the evolution of orogens.

In the sections that follow we will first outline the seminal work on metamorphism, its characterisation, interpretation in terms of tectonics, and thermal modelling that sets the scene for the current approaches to understanding metamorphism in orogeny. We will then go on to describe these approaches and their outcomes in terms of *P–T*–time records, linking these to the underlying processes, chemical, physical and tectonic, that inform and constrain models for metamorphism in orogenic systems.

General characterisation of metamorphism

Mapping metamorphism: from isograds to facies

The first attempts to systematically investigate and understand metamorphic rocks were undertaken nearly in parallel in Scotland and Scandinavia, sites of former great orogenic belts. In each case the breakthroughs were made by focussing on field mapping of the distributions of identifiable minerals in specific rock types. From 1893 George Barrow mapped mineral zones developed in metamorphosed shaley rocks (pelites) in the Dalradian to the SW of Aberdeen, Scotland (Barrow, 1893). This work was to prove seminal, even though Goldschmidt's (1915) mineralogical mapping of progressive mineral zones in the metamorphic rocks of Trondheim was carried out without reference to it.

Barrow mapped the slates, phyllites and schists of Aberdeenshire on the basis of the appearance or incoming of minerals missing from other parts of the mapped region. With the good fortune that most of the rocks were of similar bulk composition, he was able to map the appearance and presence of biotite, garnet, staurolite, kyanite and (rarely) sillimanite over an extensive area, referring to the metamorphism as 'regional thermo-metamorphism' and 'regional metamorphism'. He established the idea of a *metamorphic zone*, an area in which a zonal mineral (now known as an index mineral, a term also used in his paper) is present in rocks of a specified composition, and defined the beginning of such a zone as being at the 'outer limit' where the index mineral is first seen. These mapped 'outer limit' lines came to be referred to as *isograds*. Barrow (1912) inferred that the sillimanite zone in his zonal sequence was the highest temperature zone. Indeed, he referred to the 'outer limit' lines on his map as isothermals, the critical point being that the lines represented the 'point where the rocks have been raised to a sufficiently high temperature to develop the index-mineral of the zone.' This work profoundly influenced the subsequent work by Harker, who considered that Barrow's zones ('Barrovian metamorphism') were typical of regional metamorphism in general. Harker, who was well aware of the 'Buchan' zones in which cordierite and andalusite were typical index minerals, considered that the latter minerals were only present in the absence of deformation, and he termed these 'anti-stress minerals' to explain why minerals different from those in Barrow's zones were present in apparently regional metamorphic rocks (Harker, 1932).

In its first phase the study of metamorphism was dominated by the concept that pressure, or depth of burial, was the overriding control on what minerals were contained in the rocks (Grubenmann & Niggli, 1924). This was evident in the concept of depth zones – epizone, mesozone and catazone – that dominated classifications, and in Harker's 'stress' and 'anti-stress' minerals classification. Eventually the detailed analyses of mineral assemblages by Goldschmidt (1915) and Eskola (1915, 1920), and their documentation of the simple relations between such assemblages and bulk rock composition, demonstrated that both pressure and temperature must be critical controls on metamorphic mineralogy. Goldschmidt applied the phase rule, $F = C - P + 2$, where F is the number of degrees of freedom, or variance, of a system of C components containing P phases (minerals, fluids, melts), to hornfelses and later to regional metamorphic schists to demonstrate conclusively that these complex rocks did approach chemical equilibrium. In doing so he laid out much of the foundation for our current phase equilibrium approaches to evaluating metamorphic mineral assemblages, including assessment of variance and P–T fields of stability. Eskola, on the other hand, was able to evaluate the dependence of mineral assemblage on bulk rock composition in amphibolites in Finland. Based on these observations, and through integration of his amphibolite studies with those on eclogites and observations by other workers, Eskola (1920, 1929) was able to articulate the metamorphic facies concept, in which the mineral assemblage present in a metamorphic rock is a systematic function of the pressure and temperature conditions it experienced and its bulk composition.

Facies series, progressive metamorphism and evaluation of regional *P–T* variations

Turner (1948) was a strong and influential advocate of the use of thermodynamics and application of the phase rule, coupled with experimental constraints, to develop an understanding of metamorphism. By 1951 he (Turner & Verhoogen, 1951) was able to position Eskola's original facies on a *P–T* diagram, both in a relative sense and with, in some cases, absolute *P–T* estimates that are still in reasonable accord with modern constraints. Taking the idea of progressive metamorphic zones one step further, and adopting the grouping of facies into broad baric types as proposed by Miyashiro (1961, 1973), Turner redefined facies series '*in which mapped isograds mark successive steps in a prograde sequence of mineral transformations*'. His facies series were presented in the context of describing the metamorphic field gradients across regional terrains, linked to the newly calibrated aluminosilicate phase diagram. Along with Miyashiro's extensive work on baric types and the five-fold discrimination of facies series, Turner's analysis had a strong influence on the way in which metamorphism was characterised in *P–T* space. In particular, his statement italicised above, that the change from lower to higher grade metamorphic zones involved a *succession* of steps, instilled the view that the rocks in the different zones were linked by progressive heating along the field d*P*/d*T* gradient. In other words, all the high-grade rocks will have gone through the reactions, and assemblages, recorded in the lower-grade ones.

Turner's emphasis on the observable zonal gradient across a progressively zoned metamorphic belt, the 'metamorphic field gradient' had a profound effect. This gradient was visualised as reflecting the gradient of *T* with depth beneath the 'metamorphic surface' at the time of an essentially isochronous event. Recognising that isograds needed to be transformed geometrically to show their disposition with respect to a surface at the time of metamorphism, Turner went on to infer that isograd pattern geometry is dependent on thermal gradient and the inclination between isotherms and pressure. The germ of the idea that the field gradient was not a true thermal profile was inherent in the reference frame being the 'metamorphic surface' rather than the Earth's surface, though the idea that this 'metamorphic surface' may evolve with time was not considered. In defining and refining facies series to tie in with zones of progressive metamorphism Turner added considerable 'flesh' onto the bones of Eskola's facies concept, providing the *P–T* diagrams that illustrated the possible conditions of most facies. By the time of his 1980s book (Turner, 1981) many of the facies were reasonably well established in terms of their *P–T* regimes. Even the blueschist or glaucophase schist facies which, prior to the work of de Roever (1955, 1964), Ernst (1971, 1973, 1975), and Miyashiro (1961, 1973) in the 1960s and 1970s, had not generally been accepted as a 'true' facies, was represented on his facies *P–T* diagrams.

Turner's analysis of metamorphic field gradients was mainly focussed on assessing how accurate and precise the *P–T* arrays could be, rather than on considering what the meaning of the arrays might be in terms of the relations between different points on them. Following an exhaustive analysis of all the *P–T* arrays available, Turner (1981) concluded with the three 'typical' metamorphic thermal gradients, Blueschist, Barrovian and Buchan, that essentially correspond to the high-*P*/low-*T* (high *P*/*T*), medium-*P*/*T* and low-*P*/high-*T* (low *P*/*T*) 'shorthand' groupings used today (Fig. 7.2).

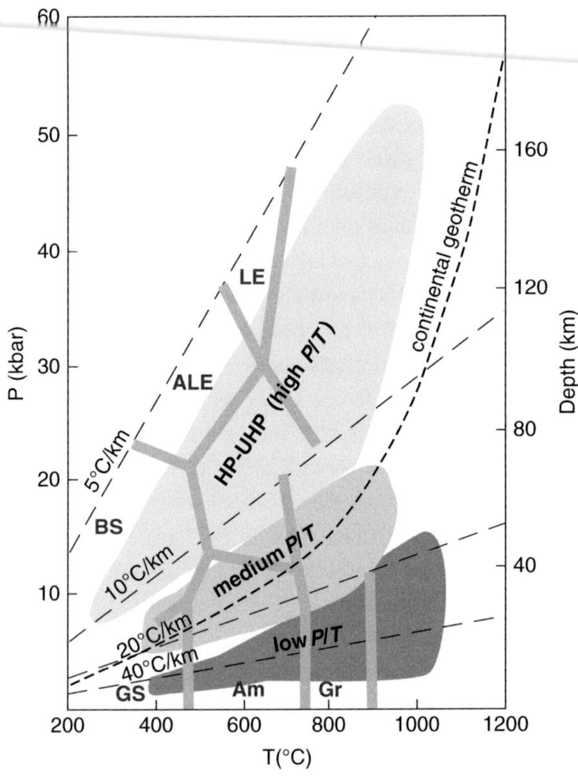

Figure 7.2 Pressure–temperature (*P*–*T*) diagram showing the broad divisions of metamorphism in terms of the facies series proposed by Turner (1981) and modified in the light of recent work on UHP and UHT metamorphism (Brown, 2007). The HP–UHP (high *P*/*T* facies series), medium *P*/*T* facies series, and low *P*/*T* facies series are distinguished by shading. A model continental conductive geotherm corresponding to a surface heat flow of 70 mW/m^2 lies to the high-*T* side of the HP–UHP metamorphic field and at significantly lower *T* for a given depth than low *P*/*T* metamorphism and much of the medium *P*/*T* field. All other lines, fields and symbols as in Fig. 7.1.

Early perspectives on metamorphism in relation to tectonics

Miyashiro (1961, 1973) was one of the first metamorphic geologists to put metamorphism into a more modern, dynamic, context. He presented an integrated overview of the possible relationship between metamorphism and tectonics, including but not exclusively focussed on his celebrated 'paired metamorphic belts' concept, tied to subduction zones and magmatic arcs. He distinguished, for example, belts found at continental margins, some of which were 'paired' in the sense that there were high-*T*/low-*P* and high-*P*/low-*T* facies series. He contrasted belts beneath 'ordinary island arcs' as in Japan, those beneath 'reversed arcs', and those belts related to continental collision. For the latter group he recognised that accreted belts of high-*T*/ low-*P* and high-*P*/low-*T* types could occur in the collisional orogen, potentially overprinted by metamorphism associated with the terminal collision itself. Miyashiro also reflected on rates of plate convergence and their consequences for the style of metamorphism. He proposed that secular variation in baric type (meaning *P*/*T* regime) to produce Phanerozoic high-pressure belts (Mesozoic to Cenozoic in his examples) reflected thicker oceanic plates, steeper descending slabs and more rapid or colder continental underthrusting. Recognising that apparently unpaired metamorphic belts did exist, he proposed that this might reflect preservation, level of exposure or alternatively an effect of later juxtaposition.

Miyashiro (1973) also showed considerable insight into the issue of the exhumation of high-pressure rocks, pointing out their close association with faults. Noting that the upper surface of a downgoing slab would be a fault zone, he speculated that this might represent one bounding surface of a high-*P* metamorphic belt. Further noting that the high-*P*/low-*T* metamorphism would mainly occur in the downgoing oceanic slab and its cover sediments, he suggested that such rocks might be added 'to the continental side' by these faults. In other words, he considered the possibility of footwall collapse, and suggested oceanwards slab retreat, long before the ideas of 'slab roll-back' and 'subduction channel' were articulated in the modern literature. His contributions, like those of Ernst (1973, 1975) on blueschists and their relation to subduction zones, were fundamental to developments in the 1970s that have led to the present views of metamorphism in mountain building.

Zwart (1967, 1969) drew attention to differences in the metamorphic patterns developed in orogenic belts, identifying a 'duality' of metamorphic belts in Europe that reflected differences in orogenic processes. Largely on the basis of observations in Europe and North America, he proposed three types that he and subsequent workers related to tectonic setting and style. His *Hercynotype* (back-arc basin type) orogen was typified by shallow, low-*P* metamorphism with temperature gradients leading to narrow *T*-dependent metamorphic zones. This type was also characterised by an abundance of migmatites and granites and an absence of ultramafics and ophiolitic rock types, very broad high-*T* metamorphic zones and limited evidence for any exhumation near the metamorphic peak. Examples of the Hercynotype orogens include not only the Hercynian itself but possibly also the Lachlan Fold Belt. His *Alpinotype*, or ocean trench related, orogens (now also referred to as 'cold orogens') featured high-*P* metamorphic zones but included metamorphism under several facies, including medium-*P* conditions. Granites and migmatites were rare or of only minor extent, whereas meta-ophiolites and ultramafic rocks might be present and volumetrically important. These *Alpinotype* orogens are elongate and relatively narrow with a high aspect ratio, and correspond to 'cold' orogenies as defined today. Nappe structures are well developed at several structural levels and have been exposed because of rapid (i.e. tens of millions of years) uplift and exhumation. His *Cordilleran* (arc) type orogens were dominated by calc-alkaline igneous rocks. In this type there is a general lack of migmatites, and of ophiolites and abyssal sediments. The low-pressure metamorphism is on a low geothermal gradient (i.e. high-*T* at shallow depths).

Based on his combined metamorphic-structural studies in the Pyrenees, Zwart (1962) recognised polymetamorphism, the overprinting of at least two distinct metamorphic events, and reasoned that this could in some cases occur during one orogeny. He discussed 'plurifacial metamorphism' and even illustrated his ideas with schematic *P*–*T* loops. In the case of the Pyrenees, the two metamorphic events identified were also of very different ages and related to entirely different orogenies (the Hercynian/Variscan and the Pyrenean), rather than being two episodes during the thermal evolution in one orogeny – in which case the term polyphase metamorphism would more correctly be applied.

Metamorphism and thermal modelling

Although not the first to consider how the thermal structure of the crust and lithosphere could impact on and be reflected in tectonics – being preceded by the thermal modelling of subduction by Oxburgh and Turcotte (1970) and others – the work of England and Richardson (1977) was seminal. It integrated the internal thermal driving forces and redistribution of heat associated with the movement of rock masses with the effects of erosion. This explicit linkage between deep Earth and surface processes provided the first step in understanding how mountain belts might evolve and, as part of that evolution, how metamorphic rocks might be exhumed. The key breakthrough in this was consideration of a mechanism for exhumation. The lithosphere-scale thermal models proposed for thrusted regions (Oxburgh and Turcotte, 1970) could demonstrate that perturbed geotherms would result and lead to anomalously low-T crust at depth, and produce temperature–depth (dT/dz) profiles in which T decreased with depth. However, in the absence of an exhumation mechanism these models could not explain how these deep, cold rocks would get to the surface, let alone how the metamorphic record could be retained in the rocks if they did manage to reach the surface.

The critical conceptual advance arising from the modelling by England and Richardson (1977) remains essentially valid: the metamorphic field array, or piezothermal array, is the locus of maximum T points attained by rocks which travelled along P–T paths that are likely to be at a high angle (in P–T space) to the preserved array itself. In the thickening model the paths traverse to higher pressures first (i.e. burial dominates) and then heating occurs as the rocks approach the surface, so that the peak T (T_{max}) is attained after the peak pressure (P_{max}), mapping out a 'clockwise' P–T evolution with time (Fig. 7.3a). In this model the deeper, higher-T rocks reach their preserved peak-T condition later (i.e. at a younger age relative to the present) than those shallower rocks that are heated for a shorter period of time before cooling takes over with continued exhumation.

The idea that isograds might not be isochronous, and that metamorphic rocks might record P–T histories that do not correspond to the final field array dT/dz gradient, had been presented somewhat earlier by Zwart (Zwart, 1962) on the basis of field and petrographic observations. However, the impact of England and Richardson (1977) was profound as it changed the view of the importance of metamorphism in tectonics. Understanding the record of metamorphism became central to the development of models for plate behaviour, including subduction zone evolution and collisional orogeny. The concepts introduced also changed the meaning, significance and use of isograds. Whereas isograds in progressive metamorphism had generally been treated as broadly equivalent in age, the England and Richardson (1977) modelling required the higher-grade isograds and mineral zones to be younger, by several million years, than the lower-grade ones. Isograds could no longer be

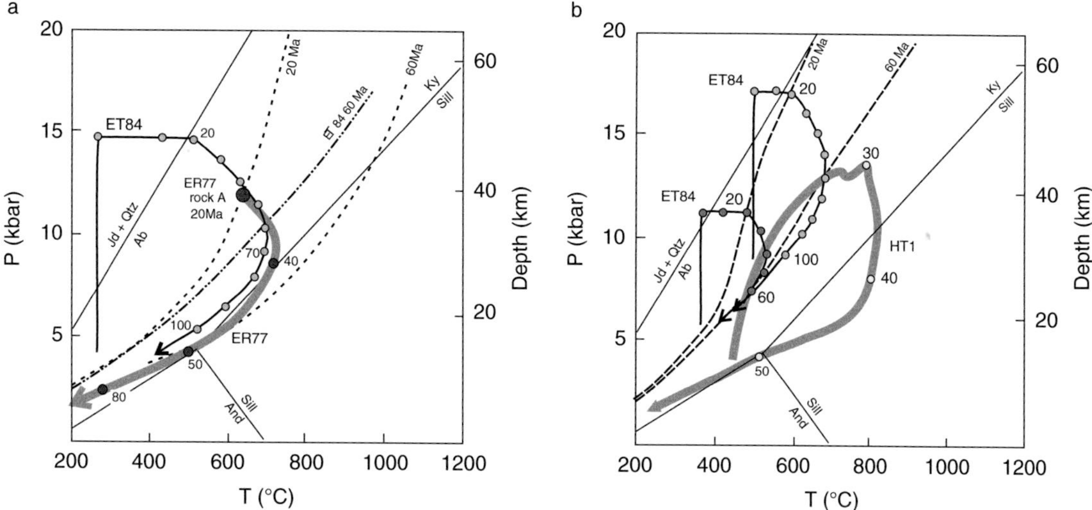

Figure 7.3 a, P–T diagram depicting a selection of P–T paths modelled for deep-seated rock packages in the classic one-dimensional overthrusting models of England and Richardson (1977) [ER 77] and England and Thompson (1984) [ET 84]. Rock package A in the England and Richardson (1977) model traverses on a 'clockwise' P–T path through the upper amphibolite facies, crossing the kyanite = sillimanite reaction and maintaining $T > 500$ °C for some 40 Ma, with the 'peak' attained at 33 Ma after thrusting. The crustal-doubling overthrust model of England and Thompson (1984) yields a P–T path of similar form and P–T position, but only for rocks taken to some 50 km, and with the T peak attained only at *c.*60 Ma after thrusting. Mineral equilibria positions are as in Fig. 7.1. Dashed lines show the transient geotherms for 20 Ma and 60 Ma after thrusting in the England and Thompson (1984) model. b, P–T diagram showing P–T paths modelled for two deep-seated rock packages in the one-dimensional homogeneous thickening model of England and Thompson (1984), compared with a mid-crustal rock incorporated deep within a ductile channel in the coupled two-dimensional HT1 model of Beaumont *et al.* (2004) and Jamieson *et al.* (2004). In all cases the rocks traverse along 'clockwise' P–T paths, but in the quite 'cold' England and Thompson (1984) model these paths always lie in the kyanite field, whereas in the HT1 case the rock package crosses into the sillimanite field and, for the selected package, achieves granulite facies P–T conditions for a considerable period of time ($T >$ 800 °C for *c.*16 Ma). Filled circles on the P–T paths represent 10 Ma intervals after initiation of the model (labelled in the Jamieson *et al.* (2004) case). Mineral equilibria positions are as in Fig. 7.1. Dashed lines show the transient geotherms for 20 Ma and 60 Ma after thickening in the England and Thompson (1984) model.

treated as fossil markers that define the P–T framework of metamorphic rocks in orogenic belts. Instead they had to be considered, along with evidence for the P–T paths of the rocks, within a dynamic frame of reference. The subsequent emphasis on P–T–time paths as records of metamorphism that constrain the underlying tectonothermal processes was given impetus by the England and Richardson (1977) modelling.

In their development of the one-dimensional modelling approach England and Thompson (1984) coupled the thermal modelling with a more detailed analysis of the metamorphic consequences (Thompson & England, 1984). They examined both homogeneous thickening and crustal doubling by overthrusting, allowing an arbitrary 20 Ma time lag to erosion after thickening. They assumed a constant mantle heat flux into the base of the crust, and all radiogenic heat production concentrated in the top 15 km of crust. The mantle heat flux and radiogenic heat production values were chosen to give surface heat flows between 45 and 75 mW/m^2. Both types of model generated clockwise *P–T* loops with significant decompression and long time intervals (10–50 Ma) spent on the erosion-controlled decompressional portions of the paths (Fig. 7.3a,b). Moreover, rocks taken to greater depths traversed 'looped' *P-T* paths for longer periods of time than those at shallower depths, so that the higher-grade, deeper-buried rocks would give younger metamorphic peak ages, and cooling ages (Thompson & England, 1984). However, as noted by Stüwe (2002), if no deep-crustal lag period is included in the model the resultant *P–T* paths do not 'loop' from maximum *P* to maximum *T*, but instead form tight 'arrowhead' type paths in which little heating accompanies decompression.

The one-dimensional models predicted that for a number of input thermal diffusivities, densities, basal heat fluxes and erosion rates, high-*P*/low-*T* metamorphic rocks of the blueschist and eclogite facies would form early, and over a large *P* interval. However, with thermal relaxation these would inevitably be overprinted by higher-*T*/lower-*P* metamorphic assemblages as the *P–T* paths traversed into the general region of Barrovian metamorphism, and so the preservation of the HP metamorphic rocks would be unlikely. England and Thompson (1984) noted that their orogen evolution could be terminated by rapid extensional collapse, leading to elevated *T* at depths of less than 30 km and *P–T* paths involving near-isothermal decompression near the metamorphic peak (ITD), terminated by a phase of near-isobaric cooling (IBC). Similar 'fast exposure' *P–T* paths have been considered essential to the formation of several IBC-type granulite terrains (Harley, 1989).

Apart from their inability to expose deeply buried rocks in one evolutionary history, these models suffer from an additional weakness. The assumption of instantaneous thickening deformation is not valid if the timescale of orogenesis is similar to that of the overall thermal evolution. In the England and Thompson (1984) models the principal deformation event pre-dates the thermal maximum for most rocks by between 20 and 60 Ma. However, in many orogenic belts geochronology now shows that the timescale of metamorphism and decompression is much shorter than that required by these simple models. Furthermore, in many examples the principal phases of peak mineral growth are broadly syn-kinematic and occur over time intervals of less than 5–10 Ma. These inconsistencies point to the importance of scale in the models, as this affects the characteristic thermal time constants and the Peclet numbers relevant to the orogenic process. Thinner thrust sheets and sequential accretion of crustal blocks would lead to shorter timescales for the metamorphic events in any one location, and allow diachroneity that is not simply depth or grade-related.

Subsequent thermal models of orogenesis and metamorphism have incorporated the temperature-dependent changes in the rheology of crustal materials, and the coupling of this with changes in physical properties resulting from metamorphic reactions, including melting. The extension of this into coupled two-dimensional thermomechanical models allows consideration of the spatial evolution of regions of accreted and thickened crust in response to changes in the rates and geometries of material accretion into the orogenic belt. Finite element and difference modelling on many thousands of nodes in two-dimensional grids has become the principal route to developing better models that either attempt to replicate the metamorphic *P–T–t* records of orogens such as the Himalaya (Beaumont *et al.*, 2001, 2004; Jamieson *et al.*, 2004), Grenville Province (Beaumont *et al.*, 2006, 2010; Jamieson *et al.*, 2010) or Alps (Stöckhert & Gerya, 2005), or attempt to forward model possible *P–T–t* paths that might result from subduction–accretion and collision (Fig. 7.3b) for various sets of assumptions in terms of plate coupling, strain-weakening, thermal structures and reaction-dependent rheologies (Beaumont *et al.*, 2006, 2009; Faccenda *et al.*, 2009; Gerya *et al.*, 2002). These complex models have required formulation of the partial or complete coupling of the thermal and mechanical behaviour of an evolving system, and solving for mass conservation at time steps. Energy balance and physical material transfer governed by groups of flow laws, and including the effects of grain coarsening, melt production and fluid transfer, are applied to each node.

The two-dimensional thermomechanical models now consider a wealth of variables and are applied to a great variety of tectonic scenarios. With respect to recent orogenic systems these models can be divided in terms of their principal focus:

1. Those that consider the *P–T* evolution and material transfer in the subduction zone or subduction–accretion channel during accretion, with or without coupling to an adjacent wedge of material (i.e. critical wedge) but not culminating in collision, and
2. Those that focus on the collisional evolution, starting either with a wedge-buttress geometry with imposed properties or with the development of an accretionary wedge then evolving into collision.

In these models the behaviour in the collisional phase is determined by such features as the extent of layering of the crustal component in terms of long-term strength and composition, and hence potential for flow and deformation partitioning with depth; lateral strength variations, for example reflecting age of crust, and how these strength variations change over time with metamorphic reactions; the onset of partial melting, and the loss of mantle lithosphere during or following crustal thickening. These models and their variants will be discussed in subsequent sections of this chapter with reference to their application to the metamorphic histories and *P–T*–time paths recorded by metamorphic rocks grouped according to their *P–T* conditions and generalised thermal gradients as well as their contexts in terms of orogen evolution.

Quantification of metamorphism: from mineral assemblages to *P–T* diagrams

Composition-assemblage diagrams and projections

Mineral assemblages are the 'currency' of metamorphism. Understanding the detailed relationships between mineral assemblages, rock composition and the key variables of metamorphism, pressure, temperature and fluid composition enables those conditions to be quantified. Mineral facies provided only a broad brush outline of what mineral assemblages are possible for quite broadly defined rock types (e.g. metabasites, pelites, calcsilicates), and could not predict what *precise* mineral assemblage one would see in a specific rock composition over a small range of P–T conditions. For example, amphibolite facies metabasites may contain clinopyroxene, epidote, garnet, chlorite or orthoamphibole in addition to a calcic amphibole and plagioclase, but not all of these phases will generally be present. The key to advancing quantification of metamorphism lay in analysing mineral assemblages via the phase rule and linking this to bulk composition in a more refined manner, utilising thermodynamically valid (i.e. consistent with the phase rule; Thompson, 1955) compositional diagrams (composition-assemblage diagrams) and projections that could not only depict mineral compositions but also allow correct inference, via tieline geometries of the potential reaction relations between them. Ultimately this approach has led to the calculated phase diagrams used today, via a number of intermediate routes.

The ability to diagrammatically represent and evaluate mineral assemblages consistent with the phase rule has been central to development of a sound theoretical basis for relating mineral assemblages to P–T and other thermodynamic variables. The introduction of the AFM (Al_2O_3–FeO–MgO) projection from muscovite by J.B. Thompson Jr. (Thompson, 1957) greatly extended the utility of composition-assemblage diagrams, allowing greenschist and amphibolite facies pelites to be treated systematically for the first time. This immediately found direct application to interpretation of mineral isograds in Barrovian (medium P/T) and Buchan (low P/T) metamorphic zonal sequences (see Plates 21, 22, 23), and is a fundamental tool used today to describe metamorphic zones in many orogens, for instance Miocene inverted metamorphic zones in the Himalayas (Kohn, 2008) and Cambrian medium P/T zones related to oblique collision in the Kauko Belt of Namibia (Goscombe & Gray, 2008). Together with subsequent projected phase diagrams and related P–T–X (where X is a compositional variable) sections, the diagrammatic approach allowed multicomponent compositions and low variance mineral assemblages involving more than four phases to be represented in terms of the phase rule. Complementary work on open systems by Korzhinskii (1959) and Thompson (1955) enabled the effects of changes in bulk composition, for example by fluid ingress or melt extraction, to be represented and assessed within a thermodynamically valid framework. The treatment of fluid-bearing equilibria using the phase rule approach was added to significantly by Greenwood (1962) in his formulation of T–X_{fluid} sections, extended to melt-bearing systems in P–T–X and T–X_{H_2O} by Powell (1978), A.B. Thompson (1982) and Waters (1988), amongst many.

Petrogenetic grids and Schreinemakers' nets

Initial quantification of metamorphic conditions relied on the experimental or experimental/natural rock location in P–T of some key mineral reactions in simple chemical systems. Classic examples include Bowen's work on the progressive metamorphism of siliceous dolomites (Bowen, 1940), studies of aluminosilicate phase relationships (e.g. Richardson *et al.*, 1969), phase relations in MAS (MgO–Al_2O_3–SiO_2) by Schreyer and his co-workers (Schreyer, 1988), the breakdown of muscovite and quartz, and the reaction of albite to form jadeite and quartz at high pressure (Fig. 7.1). Whilst providing a useful general P–T framework for considering metamorphism and placing facies in context, the 'petrogenetic grids' developed using these generally unconnected reactions from differing chemical systems do not allow one to trace, for example, the mineral assemblages that might develop in restricted ranges of rock compositions evolving within the *P–T* field of crustal metamorphism. Only in the trivial case of a one-component system rock (e.g. a pure SiO_2 rock; a pure Al_2SiO_5 rock) do such individual reactions define and fully predict the *P–T* stability fields of the 'assemblages' (SiO_2: quartz versus coesite, tridymite, cristobalite and so on; Al_2SiO_5: kyanite versus sillimanite and quartz). Even in the simplest two-component system, different mineral assemblages may be stabilised in rocks of slightly different composition. For example, in the system $NaAlSi_2O_6$–SiO_2 (components chosen to correspond to the compositions of the phases jadeite, Jd, and quartz, Qtz) the formation of the intermediate mineral composition albite (Ab), $NaAlSi_3O_8$, will lead to rock compositions that have higher initial Qtz:Jd ratios producing Ab + Qz whereas those in which Qtz:Jd was initially less than 1 would produce Ab + Jd.

An alternative approach to producing 'petrogenetic grids' explicitly utilises the phase rule coupled with the principles of topology as applied to sets. This defines the geometrical relationships between groups of mineral reactions that may occur between sets of minerals that share a chemical system and may have overlapping stabilities. The topological approach, used for analysis of the aluminosilicate phase diagram (Miyashiro, 1949) but popularised by the systematic exposition of Zen (1966), is known as Schreinemakers' analysis, after the mathematician who developed its theoretical basis. The principles embodied in Schreinemakers' analysis underpin all phase diagram work, including calculated phase diagrams in complex multicomponent systems (Spear *et al.*, 1999; R.W. White *et al.*, 2001b). The beauty of this approach, soon recognised by workers who applied thermodynamics to reaction calculations, is that experimental or thermodynamic data defining one or more univariant reactions within a system of reactions that potentially share a common intersection (at an invariant point) could be used to constrain the other reaction positions and therefore allow the *P–T* space to be divided up into smaller regions. Even with limited thermodynamic data the Schreinemakers' analysis of mineral reactions enabled systematic explorations and semi-quantitative assessments of the *P–T* conditions of mineral assemblages in model pelites, siliceous dolomites and metabasites that in many respects have stood the test of time in terms of their predictions to within ± 50–100 °C and a few kbar for 'typical' rock compositions (Spear, 1993). The *P–T* grids now produced by coupling Schreinemakers' principles and its geometric outcomes with internally consistent data sets of thermodynamic data for minerals, fluids and melts form the essential framework on which more 'rock-specific' calculated phase diagrams are constructed (Powell *et al.*, 1998; Baldwin *et al.*, 2005).

Petrogenetic grids based on Schreinemakers' analysis consist of bundles of univariant reactions within defined chemical systems, which divide up *P–T* (or other parameter) space and meet at invariant points in these systems. Once again, and even in the absence of solid solution in participating minerals, these grids only show the reactions that are *possible* in the defined system, not those that are 'seen' or traversed by a specific rock bulk composition. One conceptually elegant attribute of the analysis is that, when reactions are labelled using those system phases that are 'absent' (i.e. not involved in the reaction), then it is found that those mineral assemblages that contain the phase cannot 'see' or be affected by reactions labelled with that phase absent, because their compositions lie outside the sub-system space in which those reactions occur. That is, garnet-bearing mineral assemblages within the defined chemical system are not affected by garnet-absent reactions. When solid solution (e.g. Fe–Mg in ferromagnesian minerals such as micas, amphiboles, pyroxenes, garnet, cordierite etc.; CaAl–NaSi in feldspar, scapolite and other phases) is introduced, and minerals are considered in more complex chemical systems accounting for these compositional variations, then even rocks with appropriate bulk compositions in terms of most chemical constituents will only 'see' a univariant reaction over the limited P or T segment of that reaction over which the Mg/Fe ratios of the participating minerals straddle the Mg/Fe of the rock. At other P–T positions the mineral assemblages will be at least divariant (two degrees of freedom; $F = 2$) and the changes in P–T conditions will only be recognised from comparing assemblages of higher variance.

Divariant and multivariant equilibria – from Schreinemakers' nets to pseudosections

In a major theoretical contribution to the analysis of metamorphic assemblages and their relationships with the context of the phase rule, Hensen (1971) showed how divariant and higher variance mineral assemblages obey Schreinemakers' rules, relate geometrically to univariant reactions in a systematic manner, and can be predicted to vary in terms of the Mg/Fe ratios of constituent minerals. Building on earlier advances in depicting higher variance assemblages using T–X_{Mg} and P–X_{Mg} diagrams (Korzhinskii, 1959), Hensen demonstrated how P–X and T–X loops describing mineral compositional changes in those divariant assemblages that intersect on or 'share' a univariant reaction are disposed according to Schreinemakers' principles consistent with their end-member reaction stoichiometries. Hensen went on to illustrate how these divariant assemblages involving solid solutions form 'bands' or fields in P–T space, bounded by appropriate univariant reactions, and how these bands occur in different P–T positions, and with different widths, dependent on the precise rock composition. The buffered divariant fields could also be contoured for mineral compositions (e.g. X_{Mg}) and/or mineral modes for the chosen bulk rock composition. The P–T phase diagram thus developed for a specific rock composition, a so-called 'pseudosection' (Hensen, 1971) depicts *only* those reaction equilibria and assemblage fields 'seen' by or accessible to that rock composition. These insights facilitated the design and interpretation of Hensen's experiments on cordierite equilibria in the FeO–MgO–Al_2O_3–SiO_2 system, FMAS (Hensen & Green, 1973). In order to experimentally constrain cordierite (and garnet) compositions in the assemblages Crd + Grt + Sill + Qz and Crd + Grt + Opx + Qtz, and thereby develop these as geobarometers, experiments were required on a range of bulk X_{Mg} (and two A/AFM) compositions. Hensen's 1971 theoretical analysis provided a means of optimising the experimental strategy. The pseudosection methodology has proven to be far-reaching in its use, both in experiments in complex systems (Carrington & Harley, 1995) and, with the advent of computerised phase diagram calculations, in analysis of the P–T conditions and P–T evolutions of metamorphic rocks in general (White *et al.*, 2001b).

At the present time pseudosections (calculated P–T pseudosections and phase diagrams for specified rock compositions) and their derivatives (e.g. T vs. modal % fluid),

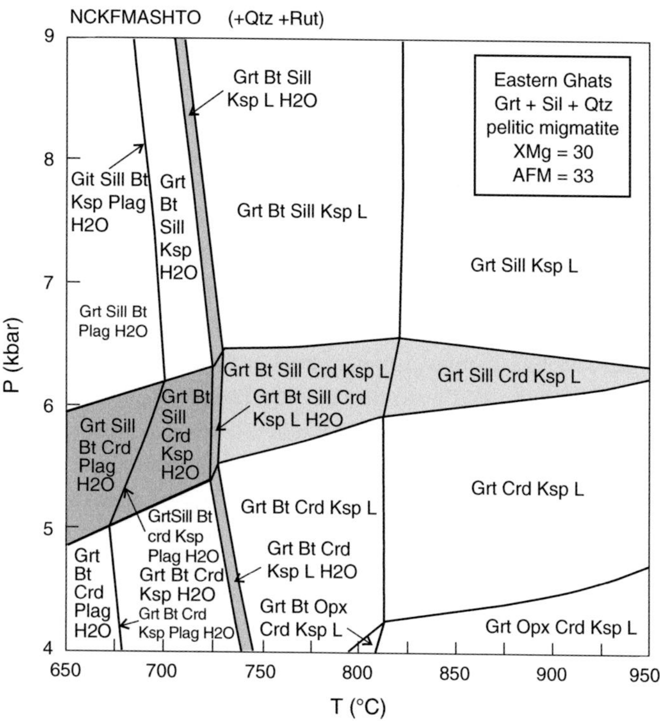

Figure 7.4 Phase diagram calculated for a specific rock composition (i.e. a pseudosection), simplified to be within the chemical system $Na_2O–CaO–K_2O–FeO–MgO–Al_2O_3–SiO_2–H_2O–TiO_2–O_2$ (NCKFMASHTO). The lightest and darkest shaded fields show the *P–T* conditions under which Grt + Sill + cordierite (Crd) coexist in this rock composition, with melt (lightest field) and without melt but with biotite (Bt) and a free volatile phase (H_2O) (darkest field). The thin fields located at 720–740 °C and near-parallel to the *P* axis show the *T* range over which H_2O-saturated melting occurs in this rock, for a given starting H_2O content.

often contoured for several mineral composition variables or modes of phases, in many cases are used as the method of choice for defining and quantifying metamorphic conditions and *P–T* histories. When combined with mineral compositional zoning information obtained from electron probe *X*-ray mapping, for example Ca–Mg–Fe zoning in garnet and amphiboles, anorthite content zoning in plagioclase or Al_2O_3 zoning in aluminous pyroxenes (Spear *et al.*, 1984; Kohn, 2003), these calculated phase diagrams can yield unrivalled constraints on the *P–T* paths along which those mineral assemblages have evolved. The principal limitations that apply to the method are the uncertainties in the thermodynamic properties of the phases, approximations in their activity–composition relationships, and the extent of preservation of mineral zoning in the light of diffusional re-equilibration.

An example of one such pseudosection, calculated on THERMOCALC (Powell & Holland, 1988) using the Holland and Powell (1998) internally consistent thermodynamic dataset and incorporating mineral activity–composition relations developed consistent with the data set, is provided in Figure 7.4. This pseudosection has been calculated in the

NCKFMASHTO system (Na_2O–CaO–K_2O–FeO–MgO–Al_2O_3–SiO_2–H_2O–TiO_2–O_2) for a rock composition that corresponds to a typical, fairly Fe-rich peraluminous pelite (X_{Mg} = 31; A/AFM = 33 in AFM projection from K-feldspar and plagioclase) from the Eastern Ghats of India. The rock is a migmatitic granulite with the preserved mineral assemblage garnet + sillimanite + plagioclase + K-feldspar + quartz + rutile (i.e. Grt + Sill + Plag + Kfs + Qtz + Rut) that coexisted with melt (now leucosome). This assemblage is stable at temperatures greater than 825 °C and pressures greater than 6.5 kbar. Contouring of the pseudosection for garnet Ca content (mol% grossular, X_{Grs}) and modal proportion further constrains *P–T* to 6.8–7.5 kbar at 830–850 °C in this example. Furthermore, the textures in this rock (e.g. inclusions of quartz, sillimanite and rutile in Grt; coronas of late garnet and late-stage biotite but a lack of cordierite) clearly show that the *P–T* evolution did not traverse the lower-pressure fields in which cordierite would be stable. This pseudosection is a simple one, dictated by the rock composition. Far more complex pseudosections may result for magnesian pelites under the same *P–T* conditions because of the enhanced stability of cordierite with bulk rock X_{Mg}, along with the competing effects of stabilisation of other minerals (e.g. spinel, sapphirine) at high-*T* especially if SiO_2 is low and quartz exhausted with high percentages of melting (Kelsey *et al.*, 2005).

Geothermobarometry

The calculation of *P* and *T* based on the compositions of coexisting mineral phases in metamorphic rocks and mantle peridotites, geothermobarometry, provides a complementary approach to estimation of metamorphic conditions that is still widely used and will remain so as new experimentally determined geothermometers and geobarometers become available. With a longer history of application than calculated phase diagrams, geothermobarometry is often referred to as 'conventional' or traditional in the sense that the methods applied do not, in general, utilise internally consistent mineral thermodynamic data optimised for all phases or even activity–composition relationships that are consistent between pairs of methods. Nevertheless, in many metamorphic regions the conventional geothermobarometers are applied, at least as a first-pass test, to give *P–T* estimates or provide evidence for spatial differences in recorded *P* or *T* – a good example being in the inverted metamorphic zones of the Himalaya.

Pressure–temperature estimates derived in conventional geothermobarometry rely on finding the *P–T* intersection of at least two different equilibria affecting coexisting minerals in a mineral assemblage. Ideally one or more of these equilibria is dominantly temperature-sensitive, with a small molar volume change relative to entropy change of reaction, whereas others are pressure-sensitive, with large relative molar volume changes of reaction. Solving these equilibria to deduce a specific *P–T* condition of course assumes that the mineral compositions are in all respects those preserved simultaneously at that condition, and that some compositional attributes controlling the results from one method, for example a thermometer, have not changed further through diffusion. The potential for offset of one equilibrium compared with another (Frost & Chacko, 1989; Harley, 1989) is particularly important at high temperatures, and can lead to *T* underestimates if not accounted for by back-diffusion calculations that correct for retrograde cation exchange (Pattison *et al.*, 2003).

Those equilibria with small d*V*/d*S* include the exchange equilibria in which cations are simply partitioned between two phases, for example Fe–Mg exchange between garnet and other more magnesian minerals such as biotite, amphibole, pyroxenes and cordierite. These major element exchange thermometers, usually developed and calibrated in independent experimental studies (Ferry & Spear, 1978; Ellis and Green, 1979; Perchuk & Lavrent'eva, 1983), are still in widespread use and in many instances provide essential *T* information consistent with phase diagram calculations (e.g. Harley, 1998b, 2008). Indeed, garnet–clinopyroxene Fe–Mg exchange thermometry (e.g. Ellis & Green, 1979) is often one of the few methods available to estimate temperatures for classic eclogite facies metabasites. Other major element thermometers include those based on the compositions of phases that lie on either side of a compositional solvus, for example muscovite-paragonite, coexisting orthopyroxene and clinopyroxene, and in the ternary feldspar system. The latter method has been applied successfully, with due caution, to high-temperature and ultrahigh-temperature granulites (e.g. Hokada, 2001).

Those equilibria with large relative d*V*/d*S*, and which can therefore be useful as geobarometers, include divariant or multivariant net-transfer reactions in which one or more mineral is consumed whilst others are produced. In simple chemical systems these reactions may be univariant, such as the reaction of jadeite + quartz to produce albite (Fig. 7.1) or of grossular garnet + aluminosilicate + quartz to produce anorthite. However, in complex systems such reactions are equilibria between all the solid solutions in the relevant assemblage, and are offset in *P*–*T* space from the end-member reaction. Knowledge of the activity–composition relations of the participating minerals, the thermodynamic parameters for the governing reaction, and compositions of the minerals preserved in the rock then allows for the calculation of pressure at given temperatures.

Popular and useful geobarometers include those that utilise the assemblages noted above, omphacite + plagioclase + quartz and garnet + aluminosilicate + plagioclase + quartz, along with many others involving garnet + pyroxenes, amphiboles or micas + plagioclase + quartz in metabasites and intermediate rocks, or garnet + cordierite + sillimante + quartz in high-*T* pelites (see Spear, 1993, for detailed discussion). The key attribute in all cases is that the reactions have large relative volume changes that drive changes in the modal proportion of high-density minerals (garnet, omphacite) compared with low-density minerals (feldspar, cordierite) as pressure changes. An example of a geothermobarometer that only involves two minerals is the Al_2O_3 content in orthopyroxene coexisting with garnet. In this case Al_2O_3 in orthopyroxene increases with increasing *T* and decreasing *P* (e.g. Harley & Green, 1982), governed by a coupled exchange of Al + Al for Mg + Si that requires a net transfer of the garnet component into orthopyroxene and hence involves a significant volume change.

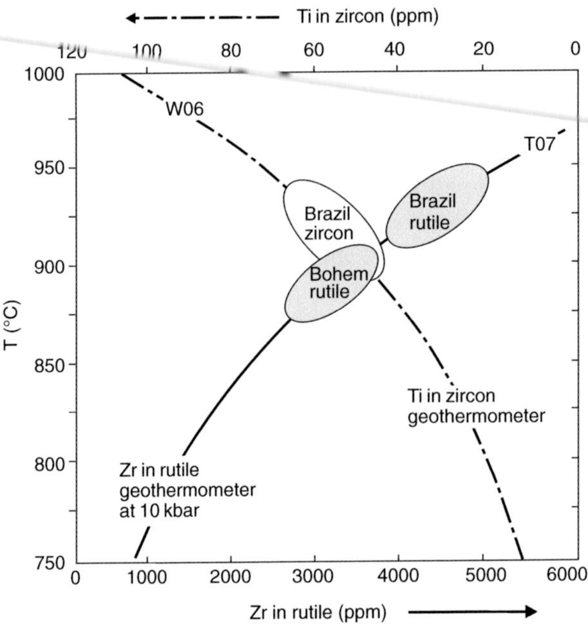

Figure 7.5 Trace element geothermometers applicable to HP–UHP and HT–UHT mineral assemblages. Temperature is plotted here against the Ti content (in ppm) of zircon coexisting with rutile + quartz, based on the calibration of Watson *et al.* (2006) [W06], and against the Zr content of rutile coexisting with zircon and quartz, based on the calibration of Tomkins *et al.* (2007) [T07] at 10 kbar. High-Zr rutiles from the Bohemian Massif and Baro Alto complex of Brazil (Zack *et al.*, 2004) yield temperatures of 870–950 °C that are consistent with other HT–UHT indicators. High-Ti zircons from the Anápolis–Itauçu Complex of Brazil (Baldwin *et al.*, 2007; Baldwin and Brown, 2008) yield minimum temperatures of 890–940 °C that are consistent with UHT P–T conditions inferred from silicate mineral equilibria.

Despite the prevalence, and intrinsic superiority, of the pseudosection or calculated phase diagram approach, geothermobarometry is still used in calculation of P–T conditions and gradients in orogenic belts, and we would advocate that appropriate major element geothermobarometers continue to be used and tested against the available mineral data and evaluated in the light of detailed petrological, textural and mineral zoning information.

The 'conventional' geothermometry described above is now being complemented by new methods based on the distribution of trace elements between accessory mineral and major phases (e.g. Y thermometry based on monazite, xenotime and garnet: Kohn *et al.*, 2005) and, in particular for extreme UHT and UHP cases, 'single mineral' trace element thermometers such as Ti in zircon coexisting with rutile and quartz (Watson *et al.*, 2006), or Zr in rutile coexisting with zircon and quartz (Zack *et al.*, 2004; Tomkins *et al.*, 2007) (Fig. 7.5). These not only open up new avenues for obtaining robust temperature estimates in some orogens, but also enable those temperatures to be tied to time when the accessory minerals are dated and so provide valuable T–time information constraining the P–T–time paths and thermal histories (Pyle & Spear, 2003; Möller *et al.*, 2003; Kohn *et al.*, 2005; Baldwin & Brown, 2008).

Crustal melting and orogeny

Melting processes and temperature conditions

Partial melting, melt segregation and migration are intrinsically linked with and related to the metamorphism of continental crust in accretionary and collisional orogenic belts, provided that sufficient heat is generated to achieve high-grade metamorphism and anatexis. Melting during orogeny may be promoted by the influx of aqueous fluids, yielding H_2O-saturated granitic melts, but this will be limited both by the availability of H_2O and the physical accessibility of fluid, for example through fracture networks (Sawyer, 2001; Brown, 2004). The potential to produce large volumes of such melts, at 650–750 °C, is limited by the melt H_2O contents – H_2O-saturated melts at pressures greater than 5 kbar contain more than 10 wt% H_2O. On the other hand, melting at higher temperatures in the range 800–1000 °C is possible without the influx of aqueous fluid through 'dehydration-melting' reactions. These reactions involve the breakdown of volumetrically abundant hydrous phases such as the micas (muscovite, biotite) in pelites and amphibole in mafic rock types to produce H_2O-undersaturated melt (e.g. 3.5–2 wt % H_2O) and anhydrous or low-H_2O (e.g. cordierite) mineral assemblages. Examples of simplified dehydration-melting reactions that may occur in pelites and quartzofeldspathic rocks include:

$$\text{Muscovite} + \text{Quartz} = \text{Sillimanite} + \text{Melt} \pm \text{K-feldspar}$$
$$\text{Biotite} + \text{Quartz} = \text{Orthopyroxene} + \text{Melt} \pm \text{K-feldspar}$$
$$\text{Biotite} + \text{Sillimanite} + \text{Quartz} = \text{Garnet} + \text{Melt} \pm \text{K-feldspar}$$

and

$$\text{Biotite} + \text{Sillimanite} + \text{Quartz} = \text{Cordierite} + \text{Melt} \pm \text{K-feldspar}$$

The loss of biotite in traversing from about 720 °C to 820 °C across the *P–T* pseudosection of Figure 7.4 is a consequence of multivariant dehydration melting reactions analogous to the last two simplified reactions listed above.

The principles and systematics underlying such 'dehydration' or 'water-undersaturated' melting have been described and analysed in depth by several workers (Powell, 1978; Thompson, 1982; Waters, 1988; Stevens & Clemens, 1993). The dehydration-melting or vapour-absent melting reactions involved have been investigated experimentally in simple and complex systems, and melt fertility of the principal crustal rock types, including greywackes, shales and hydrated mafic volcanics, established over the melting interval transected by high-grade metamorphism. There are considerable complexities arising from the melting being multivariant and therefore varying subtly between similar rock types with differing Fe/Mg, Ca/Na+K and $Al_2O_3/(CaO + FeO + MgO)$ ratios (Clemens & Vielzeuf, 1987). Nevertheless, it is well established that significant partial melting forming 5–20% leucogranitic or granitic (from pelites) to granodioritic and tonalitic (from greywackes and mafics) melts begins in the upper amphibolite facies (i.e. sillimanite + K-feldspar zone) and is common in the granulite facies. The dehydration-melting reactions generally have positive d*P*/d*T* slopes, and as a consequence may proceed with an increase in temperature and/or a decrease in pressure, for example along a looping *P–T* path beyond the maximum pressure, or on steep near-isothermal decompression paths following attainment of the thermal peak.

Relationships between melt water content and temperature, melt viscosity, H_2O content and chemistry, and melt production temperature have been documented experimentally and modelled in detail (Vielzuef & Holloway, 1988; Patiño-Douce & Johnston, 1991; Stevens & Clemens, 1993; White *et al.*, 2001b) for different generic rock types (pelites, greywackes, impure litharenites) and governing reactions. These generally demonstrate that the volumes of melt increase from a few percent (e.g. 5–10 volume %) to 20–40 volume % over the temperature interval 800–1000 °C for common rock types. From this it is inferred that temperatures above 800 °C are required for enough melt production (about 7% by volume) to cause wetting of residual mineral grains, development of melt pocket connectivity, and finally melt transfer and segregation (Sawyer, 2001). Melt loss may ensue from the anatectic high-*T* crustal region through ephemeral conduits that develop and then close during deformation.

Migmatites and migmatite terrains

Migmatites are, in the simplest terms, 'mixed rocks' composed of at least two textural components: a gneissose to schistose component and a granular, non-foliated component (Sederholm, 1907). The precise relationships between these two components, the presence of other textural and compositional variants and components, and the extent to which any granular, non-foliated 'leucosome' component can be representative of a melt have all been the subjects of considerable debate, but it is generally agreed that migmatites in most cases record the complex processes involved in melt production, segregation and extraction in the crust (Sawyer, 1994, 2001; Brown, 2004).

The record of former melt-bearing conduits is preserved in migmatite terrains as veins, lenses and sheets of trapped leucosomes that are often cumulate in chemistry, depleted in those components that escape with the final melt fraction. Further evidence for melting includes transgressive veins composed of ferromagnesian and aluminous minerals interpretable as the peritectic products associated with the melting reactions (e.g. garnet, sillimanite, cordierite) or phases formed through melt-wall rock reaction (Plates 24–29). At lower percentages of melting (<20% of likely melt) the migmatites, on scales of metres to hundreds of metres and more, largely preserve the structural geometries imposed by the gneissose component, even though on all scales there may be veining and rafting. These migmatites are termed metatexites. At larger percentage of melting this structural coherency is mostly lost, giving rise to migmatites known as diatexites.

The loss of a few volume % to several tens of volume % of melt from many high *T*/*P* rocks (granulites (HT) and ultrahigh-temperature (UHT) gneisses), and some medium *P*/*T* rocks (E-HPG gneisses), is inferred from the preservation of pristine grains of the anhydrous minerals formed in prograde dehydration-melting reactions (Sawyer, 1994; White *et al.*, 2001b). This has been verified by measurements of H_2O contents of pristine cordierites in 750–850 °C migmatites from low- to medium-*P* granulite terrains. The recorded H_2O contents (0.7–1.2 wt%) have been shown from experiments to be consistent with equilibration with H_2O-undersaturated melts formed during metamorphism near peak P–T conditions (Harley & Carrington, 2001).

Partial melting, migmatites and the strength and behaviour of orogens

In order for melting to become an important process throughout the orogeny, facilitating the development of large-scale weak zones, the deeply buried rocks must attain the thermal window for melting, outlined above. This either requires incubation following accretion and thickening for some 10–20 Ma or more of self-heating, or the input of additional heat through, for example, delamination of the mantle lithosphere. The latter has been proposed for the Betics, where convective removal of lithosphere may have rapidly increased deep crustal temperatures by over 100 °C within a time interval of <2 Ma at 20 Ma (Platt *et al.*, 1998, 2003b).

Generalising from observations in the Shuswap Metamorphic Core Complex of Canada, Naxos in the Aegean, and other migmatite-cored or migmatite-dominated terrains, Vanderhaeghe (2009) has proposed that the onset and geometry of flow accentuated or triggered by the presence of partial melts impacts strongly on the evolution of large orogenic belts. The most important consequence of partial melting is the resultant lowering of crustal strength and viscosity, which may be critical to the initiation of mid- and deep-crustal flow in 'hot orogens'. Strength and viscosity are proposed to decrease in two stages. For melt proportions of less than about 20 volume %, in which a solid framework still exists, rock strength decreases by about 2–3 orders of magnitude. Beyond about 40% melt, solid connectivity is lost and the material is essentially a melt with suspended crystals; strength is decreased by nearly 10 orders of magnitude. Vanderhaeghe (2009) also suggests that the flow regime of rocks changes along with the marked decrease in viscosity. While a broadly continuous solid network is present the partially molten rock is constrained to flow by laminar flow, whereas once the solid continuity is lost flow can be turbulent.

It can be argued that metatexites represent partially molten rocks in which a continuous solid framework is still present, whereas diatexites are akin to 'dirty' magmas with abundant solids in suspension within melt. A corollary of this is that flow in metatexites is laminar and that these cannot represent material that has been transported (unless *en masse*), whereas diatexites have the propensity to undergo turbulent flow, driven by buoyancy. In several migmatite terrains exhumed from deep in the cores of older orogens, the efficiency of buoyancy is suggested by the presence of sheets of mobilised former melt trapped structurally above, and in some cases connected via veins and dykes to, underlying metatextic migmatites (e.g. Shuswap Complex, Canada: Vanderhaeghe *et al.*, 1999). Turbulent flow in the diatexites is recorded by the heterogeneous distribution and orientations of disrupted or entrained gneissose material within them (e.g. diatexites of Mt Stafford, Central Australia: Collins & Vernon, 1991).

Melting affects rock behaviour on all scales. At the microscopic scale, surface energy minimisation determines melt geometry and hence the connectivity thresholds that allow leucosome formation by melt migration. Hydrofracture and tip propagation in front of connected leucosome pockets promote dyking to form vein networks that may act as conduits. These in turn may drain the sites of partial melting in order to feed intrusive bodies elsewhere, on scales of metres to kilometres (Brown, 2004). The transition to buoyancy-driven segregation and flow of melt, firstly as diatexites and then as cleaner magmas, may drive larger-scale flows in the deep, hot crust, and such flow may be channelled in various geometries (Vanderhaeghe, 2009). An example of three-dimensional horizontal flow would be that driven by gravity outwards to the boundaries of an orogenic plateau, as assumed in most modelling of channel flow in 'hot orogens' (e.g. Beaumont *et al.*, 2004).

The *P–T* realm of metamorphism: the current view

Before embarking on an analysis of the main types of metamorphism and their significance, informed by modelling, in orogen development, it is useful to briefly summarise the results of the vast amount of work that has been accomplished in characterising the *P–T* conditions of metamorphic belts associated with orogenic systems. Brown (2007, 2009) has provided a useful framework for this, which is largely followed herein. This framework builds on the three-fold grouping of metamorphic regions into high *P*/*T* (low *T*/*P*; Blueschist), medium *P*/*T* (e.g. Barrovian) and low *P*/*T* (high *T*/*P*; e.g. Buchan) types identified by earlier workers (Miyashiro, 1973; Turner, 1981), but extends that grouping into the highest-*P* and -*T* regimes that embrace extreme metamorphism (Figs. 7.1, 7.2).

The field of *P–T* conditions preserved by metamorphic rocks in orogens, the 'metamorphic realm' constrained by the modern methods described above (e.g. Brown, 2009), is presented in Figure 7.6. Blueschist, blueschist-eclogite and UHP metamorphic regimes, associated with subduction and terrane accretion, range from 300–1000 °C at 10–55 kbar within a broad field bounded by thermal gradients (dT/dz) of about 5 °C/km and 10 °C/km. 'Eclogite–high pressure granulite' type metamorphism (E-HPG, Fig. 7.1), present in collisional systems such as the Variscan of Europe (O'Brien & Rötzler, 2003), is an extension of Barrovian metamorphism in terms of thermal gradient (dT/dz 10 to 25 °C/km) but includes deep crust that experienced peak-*T* of at least 900–950 °C. Low *P*/*T* metamorphism includes Buchan type but extends to higher temperatures at thermal gradients of 25 to 50 °C/km. This field of metamorphism includes low-pressure metamorphism of the upper crust in extension–contraction systems (e.g. Lachlan Fold Belt: Collins & Vernon, 1992), superstructures of vertically layered orogens (e.g. Grenville Province: Rivers, 2009) or magmatic rocks and intercalated sediments in arc and back-arc settings (Miyashiro, 1973; Bohlen, 1991; Brown, 2007). It also includes granulite and ultrahigh-temperature (UHT, Fig. 7.1) metamorphism at 800–1100 °C and 20–45 km depths that is potentially produced in the deeper levels of large and long-lived collisional orogens (Harley, 1989, 1998b, 2008; Brown, 2007; Kelsey, 2008; Jamieson *et al.*, 2010).

The thermal gradients that characterise these metamorphic groupings do not correspond to any specific steady-state or transient and evolving thermal condition. This is obvious from the fact that the metamorphic rocks in each group track out *P–T* paths along which the recorded *P–T* conditions occur. It is also the case that the higher temperature sub-groups

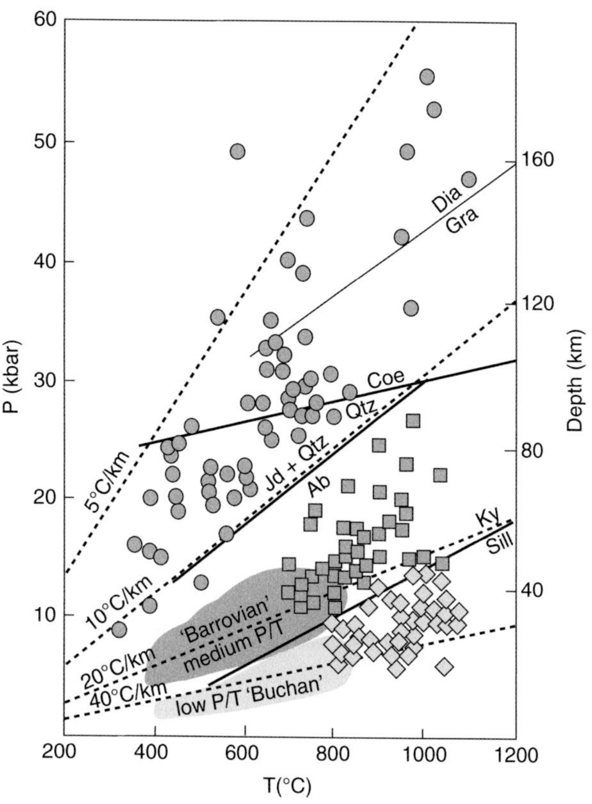

Figure 7.6 The realm of metamorphism, showing *P–T* estimates for groups of metamorphic belts and areas compiled by Brown (2007, 2009), supplemented by additional *P–T* points from Hacker (2006), Harley (2008), Kelsey (2008) and O'Brien (2008). Positions of the principal mineral reactions are from Brown (2007). Shaded fields for low *P/T* (Buchan) and medium *P/T* (Barrovian) metamorphic facies series are as for Fig. 7.1. Symbols are as follows: filled circles, HP–UHP metamorphic belts; filled squares, E-HPG (eclogite–high-pressure granulite) metamorphic occurrences; filled diamonds, HT-UHT (granulite and ultrahigh-temperature) metamorphism.

within each broad grouping do not generally traverse through the *P–T* conditions recorded in their lower-*T* counterparts, and that a number of tectonic settings can produce metamorphic rocks in the same thermal gradient grouping.

Pressure–temperature (*P–T*) paths and pressure–temperature–time (*P–T–t*) paths

The determination or constraining of *P–T* paths, and ultimately *P–T*–time paths, for metamorphic rock units is a major goal of metamorphic studies in orogenic systems. This emphasis arose from the recognition, both from the thermal modelling considerations introduced by England and Richardson (1977) and the earlier metamorphic-petrographic studies by Zwart (1962, 1967, 1969) and others, that regional metamorphic rocks do not simply record peak mineral assemblages along the gradients mapped out by the metamorphic field array, but instead record segments or points along paths that reflect, in simple terms, the interplay between burial, heating and exhumation with time. As a consequence, *P–T* paths have become key pieces of evidence to constrain tectonic and thermomechanical models for the behaviour of orogens, especially when combined with geochronological data that can constrain rates of *P–T* change.

It is important to realise that any *P–T* path depicted for a metamorphic unit is generally only constrained at a few points along its length – the 'path' being filled in by interpolation on the assumption that the determined *P–T* points or segments are linked. The construction of a *P–T* path for any rock suite involves interpreting the *P–T* information from geothermobarometry and calculated phase diagrams in concert with mineral zoning information (Plate 30) that may define *P–T* vectors, and with mineral inclusion, pseudomorphic, and spatially organised corona and mineral intergrowth microtextures (Plates 31–33) that can be used to infer reactions and hence the sense of *P–T* change. Optimally a *P–T* path will be constrained by several such reaction textures, organised in terms of relative time by their relations to structural markers and tied to a number of calculated *P–T* conditions. A rare example of one reaction texture overprinting another and thereby allowing relative timing to be established is given in Plate 33. An important problem to address in any case study is that reaction textures can in principle be generated during polymetamorphism in which the second metamorphism is completely unrelated in time and tectonics to the first one (e.g. Zwart, 1962; Harley, 1992), as distinct from textures generated during polyphase metamorphism occurring on a continuous *P–T* path relating to one orogeny.

The recognition and interpretation of reaction textures is of particular importance, but is fraught with difficulty especially where the textures are complex and spatially variable. The simplest and most easily interpretable textures are those in which single phases are pseudomorphed by their alternative polymorphs, for example coesite by quartz (in UHP rocks) and kyanite by sillimanite (Figs. 7.2, 7.6; Plate 31). Multiphase pseudomorphs that are isochemical with the reactant mineral are also relatively easy to interpret, for example jadeite + quartz replacing albite or Fe–Mg garnet replaced by intergrowths of orthopyroxene + cordierite + spinel. Highly organised, spatially controlled symplectite and corona textures formed on and between pre-existing reactant minerals may also be readily interpretable (e.g. garnet + clinopyroxene reacting to intergrowths of orthopyroxene + plagioclase in granulite metabasites (Plate 34); garnet + sillimanite separated by cordierite + sapphirine moats and intergrowths in granulite pelites (Plate 32)), especially in concert with mineral chemical zoning patterns, but the role of chemical potential gradients must be considered in their analysis (Spear, 1993; Harley, 1998c). In order to calculate *P–T* conditions or phase diagrams appropriate to such reaction textures, which inherently are disequilibrium features, it is critical to be able to define the microdomains over which equilibrium was approached (e.g. Carmichael, 1969; Powell, 1978) and identify whether fluids or melts were involved in the local equilibria (Ferry, 1976; Sawyer, 1999).

There are now a wealth of *P–T* paths documented, with varying degrees of precision and uncertainty, from metamorphic rocks in orogenic belts. Selections of these are presented in subsequent sections of the chapter. The *P–T* paths are mostly characterised by increasing *P* and low-*T* preceding the region of peak *P* and *T*, which are not necessarily coincident, followed by decreasing P at more elevated T, with varying degrees of cooling. These 'clockwise' P–T paths may be 'hairpin' in shape, with maximum P being recorded at or near the maximum T, or broad with the maximum P attained earlier than peak T. The post-peak T paths are generally referred to as 'near-isothermal decompression' paths if cooling is minor compared with the decrease in P (ITD: Harley 1989; $dP/dT > 25$ bar/°C) or 'decompression-cooling' paths if dP/dT is lower and near-parallel to reactions like the kyanite–sillimanite transition, as in the SW Grenville Province of Ontario (Rivers, 2009). Post-peak P–T histories dominated by cooling at depth in the crust of an orogen, with low dP/dT (<5 bar/°C) and consequent production of dense minerals such as garnet in reaction coronas on higher-T minerals, are referred to as 'near-isobaric cooling' (IBC) P–T paths (Ellis *et al.*, 1980; Harley, 1989; Bohlen, 1991). The prograde P–T paths of the high-T upper amphibolite to granulite facies metamorphic terrains that preserve such post-peak IBC histories may be 'clockwise', and preceded by earlier ITD (Harley, 1992). Alternatively, as argued by Bohlen (1991) and Sandiford and Powell (1991), many high-T IBC belts may have followed 'counter-clockwise' prograde paths involving limited burial at high-T followed by cooling at the burial depths attained – for example in a zone of magmatic accretion.

Further discussion of the P–T records and P–T–time evolutions of metamorphic regions follows in the subsequent sections of this chapter, in which the metamorphic records are considered along with relevant geochronology and structural context to define P–T–time histories for the main styles of metamorphism preserved in orogens.

Blueschist–eclogite and UHP metamorphism

Background and *P–T* domain

Eclogites were amongst the first metamorphic rocks to be recognised and systematically documented, for example by Eskola (1915), but for at least 50 years following their identification their status not only as a facies but also as P–T-sensitive rocks was questioned. This controversy arose partly because many of the reported eclogites were found as local 'pips', boudins or relicts in amphibolite facies rocks (e.g. Norway) or 'knockers' in disorganised low-T glaucophane schists as in the Franciscan. De Roever (1955, 1964) and Miyashiro (1973) recognised their importance, and in this were supported by Ernst (1971, 1973) and others. However, the critical evidence leading to the recognition of HP and UHP metamorphism as central to orogenesis was the identification of blueschists and eclogites in structurally coherent or extensive slices in the Alps (Dent Blanche Nappe, Zermatt–Saas, Sesia Lanzo zone, Adula Nappe), New Caledonia, Cyclades (e.g. Syros) and Caribbean (Isle Margarita). In some of these cases isograds of progressive metamorphism could be mapped – for example lawsonite to epidote glaucophane schist – and blueschist to eclogite transitions documented (Miyashiro, 1973; Turner, 1981). A vitriolic debate about the '*in situ*' versus 'exotic' nature of the so-called 'country rock' eclogites in the Caledonian Western Gneiss Region of Norway typified the lack of clarity on the issue: several workers regarded the

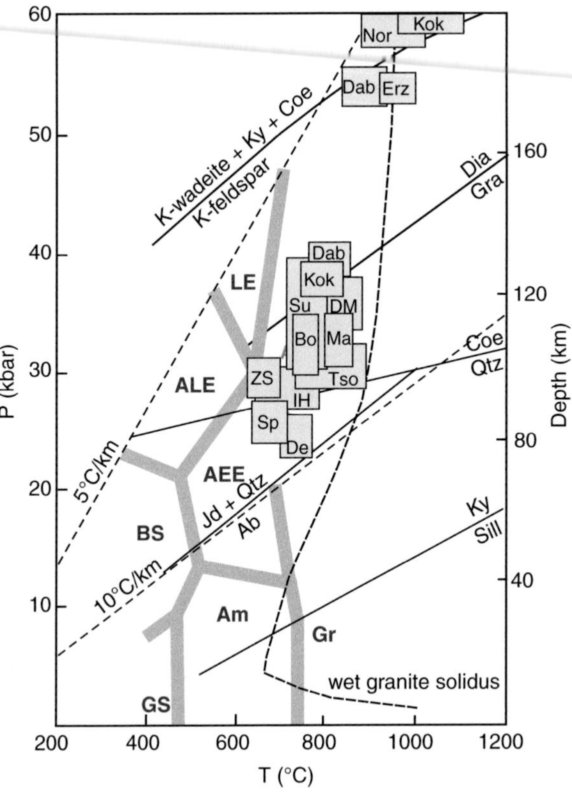

Figure 7.7 *P–T* estimates for selected HP–UHP metamorphic belts and occurrences, based on the compilations of Chopin (2003), Hacker (2006), O'Brien (2008) and Zhang *et al.* (2009). Named areas as follows: Bo, Bohemian Massif, Czech Republic; Dab, Dabie–Shan, China; De, D'Entrecasteaux Islands, Papua New Guinea; DM, Dora Maira, Italy; Erz, Erzgebirge, Germany; IH, Indian Himalaya; Kok, Kokchetav Massif, Kazakhstan; Ma, Mali; Nor, Western Gneiss Region (Nth part), Norway; Sp, Spitsbergen; Su, Sulawesi; Tso, Tso Morari dome, Himalayas; ZS, Zermatt–Saas Fee, Switzerland. Granite minimum melting curve and K-wadeite reaction are from Hacker (2006), other reactions and facies positions are from Brown (2007), as in Fig. 7.1.

eclogites as exotic, probably of mantle derivation and present as tectonic 'pips', whereas others pointed out the systematic spatial variation in calculated *P–T* estimates obtained from these eclogites and argued that they were relics that showed that the whole terrain had been to high pressures. The discovery in the early 1980s of coesite in crustally derived pyrope-quartz rocks in the Dora Maira area of the Italian Alps by Chopin (Chopin, 1984) and in eclogite from the Western Gneiss Region (Smith, 1984) greatly extended the pressure range of metamorphism – leading to the distinction of UHP metamorphism and eventually the remarkable discoveries of coesite–eclogite UHP terrains and diamond-bearing UHP rocks in orogenic belts (Fig. 7.7). This also presented a major new challenge to tectonic models for mountain building and in particular the mechanisms of material transfer and exhumation during subduction, accretion and collision (Chopin, 2003; Hacker, 2006).

The diagnostic minerals and mineral assemblages of HP/UHP metamorphism (Plates 35, 36) allow construction of at least three 'subfacies' within the eclogite facies (Figs. 7.1, 7.7). At the

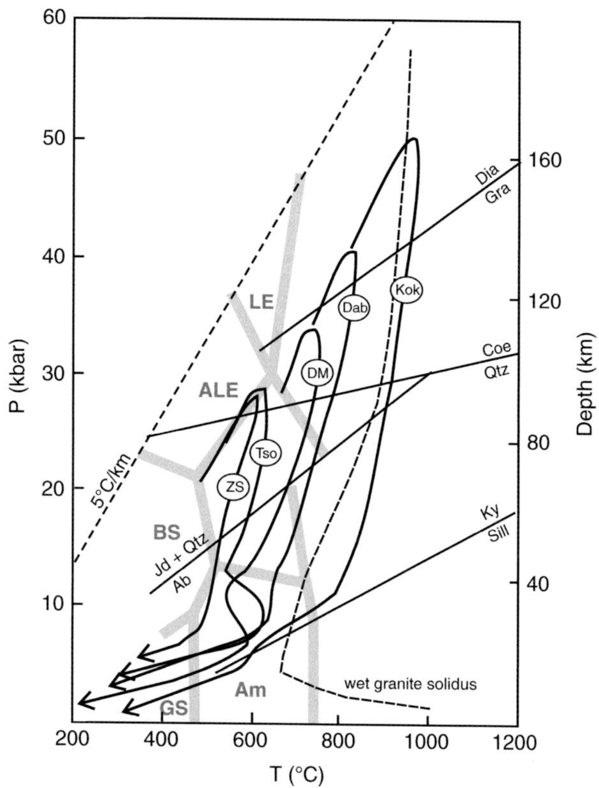

Figure 7.8 Selected *P–T* paths for HP–UHP belts. Dab, Dabie–Shan (Hacker, 2006; Hacker *et al.*, 2006; Zhang *et al.*, 2009); DM, Dora Maira (Schertl *et al.*, 1991; Rubatto and Hermann, 2001; Chopin, 2003; O'Brien, 2001); Kok, Kokchetav Massif, Kazakhstan (Sobolev and Shatsky, 1990; Hermann *et al.*, 2001; Hacker *et al.*, 2003); Tso, Tso Morari dome (de Sigoyer *et al.*, 2004; Beaumont *et al.*, 2009); ZS, Zermatt–Saas Fee, Switzerland (Angiboust *et al.*, 2009). Granite minimum melting curve and K-wadeite reaction are from Hacker (2006), other reactions and facies positions are from Brown (2007), as in Fig. 7.1.

low-temperature end of the spectrum, often forming a high *P/T* facies series in association with prehnite–pumpellyite and lawsonite blueschist facies rocks (metagreywackes and metabasites), are the lawsonite eclogites. These, and lawsonite blueschists, principally occur in subduction-accretion complexes such as those exposed in New Caledonia, Syros, Sulawesi and Venezuela (see Schreyer, 1995; Chopin, 2003 for overviews) but are also found trapped as thrust-bound nappes in 'cold' collision belts, well exemplified by the 44 Ma Zermatt–Saas meta-ophiolite of the European Alps (Angiboust *et al.*, 2009) and the 55 Ma Tso Morari HP/UHP complex of India (de Sigoyer *et al.*, 2004). Peak temperatures, attained coincident with peak pressures of 20–30 kbar on 'hairpin' type *P–T* paths, generally are in the range 520–580 °C in these 'cold' eclogites, and the critical UHP indicator coesite is only occasionally recognised. Post-peak or 'retrograde' *P–T* paths for these high *P/T* terrains are steep in terms of d*P*/d*T* and traverse down-pressure nearly isothermally into and through the epidote blueschist/transitional hydrous eclogite fields at *c*.500 °C to about 10 kbar (Fig. 7.8). These paths then are often characterised by cooling with decompression to 4–5 kbar in the greenschist facies (400–500 °C), and consequent overprinting by albite + chlorite + clinozoisite assemblages in the case of metabasites. In some cases, for example Tso Morari and in the Western Gneiss Region of Norway, the decompressional *P–T* path intersects the epidote amphibolite facies and features a thermal excursion to over 600 °C at 8–10 kbar that is recorded in strong overprinting of the HP assemblages by epidote–actinolite or plagioclase–hornblende assemblages (Fig. 7.8).

Some of the most comprehensive work on eclogite *P–T* records and paths has been in the western Alps. As explained in Chapter 5, the Western Alps are an amalgam of Cretaceous and Eocene orogens. NNE-directed convergence and oblique subduction in the late Cretaceous produced HP/UHP metamorphism in the Sesia Zone, where 70–65 Ma HP metamorphism of eclogitic micaschist was followed by its partial exhumation and juxtaposition with supracrustal schists by 65–60 Ma. UHP metamorphism occurs in both continental (Dora Maira) and oceanic-derived (Zermatt–Saas Fee ophiolite) units of the Axial Zone, at much younger ages than originally proposed based on mica Ar–Ar ages. This *P–T* evolution is episodic, with several diachronous HP/UHP events of more limited spatial extent, each terminated by rapid exhumation on steep d*P*/d*T* paths (O'Brien, 2001; Beltrando *et al.*, 2010).

A detailed prograde to retrograde *P–T* evolution has been documented for the Zermatt–Saas ophiolite by Angiboust *et al.* (2009), who claim it to be the deepest continuous slice of exhumed HP oceanic lithosphere. This is a 'cold' lawsonite eclogite slice, with assemblages of omphacite-garnet ± phengite ± epidote ± lawsonite ± glaucophane in metabasalts and garnet–chloritoid-talc ± lawsonite ± phengite in metamorphosed hydrothermallly altered basalts. These eclogites record peak *P–T* conditions of 23 ± 1 kbar and 540 ± 20 °C over most of the Zermatt–Saas slab. Markedly zoned garnet porphyroblasts coupled with replacive textures such as paragonite + epidote pseudomorphs after lawsonite and subsequent fabrics involving epidote + paragonite + chlorite + albite define a clockwise 'hairpin' *P–T* path featuring near isothermal decompression from the peak conditions to about 8–10 kbar and 480–520 °C, followed by a cooling-dominated terminal path in the mid crust (Fig. 7.8). The *P–T* conditions and path are similar for nearly all of the Zermatt–Saas ophiolite, the detachment and exhumation of which may have been triggered by the arrival of continental crust into the subduction zone. Only the tectonically bounded UHP lens or sheet at Lago di Cignana, from which coesite has been reported, was taken to deeper levels. This may have been a detached frontal extremity of the Zermatt–Saas slice, or have been a separate slice, prior to exhumation.

Ranging up to rather higher pressures in the range 35–55 kbar, but still lying on a d*T*/d*z* gradient of about 6–7 °C/km for the peak assemblages and so preserving equilibration at temperatures from 750 to perhaps 950 °C, are the HP/UHP terranes that lack lawsonite but instead contain epidote-bearing assemblages (Fig. 7.7). These include epidote–glaucophane–omphacite–garnet transitional eclogites in several Alpine cases (e.g. Sesia Zone), as well as epidote eclogite sensu stricto (e.g. Norwegian Caledonides, Dabie–Sulu). The first-order diagnostic mineral for UHP metamorphism is coesite, the high-*P* polymorph of quartz. Coesite has now been identified in perhaps 22 different HP/UHP terrains globally, ranging back to late Proterozoic in age. These coesite-bearing UHP terranes range in area from large belts such as the Dabie–Sulu belt of China (30,000 km^2: Zhang *et al.*, 2009),

Western Gneiss coesite province (22,000 km^2: Carswell *et al.*, 2003) and Kokchetav Massif of Kazakhstan (1200 km^2: Sobolev & Shatsky, 1990) to less extensive slivers or slices such as Dora Maira (150 km^2: Chopin, 1984; Schertl *et al.*, 1991), parts of the Bohemian Massif, and Tso Morari in the Himalayas (de Sigoyer *et al.*, 2004). Diamonds have now been recorded from the Dabie–Sulu, Western Gneiss UHP sub-area, Sulawesi, Erzgebirge, Rhodope, and Quinling UHP terrains (Chopin, 2003). The UHP domains in most cases are bounded by tectonic contacts with either thrust or extensional geometries in present orientation. Significant metamorphic pressure 'gaps' are evident between the UHP units and their overlying (and underlying) rock units – at Dora Maira this gap is of the order of 14–17 kbar, corresponding to 45–55 km of 'missing' structural thickness (Chopin, 2003), whilst in the Tso Morari case the pressure gap between the upwardly domed UHP unit and the overlying nappe unit is at least 16 kbar, and may be as much as 30 kbar if a reported microdiamond occurrence in the UHP nappe is verified from further localities.

The *P–T* conditions of UHP metamorphic belts are constrained, in most cases, by several mineral equilibria independent from the basic but essential observation of coesite as inclusions in garnet and other phases (omphacite, carbonates) or rarely as a matrix mineral. Figure 7.7 shows the *P–T* estimates for several UHP and HP metamorphic areas found in recent and older orogens, together with some of the constraining equilibria. The Dora Maira example, constrained rather well by mineral equilibria within pyrope–coesite and jadeite–kyanite–coesite assemblages, preserves 35 Ma UHP metamorphism at 34 ± 2 kbar and 700–750 °C, followed by a post-peak decompressional *P–T* path traversing over 70 km prior to significant cooling (Fig. 7.8). The question of how fast this near-isothermal decompression, typical of the UHP terrains, occurred is one that is critical to models of their genesis and exhumation.

Timescales of HP/UHP metamorphism and rates of exhumation

Significant effort has been made to determine the ages and timescales of the UHP metamorphism in Alpine (e.g. Dora Maira), Himalayan (Tso Morari) and Cambrian (Kokchetav) case studies. The approach has been to date accessory minerals that coexist with the UHP minerals, or in optimal cases preserve inclusions of coesite or diamond, and accessory minerals in lower-*P* overprinting assemblages, and complement these data with Ar–Ar ages on micas, Lu–Hf or Sm–Nd isochrons from garnets, and other methods to define temperature–time (d*T*/dt) gradients and from these and the *P–T* paths constrain exhumation rates, d*z*/d*t*. Gebauer, Rubatto, Hermann and others (de Sigoyer *et al.*, 2004; Hermann *et al.*, 2001; Rubatto & Hermann, 2001; Carswell *et al.*, 2003; Hacker *et al.*, 2003, 2006) have applied this type of detailed and integrated approach to several terrains to show that exhumation rates in each case had to be 'fast' – of the order of centimetres per year.

There are problems in precision and accuracy inherent in defining d*T*/d*t* and d*z*/d*t* from the geochronology, as the reported ages often have uncertainties of ±2–5 Ma that are of similar magnitude to the apparent timescale of post-peak decompression and exhumation from UHP to mid-crustal depths. For example, in the Tso Morari case zircon U–Pb geochronology yields an age of 55 ± 5 Ma for UHP assemblages (de Sigoyer *et al.*, 2004) compared with 48 ± 5 Ma for 12 kbar amphibolite overprinting that occurred at temperatures still greater than the closure temperatures of several chronometers. The high-T decompression could, therefore, have occurred over a period of 1 to 15 Ma, and dz/dt accordingly could have been between 6 and 0.4 cm/a, in the absence of additional chronometer information. In most cases the age and P–T information are further complicated by the evidence, seen in reaction textures in medium-P assemblages that overprint the UHP ones, for at least two-stage exhumation. In Dora Maira precise U–Pb ages on zircon included in UHP garnet gives an age of 35.4 ± 1 Ma for metamorphism at a burial depth of *c*.110 km. This age is supported by a UHP-titanite age of 35.1 ± 0.9 Ma (Rubatto & Hermann, 2001). Later titanite, formed in equilibrium with overprinting assemblages formed at only 35 km, gives an age of 32.9 ± 0.9 Ma. These results, and consistent garnet Lu–Hf ages, constrain this 'first phase' exhumation rate to *c*.3.4 cm/a. Similarly impressive constraints on dz/dt are obtained from the Kokchetav Massif (45 kbar, 950 °C) UHP terrain, possibly the most extreme UHP terrain recognised. The first zircon data obtained from the microdiamond-bearing metamorphosed continental metasediments defined an age of 530 ± 7 Ma (Sobolev & Shatsky, 1990). Further detailed zircon U–Pb dating by Hermann and co-workers (e.g. Hermann *et al.*, 2001) has revealed a more complex history that refines the timescale of UHP. Metamorphic zircons are polyphase, with pre-UHP cores overgrown by zones that contain inclusions of diamond, omphacite, coesite and Ti-phengite and yield a peak-UHP age of 525 ± 5 Ma. Distinctive overgrowths that occur with biotite + garnet + plagioclase and preserve flat HREE patterns yield ages of 528 ± 8 Ma for a granulite overprint. Amphibolite facies zircons, depleted in Eu but high in HREE due to garnet breakdown, yield similar ages of 526 ± 5 Ma, and Ar–Ar mica ages indicate cooling to *c*.300 °C by 515 ± 5 Ma. Taken as a whole, these results allow a maximum of 6 Ma for decompression from the peak UHP conditions (at 150 km) to granulite and amphibolite conditions (at 30 km) – requiring dz/dt of at least 2 cm/a and confirming fast exhumation rates of UHP rocks in orogenic belts (Fig. 7.9).

Extremely young UHP metamorphic assemblages (24–30 kbar, 870–930 °C) have been recorded from Fergusson Island in the D'Entrecasteaux Islands of Papua New Guinea. The HP/UHP ophiolitic metabasite unit in this example is in faulted contact beneath much lower-grade and essentially unmetamorphosed sediments, in a geometrical relationship similar to core complexes. Ar–Ar cooling ages of 4–2 Ma coupled with the zircon evidence for UHP and HP metamorphism at 8 Ma demonstrate the rapid exhumation of these UHP rocks, initially driven by buoyancy and evolving into unroofing beneath an extensional detachment. The exhumation path here is undeniably fast – dz/dt in the range 1–3 cm/a is suggested by the exposure of these rocks at the Earth's surface in recent times (Baldwin *et al.*, 2004).

Models for UHP rock exhumation

Several workers have developed two-dimensional coupled thermomechanical models of subduction channels, defined as finite-width active zones of material transport located at the upper interface of a downgoing slab with its overriding plate. Gerya and co-workers (Gerya *et al.*, 2002) have considered the subduction channel and its adjacent or overlying

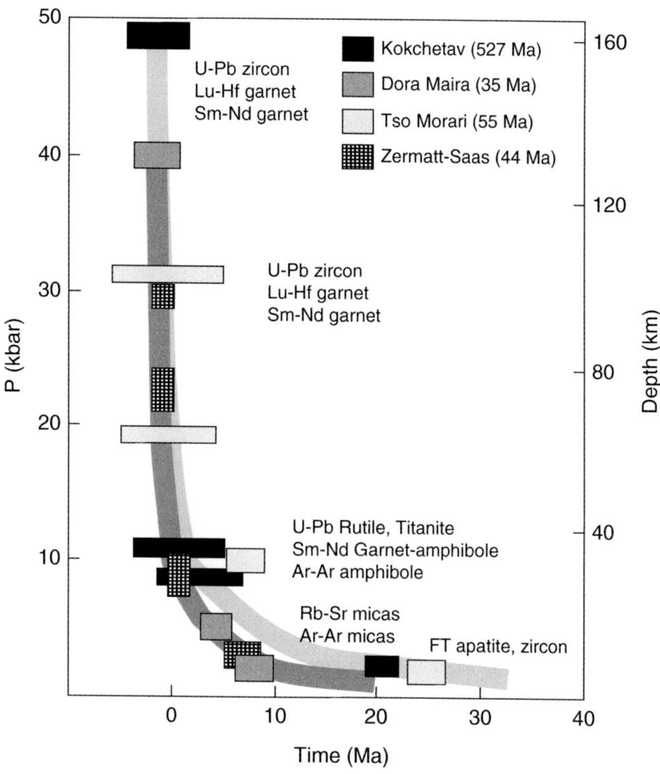

Figure 7.9 Depth vs. time records for selected HP–UHP belts. Sources as for Fig. 7.8. The geochronology is referenced in each case against the established age of peak-*P* metamorphism, obtained from high-*T* geochronometers (e.g. zircon U–Pb) as labelled on the diagram. These reference ages are noted in the key. Depth has been calculated from the *P–T* path using the pressures at those temperatures in the ranges of the closure temperatures for the lower-*T* mineral chronometers, or more directly from *P–T* conditions established for overprinting mineral assemblages (e.g. Tso Morari, Dora Maira).

orogenic wedge in scenarios prior to any collision and also evolving towards and into collision (Figs. 7.10, 7.11). Material is fed into the subduction channel from either or both of subduction-erosion of the active margin plate and the in-feed of sediments (oceanic and accretionary prism). In their models, depending on velocity and subduction dip, four domains or evolving zones are developed within about 10 Ma of subduction initiation. Oceanward, near the trench, is the low-grade (low *P* and *T* – zeolite/prehnite–pumpellyite facies) accretionary complex. Inboard of this is a wedge of nappes of variable provenance and scales of interleaving, followed by a zone comprising either a megascale 'melange'or stacked set of HP and UHP slices extruded from the channel (Gerya *et al.*, 2008). Finally, stranded on the upper plate, is a lid or backstop/lid zone in which the metamorphism is largely of low *P/T* type or in which rocks have not experienced significant burial during deformation (Stöckhert and Gerya, 2005). It is argued by Gerya that the presence of UHP

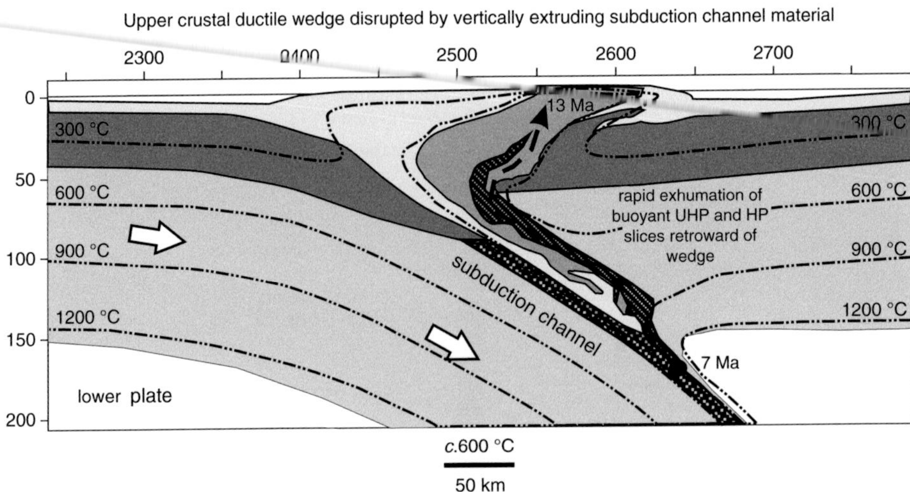

Figure 7.10 Snapshot of the results of coupled two-dimensional modelling of subduction and accretion by Gerya and co-workers (Gerya *et al.*, 2002, 2008). Intense deformation in the accretionary/orogenic wedge overlying the subduction zone coupled with vertical extrusion of low-viscosity material from deep within the subduction channel leads to a retro-wedge zone of stacked HP–UHP nappes and slices. This snapshot shows the lithospheric cross-section at 13 Ma following incorporation of low-viscosity material into a subduction channel. The 13 Ma isotherms are strongly disturbed in the ductile wedge, driven upwards by the buoyant extrusion of the HP–UHP packages over a 6 Ma time period after reaching peak-*P*. An approximate depth–time path for one UHP rock package, from 7 Ma to the snapshot time of 13 Ma, is shown (dashed path). Published by permission of Elsevier.

continental material as slices does not necessarily indicate collision, as these could be sourced from thinned crustal basement captured from the downgoing slab or from deeper continental crust tectonically stripped from beneath the upper plate. It is further suggested that the amount of UHP continental material is so small in cases such as the Alps that its derivation from subducted microcontinental slices is less likely than from subduction-erosion of the upper plate, which can also source mantle peridotite slivers and pips.

An important feature of these models is the allowance for mantle wedge hydration and serpentinisation, which modifies mantle density and strength. Another feature is the adoption of layered viscosity approximations to distinguish different types of input material (e.g. soft sediments vs. harder older crust). The strongly *T*-dependent flow laws used can lead to intense distributed deformation, with overturn of units on a 100 km scale, and 'marble-cake' mixing of hydrated, hybridised and melt-infiltrated mantle materials, with some subduction component, at 50–70 km beneath the fore-arc (Gerya *et al.*, 2008). The viscosities control whether the large-scale configuration consists of stacked nappes and coherent units or melange-like mixtures of rocks. Whatever the structure produced during this accretionary phase, it can in principle be 'locked in' prior to any terminal collision, producing a structural and metamorphic template that is then disturbed and re-oriented by subsequent collision-related deformation and metamorphic events.

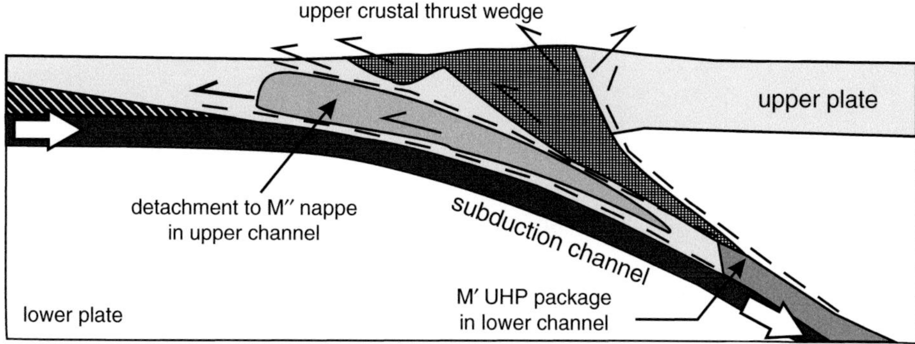

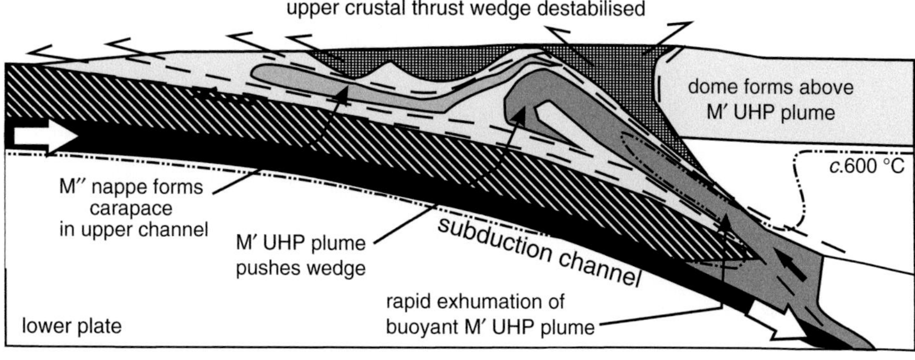

Figure 7.11 Processes involved in the exhumation of HP–UHP rocks according to the models of Beaumont *et al.* (2009) and Warren *et al.* (2008). In a, the combination of relatively thin subduction channel and strong crustal materials being translated down the channel leads to the UHP rocks remaining attached to the downgoing slab, whereas nappes detach in the upper channel and are expelled upwards and to the pro-side of the overlying wedge (M″ nappe of HP rocks). In b, the thermal weakening of the material transferred down subduction zone leads to its detachment and extrusion as a plume (UHP M′ package) back up the channel to catch up with nappes in the upper channel (e.g. M″) and hence slow down in ascent rate and cause doming beneath and pro-ward of the overlying wedge. Redrawn and published by permission of Elsevier.

The model metamorphic *P–T* paths generated for material that approaches and is taken down the subduction channel are generally characterised by rapid near-isothermal decompression (ITD) following 'hair pin' peak *P–T* conditions after prograde burial, or subsequent to a period of deep-seated (HP to UHP) cooling (IBC) before entrainment in the channel and return to the overlying deforming wedge. Some material is transported to 100 km or deeper, and then is transported back up the topmost 1–4 km wide channel zone by buoyancy-driven forced return flow, perhaps activated by slab breakoff, over a total time period of *c.*20 Ma. The standard model in this case yields peak-*T* of only 500–600 °C at 30 kbar, lower by 100 °C or more than recorded by several of the HP/UHP rocks for

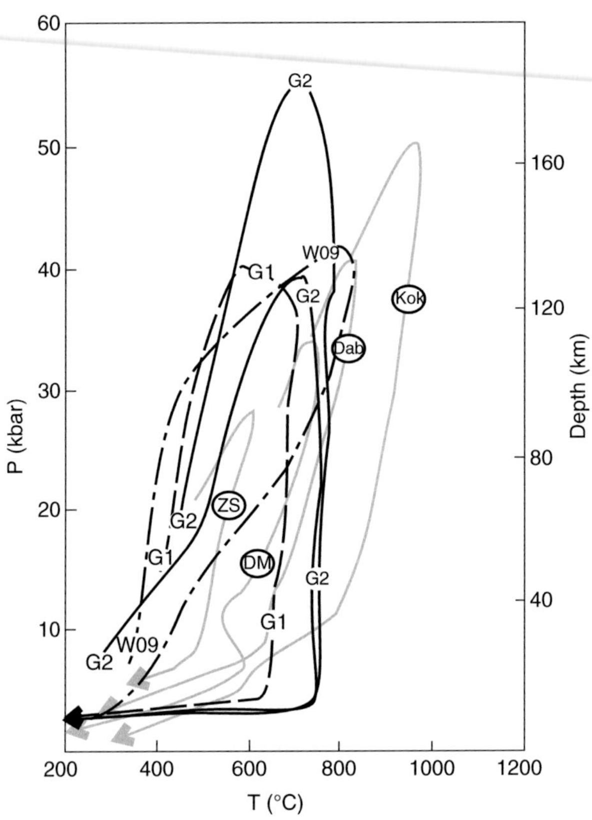

Figure 7.12 Comparison of P–T paths determined for selected HP–UHP regions (from Fig. 7.8) with model P–T paths for selected rock packages in the models developed by Gerya *et al.* (2008) [G1 and G2] and Warren *et al.* (2009) [W09]. The G1 path is for a package taken to 40 kbar in the Geyra *et al.* (2008) reference model, in which the effects of shear heating are not added into the model. The two G2 paths are for rocks taken to *c.*40 kbar and 55 kbar in the Gerya *et al.* (2008) model that includes a component of shear heating. Comparison of G1 and G2 paths shows a T difference of about 100 °C for $P < 30$ kbar during exhumation. The very rapid exhumation rates in these models leads to higher-T at low-P when compared with natural cases. The W09 model P–T path reaches P–T conditions compatible with UHP areas but the dP/dT slope of the exhumation path is only moderate, leading to lower-T at 10–20 kbar in the model when compared with the natural cases.

these pressures (see Fig. 7.12 later). The introduction of shear heating, or viscous dissipation, as an additional and effective heat source for metamorphism in the thin channel leads to higher-T conditions at pressures within the UHP range (Gerya *et al.*, 2008).

Notwithstanding the absolute temperatures attained, these subduction channel models are very effective in yielding situations in which rock packages that are adjacent now (or at the end of the model) have very different P–T paths and peak-P conditions, consistent with field observations. A further outcome of this model is that UHP rocks, generated from material fed into the channel early in the subduction phase, may return near to the surface early, episodically and prior to any collision. This type of history matches well with the records of HP and UHP rocks in active accretionary belts such as Indonesia (e.g. Sulawesi).

Stronger upper-crustal lids on the continental side of the channel, or retro-dipping backstops, prevent the rapid uprise of material and its exposure. In some test cases this leads to the rocks previously taken down to 100 km or more finally residing in the middle crust for a period of thermal incubation and overprinting, unless further exhumation is promoted by subsequent collision. Units that remain resident in the middle crust would have their HP/UHP assemblages largely overprinted by Barrovian-type medium *P*/*T* mineral assemblages, as observed in some of the HP and UHP case studies referred to above.

The modelling of *P*–*T*–time–deformation records in the subduction channel has also been considered recently by Warren *et al.* (2008) and Beaumont *et al.* (2009), with particular reference as to whether slab break-off is an essential pre-requisite for buoyancy-driven return flow (Fig. 7.11). These models begin with an established subduction zone being fed with material carried on the downgoing plate. The factors critical to determining whether buoyancy-driven exhumation occurs are the magnitudes of down-channel shear-coupling and strain-weakening, which allows downward traction to decrease and buoyancy to take over. As the amount of strain-weakening is mainly controlled by the thermal state, composition and mineralogy of the channel material, the time at which the balance turns in favour of buoyancy will be linked to the type of crust subducted. If this is the case then slab break-off, as for example proposed and applied by Chemenda *et al.* (1995), is not required as a trigger for return flow up the channel (Warren *et al.*, 2008).

In the scenario examined by Warren *et al.* (2008) the subducting continental crust has an assumed input geometry comprising a thinned and hotter margin which arrives earlier, and a stronger interior and lower crust which is carried into the system later. Progressive decoupling of strain-weakened crustal material then leads to the development of at least two packages or slices of continental crust that reflect the imposed input properties. These packages are taken down the subduction channel to differing depths, detach, and return up the channel with differing histories. The leading edge, continental margin mid-crust package (termed M′) descends to deeper levels where it experiences UHP metamorphism before its detachment is promoted by strain-weakening. A package (M″) that arrives later and is stronger, being from the more interior part of the continental margin, detaches at shallower levels to form a nappe or slab above the M′ package. The latter forms a buoyancy-driven ductile plume or diapir, which eventually drives the overlying HP M″ package upwards. The result is that a UHP slice is finally left situated structurally below an overlying HP nappe, with a significant *P* gap reflecting the differential motion between M′ and M″ during the syn-subduction 'plume' stage. The evolution takes place over a 6–10 Ma time interval following entry of the continental crust into the subduction channel. The material that is exhumed first is the M″ HP nappe that forms a carapace on the M′ UHP unit. This carapace is exhumed by about 10 Ma after incorporation into the subduction channel, having been taken to *P*–*T* conditions in the range 24–28 kbar and 500–550 °C. The underlying M′ package, taken in this model to 41.5 kbar and 800 °C, is exhumed within a further 2 Ma. The averaged exhumation rates are therefore 2.2–4.9 cm/a over this 10–12 Ma time interval. The model generates clockwise-hairpin *P*–*T* paths that are in most respects compatible with those determined from HP/UHP terrains (Fig. 7.12). There is a trade-off between peak-*T* and strain-weakening: greater and earlier strain-weakening causes rocks to return from shallower depths, and even if the rocks get to 30 kbar the peak-*T* still does not go beyond 520 °C. If no strain-weakening occurs then most material does not detach until much deeper in the mantle, at which depths it may underplate the whole lithosphere, as is seen in some of the models presented by Gerya and co-workers. With strain-heating as an additional input, the P–T paths are displaced to higher-T and exhumation occurs at about 7–9 Ma after arrival of the continental material into the subduction channel. In this case temperatures at 20–35 kbar are greater than 650 °C, and ITD follows the peak P–T condition (Fig. 7.12). The effect of any slab break-off during the subduction-collision phase is to decrease the slab angle and facilitate widening of the subduction channel and hence incorporation of mantle material into the suite of packages, but it is not a pre-requisite to exhumation.

The Warren *et al.* (2008) model derives its UHP continental crust from an impinging continent arriving at the subduction-accretion zone, rather than from subduction-erosion. Of course, both may be possible at different stages of the evolution of the subduction zone and depending on the width of the intervening ocean prior to collision. Old continental interior material that arrives later than the packages discussed above will only detach at >700 °C. One driver for exhumation, at a later stage of collision when strong crust enters the subduction zone, is 'plunger extrusion in which, like a piston, a stronger (i.e. less strain-weakened) sheet of material pushes downwards and causes adjacent/overlying weaker material to be extruded upwards along its margin, giving two-way flow (Warren *et al.*, 2008).

Beaumont *et al.* (2009) have applied these concepts to the Tso Morari UHP domain, using the observed geology, geometries of units and P–T–t paths as constraints. An observation central to their modelling is that the UHP belt occurs in a structural dome as a sheet bounded above and below by lower-P rocks. Whilst some of these 'lower-P' units may also be HP, the flanking units are low-grade accretionary wedge and/or upper crustal sediments. The UHP rocks are spatially associated with ophiolites near the suture, and are bounded by nappes across ductile shear zones characterised by thrust-sense movement. Steep UHP structures are overprinted by shallow amphibolite facies ones. Extensional structures only occur at the highest structural levels, and are syn- to post-UHP exhumation in relative age. Beaumont *et al.* (2009) argue that these features point to a UHP plume ascending up the subduction channel to interact at lower P with the overlying accretionary wedge, with or without an intervening HP carapace package, to form a dome within 11 Ma of the initial collision. At Tso Morari, collision at 55 Ma yields an age of 53–50 Ma for UHP at 28–30 kbar and 620 °C. Extension of the nappe occurred at *c*.48 Ma and cooling to <300 °C by 40 Ma. The modelling, consistent with this P–T–t record, gives a time span of about 1 Ma at peak UHP conditions, and exhumation in only 1–2 Ma at 3–6 cm/a.

Collision and medium *P/T* metamorphism: Barrovian type metamorphism

Barrovian facies series and the Himalayan case study

Barrovian type (medium P/T) metamorphism, long regarded as 'typical' of orogenic belts, has been extensively described and evaluated in terms of its metamorphic mineral zones and assemblages, piezothermal arrays and P–T conditions and paths for many examples around the globe – including in the original Barrow's zones themselves, in the Acadian of east USA and Lepontine metamorphism of the Swiss and Italian Alps. However, nowhere has this style of metamorphism been more intensively studied, and its implications for orogen evolution debated, than in the Himalayas (Plates 21, 22, 23).

Deep subduction (80–160 km) of the leading edge of the Indian plate occurred from between 55 and 45 Ma (de Sigoyer *et al.*, 2000), producing the relatively scarce HP and UHP metamorphic remnants present in the northern Himalaya. The principal metamorphic feature of the Indo-Asian collision, however, is exposed as a 50–100 km wide and 1500 km long belt south of and beneath a major normal-sense shear zone system, the South Tibetan Detachment (STD). This large-scale regional metamorphic belt is present in two tectonically bound slabs of Indian crust – the Greater Himalayan Sequence (GHS) (also known as the High Himalayan Crystallines, HHC) and the Lesser Himalayas (LH). The metamorphism is young in age, laterally extensive and well exposed in an active collision zone. Metamorphism, though complex, started with broadly Barrovian kyanite–sillimanite grade metamorphism at 34–32 Ma (Searle *et al.*, 2006) and continued with a high-T overprint at 24–17 Ma in the GHS, and may be younger (14–4 Ma) southwards. The metamorphic zones are little disturbed in their geometry and disposition by post-orogenic events, and – most spectacularly – inverted in their preserved thermal profile. Inverted metamorphism refers to the geometry of the metamorphic field gradient or piezothermic array, and is the situation in which the highest-grade metamorphic rocks occur in the structurally highest field position. In other words, the deepest rocks are 'cooler' or preserve lower-grade assemblages than the shallower rocks.

The key metamorphic data that inform the thermal models for the evolution of the Himalayan orogenic front include the metamorphic field gradient and its calculated P–T distribution, the P–T paths obtained from rocks within the GHS, in the MCT zone and beneath it, and the distribution of peak-T and cooling ages for rocks at all structural levels. There are several well-documented sections for which such data are available (Searle *et al.*, 1992, 2006; Goscombe *et al.*, 2006; Kohn, 2008). Noting that there are some significant variations (Searle *et al.*, 1992), a generalised metamorphic cross-section in the India to Nepal region is as depicted in Figure 7.13 (Goscombe *et al.*, 2006; Kohn, 2008). This is complemented below by a consideration of the metamorphic and geochronological evidence available from sections located along the Langtang Valley and near Mount Everest.

As described in a previous chapter, the famous Main Central Thrust lies broadly in the vicinity of the boundary between the GHS and LH but does not necessarily coincide with this protolith boundary. Going up-section from beneath the MCT, the metamorphic field gradient or piezothermal array is structurally inverted. A Barrovian zonal sequence is seen, with chlorite, biotite, garnet and kyanite zones exposed over a structural thickness of 3–4 km in the LH as the MCT is approached. Barrovian-style kyanite zone metamorphism (Ky + Musc) occurs in the GHS at the MCT, and the apparent grade of the GHS increases upwards into sillimanite–muscovite and eventually sillimanite–K-feldspar migmatite zones within the GHS from 2 km to 15 km structurally above the MCT. At Langtang the peak P–T conditions in the garnet zone (LH) are 8 ± 1 kbar and 550 ± 50 °C. Within the kyanite zone the preserved pressures vary between 8 and 12 kbar, but with considerable geobarometric uncertainties, and T generally increases up section from 600 °C to 700 °C

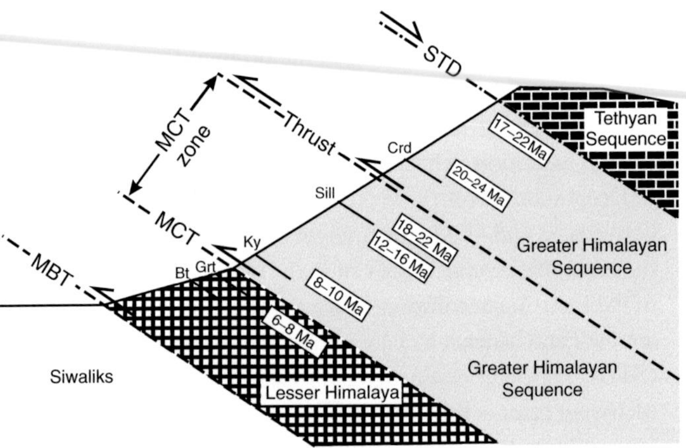

Figure 7.13 Generalised schematic section across the metamorphic zones of the Lesser Himalaya and Greater Himalayan Sequence, from the Main Boundary Thrust (MBT) to the South Tibetan Detachment System (STD), modified from Goscombe *et al.* (2006). The section depicts the general position of the biotite, garnet, kyanite and sillimanite, in isograds within the generally inverted metamorphic zonal sequence, and the first appearance of cordierite high in the GHS. Age information on the termination of high-*T* metamorphism (numbers in rectangles) is from the compilations of Goscombe *et al.* (2006) and Kohn (2008). Redrawn and published with permission of Elsevier.

and to perhaps 750 °C. At higher structural levels, within the GHS, peak temperatures are maintained at >700 °C and may be as high as 825 ± 25 °C in the sillimanite–K-feldspar migmatites that occur within the top 5 km of the section beneath the STD. Pressures corresponding to these peak-*T* conditions decrease from >9–10 kbar to perhaps 5–6 kbar across the 15 km of structural height in the GHS (Fig. 7.14). This *P–T* distribution means that the metamorphism below and into the MCT is apparently inverted in both *T* and *P*, with uniform d*T*/d*z* of 20 °C/km, and that above the MCT zone inverted only in terms of *T*, with d*T*/d*z* increasing up to 40 °C/km on an increasing-*T* excursion away from the classic 'Barrovian' medium *P/T* greenschist to amphibolite facies series, towards a 'Buchan' style low *P/T* metamorphism within marginal granulite facies (Kohn, 2008).

High-level leucogranites in the upper parts of the GHS formed by muscovite-controlled dehydration melting of GHS protoliths, requiring peak *T* of > 700 °C. These leucogranite melts segregated, migrated and came to be emplaced at upper structural levels close to the STD, coeval with south-directed thrusting along the MCT zone at 22–18 Ma (Reddy *et al.* 1993; Massey *et al.* 1994; Fraser *et al.*, 2000). High strain, syn-sillimanite, shear zones and ductile thrust surfaces displacing sillimanite-zone over kyanite-zone rocks within the GHS indicate deep-seated top-down, south-directed stacking and shearing, supported by Kohn's (2008) analysis. The overall averaged d*P*/d*z* gradient of 0.27–0.35 kbar/km calculated across the GHS may therefore be coincidental. It does not actually reflect a true lithostatic load but instead reflects the presence of early thrust-like tectonic breaks combined with ductile thinning of GHS slices (Fraser *et al.*, 2000; Kohn, 2008), associated with high-*T* metamorphism that overprints an earlier, 35–31 Ma, kyanite-grade one.

The *P–T* paths of schists from the LH and GHS have been constrained using mineral inclusions, modelling of zoning patterns in garnet, and calculated *P–T* phase

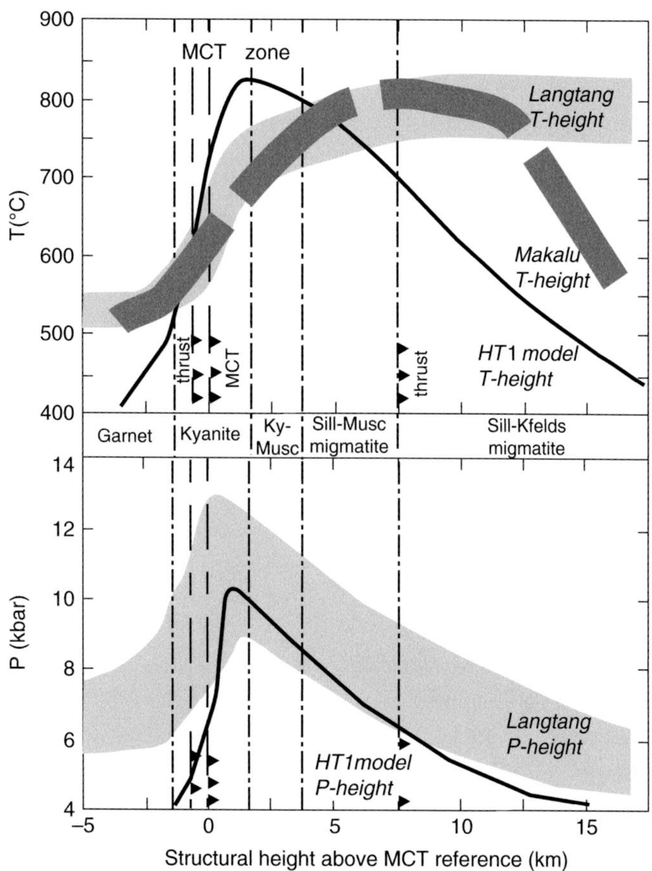

Figure 7.14 Variations in metamorphic *P* and *T* with approximate structural height above the MCT (or its inferred correlative), compared with predictions from the HT1 channel flow model of Beaumont *et al.* (2004) and Jamieson *et al.* (2004), modified from Kohn (2008) and with additional *T*-height data from Goscombe *et al.* (2006) ('Makalu' curve). Metamorphic zones going up-section are named according to the isograds documented by these authors. The maintenance of high metamorphic *T* to high structural levels (Langtang section – Kohn (2008)) differs from the predictions of HT1. Estimated metamorphic pressures in general are greater than those predicted in the HT1 model but follow a similar trend with structural height. Published with permission of the Geological Society of America.

diagrams. At lower structural levels, beneath the MCT zone, the paths are clockwise and 'hairpin' in style, with the retrograde *P–T* path tracking back along similar low-*T* and moderate-*P* conditions to the prograde path, as shown in Figure 7.15. Within the GHS the paths associated with the high-*T* (sillimanite zone) metamorphism are dominated by a heating-cooling phase which is at high-*P* (10–12 kbar) near the MCT and at lower-*P* in the structurally higher migmatite zones. The post-peak IBC recorded in these GHS zones does not appear to have been preceded by any decompression.

The final line of metamorphic *P–T–t* evidence bearing on models for Himalayan orogenesis is the preserved age structure at the orogenic front – the time and timescale

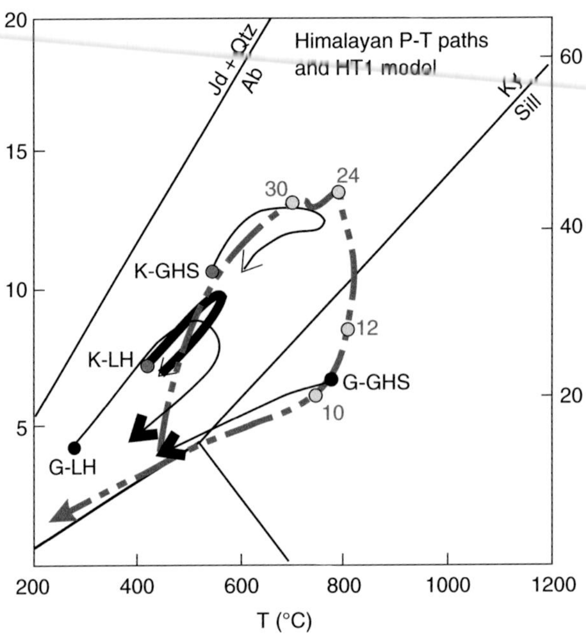

Figure 7.15 Selected P–T paths recorded in Greater Himalayan Sequence (GHS) and Lesser Himalaya (LH) schists in the vicinity of the Main Central Thrust, compared with a typical GHS P–T path (dashed line labelled with ages in Ma, present day being zero) produced by the HT1 model (Jamieson *et al.*, 2004). P–T paths prefixed K- are from Kohn (2008), and relate to the Langtang section. P–T paths prefixed G- are from Goscombe *et al.* (2006) and refer to the Makalu section. None of the documented P–T paths appear to record the extensive P–T excursion and ITD phase produced in the HT1 model.

of high-T metamorphism and syn-metamorphic deformation/thrusting, and the cooling ages of rocks at different structural heights (Fig. 7.16). In this summary we are specifically referring to the 'M2' Himalayan metamorphism and not the occasionally preserved earlier (34–32 Ma) prograde metamorphism at *c*.650–680 °C and 7–8 kbar recorded in zoned garnets and early kyanite-sillimanite fabrics in some GHS rocks. Monazite and zircon U–Pb geochronology for the 'M2' and possibly later mineral assemblages indicate a structurally downward and outward younging in the timing of the end of high-T metamorphism and related deformation. In the upper 10–15 km of the GHS the metamorphism associated with sillimanite occurred at 23–17 Ma. Cooling to 300–400 °C (Ar–Ar estimates) occurred by 17–14 Ma, at rates of *c*.40 °C/Ma. Near the MCT itself, or within the broader MCT zone in the lower GHS, ages for syn-thrusting metamorphism vary from 16 Ma to 8 Ma. At Langtang, displacement on the MCT has been dated at 16 ± 1 Ma whereas syn-metamorphic displacements on structurally lower thrust surfaces are proposed, based on monazite U–Pb ages, to have occurred at 10.5 ± 0.5 Ma (Kohn, 2008). In the frontal part of the system, below the MCT within the garnet-zone LH sequence and Lesser Himalaya Duplex, movement occurred from 14 Ma to perhaps 3.5 Ma, though whether the younger ages actually reflect regional-scale metamorphism at *c*.500 °C or the fault-controlled access of hot fluids is the subject of debate (Searle *et al.*, 2006; Kohn, 2008).

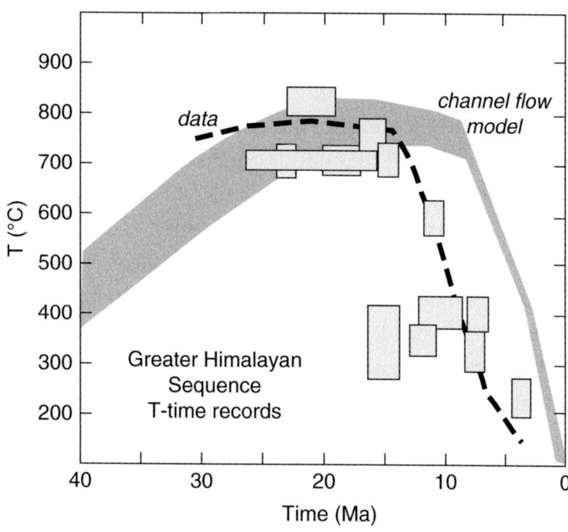

Figure 7.16 Temperature–time cooling paths determined for the Greater Himalayan Sequence based on U–Pb dating of monazite and zircon (peak and near-peak data), and mineral chronometers with different closure temperatures (e.g. mica and feldspar Ar–Ar) to determine cooling ages. Note that cooling from high-*T* conditions appears to have occurred significantly earlier (by 5–10 Ma) than predicted by the HT1 (Jamieson *et al.*, 2004) channel flow model. Modified from Kohn (2008) and with additional data from Searle *et al.* (2006). Published with permission of the Geological Society of America.

It should be noted that the description of the inverted metamorphic zones given above is generalised for the region from the central Indian Himalayas to Nepal, and is not necessarily representative of the entire strike-length of the Himalaya between its eastern and western syntaxes. A notable variant on the geometry is described from Zanskar, in Pakistan (Searle *et al.*, 1992). Here the inverted gradient lies above the MCT, which occurs low down in the structural cross-section. The structure has been interpreted by Searle *et al.* (1992 and subsequent papers) as one involving overfolded isograds, wedged between a lower compressional shear zone (the MCT) and a structurally higher extensional one, the Zanskar Shear Zone (equivalent to the STD). This antiformal megafold is cored by the highest-grade migmatites, and grades decrease structurally upwards beyond this highest-*T* zone, towards the upper extensional shear zone. Recent models have re-interpreted the geometry as one of an extruding channel of hot middle crust, with a cool upper margin adjacent to the STD, leading to a geometry of folded 'fossil' isograds (24–20 Ma old) overprinted downwards and outwards (southwards) as extrusion propagated outwards with time (to *c*.16 Ma) and broadened the extruding zone (Searle *et al.*, 2006).

Barrovian collisional metamorphism and inverted zones: models for orogen development

Several models have been proposed to explain the inverted metamorphic sequences of the Himalayas. The original was the classic 'hot iron' model of Pecher and LeFort (1986), in which syn-metamorphic thrusting of a hot slab over a cold one, on the MCT, was proposed to cause heating of the latter and retrogression of the former. Heat transfer across the moving boundary zone would, in this model, produce a continuous but inverted gradient in *T* coupled with a continuous gradient in *P* that correlates with structural height, from below the boundary (MCT) and into the upper slab. A variant of this model added the effects of shear heating on the boundary in order to produce the elevated temperatures at the MCT (600–650 °C) required by the metamorphic mineral assemblages.

Two other models, proposed early in the evolution of ideas on the inverted zones, include post-metamorphic deformation of the metamorphic zones, either by refolding and overturning or by thrust-controlled imbrication. In the case of refolding and overturning, the post-metamorphic inversion of the zones initially formed in a constant dT/dz field gradient would, in the absence of further equilibration, lead to increases in metamorphic *P* as well as *T* upwards (i.e. structural height would correlate negatively with preserved metamorphic pressure). In the case of post-metamorphic stacking or imbrication, in which slices of different grade rocks are re-stacked into an apparent hot-over-cold sequence, discontinuities in *P-T* conditions would be expected across the post-peak contacts.

Recognition that none of these models adequately accounts for the observed disposition of inverted metamorphic zones in the vicinity of the MCT along the Himalayan front (e.g. Spear, 1993), coupled with the ability to investigate the thermal evolution of orogens using two-dimensional thermomechanical models, has led to currently debated models that involve lateral translation of material both into and out of the orogen or invoke temporal shifts in the locus of deformation as collision proceeds. Of these, the channel flow model, described in previous chapters, has gained considerable support and attracted equally considerable criticism. The following section considers the extent to which the channel flow model, as generally described, accounts for the metamorphic record. We also briefly evaluate how the competing model, of lateral extrusion in a widening critical wedge, fits with the evidence preserved in the metamorphic rocks exposed on this orogen flank.

Channel flow and Himalayan metamorphic zones: timing and *P–T* paths

Channel flow is the lateral tunnelling of a low-viscosity middle crust under the influence of gravity (see Chapter 6). In the Himalayan case it is postulated to have accommodated the eastward growth of the Tibetan plateau and enabled dissipation of excess gravitational potential energy by extrusion of a mid-crustal channel southward, facilitated by flank erosion (Jamieson *et al.*, 2004). Transient extension of the Tibetan upper crust in the mid-Miocene is considered in modified models (Jamieson *et al.*, 2006) to have caused the formation of a subsidiary channel at that time that facilitated the north Himalayan gneiss domes. The main channel is proposed by Jamieson *et al.* (2006) to have become active once again in the late Pliocene, possibly stimulated by rise of the monsoon and extreme focussed erosion.

For channel flow to initiate there needs to be a thermal incubation period of perhaps 20 Ma during which large portions of the crust can self-heat, depending on the amount of heat production in the crust involved. (In the model 'hot' orogen used by Beaumont *et al.* (2004) and Jamieson *et al.* (2004), heat production in the upper crust is 2 $\mu W/m^3$ and in the lower crust 0.75 $\mu W/m^3$. The vertical heat flow at the base of the crust, Q_{moho}, is

Homogeneous Channel Flow HT1 (Jamieson *et al.*, 2004)

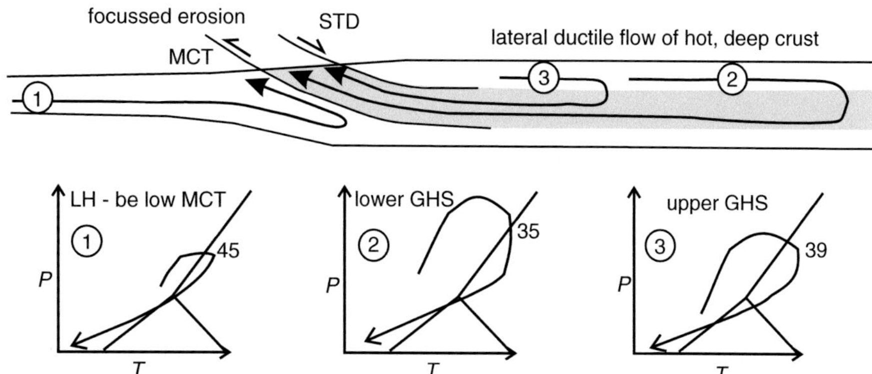

Channel Flow with extruded dome HT111 (Jamieson *et al.*, 2006)

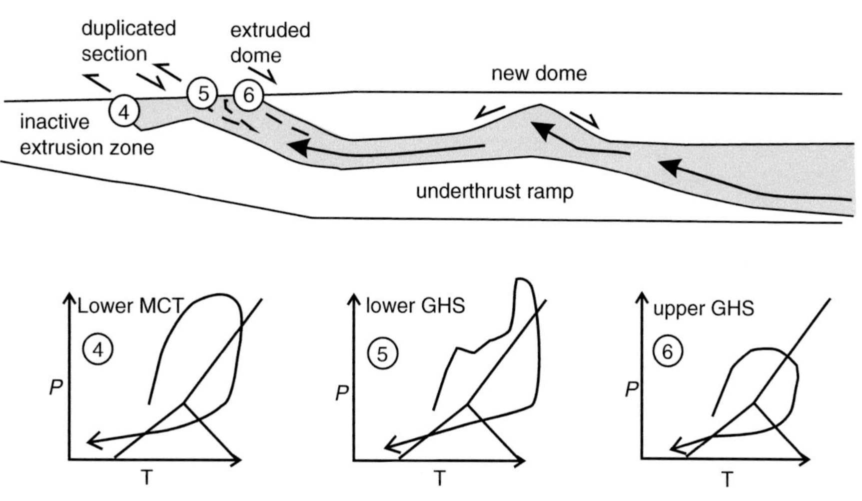

Figure 7.17 Channel flow models and their *P–T* path predictions, based on the original HT1 model of Jamieson *et al.* (2004) and the modified two-stage model investigated by Jamieson *et al.* (2006). Flow paths of rocks incorporated into the ductile, melt-bearing, channel are shown by the arrowed lines. Material points 1–3 and 4–6 at different structural heights in the two models are shown, each with corresponding *P–T* paths referenced against the aluminosilicate phase diagram. In all cases the *P–T* paths describe extensive 'clockwise' loops, and most involve significant decompression through 2–8 kbar at $T > 600$ °C. For paths 1–3 (HT1 model) the age (in Ma) at which peak-*T* is attained is shown by the small numbers. These are the times elapsed since the start of the model, which is 54 Ma in HT1. Hence, in the LH (path 1), 45 Ma corresponds to an age of 9 Ma, and in the GHS (paths 2 and 3), peak times of 35 Ma and 39 Ma correspond to geological ages of 19 Ma and 15 Ma. Redrawn and published with permission of John Wiley & Sons.

20 mW/m^2, T_{moho} is 704 °C and surface heat flow 71.25 mW/m^2). Hence, if collision occurred at 55–50 Ma then only by 35–30 Ma, at the earliest, could a low-viscosity mid-crustal channel have developed. In the standard model (HT1: Beaumont *et al.*, 2004; Jamieson *et al.*, 2004) mid-crustal material that goes into the convergent zone early is

translated into the orogen for 18 Ma (90 km of displacement at 5 cm/a), along with the advancing subduction boundary. At appropriate depths, and provided T >700 °C, this material, represented by GHS in the Himalaya, would be entrained in the low-viscosity zone at depth and then flow outwards to be juxtaposed over and against newly accreted material of the LH at the orogen flank, to form inverted metamorphic zones in the latter unit. The residence time in the orogen is longer for the metamorphic rocks in the channel, leading to higher-T metamorphic records for rocks within the GHS (Fig. 7.17).

In the standard model (HT1: Jamieson *et al.*, 2004) the melt-promoted channel is fully developed 12 Ma after the initial 20 Ma incubation period. It tunnels forward under most of the thickened crust, and eventually much of the deeper crust in the orogen core is at >800 °C, by about 21 Ma before the present. After a further 10 Ma (i.e. 42 Ma after the start of the model) the channel is extruded at the topographic gradient frontal to the mountain belt (i.e. at 12 Ma before present). Once incorporated into the channel, rocks experience their maximum pressure, P_{max} (up to 13 kbar) at 28–26 Ma, whereas their T_{max} occurs in the range 24 Ma to 12 Ma depending on position in the channel. In the model the resultant P–T paths for rocks within the channel are clockwise and involve several kbar of ITD (Fig. 7.17). Cooling of the rocks within the channel, and now at the mountain front, proceeded since 10 Ma and rapid exhumation of the channel rocks since 6 Ma.

Problems with channel flow as a model for the Himalayan metamorphism

Several workers (e.g. Goscombe *et al.*, 2006; Kohn, 2008) have developed detailed critiques of the channel flow model as applied to metamorphism of the Himalayas. The key problems that emerge from comparison of the model (HT1: Jamieson *et al.*, 2004) with the field data are:

1. Channel flow predicts a strong decrease in peak T with structural height within the channel, from 800 °C to 500 °C, concomitant with the decrease in metamorphic pressures from 10 to 4 kbar. This is not consistent with the P–T data outlined above for the GHS, in which rather high-T conditions are preserved irrespective of structural height (Goscombe *et al.*, 2006; Kohn, 2008).
2. P–T paths in the channel flow model are clockwise and involve significant high-P to high-T loops. For example, model P–T paths in the GHS are predicted to involve significant ITD from the kyanite field (12 kbar) into the sillimanite field (10–7 kbar) at 800 °C, with terminal cooling then occurring at <6 kbar (Figs. 7.14, 7.15, 7.17). The lack of an extensive ITD record in most or all GHS rocks prior to cooling is hard to reconcile with the model.
3. Finally, the 'end-peak' age structure appears to be inconsistent with simple channel flow. The model predicts similar, and young, ages for cooling from peak to near-peak conditions for both the LH and GHS. The recorded ages for cooling from peak to near-peak conditions in the GHS are, however, 5 to 10 Ma older (Fig. 7.16). Cooling in the GHS above and within the MCT zone occurred some 5 to 8 Ma too early: according to Kohn (2008), by the time syn-metamorphic deformation was being experienced below the MCT, at <10–5 Ma and 500–550 °C, the GHS rocks of the channel had already cooled

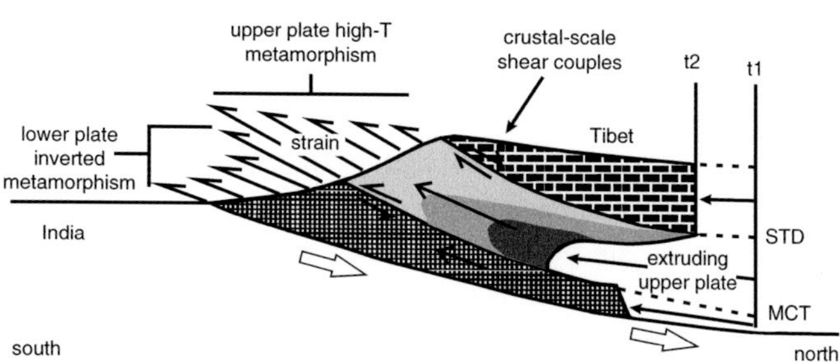

Figure 7.18 Syn-metamorphic ductile extrusion model for development of the inverted metamorphic zonation in the vicinity of the MCT, modified from Goscombe *et al.* (2006). This model incorporates downward and outward variations in strain associated with displacement between the STD at the top of the deforming section, and a southward-propagating MCT zone at and in the base of the structural section. Increasing grade of metamorphism in the GHS is shown by darker shadings. In this conception of the metamorphism there is a decrease in P and T in the upper part of the GHS section, close to the STD. Displacement of the metamorphic zones in the GHS at time t2 is *relative* to that at time t1, and does not indicate total displacements of the rock units across the whole Himalayan front. Redrawn and published with permission of Elsevier.

to below 400 °C. These timing inconsistencies provide significant problems for the generalised channel flow model, which requires longer residence times at high-T in the channel for the hotter metamorphic zones and delayed cooling to within the past 10 Ma.

As an alternative to the channel flow model, Kohn (2008) argues for a critical taper model in which wedge extrusion – lateral expulsion of the GHS and other sequences – occurs. This is conceived to arise because of a pressure gradient imposed by the narrowing of the vertical distance between the surfaces that bound the deformation. The existence of a high-T metamorphic belt in this case does not require large-scale channel flow: the lateral separation between units within and below the region of expulsion prior to wedge movement is 250 km instead of the much larger separation (1000 km) inferred in the channel flow model. Furthermore, the high-T isotherms at depth in the orogen do not tunnel pro-ward but remain essentially stationary with respect to the surface and leading edge of the orogen. This in turn leads to the high-T rocks spending less time at or near their peak-T condition. Goscombe *et al.* (2006) invoke slab extrusion involving distributed reverse movement within the slab, combined with a shear couple between the bounding zones, operating from 24 Ma to at least 13 Ma (Fig. 7.18). They suggest that the lack of any early high-P record in the rocks they examined means that the high-T rocks of the wedge cannot have been displaced up and over the structurally lower rocks. Instead, these workers suggest that continued extrusion is focussed in the lower plate itself (i.e. at lower structural levels). This is consistent with the critical wedge model, in which the disposition of metamorphic zones and P–T–t relationships are controlled by the outward (pro-ward) in-sequence propagation of thrusts, with progressively younger thrusting deformation occurring southward with footwall collapse (Kohn, 2008). This downward and outward propagation leaves its record as a set of southward-younging high-strain zones or thrusts within the GHS and LH, giving spatially discrete metamorphic events: at 24–20 Ma in the GHS; on the MCTZ and STDS at 18–10 Ma; and in the LH at 9–3 Ma.

The critical taper model as proposed by Kohn and others predicts increasing T across the MCT and into the GHS and uniformly high T to high structural levels, but also predicts high preserved P at those levels (10–12 kbar), a feature that is not observed. Searle *et al.* (2006) have also pointed out that the metamorphic significance of the young (< 9 Ma, and especially 3 Ma) monazite ages obtained by Kohn and co-workers is ambiguous, principally because they are younger than mica Rb–Sr and Ar–Ar cooling ages of 17–14 Ma that require the crustal section to be at $T < 300$ °C well before 10 Ma. Searle *et al.* (2006) accept that channel flow controlled the thermal evolution of the GHS over the period 30–17 Ma, but modify the general model to incorporate the combined effects of vertical pure shear flattening within the channel and expansion of the channel through basal thrust propagation down-section with time. At the time of writing of this contribution it is apparent that whilst channel flow cannot explain all the metamorphic P–T and P–T–time evidence from the Himalaya, it is nevertheless an important potential process, and may have been operational for an extensive period of orogen evolution (*c.*12–15 Ma) prior to localisation of deformation to lower structural levels southwards along the Himalayan front.

Collision and medium *P*/*T* metamorphism: eclogite–high-pressure granulite (E-HPG)

Background and *P*–*T* domain

Barrovian metamorphism at greenschist to upper amphibolite facies conditions, as detailed above for the Himalaya, is only part of the realm of orogenic metamorphism that falls within the medium P/T field, as distinguished by field gradients and peak P–T arrays with dT/dz in the range 15–20 °C/km. At generally higher-T, and hence P, are the high-pressure granulites and associated medium-temperature eclogites (labelled E-HPG in Fig. 7.1; Brown, 2007, 2009). These occur at temperatures of 650–1050 °C but preserve significantly lower pressures (12–20 kbar) than the HP/UHP metamorphic rocks described earlier, which lie on a much steeper dT/dz gradient.

The E-HPG group of metamorphic rocks and terrains straddles the eclogite to granulite facies boundary (Figs. 7.1, 7.6). As a result, in appropriate metabasite and intermediate bulk rock compositions the peak clinopyroxene is only rarely a diopsite–jadeite solid solution (omphacite) but more usually is one in which calcium-Tschermaks ($CaAl_2SiO_6$ or CaTs) is an important component, and in which the Jd/CaTs ratio is low (Green and Ringwood, 1967). These rock types therefore range from true eclogites containing garnet + omphacite along with amphibole, quartz and other phases but lacking plagioclase, to garnet + clinopyroxene (omphacite or Alrich diopside) + Ca-plagioclase + quartz + rutile high-P granulites. Higher-T (800–1000 °C) mineral assemblages developed in granitic bulk rock compositions involve garnet + kyanite + quartz + ternary feldspar + rutile, and those in intermediate orthogneisses may contain additional clinopyroxene or lack kyanite (O'Brien and Rötzler, 2003; O'Brien, 2008). Many bulk rock compositions that at lower-P would contain orthopyroxene lack this phase in E-HPG metamorphism because of the stability of Fe–Mg–Ca garnet. For example, the dehydration melting of biotite + quartz, which produces orthopyroxene + melt at *c.*800 °C and $P < 10$ kbar, instead forms garnet + melt at HPG pressures. Whilst lower-*T* medium *P*/*T* rocks with granitic and pelitic compositions may contain muscovite + kyanite, at $T >$ 700–800 °C in the E-HPG field muscovite (phengitic) breaks down to yield garnet + kyanite + melt, producing high-*P* leucogranites such as those at the type granulite localities of the Bohemian Massif.

Pressure–temperature–time paths

In several well-documented cases including the Bohemian Massif of the European Variscides, the southern portion of the Western Gneiss Region in Norway and the Grenville Province of Canada the general form of the post-peak *P–T* path is from HPG pressures of 14–20 kbar into lower-*P* amphibolite or granulite conditions of 7–10 kbar and 650–800 °C. Critical evidence for this includes the replacement of the high *P–T* clinopyroxenes by plagioclase + lower-Jd clinopyroxene or plagioclase + hornblende, of Fe–Mg–Ca garnet by intergrowths of orthopyroxene + spinel + plagioclase, and of garnet + clinopyroxene by orthopyroxene + plagioclase + minor amphibole. Kyanite-bearing HPG assemblages provide supporting reaction texture evidence in the form of spinel + plagioclase symplectites replacing kyanite.

There is considerable variability in the post-peak *P–T* paths of E-HPG regions in terms of the amounts of true ITD compared with decompression-cooling (Fig. 7.19), and in the

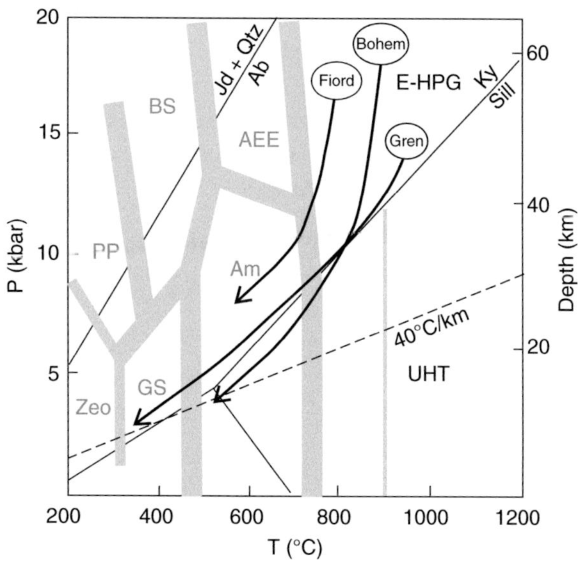

Figure 7.19 *P–T* paths for selected E-HPG terrains. Facies boundaries and fields and positions of reference mineral reactions are based on Brown (2007) and as in Figure 7.1, with the addition of the prenhite-pumpellyite facies (PP) and zeolite facies (Zeo). Three selected *P–T* paths are illustrated: Bohem, S Bohemian Massif (O'Brien, 2008; Schulmann *et al.*, 2008); Gren, Grenville Province, near allochthon boundary thrust (Rivers, 2009); Fiord, Fiordland granulites, New Zealand (De Paoli *et al.*, 2009).

timescales of post-peak decompression and exhumation – from relatively rapid (e.g. within 10 Ma of peak metamorphism) to much longer or delayed by several tens of millions of years compared with lower-*P* rocks. As a consequence a variety of models have been proposed to account for the formation and exhumation of E-HPG metamorphic rocks in orogens. These are best considered with reference to specific examples that preserve different *P–T*–time records, as detailed below.

Formation and exhumation of E-HPG metamorphic rocks

It is important to recognise that not all E-HPG metamorphism is associated with collisional orogeny. E-HPG metamorphism may occur in the deepest levels of very thick (50–60 km) magmatic arcs through transient heating of earlier-formed deep-seated intrusives by later magmatic pulses during continued arc activity. In this type of E-HPG metamorphic belt, as exemplified by the 126–116 Ma Fiordland granulites of New Zealand (De Paoli *et al.*, 2009), the E-HPG metamorphism is short-lived (*c.*10 Ma) and broadly contemporaneous with magmatism. In this example, post-peak ITD from 18 kbar and 800 °C to 14 kbar and 700–750 °C, followed by decompression with cooling into the amphibolite facies at 8–10 kbar and 600–700 °C (Fig. 7.19), has been interpreted to reflect recrystallisation during ascent contemporaneous with continued arc development (De Paoli *et al.*, 2009).

To understand how E-HPG metamorphic rocks, and also their lower-*P* granulite relatives, evolve and may be exhumed in an orogen it is worthwhile to first note that these deep-level rocks generally appear to have been formed in the core regions of large orogens (e.g. Beaumont *et al.*, 2004), where uniformly high temperatures (>700 °C) are expected to be achieved at depths of greater than 40 km because of radioactive self-heating over timescales of 20 Ma or more. Those rocks that remain trapped far within a large orogen core at the end of collision, whether in a low-viscosity channel or beneath it, will remain so unless additional processes operate that can induce their exhumation. Some possible processes and scenarios include:

1. Syn- to post-collisional gravitational spreading that removes the orogenic lid or causes it to founder, thereby allowing deeper-level rocks to approach the surface. This requires a large residual gravitational head, coupled with high enough heat production in the deeper crust to cause significant self-heating following collision. It also means that the exhumation of the deeper-level (E-HPG and G-UHT, where G is granulite) rocks will be accompanied by the formation of sub-horizontal ductile fabrics, broadly synchronous with extensional structures developed in overlying lower-*T* rocks.
2. Break-up of the layered and strength-related superstructure/infrastructure relationship in the crust by a change in tectonic forces and boundary conditions or by buoyancy-driven internal re-organisation. Examples of these 'triggers' may include the arrival of a strong crustal indentor, the frontal 'steamroller' effect of a distal collision, and bouyancy-driven doming and extrusion of lower-density hot rocks (e.g. melt-bearing felsic crust) beneath higher-density nappes.

These processes need not be mutually exclusive. The following section gives examples of how such processes may have been involved in the evolution of E-HPG rocks in two

Phanerozoic orogens – the comparatively 'hot' European Variscides as exposed in the Bohemian Massif and the 'colder' Caledonian Orogen as observed in southwestern Norway and Greenland. This is followed by consideration of the Mesoproterozoic Grenville orogen in Canada, for which detailed thermomechanical models have been developed to explain the long-lived *P–T*–time paths of several E-HPG and granulite facies domains and relate these to the tectonic architecture of the orogen.

Bohemian Massif – rapid buoyancy-driven vertical extrusion of HPG?

The Variscan Orogen of Europe is a large hot orogen formed through the collision of Laurussia with Gondwana and its microcontinental forerunner fragments (O'Brien, 2008). The E-HPG metamorphic rocks of the Variscan orogen in the Bohemian Massif are typified by gneisses of upper crustal derivation, metamorphosed at 16–20 kbar and 850–1000 °C at 340 Ma, which are characterised by steep fabrics, HP migmatisation and the formation of garnet + kyanite leucogranulite (O'Brien, 2008). Intercalated with these are eclogite facies rocks taken to coesite conditions of 28–32 kbar and 900 °C, and garnet peridotite bodies. A plausible way to produce this assemblage of diverse rocks with distinctive *P–T* records is by deep continental subduction terminated by delamination of the lowermost subducted crust and its lithospheric mantle, leading to underplating of the overriding plate by the remaining upper units of the subducted continental crust (O'Brien, 2008). This would give rise to a 'hot' orogen consisting of thickened crust, the deepest parts of which would be dominantly felsic, of low density, and potentially characterised by high heat production. This high heat production, possibly in concert with high mantle heat advection, may be responsible for the generation of hot ultrapotassic magmas ('durbachites') essentially synchronous with peak metamorphism and initial decompression in the Bohemian Massif (Schulmann *et al.*, 2008).

In the southern Bohemian Massif post-peak ITD and decompression with cooling of the hot (850–1000 °C) HPG rocks occurred within 10 Ma of the metamorphic peak, at an average d*z*/d*t* of 0.3 cm/a (Fig. 7.19). This occurred despite the rocks forming the lower part of the orogen. Schulmann *et al.* (2008) have argued that vertical extrusion of the buoyant, gravitationally unstable HPG rocks caused this initial exhumation to 7–10 kbar at 800 °C, possibly in combination with horizontal compression that stimulated folding, growth of domes and subsidence of intervening higher-level rocks. It is further suggested that in some parts of the orogen lateral flow ensued at these pressures producing strongly recrystallised 7–10 kbar granulite (d*T*/d*z* = 20–30 °C/km) assemblages which were then transported in hot fold nappes over a rigid continental indentor by *c*.330–325 Ma. It is proposed that the upper crustal rocks formed a carapace or 'lid' during the entire orogenic evolution, experiencing only minor burial with heating (to 9 kbar and 700–750 °C) prior to limited decompression perhaps related to gravitational spreading of the orogen subsequent to the initial buoyancy-driven high-*T* vertical extrusion phase of crustal re-organisation.

Caledonian E-HPG in Norway and Greenland

In the Western Gneiss Region of Norway the idea of extensional unroofing has been in vogue since the 1980s, when it was recognised that the lower boundaries of the Devonian basins such as Sogndal were high strain zones decorated with HP relics and lower-P mylonites (Andersen and Jamtveit, 1990). Milnes and Koyi (2000) further argued for earlier ductile rebound of the deep Caledonian orogenic root, composed of strongly reworked Baltic shield basement, prior to its final exhumation beneath the extensional detachment zones represented by the sub-Devonian basin high strain zones. In this interpretation, lateral spreading and extrusion of the E-HPG deep crust occurred beneath a little-deformed carapace comprised of less reworked Baltica basement transected by anastomosing shear zones. Late-Caledonian (Scandian) eclogite layers and lenses, formed at 425–400 Ma, underwent ITD from >15 kbar to 10 kbar at 600–650 °C, experiencing marginal hydration to amphibolite under the latter *P–T* conditions. This phase of decompression did not exhume the E-HPG rocks, but produced a tilted structural and metamorphic depth section. Further extension partitioned into detachment horizons that cut through the tilted basement gneiss section ensued from 405–400 Ma, displacing upper crust and allochthonous nappes down against the variably 'Caledonised' deeper crust.

McClelland and Gilotti (2003) have described the structural setting, *P–T* conditions and exhumation of E-HPG gneisses from the western side of the Caledonian Orogeny in east Greenland. An upper unit of low-grade and sedimentary rocks lies above an extensional detachment considered analagous to the STD in the Himalaya. Beneath this is a middle-crust unit comprised of medium-*P* and low-*P* migmatites that host leucogranite sheets and granite plutons, formed largely synchronously with high-level extension at 430–424 Ma. Underlying these mid-crustal migmatites, and separated from them by a high-strain zone interpreted as a detachment, is the deep-crustal basement unit that preserves the E-HPG mineral assemblage Grt + Cpx + Amphibole + Qtz + Rut + Plag in orthogneisses and Grt + Ky + Kfs + antiperthite + Qtz + Rut in metasediments. Collectively these yield *P–T* conditions of 14–16 kbar and 800–850 °C.

The most significant feature of these east Greenland E-HPG rocks is the timing of their peak metamorphism, constrained by zircon U–Pb dating of high-*P* leucosomes to be 403 ± 5 Ma. This age, some 20 Ma later than migmatisation and leucogranite formation in the overlying middle-crust unit, demonstrates that the deep-crust continued to evolve to high-*T* within the orogen core long after shallower-level rocks were exhumed through syn-orogenic extension. The E-HPG metamorphic rocks in this instance required a later phase of extension or extrusion to unroof and juxtapose them against the significantly lower-*P* rocks of the orogenic lid and middle crust.

The Grenville Province: heterogeneous ductile flow and exhumation of E-HPG rocks and deep orogen interiors

The Grenville Province of western Ontario is a well-studied orogen for which an abundance of metamorphic and structural data is available, complemented by deep-crustal seismic reflection imaging that provides for an unrivalled insight into the crustal architecture of the Grenville orogen front and interior (Rivers, 2009). From northwest to southeast the gross crustal structure has been divided into three belts separated by low-angle thrust or detachment zones. Unmodified Superior Province basement in the northwest is overthrust in the Grenville Front

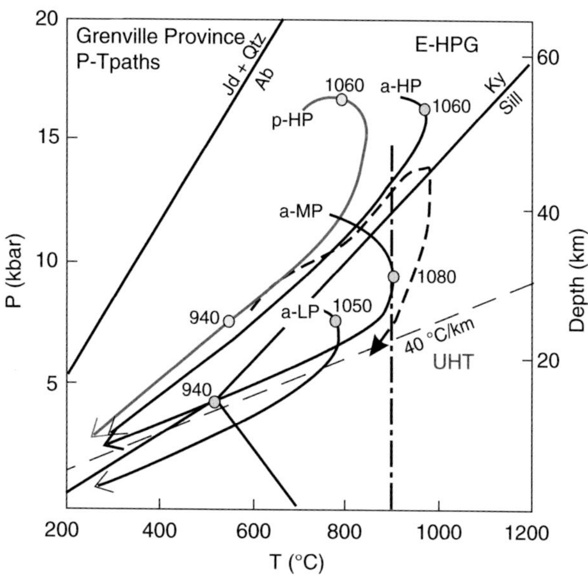

Figure 7.20 *P–T* paths for selected domains within the E-HPG Grenville Province in Ontario, Canada (Rivers, 2009). Positions of reference mineral reactions based on Brown (2007) and as in Figure 7.1. Four representative *P–T* paths are shown: p-HP, parautochthonous high-pressure (HP) domains; a-HP, allochthonous HP domains; a-MP, allochthonous medium-pressure domains; a-LP, allochthonous low-pressure domains. Ages (in Ma) along these paths, derived from U–Pb mineral dating and closure-*T* chronometers, are shown in italics. A model path produced from the GO thermal modelling of Jamieson *et al.* (2006) and Jamieson and Beaumont (2011) is shown by the dashed *P–T* path. This corresponds to *P–T* path 3, for an orogen core nappe thrust over a rigid indentor, in Fig. 7.21.

Tectonic Zone (GFTZ) by reworked slices of equivalent basement in the parautochthonous belt. This belt is in turn overlain to the southeast by high-grade gneisses within tectonically bound and strongly deformed deep-crustal slices in the polycyclic allochthon belt, and by lower-*P* and less deformed units of the orogenic lid or upper crust. The thrusts within the GFTZ collectively amount to only a few tens of kilometres lateral displacement, whereas far more displacement (>100 km) is inferred to have occurred on the lower boundary to the allochthonous belts, the Allochthon Boundary Thrust, which has a complex history of re-activation (Rivers, 2009). Metamorphism in the GFTZ (at 6–12 kbar and 700–800 °C) occurred late, at 1000–980 Ma in the 'Rigolet' phase of the Grenville orogeny, whereas E-HPG and lower-*P* granulite metamorphism in the allochthonous belt is of Ottawan age (1090–1020 Ma), overprinted by Rigolet phase amphibolite facies assemblages in places.

Long-lived decompression-cooling *P–T* paths are preserved in E-HPG (12–14 kbar, 860–960 °C) and medium-*P* amphibolites and granulites (8–10 kbar, 700–800 °C) that occur in the thrust- and shear-zone bound allochthonous accreted domains (Fig. 7.20). High-*P* granulites formed in these rock packages from the deep interior of the Grenville orogen experienced ITD ($dP/dT = 40$–60 bar/°C) or decompression-cooling (15–20 bar/°C) paths through significant *P* intervals (*c.*6 kbar), and from geochronology resided at $T > 850$ °C for over 50 Ma. The partitioning of heterogeneous deformation into the mid- and deep-crust leads to the presence of a shallower, long-lived orogenic 'lid' or superstructure that undergoes

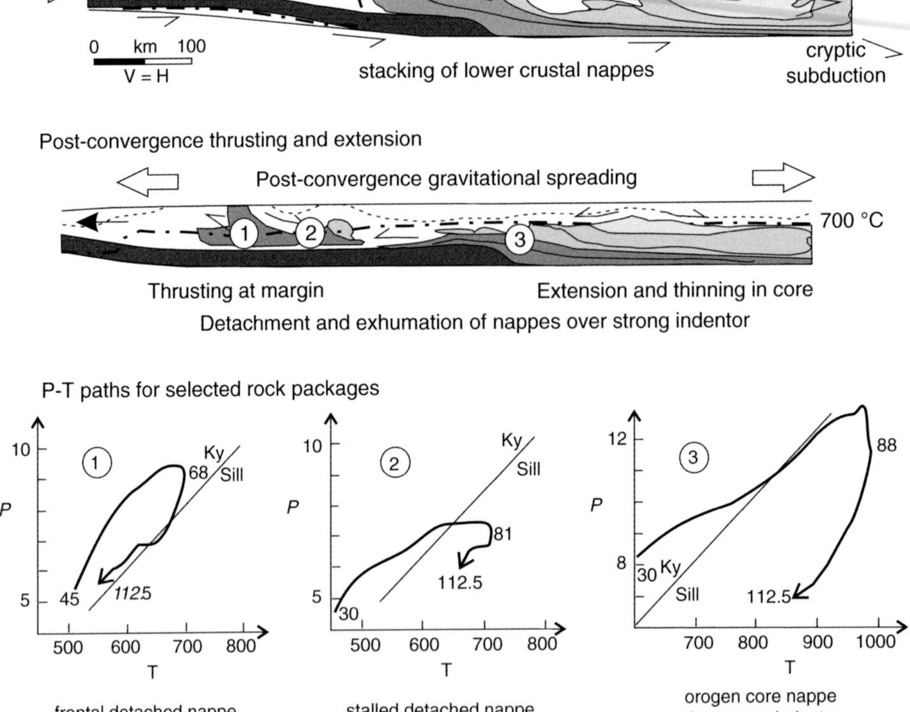

Figure 7.21 Crustal sections at two time slices in the GO-ST87 model of Jamieson *et al.* (2010) as summarised by Jamieson and Beaumont (2011), developed for slow convergence and for application to the Grenville Province of Ontario, where laterally heterogeneous crust is incorporated into the orogen. The 700 °C isotherm is shown for both the syn-convergence and post-convergence cases. Three *P–T* paths are shown for rock packages that have experienced the syn-convergence thickening followed by post-convergence lateral spreading. The rock packages are from the late-thrusted frontal zone of the orogen (path 1), within a stalled nappe (path 2), and in a deep-crustal nappe thrust forward over a deep rigid indentor (path 3). Simplified and modified from Jamieson and Beaumont (2011). Redrawn and published with permission of John Wiley & Sons.

heating to 600 °C and cooling at $P < 6$ kbar without significant changes in P throughout much of the orogeny. Metamorphism in this 'lid' is therefore of low P/T type, in contrast to the E-HPG metamorphism recorded in the tectonically underlying allochthons of the polycyclic belt.

Beaumont *et al.* (2006) and Jamieson *et al.* (2010) have applied their coupled thermo-mechanical modelling of large-scale orogens to the Grenville Province of eastern Canada, with important modifications from their 'HT' models introduced to account for its imaged structure, long-lived high-T history, and record of the two distinct but related tectonothermal phases (Jamieson & Beaumont, 2011). The derived 'GO' (Grenville Orogen) model uses a convergence rate of 2 cm/a to account for the involvement of old, strong crust which is heterogeneously deformed into high-T ductile fold nappes (Fig. 7.21). The orogen becomes thick, hot and ductile over a long time period because of the slower convergence rate, such that $T > 700$ °C occurs in the deeper crust only after about 40 Ma of collision. Ductile deformation assisted by melt-weakening may be initiated, but mid-crustal channel flow is inhibited by the low rates of melt production and, depending on rock type, the infertility of the incoming old and dry crust. The limited melt-weakening requires that any outward flow of the orogen is dominated by tectonic drivers rather than being gravitation-controlled. Crustal units deep in the orogen are stacked by ductile thrusting, coeval with flattening by ductile thinning. Strong incoming crustal units could become underthrusting indentors, causing upwards ramping, thrusting and lateral expulsion of the weaker slices or blocks already within the orogen. The lower crustal rocks forced upwards and outwards over such a rigid indentor would experience decompression and exhumation if near the flank of the orogen, or if erosion is promoted above the indentor region.

By the end of the collision and thickening stage in the reference 'GO' model (Jamieson & Beaumont, 2011) the 700 °C isotherm under the orogen plateau occurs at 15–20 km below the surface, giving an effective dT/dz of 35–44 °C/km at that depth. Much of the orogen core would be in granulite conditions, and for a considerable period of time (several tens of Ma). When convergence stops these orogen interior granulites will be initially trapped within the orogen, but after a period of some 25–30 Ma of continued radioactive self-heating may become weak enough to flow outwards over a strong indentor (e.g. the Superior Province cratonic crust). This second phase of reverse-sense movement is proposed to produce post-peak thrusting and associated metamorphism some 25–30 Ma younger than the main collisional phase. The post-collisional outward flow could result in 25% thinning of the orogen interior coeval with thrusting in the flanks and foreland. Slow cooling with decompression may ensue with post-collisional extension up to 50 Ma after the end of the collision itself.

This model ('GO' models in Beaumont *et al.*, 2006; Jamieson & Beaumont, 2011) successfully accounts for many of the features of the Grenville Province (e.g. Rivers, 2009), and is consistent with the heterogeneous ductile flow of deep-crustal slices during lateral spreading following the end of convergence. The model provides a good explanation for the overprinting of granulite facies mineral assemblages of Ottawan age (1090–1020 Ma) by amphibolite and lower grade assemblages in the Rigolet phase at 1000–980 Ma, occurrence of medium P/T Rigolet metamorphism in the GFTZ, and general characteristics of the P–T–time paths in the allochthonous domains. Moreover, the modelling applied provides insights into possible mechanisms for the eventual exhumation of the long-lived deep crust of orogen interiors, of relevance not only to the E-HPG rocks described in this section but also granulites in general.

Low P/T metamorphism: granulite and UHT metamorphism in orogeny

Background and *P–T* domain (low *P/T* and UHT)

Granulite metamorphism is defined by the incoming of orthopyroxene in metabasites, by dehydration or melting reactions involving the consumption of hornblende. In an overview of granulite metamorphism that demonstrated its wide range of P–T conditions and P–T paths, Harley (1989) emphasised the well known P-related divisions of granulites based on the incoming of garnet in metabasites, and complemented this with subdivisions into lower- and higher-T groups based on assemblages in pelites. Under low- and medium-P granulite facies conditions pelites lack muscovite, and are sillimanite–K-feldspar gneisses and migmatites containing additional biotite, garnet, cordierite, Ti-phases or (at lowest-P) spinel depending on bulk rock composition. At 750–900 °C, the typical assemblages in pelites involve combinations of garnet and/or cordierite + quartz + K-feldspar with sillimanite or orthopyroxene, and those in metabasites are the classic 'two pyroxene' granulites that may contain additional garnet if P is high enough (e.g. >7 kbar).

Migmatitic structures and microtextures are common in granulite facies pelitic gneisses, reflecting the extensive progress of fluid-deficient dehydration-melting reactions controlled by biotite, for example the simplified model reaction (Carrington & Harley, 1995):

$$\mathrm{Bt + Sill + Qtz = Grt + (Crd) + Kfs + melt}$$

At $T >$ 800–900 °C the proportion of leucogranitic melt produced by the melting reactions is significant (10–20 volume %) and leads to important changes in the strength and viscous behaviour of the hot crust, as detailed in the section on crustal melting and orogeny (Plates 24–28). Migmatitic features are also seen in metabasites under these P–T conditions, though generally a lower percentage of melt is present in these rocks.

Beyond 900 °C and at pressures between 6 and 12 kbar is the field of UHT metamorphism (Harley, 1998b, 2008; Brown, 2007, 2009; Kelsey, 2008; Fig. 7.22). UHT has been known since the 1960s, firstly by workers in the former USSR and then through discoveries in Antarctica, Canada and India. However, its wider recognition as a major form of metamorphism associated with orogeny came only in the 1980s–1990s, mostly as a result of the detailed and comprehensive evidence presented on UHT P–T conditions in the Napier Complex of Antarctica by Ellis *et al.* (1980) and subsequent workers, as summarised by Harley (1989, 1998b).

The first-order diagnostic mineral assemblages for UHT metamorphism (Plates 37, 38) are mainly recognised from pelites (Fig. 7.22). The classic indicator is sapphirine + quartz (Spr + Qtz), now regarded as indicating $T >$ 990–1020 °C at P of 7–11 kbar (Kelsey *et al.*, 2004; Kelsey, 2008) so long as the sapphirine has low Fe^{3+} contents. The highest-T Spr + Qtz assemblage recorded is a Spr + Opx + Qtz granulite from the Archaean Napier Complex, in which the 12.5 wt% Al_2O_3 orthopyroxene implies peak-T of >1070 °C and perhaps 1120 °C (Harley, 2008). Orthopyroxene + sillimanite + quartz (Opx + Sill + Qtz) in which the orthopyroxene is highly aluminous (Al_2O_3 > 8–9 wt%), and in particular Opx + Sill + Qtz with additional K-feldspar (Harley, 2004; Kelsey, 2008), is indicative of $T >$ 900 °C at P of 8–13 kbar. Grt + Opx + Sill ± Qz assemblages from UHT domains in Antarctica (e.g. Mather Supracrustals: Harley, 1998c) preserve highly magnesian garnet (X_{Mg} = 60–70) and aluminous orthopyroxene (Al_2O_3 = 11 wt%) that lead to peak P–T estimates of 12 kbar and 1030 °C (Harley, 1998c, 2008). Similarly extreme peak-T conditions are deduced for Grt + Opx + Sill + Qtz and Grt + Opx + Spr + Sill assemblages from a number of UHT areas and domains worldwide – including In Ouzzal,

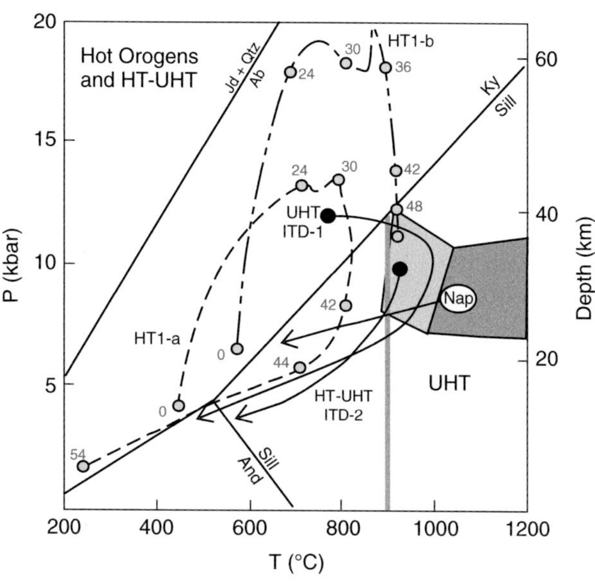

Figure 7.22 Some typical P–T paths for HT–UHT terrains (solid lines), compared with modelled P–T paths for two rock packages in the hot orogen channel flow model HT1 of Jamieson *et al.* (2004) (HT1-a, HT1-b). Two types of HT–UHT ITD paths are distinguished: ITD-1, where extensive decompression at $T > 900$ °C is followed by cooling at 5–7 kbar; and ITD-2, where decompression is accompanied by cooling so the rocks do not remain in the UHT field for much of the decompression history. Note that whilst path HT1-a exhibits extensive ITD over a similar range in P to the HT–UHT–ITD paths, the T is lower by some 100–150 °C. Model path HT1-b, on the other hand, attains UHT conditions, but only subsequent to its decompression from 20 kbar to *c.*11 kbar, at which point it then resides and would cool in the deep crust rather than undergo further ITD in the HT1 model. The light stippled field is the stability of orthopyroxene + sillimanite + quartz; the heavier patterned field is the stability field of sapphirine + quartz.

The path labelled Nap is an example of a UHT–IBC (near-isobaric cooling) path from the Archaean Napier Complex of Antarctica (Ellis *et al.*, 1980; Harley, 1989, 1998b). The stippled field is that of the UHT mineral assemblage Opx + Sil + Kfelds + Qtz + Melt, and the cross-hatched field is that of the key assemblage sapphirine + quartz (Spr + Qtz).

Algeria (Ouzegane and Boumaza, 1996), Anapolis, Brazil (Moraes *et al.*, 2002), Palni Hills, Southern India (Raith *et al.*, 1997) and the Kontum Massif, Vietnam (Osanai *et al.*, 2004), to note just a few of those reviewed by Kelsey (2008).

Further evidence for UHT may include ternary feldspar in pelitic gneisses, osumilite + orthopyroxene + garnet in pelites, and Fe-Mg pigeonite in metaironstones (Harley, 2004, 2008). Key mineral chemical evidence now also includes high Ti contents in zircon (>50 ppm) and quartz (>200 ppm) coexisting with rutile, and correspondingly high Zr in rutile (>3700 ppm) (Baldwin *et al.*, 2007; Tomkins *et al.*, 2007; Harley, 2008; Kelsey, 2008). The occurrence of spinel + quartz may provide supporting evidence, but is not diagnostic in itself because of the very significant P–T shifts associated with minor constituents such as Zn, Fe^{3+}, Cr and V in spinel, and the expanded stability of spinel + quartz at low-P ($P < 5$ kbar) (Waters, 1991).

UHT has now been identified in 46 different high-*T* regions globally, ranging up to mid Archaean in age. These UHT occurrences range in area from large belts such as the 10,000 km^2 Napier Complex of Antarctica (Ellis *et al.*, 1980), the Eastern Ghats (India), Wilson Lake (Labrador), In Ouzzal (Algeria) and the Arequipa Complex (Peru), to horizons, slivers and rare boudins and relict bodies on scales of only metres to kilometres, for example at Anapolis, Brazil (Moraes *et al.*, 2002) and St Maurice (Quebec) within otherwise unexceptional granulites. In many, though not all, examples the broader setting of the UHT rocks is one of strongly deformed low *P*/*T* mid- to deep-crustal high grade rocks associated with major orogenic events (Harley, 2008).

Pressure–temperature–time paths in G-UHT

The *P*–*T* records of granulites are constrained by assemblages, calculated phase diagrams in complex chemical systems, and geothermobarometry as described in earlier sections of this chapter. The high-*T* conditions, however, limit the retention of prograde mineral zoning and may even modify the compositions of minerals participating in post-peak reaction textures (Harley, 1989; Spear, 1993). Nevertheless, examination of reaction textures coupled with the calculated phase diagram approach and application of new trace element geothermometers (e.g. Zr in rutile) often allow tight constraints to be placed on post-peak *P*–*T* paths.

Most granulites are typified by ITD and decompression-cooling paths, although a smaller but significant number may be characterised by near-isobaric cooling (IBC) paths (Harley, 1989; Brown, 2007; Fig. 7.22). Some examples of IBC granulites appear to have formed at only 3–6 kbar within low-*P*/high-*T* belts, perhaps associated with extension–contraction orogenesis in which collision has not occurred (e.g. central Australia: Collins & Vernon, 1991), produced in the deeper levels of orogenic 'lids' (e.g. central Grenville Province: Rivers, 2009), or formed as a result of mid-crustal magmatic accretion in arc or rift settings (Bohlen, 1991). However, IBC, or at least a *P*–*T* evolution dominated by cooling, is also recorded by garnet-forming reaction textures in some deeper-level granulites (6–10 kbar) that, on the basis of their deformation fabrics and ages, constitute the mid- to deep-crust during orogeny (Fig. 7.22). In some cases, for example parts of the Grenville Province, the 530 Ma Highland Complex of Sri Lanka, 990–930 Ma granulites from the Eastern Ghats of India and the Rayner Complex of Antarctica, the 649–634 Ma Anápolis–Itauçu Complex of Brazil, and the 2740–2500 Ma Scourian granulites of the Lewisian of northwest Scotland, this cooling can be shown to have followed earlier near-peak to post-peak decompression. The classic example of G-UHT metamorphism followed by IBC is the Archaean Napier Complex, Antarctica, which will be returned to in Chapter 12.

Returning to the dominant ITD and decompression-cooling *P*–*T* paths, these have been documented from reaction textures in both 'normal' 750–900 °C granulites and UHT granulites. As with all interpretations of reaction textures, it is important to determine, if possible, the ages of the reactant and product assemblages in order to decide whether the textures reflect a single *P*–*T* path or the superposition of two unrelated *P*–*T* events – as discussed earlier in this chapter.

The key reaction texture and mineralogical lines of evidence for ITD and decompression-cooling are generally similar for mafic and intermediate rock compositions, but differ for pelites and aluminous rock types in the normal granulite and UHT cases. In mafic and intermediate granulites typical textures include the replacement of garnet + clinopyroxene (aluminous augites) + hornblende by intervening layered coronas of orthopyroxene + plagioclase + ilmenite, or of garnet + quartz by orthopyroxene + plagioclase, accompanied by marked increases in the Ca-content of plagioclase. In pelitic rocks equally common textures include the formation of cordierite and/or plagioclase as thin rims on garnet + sillimanite + quartz, and orthopyroxene + cordierite on garnet with quartz. Calc-silicate rocks also preserve excellent textures that may be indicative of ITD or decompression-cooling, such as the replacement of grossular garnet + calcite + quartz by wollastonite + scapolite. Many of these reaction textures can be constrained from calculated phase diagrams (pseudosections) to reflect decompression through the *P–T* window 9 to 5 bar and 700–860 °C in the case of the common granulites, on d*P*/d*T* gradients of 40–25 bar/°C.

ITD and, in some cases, decompression-cooling *P–T* paths, collectively referred to here as 'UHT-ITD' paths, are recorded in aluminous and magnesian UHT rocks by spectacular reaction textures principally involving the formation of two- and three-phase symplectites at the expense of Mg-rich garnet. Indeed, on the basis of these textures post-peak UHT-ITD *P–T* paths are inferred for some 90% of the UHT areas so far recognised. The classic textures include the partial pseudomorphing of garnet by sapphirine + aluminous orthopyroxene + sillimanite, sapphirine + aluminous orthopyroxene + cordierite or spinel + aluminous orthopyroxene + cordierite; breakdown of garnet or orthopyroxene with sillimanite (or, in some instances, kyanite) to form sapphirine + cordierite; the very rare breakdown of garnet + sillimanite to form sapphirine + quartz; and garnet-corundum textures involving the formation of sapphirine + sillimanite or cordierite.

These are but a few of the many reaction textures described in detail from several UHT areas and localities (Harley, 1998b). In a number of cases the prograde evolution of the UHT-ITD granulites can be deduced based on inclusions of biotite + sillimanite in porphyroblastic garnet, succeeded by kyanite as inclusions and porphyroblasts, now pseudomorphed by sillimanite aggregates. These observations have been used to constrain prograde *P–T* paths traversing from 7–8 bar and 750–800 °C into the kyanite field at 10–12 kbar and 850–950 °C, followed by thermal excursion to peak-*T* conditions within the sillimanite field, at 9–12 kbar and 950–1050 °C depending on the UHT occurrence (Fig. 7.23). In detail the post-peak UHT-ITD *P–T* paths show some variation over the *P* range of interest (12–6 kbar). Some UHT occurrences (e.g. Mather supracrustals: Harley, 1998c, 2008; Palni, southern India: Raith *et al.*, 1997) are characterised by true ITD (d*P*/d*T* of 40–80 bar/°C) to about 8 kbar at 980–1050 °C, followed by near-isobaric cooling (d*P*/d*T* of 4–8 bar/°C) to 700–750 °C at *c*.5–6 bar. Others appear to be characterised by decompression with cooling, on post-peak *P–T* paths traversing from 9–12 kbar and 950–1050 °C to 5–7 kbar and 750–800 °C with average d*P*/d*T* of 20 bar/°C.

The key attributes of these varied UHT-ITD paths, irrespective of the details, are that they all traverse forward and back across the G-UHT metamorphic domain at pressures that would correspond to crustal depths of between 45 and 20 kilometres, and all of this

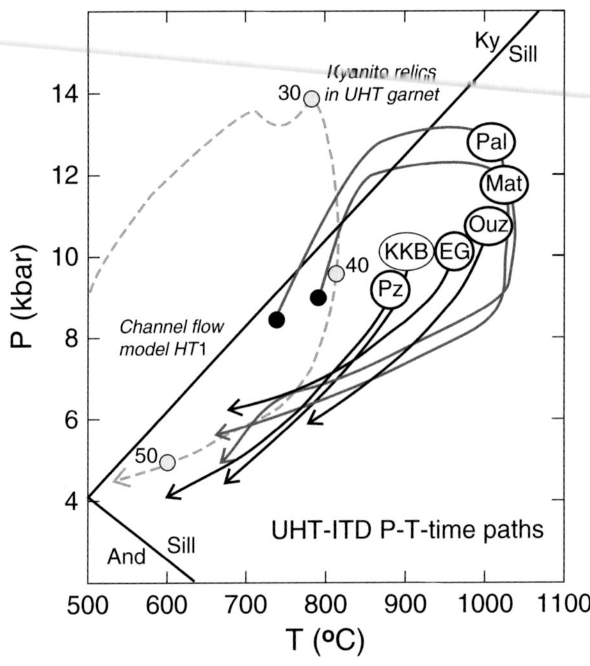

Figure 7.23 Examples of *P–T* paths for HT–UHT terrains characterised by ITD or decompression with cooling. The selected *P–T* paths are: EG, Eastern Ghats belt, India (980–930 Ma) (Kelsey, 2008); KKB, high-*T* portion of the Kerala Khondalite Belt, southern India (560–525 Ma) (Harley, 2008; Kelsey, 2008); Mat, Mather Supracrustals, Rauer Islands, Antarctica (590–530 Ma) (Harley, 1998c, 2004, 2008; Kelsey *et al.*, 2005); Ouz, In Ouzzal, Algeria (Ouzegane & Boumaza, 1996; Pal, Palni Hills, southern India (545 Ma) (Raith *et al.*, 1997); Pz, Prydz Belt, Brattstrand Bluffs, Antarctica (545–530 Ma) (Harley, 1998b).

P–T evolution lies at $T > 800$ °C. Any modelling of orogen evolution to produce UHT rocks, whether in collisional or accretionary orogens, must be able to account for these *P–T* paths and their timescales. We now briefly outline evidence for the timescales of G-UHT metamorphism before going on to consider models that may account for UHT metamorphism.

Timescales of G-UHT metamorphism: orogenic and other settings

The high temperatures attained in G-UHT metamorphism render most lower-*T* chronometers of limited use in constraining the timescales of peak and near-peak metamorphism and the post-peak *P–T* evolutions of many terrains. Exceptions to this would include rapidly evolved G-UHT areas in which the lower-*T* chronometers (e.g. Ar–Ar mica chronology) yield ages close to those obtained from accessory minerals formed at or near peak conditions. Even in those cases, however, the definition of the age of the metamorphic peak may not be straightforward because of the propensity of the zircon and monazite to form at different times and respond to a variety of reaction-, melt- and fluid-related processes in G-UHT rocks, as outlined in Chapter 4.

Long-lived residence times at peak and near-peak conditions have been demonstrated for a number of G-UHT characterised by post-peak IBC or by cooling with decompression. Several allochthonous crustal segments within the Grenville Province, including the Adirondack Highlands, record long-lived decompression and cooling from peak *P–T* conditions of 7–10 kbar and 800–900 °C at 1080–1050 Ma to *P–T* near the aluminosilicate triple point at 1000–950 Ma (Rivers, 2009). This corresponds to about 100 Ma of deep to mid-crustal residency and averaged cooling rates of 3–4 °C/Ma, and time spans at $T >$ 800 °C (Δt_{800}) in the range 30–50 Ma. Similarly long-lived orogen residency is inferred for G-UHT rocks of the central Eastern Ghats of India and potentially correlative granulites in the Rayner Complex of East Antarctica, which record high-*T* metamorphic events from 1000 Ma through to 930–920 Ma and may have been at >800 °C for some 80 Ma. Perhaps the longest deep-crustal residence time for any G-UHT terrain is that indicated for the Archaean Napier Complex of east Antarctica (Kelly & Harley, 2005; Harley *et al.*, 2007). Zircons that formed at 2590 Ma, 2550 Ma, and 2520–2510 Ma in partial melt leucosomes and through high-*T* metamorphic reactions at 7–10 kbar in the Napier Complex imply $T >$ 900–1000 °C for some 80 Ma in this IBC terrain. Younger zircons (2480–2460 Ma) produced through subsequent reactions associated with garnet corona formation, rutile recrystallisation and localised hydration reactions at 800–750 °C indicate a Δt_{800} of some 100 Ma, and imply average cooling rates of some 3 °C/Ma over this time interval.

The timescales and high-*T* residence times of those G-UHT areas characterised by steeper post-peak *P–T* paths, including the UHT-ITD granulites, appear to vary between 50 Ma and 10–15 Ma. Some of the better-documented examples include those granulites and UHT-ITD rocks formed in the latest Neoproterozoic to Cambrian, 'Pan African' high-*T* belts of southern India (e.g. Madurai Block, Kerala Khondalite Belt), Sri Lanka Highlands, Madagascar and east Antarctica (Prydz Belt, Lützow–Holm Belt). Zircon and monazite ages from leucosomes, migmatites and within UHT minerals in these areas suggest continuous G-UHT metamorphism at $T >$ 800 °C over periods of 20–40 Ma from *c.*560–545 Ma to 520–510 Ma, followed by relatively rapid cooling within the subsequent 10 Ma. In general these areas appear to be characterised by long-lived G-UHT metamorphism terminated by ITD or decompression with cooling through 20 km at post-peak exhumation rates of 1–2 mm/a – rates that are comparable to those obtained in modelling of orogen flanks and within-plateau domes in the hot orogen (HT) models of Jamieson *et al.* (2004) and Jamieson and Beaumont (2011).

Short-lived G-UHT metamorphism has also been documented. For example, the Anápolis–Itauçu Complex of Brazil has been shown to have experienced relatively short-duration UHT metamorphism followed by UHT-ITD and cooling over the 15 Ma time interval from 649 Ma to 634 Ma (Baldwin *et al.*, 2005, 2007). This rapid evolution has been interpreted to reflect thickening of hot back-arc crust as a consequence of arc–continent collision (Baldwin & Brown, 2008). The UHT-ITD supracrustals of the Mather Paragneiss in the Rauer Group of east Antarctica may also record a short-lived UHT event, at 590–580 Ma, that may reflect accretion onto the leading margin of an Archaean craton prior to later collision with a Mesoproterozoic block in a distinct orogenic episode that partially overprinted the UHT record at 545–515 Ma. Other examples of short-lived G-UHT metamorphism, with associated ITD, that may reflect the rapid transfer of heat

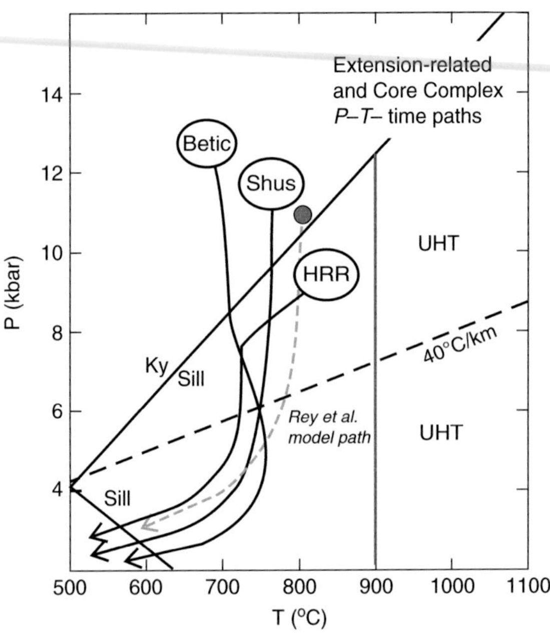

Figure 7.24 Examples of *P–T* paths for terrains characterised by late-stage extension. Convective removal of mantle lithosphere is inferred to have actived extensional unroofing of the deeper crust in the Betic–Rif orogen, leading to ITD and decompression with heating (Platt *et al.*, 1998, 2003). *P–T* paths are shown for two Cordilleran Core Complexes: the Shuswap complex of Canada (Shus) (Vanderhaeghe *et al.*, 1999; Whitney *et al.*, 2004) and the Humboldt–Ruby Range complex (HRR) (Rey *et al.*, 2009). The dashed *P–T* path is that modelled by Rey *et al.* (2009) for fast extension, as described in the text.

promoted by extension and magmatic underplating, or the accretion of thinned crust on thinned lithosphere, may include the Palaeozoic Gruf Complex, located within the Alps, and the Palaeoproterozoic South Harris Complex of Scotland.

An example of the rapid attainment of granulite *P–T* conditions (5–7 kbar and 850 °C) and their evolution on steep ITD paths over a short time interval (*c.*15 Ma) in an orogenic setting is provided by the the Betic Cordillera (Platt *et al.*, 1998). The Betic–Rif orogen preserves nappes of both high-*P*/low-*T* schists (the Nevado Filabrides) and low-*P* /high-*T* schists and gneisses, along with fragments and sheets of mantle-derived peridotite such as the 'hot' Ronda peridotite which itself preserves an envelope of high-*P* to high-*T* contact metamorphic rocks. Platt *et al.* (2003) have argued for coeval attainment of peak temperatures (650–850 °C) across a large volume of the middle to deep crust of the Betic–Rif orogen at 22–18 Ma. This rapid and regionally extensive heating and contemporaneous decompression (Fig. 7.24) is considered to have been initiated by the rapid convective removal of the lithospheric mantle root beneath the Betic–Rif orogen at *c.*27–28 Ma. The observation of reflective deep crust above a horizontal Moho in spite of the complex and topographically varied crustal superstructure of the Betics is consistent with deep crustal flow and Moho renewal following this lithosphere removal. Removal of the lithosphere is argued to have triggered upper crustal extension above domes within which rapid, probably melt-assisted, vertical transfer of the deep crust occurred. The rapid formation and evolution of granulites and other low-*P*/high-*T* rocks over the period 28 Ma to 14–12 Ma is entirely post-collisional, occurring some 25 Ma after crustal thickening and the HP metamorphism recorded in the higher structural level nappes. The critical factors enabling high-*T* metamorphism at relatively shallow crustal levels in this orogen are inferred to be the advection of heat from the underlying mantle following convective removal of the lithosphere below 70 km depths and the extensional spreading of the remaining crust as it heats up in response to the increased basal heat flux.

Hot orogens and G-UHT metamorphism

As is evident from the discussion of the timescales of G-UHT metamorphism, there is considerable debate as to the tectonic settings of granulite metamorphism, particularly with respect to the importance of granulite formation related to collisional orogeny versus formation within arc-backarc accretionary settings where lithospheric thinning and magmatic underplating may add significantly to the thermal budget (Ellis, 1987; Harley, 1989; Bohlen, 1991; Sandiford & Powell, 1991; Collins & Vernon, 1992). A good case in point is provided by the very young Cenozoic granulites exposed in the Hidaka HT/UHT metamorphic complex of Hokkaido, Japan (Kemp *et al.*, 2007). These formed at 19 Ma through the metamorphism of 41–37 Ma arc-related gabbros and tonalities and 80–50 Ma accretionary complex sediments. In this case, metamorphism at *P–T* conditions of 8 kbar and 900 °C is attributed to heating through back-arc extension, magmatic under-accretion and associated lithospheric thinning (Kemp *et al.*, 2007). This thermal pulse was short-lived, being terminated by subsequent arc–arc collision and obduction which led to the rapid exhumation of the granulites on a *P–T* path involving ITD to 3.5 kbar at $T > 800$ °C and then cooling in the upper crust to 500 °C by 15 Ma. The clockwise *P–T* path produced in this extension–accretion setting is similar to those deduced for many granulites from orogenic belts, and so is not diagnostic of a particular tectonic setting (Brown, 2007). However, the residence time of the Hidaka granulites at $T > 800$ °C is very short (2–6 Ma), and there is a short time-gap between the age of the high-*T* metamorphism and the ages of the magmatic and sedimentary protoliths. This short-duration high-*T* metamorphism of recently formed protoliths potentially contrasts with the generally longer-duration G-UHT metamorphism of varied lithologies of different ages that might be expected in 'hot' collisional orogens (Beaumont *et al.*, 2006; Jamieson *et al.*, 2004, 2010; Jamieson & Beaumont, 2011).

The HT1 model of Beaumont *et al.* (2004) illustrates the potential for approaching G-UHT conditions in large hot orogens, although it does not replicate the extreme *T* conditions of 1000–1050 °C recorded in many of the UHT-ITD cases described above. For convergence rates of 5 cm/a, continued collision coupled with radioactive self-heating of the thickened continental crust leads to melt-induced weakening some 20–25 Ma following initial collision. Rocks that remain buried by the end of the model (54 Ma), trapped at that time in or beneath a melt-weakened mid-crustal channel at *c*.40 km depths (points E1 and E2 in Jamieson *et al.*, 2004, figure 11), may, by 54 Ma, have experienced clockwise *P–T* paths that include an excursion into transient high-*P* conditions within the

E-HPG field prior to ITD into granulite conditions. In this model those rocks starting at 25 km depths (6–7 kbar) undergo prograde burial, reaching their maximum pressures (P_{max}: 18–19 kbar; *T*: 780–870 °C) some 22–34 Ma after incorporation into the orogen. If these rocks are then translated outward towards the orogen flank, within a zone of lateral or oblique ductile flow, they may evolve to lower-*P* on ITD paths through 7–8 kbar over periods of 6–18 Ma. The final rock products by 54 Ma in this case would be granulites, with preserved *P–T* conditions of 10–11 kbar and 860–930 °C (Fig. 7.20). These G-UHT rocks may or may not record their prior higher-*P* history and much of the ITD from E-HPG conditions, depending on the extents of recrystallisation and strain partitioning accompanying the deep-crustal ductile flow and given the amount of time the rocks are resident at $T > 800$ °C (24 Ma) and $T > 900$ °C (18 Ma for some material points).

As with the E-HPG rocks discussed in the previous section, the fate of the granulites left within the orogen on cessation of convergence depends upon what further processes then act on the orogen to redistribute its mass, and what events occur that affect the orogen long after it has ceased to evolve. On cessation of convergence the orogenic structure may relax through gravitational spreading, leading to post-collisional, post-peak, extension and thinning in the orogen core coupled with coeval foreland-directed deformation on the orogen flanks. However, as considerable lateral motion (e.g. by channel tunnelling) will have already occurred during collision, and mass will have been lost through exhumation at the flanks, the total extension in the core is likely to be limited, with Jamieson and Beaumont (2011) estimating only 10% thinning. The thermal consequence of this is that granulite and UHT (>900 °C) metamorphism will continue in the deep orogen core for several tens of millions of years after collision, followed eventually by cooling accompanied by moderate decompression through 1–2 kbars. The total time that these granulites spend near their peak *T* conditions (>800 °C) will therefore be of the order of 40 Ma. A further consequence of this is that HT and UHT rocks deep within the orogen core will remain trapped there, possibly undergoing long-term IBC at depth, until exhumed accidentally by a subsequent, completely unrelated, tectonic event – for example a later orogeny or continental rifting – or entrained as xenolithic material in eruptive alkali basalts or kimberlites.

The analysis presented above illustrates that the large, hot orogen models are capable of generating UHT conditions at 10–12 kbar, but only after prolonged residence in the deep crust and following extensive decompression from E-HPG conditions (Jamieson & Beaumont, 2011). The models do not produce the thermal excursions at 8–12 kbar documented from UHT-ITD rocks, and do not duplicate the *P–T* paths now seen as typical of these UHT-ITD areas, in which most of the *P–T* path occurs at $T > 800$ °C and within the *P* window 6–13 bar, over time periods of 15–80 Ma. None of the currently recognised UHT-ITD areas preserve any evidence requiring their origin in doubly thickened crust, but instead indicate that additional heat contributions are required at crustal or lithospheric depths of 40–50 km, rather than at 70 or 100 km, in order to generate the *P–T* conditions and paths. An obvious additional heat contribution is heat advected from asthenospheric mantle following the convective removal of lithospheric mantle, as advocated for the Betics (Platt *et al.*, 2003b). However, it is possible that other additional heat sources, not yet integrated into the large, hot orogen models, may contribute to the formation of G-UHT rocks in orogens, particularly in older mountain belts.

Recent experimental work that indicates lower thermal diffusivities for crust rocks at high temperatures (Whittington *et al.*, 2009) suggests that, once attained, high temperatures will be maintained during crustal thickening (i.e. it will retain its heat over long time periods). Given this, if radiogenic heating is enhanced through enrichment of deeper crust in heat-producing elements, for example through concentration of accessory minerals in residual material after extraction of leucogranitic melts, then it could be argued that the deep crust might potentially reach temperatures of 900 °C or more without the need for the addition of advective heat from the mantle (McKenzie & Priestley, 2008). The magnitudes of heat production required are of the order of 3–4 $\mu W/m^3$, significantly higher than estimates for many crustal rocks (0.5–2 $\mu W/m^3$) but not anomalous when compared with heat productions measured from some migmatitic gneisses, especially when those heat productions are back-calculated to the age of metamorphism in older belts.

Thermal excursions into UHT conditions (>900 °C, 8–12 kbar) may, as a consequence, be feasible for residual, high heat-production crust trapped deep within a large orogen interior, providing its residence time in the orogen is of the order of 20–50 Ma. The requirement for high radiogenic heat production is relaxed somewhat when the additional contribution of mechanical heat production, shear or viscous heating during deformation of relatively strong crust, is considered. Burg and Gerya (2005) have estimated that prior to the onset of extensive melting and thermal weakening in a collision belt, viscous heating could contribute between 0.1–1 $\mu W/m^3$ to the heat budget, and provide a major 'thermal primer' in the region of mid-crustal high strain zones. On the other hand, the effects of melt production later in orogen evolution may to some extent counteract heating via radiogenic and mechanical sources because of the latent heat of melting acting as a 'heat sink'. Systematic measurements of heat-producing elements in metamorphic rocks and mapping of any spatial associations between UHT domains and high strain zones are required in order to help evaluate whether UHT metamorphism in orogens is a consequence of factors internal to the crust (radiogenic and mechanical heat production) or external to it (e.g. lithosphere removal).

Gravitational spreading, core complexes and extensional metamorphism

The Betic Cordillera, briefly described in the previous section, may provide one example of the gravitational collapse and extension of a collisional orogen, leading to the exhumation of deeper-level gneisses (Platt *et al.*, 1998). In that example, the trigger for post-collisional extension is suggested to be the convective removal of subjacent mantle lithosphere, which results in a buoyant crustal section that then extends in order to approach a state of gravitational equilibrium (Platt *et al.*, 2003). More generally, it has been suggested by many authors that gravitational spreading or collapse is a natural progression in orogeny when collision stops and if a large mass of topographically elevated, buoyant crust remains and is able to flow at depth (Rey *et al.*, 2001; Vanderhaeghe *et al.*, 2003; Rey & Houseman, 2006). Vanderhaeghe (2009) and others argue that horizontal and heterogeneous domal

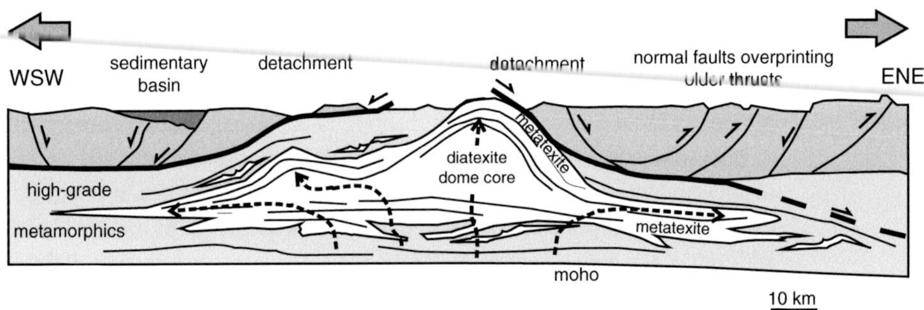

Figure 7.25 Conceptual crustal cross-section of a migmatite-cored domal metamorphic core complex, based on the Thor-Odin dome of the Shuswap MCC (Vanderhaeghe *et al.*, 1999). Modified from Rey *et al.* (2009). Published with permission of Elsevier.

flow of the hot middle and deep crust will occur contemporaneously with extension in the upper crust that is accommodated by episodic failure on listric normal faults that sole into detachment zones above the doming or exhuming ductile middle crust. This geometry is similar to the situation described in this chapter with reference to relatively 'cold' (600–650 °C) E-HPG basement in the southern Western Gneiss Region of Norway (Andersen & Jamtveit, 1990). However, where the middle crust is hot, well within the T range appropriate for melting ($T > 700$–800 °C), it is likely that heterogeneity in upper crustal extension will facilitate localised vertical extrusion and exhumation of metatexites cored by domal diatexites, as observed in the Cordilleran Core Complexes of NW USA and Canada, parts of the Variscan of Europe, and elsewhere (Zwart, 1963; Vanderhaeghe *et al.*, 1999; Whitney *et al.*, 2004; Rey *et al.*, 2009).

Metamorphic core complexes are exposures of ductile, medium- to high-grade middle to lower crust exhumed from beneath overlying brittle and significantly lower-grade or unmetamorphosed, highly extended, upper crust (Fig. 7.25). In the classic geometry described by Coney (1980) and amplified on by subsequent workers, the hanging wall superstructure of normally faulted thinned upper crust is juxtaposed against the footwall metamorphic core complex on a low-angle detachment characterised by strong mylonitic fabrics, across which there are marked gradients in strain (decreasing intensity of mylonitic fabric downwards) and metamorphic temperatures (rapid increase in recorded T with depth over a narrow structural thickness).

Exhumation of the hot footwall metamorphic core complex may be driven entirely by the extensional removal of overlying crust. However, when the deep crust is at high enough temperature to undergo melting the associated decrease in viscosity and potential for buoyancy-driven upward extrusion of melt-rich migmatites may facilitate the development of migmatitic metamorphic core complex domes. These are characterised by metatexite migmatites that beneath the detachment have shallow foliations and sub-parallel leucosome segregations. Where the upper crust has been completely removed, domal migmatitic structures, in which core regions of diatexite are mantled by metatexite and evidence is present for syn-anatectic flow, may occur on kilometre scales, as typified by the Shuswap Metamorphic Core Complex of Canada (Vanderhaeghe *et al.*, 1999; Vanderhaeghe, 2009) (Fig. 7.25). Structural features associated with these domes are concentric foliations, down-dip mineral lineations, asymmetric cascading folds consistent with the upward relative motion of the diatexite core, families of elliptical domes of different scales, and diapirs. These all point to gravitational instability as the principal driving force, facilitated by the marked density difference between melt-rich diatexite and the melt-drained envelope of metatexite and, possibly, restite. It is considered that buoyancy acts in concert with lateral pressure gradients to allow vertical escape of high percentage melts through the metatexite envelope (Vanderhaeghe, 2009).

The metamorphic record of extension and core complex evolution depends to a large extent on the rates of extension and position of mid-crustal rock packages in relation to the overlying detachment when extension is asymmetric. Rey *et al.* (2009) have examined these effects, and the additional impacts of melting, for 25% extension of crust at strain rates corresponding to extension at 1 mm/a and 1 cm/a. In each case the extension facilitates focussed exhanced exhumation of the deeper crust as domes some 30–50 km across. For slower extensional strain rates, kyanite-bearing rocks initially located at 30–40 km in previously thickened crust experience decompression-cooling *P–T* paths ($dP/dT = 24–16$ bar/°C) located more or less along the kyanite–sillimanite reaction. For example, after 40 Ma rocks initially at 11 kbar and 770 °C may be exhumed to 5 kbar and 500 °C, depending on their final position in the dome beneath the detachment. At faster rates of extension (1 cm/a) that may be relevant to migmatitic core complexes such as the Shuswap (Rey *et al.*, 2009) the *P–T* paths for rocks from similar initial conditions are dominated by rapid ITD, followed by cooling (IBC) in those cases where the rock packages move laterally in the domed core complex region after this initial decompression. In this case those kyanite-bearing schists at initial *P–T* conditions of 11 kbar and 770 °C undergo ITD through 4–7 kbar at $T > 700$ °C, producing sillimanite-bearing migmatitic metatexites and diatexites and in the most exhumed case (in the core of the dome) cooling to $T < 600$ °C and into the andalusite field, all within 4 Ma (Fig. 7.24). This type of *P–T* evolution is broadly comparable to, though not coincident with, that recorded in the Thor Odin Dome of the Shuswap Metamorphic Core Complex (Fig. 7.24). In that metamorphic core complex, migmatites were generated from 60 Ma at initial *P–T* conditions of 10.5 bar and 760–800 °C, evolving to lower-*P* on an ITD path to about 4 bar over a time period of some 8 Ma. Isobaric cooling to Ar–Ar closure temperatures of 300 °C was completed by 4 Ma, giving a 12 Ma timescale for the whole metamorphic evolution related to post-collisional extensional collapse.

Concluding remarks

This chapter has reviewed the realm of metamorphism associated with orogeny, in doing so discriminating the three main *P–T* domains now understood to be attained and traversed by metamorphic rocks formed in accretionary and collisional orogenic belts. From this review it is apparent that, although much remains to be done in order to further constrain both the prograde histories and exhumation rates of the high *P/T* HP–UHP metamorphic belts, there is a clear and consistent relationship between subduction-accretion processes and the formation and evolution of these belts. To some extent the relationship between orogen evolution, in terms of thickening of continental crust and orogenic plateau formation, and the formation and P–T evolution of E-HPG rocks and medium P/T is also well understood and explained in many respects by coupled two-dimensional models of large, hot orogens. On the other hand, whilst low P/T metamorphism can be explained in terms of gravitational spreading, magmatic heating, lithosphere delamination or accretion of hot back-arc and arc crust, there remain significant issues with forming G-UHT metamorphic belts that record thermal excursions to over 1000 °C at 7–12 kbar. The high heat productions that appear to be required for such extreme metamorphism in orogens are likely to have important consequences for orogen behaviour in the Precambrian. We will return to this theme, and the possibility of 'ultra-hot' orogens (Chardon *et al.*, 2009) in Chapter 12.

8 The erosion and exhumation of mountains

This is a large topic which covers the decay of mountain belts by erosion and more controversially, the possible link explored in Chapter 11 between mountain relief and climate. The topic has already been mentioned in Chapter 5 because it seemed appropriate to discuss the exhumation of the European Alps in the context of the recent dating of the HP metamorphism. A huge amount of literature now exists on the topic, which may be simply set out as the link that mountain building creates topography and this in turn generates precipitation as rain or snow and ice. A more accurate formulation would be that many factors are involved if the effect of topography on climate is to be more than transient. Thus mountains may amplify and modify rather than create precipitation in the first place. The relationship is well shown in the Canary Islands which are in a semi-arid belt. Low islands such as Lanzarote and Fuerteventura are semi-arid. However, on Tenerife there is a volcano of 3700 m altitude, rising straight from sea level. On the windward north side of the volcano the island is very wet, with green pastures on the coastal plain, whereas the leeward side is arid or semi-arid. As a consequence of the precipitation and thus enhanced erosion on the windward side of the island there is also a marked asymmetry in steepness of relief: the wet north side has steep relief and a distinct coastline with high cliffs whereas the leeward south side is more gently sloping. Some global climate models show that mountains, while they do not determine whether or not a monsoon occurs, tend to lengthen and enhance monsoonal periods.

The factors involved in climate include atmospheric circulation and ocean water circulation. If these are not appropriate then the mountains will not create precipitation. The English Lake District and the Scottish Highlands would not be such wet places if the global atmospheric circulation pattern in the form of the Atlantic 'westerlies' did not dominate the climate of Britain. Over the past 100 Ma temperatures have decreased globally and this has been attributed to changes in sea-floor spreading rates, or in land–sea distribution, or in the rates of out-gassing of volcanoes. Rapid uplift of the Himalayan–Tibetan region may have caused cooling and glaciation in Europe and North America during the past few Ma because of its profound effect on global atmospheric circulation, causing a greater southward penetration of Arctic air masses over North America and Europe. These are controversial topics mainly because they seek a straightforward cause and effect relationship between geological processes and climate (Fig. 8.1).

Rivers play an important role in the erosion process. River incision proceeds by plucking of material from the river bed, by abrasion and by cavitation. The river profiles and network control the relief of mountains by carving them into peaks and valleys. Changes in topography produce a rise or fall in the river base levels. Again the dynamic equilibrium between uplift and erosion may be upset by tectonic driving forces such as a change in the

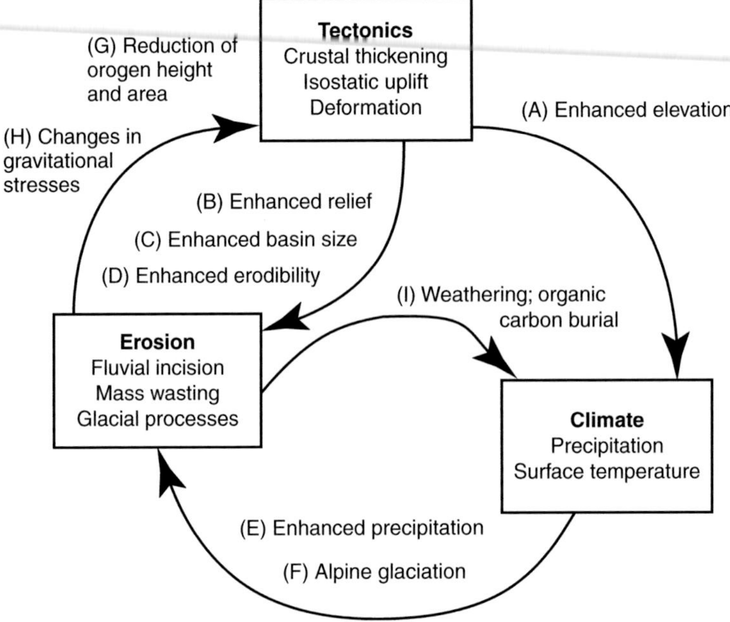

Figure 8.1 Interactions and feedbacks for tectonics, climate and erosional processes. For explanation see text. Modified from Willett *et al.* (2006), published by permission of the Geological Society of America.

rate of plate convergence across an orogen. However, it should not be forgotten that the effect of glaciers and land ice is at least as powerful; for example the western coast of South America in Chile changes markedly from a straight coastline in the north to a fjordland coastline in the south. In the south, the glaciers in the Andes act like chainsaws in cutting the fjords. All regions where glaciers have been active show the same effect.

The complexity of the interplay of various processes is also shown by evidence that in the Himalaya precipitation varies considerably between places which are only a few kilometres apart along strike but which have markedly different topography. Furthermore, Molnar (2001) questioned the simple correlation between precipitation and the amount of erosion, because an increase in aridity may produce an increase in the erosion rate if floods increase.

Mountains as barriers

Mountains influence climate because they are obstacles to global air circulation patterns; additionally, they are sources of elevated latent heat and, by changing water exchanges between continental surfaces and the atmosphere, they give rise to increased snow cover, elevated plateau or rain shadow deserts.

Under steady-state conditions removal of a landmass by erosion is balanced by isostatic uplift. The cyclical process is as follows – topography is reduced and decays by river and glacier erosion and by landslides. Surprisingly, as first pointed out by Holmes (1944) and more recently by Molnar and England (1990), isostatic factors mean that the removal of rock material by river erosion leads to an increase in the heights of mountain peaks as material is removed from the valleys. In turn, this increased elevation relative to the regional base level increases the rates of erosion by fluvial incision and transport, providing a positive feedback loop amplifying the effect.

Box 8.1 **The interplay of tectonics, erosion and climate**

Topography tends to increase localised precipitation which gives rise to increased river discharge and incision. Figure 8.1 shows the physical reaction between tectonics, surface processes and climate. The increased river incision leads to oversteepening of relief which in turn increases the risk of hillslope failure. Tectonic activity increases relief by normal faulting caused by isostatic adjustments and by folding or buckling of the crust. A link between tectonics and erosion is shown by the consequent increase in erosion and sediment yield. Accretion of new material into the orogen gives a larger area subject to erosion and therefore sediment yield is increased. Increased elevation relative to base level increases river channel gradients and transport. Topography increases precipitation and hence river discharge and incision is increased. On shorter timescales, tectonics influences erosion rates by earthquakes which trigger landslides. The amount of material that is removed from a mountainous area depends on latitude and altitude, and position with respect to the global atmospheric and oceanic circulation pattern. More locally it depends on the erodability of rocks, the erosivity of rivers, channel slope, rate of water discharge and uplift.

Climate links surface uplift and erosion rates by orographic forcing of precipitation and the onset of glaciation as mountains rise high enough for permanent glaciation to occur. The feedback of erosion on tectonics is provided by the redistribution of mass which influences gravitational stresses with respect to the geoid. Hence tectonics becomes important here because tectonic processes are strongly influenced by gravity. If erosional mass or fluxes outpace the rate of crustal thickening, the mean elevation and area of the mountains will be reduced, leading to a reduction in sediment yield and erosional flux.

Erosional events may exist in pulses or may occur by steady-state processes. The Himalaya is an example of the former because erosion or tectonic denudation there has been shown to be episodic (see Chapter 12). Similarly in the Alps the high-temperature geochronometers indicate that tectonic events are episodic although low-temperature geochronometers indicate a steady state. Both may be right.

Examples of the linkage between tectonics and erosion

The preceding account seeks to explain the balance between the tectonic processes which create topography and the erosional processes which tend to destroy it. The system tends to a steady state in which the flux of material into the orogen, along thrust sheets for example, is balanced by the erosion rate which removes material from the orogen. Brocard and van der Beek (2006), working in the Alps, showed that topographic relief is due to competition between uplift and erosion by river incision. In other words, the erosion rate is proportional to the rate of uplift of the orogen. Thus we can speak of equilibrium landforms or mature landforms which are at or near the steady-state condition.

The close linkage between tectonics and erosion is illustrated by the work of Wobus *et al.* (2003) in central Nepal where there is a spatial correlation between, on the one hand, a marked break in rock uplift rates, knick points in river profiles, present-day out-of-sequence thrusting and, on the other hand, the heaviest precipitation which occurs in a zone about 30 km south of the MCT. A knick in the river profile separates a region to the north with high precipitation and rapid uplift from one to the south with lesser uplift. Did monsoon-driven erosion focussed at the Greater Himalayan topographic front cause the faulting? Burbank *et al.* (2003) disputed this and suggest that erosion rates from apatite fission track data show no difference in erosion over a 20 km wide part of the Greater Himalaya. From this they conclude that monsoonal precipitation does not correlate with erosion.

Thiede *et al.* (2004) in the Sutlej valley in the western Himalaya showed that climate controls the orogenic structure: in particular, maximum precipitation occurs at the breaks in topography marked by the Main Central Thrust and the Main Boundary Thrust. Landslides are frequent in periods of intense monsoon.

Again, Stewart *et al.* (2008) showed that the Brahmaputra river in the eastern Himalaya cuts through an antiform which was exhumed at a rate of 3–5 mm/a for the period between 5 Ma and 1 Ma ago. In more recent times the rate has been higher, i.e. 10 mm/a (see Booth *et al.*, 2004 and Seward and Burg, 2008). Of the sand in the Brahmaputra river, 45% is from old basement gneiss found in the Brahmaputra canyon, thus only 2% of the total drainage area of the river makes a major contribution to the sediment in the river. On a larger scale, rapid exhumation in the Eastern Syntaxis has contributed *c.*20% of the total detritus in the Bengal Fan.

Finally, studies in the Pyrenees exemplify the interaction between erosion and deformation. This interaction contradicts the traditional view that erosion followed tectonism, because it is now clear that erosion is active during tectonism and indeed exerts a control on the progress of deformation. In order to demonstrate that point, Sinclair *et al.* (2006) have developed a model for the Pyrenees which incorporates several features of orogenic growth, in particular the accretionary, erosional and sedimentary flux in the orogenic system (Figs. 8.2, 8.3, 8.4). The word 'flux' refers to the idea that in dynamic wedges material is flowing into and out of the wedge, the former process being tectonic and the latter erosional. The flowing material has trajectories, and numerical models show that without erosion the trajectory paths for a parcel of rock accreted from an underthrust plate will be directed towards the overlying plate. However, these trajectories through an orogen are deflected when erosion drives the surficial removal of mass. Trajectory paths determine the evolution of orogens and are therefore a first order control on orogenesis.

The results of Sinclair *et al.* (2006) showed that small doubly vergent orogens focus deformation and erosional unroofing towards the pro-wedge when coupled to a spatially invariant erosion system. The strong coupling between erosion and deformation is greater on the pro-wedge side and this defines the strong asymmetry of the orogen. Underplating is

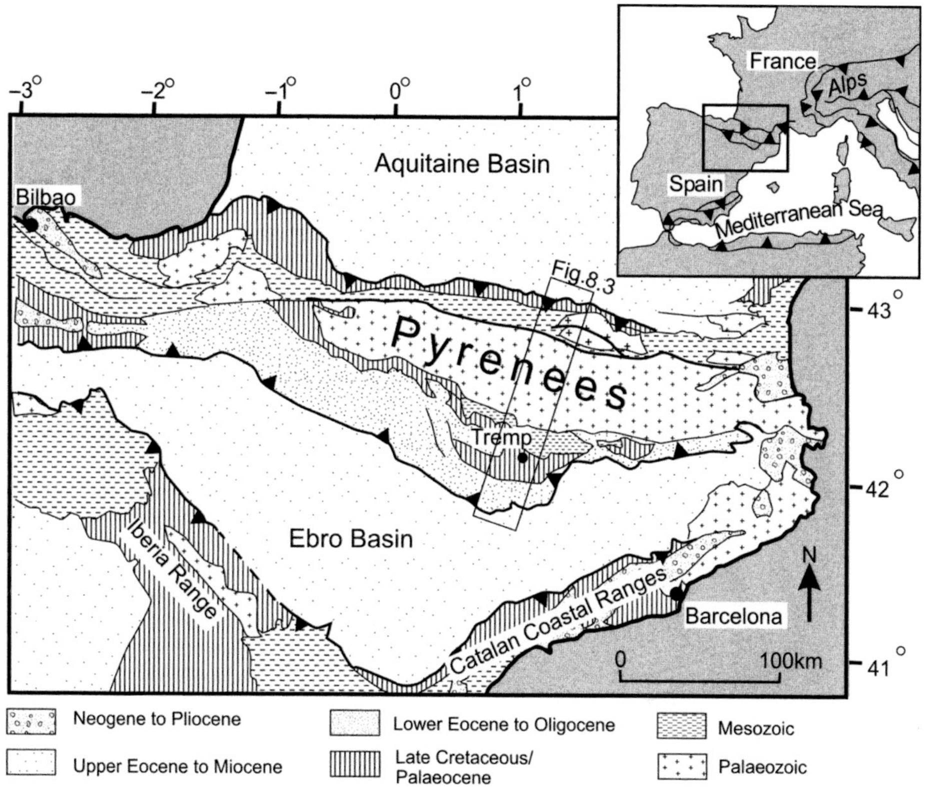

Figure 8.2 Location maps of Pyrenees showing stratigraphical and structural units. From Sinclair *et al.* (2006), published by permission of the *American Journal of Science*.

a significant factor in the Pyrenees. It is linked directly to maximum erosional efflux and goes on to determine the asymmetry of sedimentation rates in the foreland basins.

The Pyrenees formed by the convergence of the Iberian and European plates and evolved in stages over *c*.60 Ma starting in the late Cretaceous and ending in the early Miocene (Figs. 8.2, 8.3). About 165 km of convergence of the Iberian and European plates took place over this period. What interested Sinclair *et al.* (2006) was the asymmetric and doubly vergent nature of the orogen, a complexity which is in contrast to the simple asymmetric plate motion. The terms pro-wedge (that is, the part above the subducting plate) and retro-wedge (the part above the underthrust plate) come from the wedge terminology introduced by Willett (1999) for doubly vergent orogens. Another example of a doubly vergent orogen given by Willett is the Himalaya as a pro-wedge and Tibet as the retro-wedge. The Pyrenees developed as follows (Fig. 8.4):

1. Initiation of the south-verging pro-wedge in the Late Cretaceous. Sedimentation in marine basins.
2. Frontal accretion and underplating in the pro-wedge in end-Palaeocene to mid-Eocene times, accompanied by unroofing of the Axial zone, erosion and foreland basin marine sedimentation.

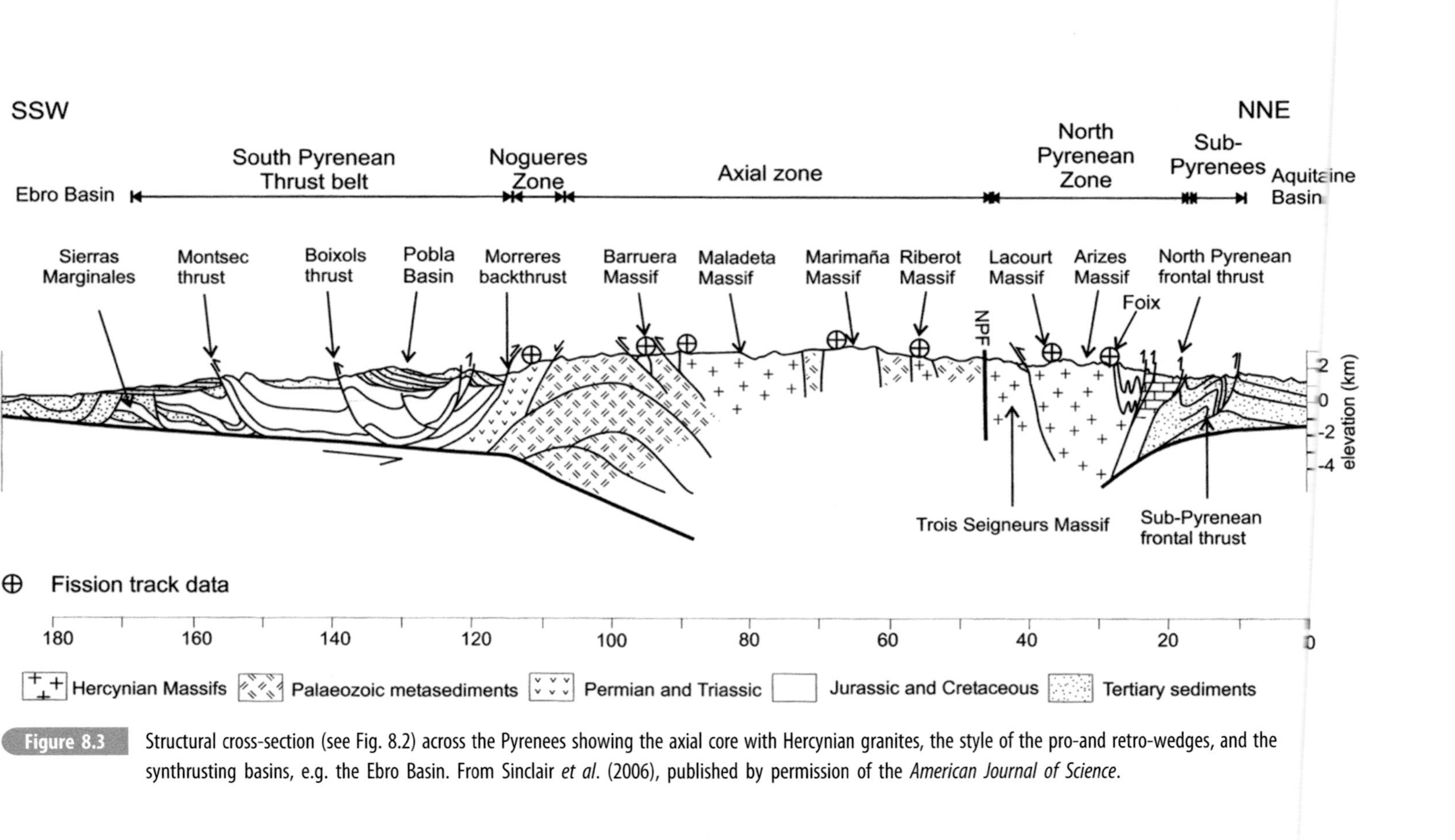

Figure 8.3 Structural cross-section (see Fig. 8.2) across the Pyrenees showing the axial core with Hercynian granites, the style of the pro-and retro-wedges, and the synthrusting basins, e.g. the Ebro Basin. From Sinclair *et al.* (2006), published by permission of the *American Journal of Science*.

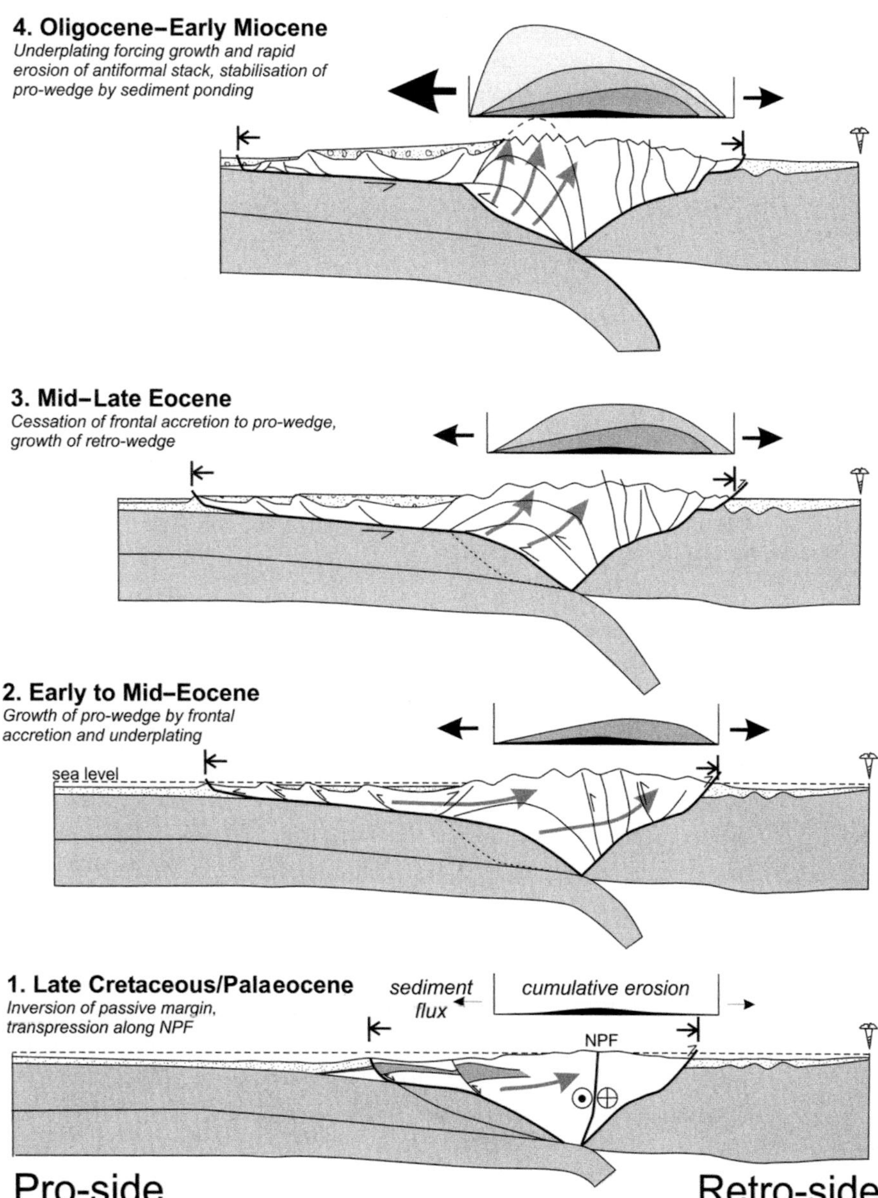
4. Oligocene–Early Miocene
Underplating forcing growth and rapid erosion of antiformal stack, stabilisation of pro-wedge by sediment ponding
3. Mid–Late Eocene
Cessation of frontal accretion to pro-wedge, growth of retro-wedge
2. Early to Mid–Eocene
Growth of pro-wedge by frontal accretion and underplating
sea level
1. Late Cretaceous/Palaeocene
Inversion of passive margin, transpression along NPF
sediment flux
cumulative erosion
NPF
Pro-side
Retro-side

Figure 8.4 Parts 1–4 show the evolution of the Pyrenees, indicating fluxes or movement of material through the orogen as shown by frontal accretion, underplating, erosion and sedimentation. NPF, North Pyrenean Fault. From Sinclair *et al.* (2006) Fig. 15, published by permission of the *American Journal of Science*.

3. Cessation of pro-wedge growth in mid to late-Eocene times; propagation of the retro-wedge was initiated with north vergence and accretion. The pro-wedge was now inactive and the foreland basins were filled to or above sea-level.

4. In Oligocene to early Miocene times there was reactivation of the pro-wedge which was induced by underplating instead of frontal accretion. This was accompanied by accelerated erosion in the Axial zone and an increased rate of sedimentation in the foreland basins seen on the southern side of the Pyrenees, e.g. the Ebro basin, and the Aquitaine basin on the north.

Sediment supply after 50 Ma was far greater on the pro-wedge side than on the retro-wedge side. As predicted by wedge theory (see Chapter 4), the close interplay of thrusting and volume loss by erosion has caused adjustments of wedge taper. The significant point here is that the pro-wedge developed before the retro-wedge. From sand box modelling, Merle and Abidi (1995) showed that synthrusting erosion promoted out-of-sequence backthrusting in orogenic wedges.

Thermochronology (low temperature, apatite and zircon fission track) and sedimentation rates in the foreland basins show a shift in time in the focus of erosional exhumation towards the pro-wedge. A broad region of unroofing existed from 60 to 40 Ma, then there was a southern shift with a period of highest erosion rates from 36–20 Ma linked to the growth of the antiformal stack in the Axial zone. The youngest apatite fission ages of 35–20 Ma record the growth of an antiformal stack due to underplating which induced vertical trajectories. There has been 6–9.5 km of erosion over the core of the Axial zone since 50 Ma, representing the greatest depth of exhumation in the Pyrenees. Mass balance in the wedge is conserved because erosional loss is accommodated by tectonic thickening to restore the taper. The material flux is greater on the pro-wedge side. The process involves a dynamic coupling between erosion and structural deformation of the pro-wedge.

During pro-grade thrusting there was advection of material towards the retro-wedge. In the evolution of the thrust stack there was a transition from dominantly frontal accretion to underplating and the growth of an antiformal stack in the inner part. This is attributed to stable sliding of the thrust stack caused by thick sediment accumulation in ponded piggy-back basins on top of the pro-wedge. The progressive incorporation of basins into the thrust stack resulted in a thickening of the wedge and an increase of the surface slope, thereby driving it into a stable configuration. This system is asymmetrical and results from a reduction in the velocity of rock particles moving from the pro-wedge front to the retro-wedge. The relations between tectonics, erosion and sedimentation in the Pyrenees are discussed further in Chapter 9.

Dating the rates of exhumation and uplift in mountain belts

There is a great amount of data on the exhumation, uplift and erosion rates in orogens, all of which emphasises that, apart from a few dramatic examples, long-term slow processes are involved. Manifestly, uplift is substantial over geological timescales, but the yaks in the Himalaya would not be aware of what was happening beneath their feet. For example the Western Alps are being uplifted by 0.5–1.0 mm/a in recent times, as determined by cosmogenic dating of uplifted river terraces. The dating relies on the production of cosmogenic nucleides such as ^{10}Be, ^{26}Al, ^{3}He or ^{21}Ne isotopes and gives a measure of the residency time on the surface. A similarly slow uplift is shown by the Himalayan data. Cosmogenic nucleides have also proved very useful when dating the timing of great landslides, rock faces and fault scarps in orogenic belts.

Another important point about mountains and climates is that orogenic growth is likely to produce a rain shadow on the leeward side of a mountain range, for example in the arid deserts in western USA east of the Sierra Nevada. It also guides drainage patterns or glaciers. Erosion may be said to be symmetrical or asymmetrical with respect to a mountain range, meaning that erosion may operate on both sides of the high peaks or only one side. Given enough precipitation the drainage pattern will cover the entire orogen up to channel heads or snow lines. But precipitation may be focussed on one part of the orogen. For example, in the Himalaya the monsoonal precipitation is strongly focussed on the High Himalayan front above an elevation of 2–3.5 km, to the extent that 80% of precipitation is limited to that small area. This focussing is interpreted by Beaumont *et al.* (2001, 2004) as a control on the rate of channel flow and extrusion (see Chapter 6).

The recognition that surface processes are important in the growth of orogens is a fairly recent development. The Keystone thrust in southern Nevada is a remarkable example of a thrust which has emerged onto the Earth's surface. Johnson (1981) showed that erosion facilitates thrust motion by reducing the load of the thrust sheet (Plates 19, 20). In orogenic wedges, erosion affects the critical wedge taper and thereby triggers a structural response in the wedge (see Chapter 4). If, owing to erosion, the critical taper cannot be maintained then wedge thickening by thrusting ensues.

The mass balance within an orogen

The above discussion establishes the point that the gross topographic shape of mountain belts is determined by a balance between tectonically induced uplift and the erosional degradation of the mountains. The balance referred to is between the input of material into thrust sheets on the one hand and the efflux of material by erosion on the other. A knowledge of these two opposing factors leads to an understanding of the mechanics of mountain building. The internal stress regime in an orogen is modified by the redistribution of mass and thus the mountains respond to spatial and temporal changes in erosion rate. Because climate is a significant control on erosion, changes in climate must affect the mechanics of mountain building (see Chapter 12).

Terminology

There are several terms used in relation to this topic and it is important to distinguish between them. *Uplift* means the raising of a surface relative to mean sea level. *Cooling rate* refers to the cooling history of a metamorphic complex (see *P–T–t* paths) determined by geochronological and thermobarometric techniques. *Erosion* refers to transportation of weathered rocks destroyed by mechanical or chemical weathering: that is, surface alteration processes. *Denudation* is the removal of rocks from their original position by river transport or landslips etc. *Exhumation* is the process whereby rocks are transported to the surface after metamorphism. Note that exhumation and cooling do not necessarily change the overall surface height. For example if erosion keeps pace with uplift then no change of surface elevation occurs.

The measurement of the rate of exhumation

This may be gained from (a) the cooling history of the rocks and minerals in the orogen, e.g. as determined by $^{40}Ar/^{39}Ar$ dating of micas, fission track geochronology and other techniques; (b) the timing and conditions of metamorphism (*P–T–t* paths; see Chapter 7) as determined by a study of geochronology and the thermobarometric histories of rocks, and (c) the sedimentary history of the foredeeps which should contain a record of the unroofing of the orogen (see Chapter 9).

Orogenic history

Traditionally orogenic history was given as pre-, syn- and post-orogenic, with the uplift and erosion being the last phase. However, a more complex history is revealed in polyphase orogenic belts such as the Canadian Rocky Mountains where thrusts have been emplaced sequentially and each thrust sheet generated uplift, erosion and foredeep sedimentation. Here we refer to the exhumations of the Alps and Himalaya in order to demonstrate this point.

Exhumation of the Alps

The unroofing of the Alps started in late Oligocene times soon after peak Lepontine metamorphism. Luth and Willingshofer (2008) described the cooling history of the Eastern Alps using Rb/Sr, K/Ar (biotite), and apatite and zircon fission track dating. These data show cooling from upper greenschist/amphibolite facies at 500–600 °C to 110 °C between 25 and 10 Ma, a cooling rate of *c*.30 °C/Ma. Cooling in the Pennine units of the Tauern Window occurred in the Oligocene to early Miocene at rates of 50 °C/Ma, continuing in the mid Miocene. The cooling was diachronous in the Tauern Window, being earlier in the east by *c*.5 Ma. In the Dauphinois zone, Crouzet *et al.* (2001) described tectonic denudation at about 24 Ma followed by faster cooling after that date. Similar *P–T–t* paths that display a cooling history can be determined in the Lepontine Alps and the Simplon area.

As well as geochronological data, evidence for exhumation is also derived from study of the history and nature of the mineral assemblages in the rocks. In the Eastern Alps the thickness of the Austro-Alpine nappes is *c*.20 km. At deeper levels a greenschist to low amphibolite facies metamorphism shows a maximum pressure of 6 kbar at 100 Ma, that is in the late Cretaceous prior to collision. Pennine nappes underlying the Austro-Alps, exposed in the Tauern Window, show early Oligocene metamorphism at 10 kbar (*c*.35 km depth) and there is some evidence of higher-pressure assemblages. Thus at least 15 km of rock was removed from near the basal contact of the Austro-Alpine nappes after HP metamorphism and before or during the Lepontine metamorphism at 40 Ma. This evidence may mean that the contact between the Austro-Alpine and Pennine nappes is a normal extensional fault which was active before or during the Lepontine metamorphism.

The uplift of the Alps may have been by (1) the buoyant return of slabs of continental crust which had been subducted into the mantle and (2) underplating of the crust accompanied by isostatic uplift and erosion. What is clear is that there has been a huge loss of rock linked to the uplift. For example, about 50 km originally covered the Sesia zone and even more cover, *c*.90 km, existed above the Dora Maira zone. The removal could have been by erosion: thus the early Oligocene–Miocene molasse basin, which is situated north of the Helvetic zone, has thick successions of arenaceous rock which were eroded from the Alps. But erosion cannot account for all the loss of material from the rising Alps because there are few examples of molasse-type basins before the Oligocene, and the Alpine flysch of late Cretaceous and early Cenozoic age is not of great volume (see Chapter 9). Tectonic denudation, in early Miocene times, involved orogen-parallel extension and served to enhance the unroofing of the Alps.

Examples of the conflicting datings are as follows. Schlunegger and Willett (1999) described a rapid exhumation of the Lepontine region at 35–20 Ma at a rate of 1 km/Ma, but whether this was by tectonic denudation or erosion is not specified. Carrapa *et al.* (2003) described $^{40}Ar/^{39}Ar$ mica dates from sediments in rivers in the Piedmont Basin of North Italy. These suggest rapid uplift at 38 Ma followed by slower cooling over a 30 Ma period. In addition, sedimentation associated with uplift occurred in pull-apart basins in the eastern Alps.

Other evidence of the rapid unroofing of the Alps comes from the foreland basins. For example, HP minerals appeared in the sediments of foreland basins in southeastern France at 30–32 Ma, close to the revised age for the Barrovian metamorphism, which again shows a rapid ascent in the lithosphere after metamorphism (Morag *et al.*, 2008). Late exhumation is shown by apatite fission track ages of late Neogene to recent age, the uplift rate then being 0.5 mm/a. Vernon *et al.* (2009) described variable late Neogene exhumation in the central European Alps, such as an increase from 750 m/Ma to 3000–1000 m/Ma at 5 Ma followed by a drop to 700 m/Ma at 5–3 Ma.

Exhumation of the Himalaya

The approach is illustrated well by studies on the eclogites found in the Western Himalaya. These rocks formed at temperatures of 580–600 °C and pressures of *c*.24 kbar which translates into depths of over 80 km (roughly, 1 kbar equals 3.5 km). The formation of the eclogite is dated at 47–46 Ma. This was followed by rapid exhumation between 46 and 40 Ma as given by study of mica, biotite and phengite ages, all of which have lower blocking temperatures than the zircons which gave the age of eclogite formation. Thus an exhumation from depths of c.80 km in a short period of <6 Ma indicates a rate of exhumation of 40 mm/a. Similar results come from the Himalaya of Pakistan and at Tso Morari in NW India, i.e. c.1–4 Ma has been estimated for the time taken to exhume the eclogites there. For example, Parrish *et al.* (2006) have dated the UHP metamorphism in Pakistan at 46.4±1 Ma, and the eclogite was exhumed to 35 km depth at or before 44 Ma. This gives an exhumation rate of 30–80 mm/a. Following this rapid exhumation, both areas show very slow rates of exhumation. The mechanism for rapid exhumation is unclear. Both erosion and faulting have been invoked. Chemenda *et al.* (2000) proposed a slab break-off and a buoyant rebound taking the eclogites to higher levels in the orogen (see Chapter 5).

Another region of spectacularly fast exhumation is Nanga Parbat in the western syntaxis. After high-grade metamorphism at 46–36 Ma, the Ar^{40}/Ar^{39} ages decrease from c.40 Ma to 5–0 Ma giving an average cooling rate of 300 °C per Ma over this comparatively long period. This indicates an exhumation rate of 0.30 mm/a. Short-term and recent datings show much higher rates occur at Nanga Parbat, with uplift rates reaching c.5 mm/a for the past 1–2 Ma, as shown by fission track annealing rates. This gives a mean erosion rate of 3–4 mm/a. These rates are not exceptional in active orogens. For example, Burbank and Beck (1991) described rapid rates of denudation of 1–15 mm/a over intervals of 0.2–1.5 Ma in the Cenozoic foreland basin in Pakistan. The interesting point about this example is that uplift rates and erosion rates are nearly balanced and the topography is in a steady state despite the uplift. Lastly, at Shisha Pangma in the central Himalaya, Searle *et al.* (1997) recorded rapid unroofing of the Himalaya at or before c.17–14 Ma when about 12 km was removed by erosion and normal faulting, giving a mean exhumation rate of c.4 mm/a (Fig. 8.5). These results indicate that the Himalaya reached their maximum elevation at c.17 Ma.

What of exhumation rates generally in the Himalaya? The earliest metamorphism in the Greater Himalaya occurred within 10–20 Ma of the India–Asia collision.

The present-day exposure of high-grade rocks in the Greater Himalaya implies the removal of over 20 km of rock. In the Zanskar area of the NW Himalaya of India, Searle *et al.* concluded that after NeoHimalayan metamorphism at c.35–25 Ma, cooling to below 350 °C occurred between 21 and 5 Ma. The top and bottom of the Greater Himalayan Sequence, which is some 10–15 km thick, show cooling rates of c.16 °C/Ma. These rates translate to a 0.53 mm/a denudation rate. After 4 Ma the exhumation rate increases to 1.10 mm/a. In the central Himalaya, $^{40}Ar/^{39}Ar$ data show that the top of the Greater Himalayan Sequence (GHS) was at c.500 °C between 35 and 25 Ma, whereas the base of the sequence was at this temperature much later, between 21 and 4 Ma (Godin *et al.*, 2001). The top part cooled to c.350 °C at 14–16 Ma whereas the base cooled to this temperature at 12–3 Ma, just as the cooling/uplift model predicts. The cooling rate for the top is c.7 °C/Ma over a period from 35 to 13 Ma, and 15 °C/Ma between 25 and 14 Ma. The base of the GHS cooled at a rate of 7–15 °C/Ma. These cooling rates imply an exhumation rate of 0.38 mm/a. Apatite fission track ages from the GHS show an erosion rate of 2–5 mm/a over the past 1 Ma (Fig. 8.6).

The exhumation rate for the MCT zone at the base of the GHS has been 3–5 mm/a since about 7–4 Ma. Rapid cooling rates of <65 °C/Ma between 5 and 2 Ma are given by zircon fission track ages and muscovite ages. From $^{40}Ar/^{39}Ar$ muscovite ages, the top of the GHS

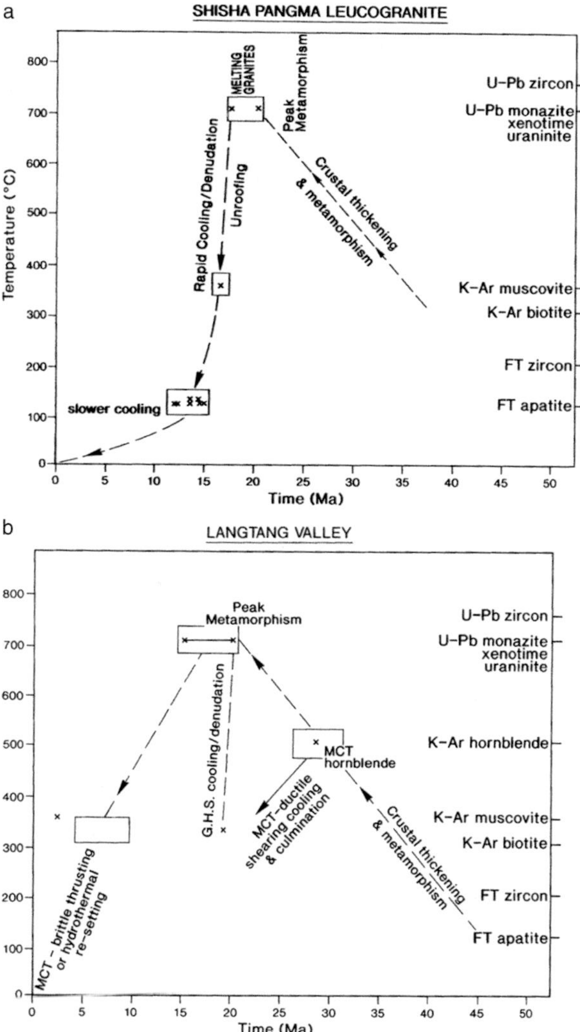

Figure 8.5 Rapid exhumation at Shisha Pangma and Langtang in the Himalaya. From Searle (1997), published by permission of the *Journal of Geology*.

was more than 10 km deep between 15 and 13 Ma; in other words the GHS was not exposed at the surface before 15–13 Ma. In the next chapter it is shown that the sediment record in the foredeep supports this conclusion.

In the Eastern Himalaya the GHS cooled below *c*.350 °C between 14 and 11 Ma, giving an average cooling rate of 29 °C/Ma and an exhumation rate of 0.97 mm/a.

Many other examples could be provided from the Himalaya, but the above gives an idea of the approach to the question of cooling, uplift and exhumation. The next chapter discusses foredeep basins which contain a record of mountain exhumation and allow tracking of the process of unroofing.

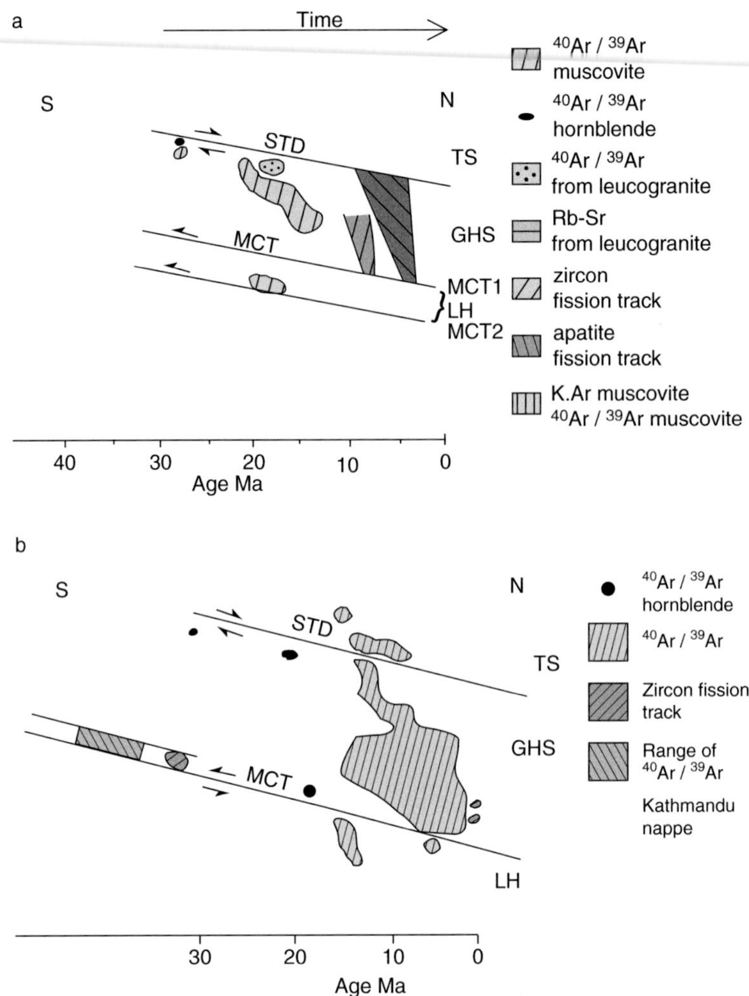

Figure 8.6 Exhumation rates in the Himalaya. a, Cooling history of the Garhwal and Ladakh Himalaya. b, Cooling history of central Nepal. The cooling in the Greater Himalaya starts earlier near the top and youngs downward. GHS, Greater Himalayan Sequence; LH, Lesser Himalaya; STD, South Tibetan Detachment; TS, Tethyan Himalaya. Cooling ages in Kathmandu nappe (external zone) are: $^{40}Ar/^{39}Ar$, 24–18 Ma, zircon fission track, 14–11 Ma. Prepared from data in An Yin (2006).

Further reading

Allen, P.A. and Homewood, P. (1986). *Foreland Basins*. International Association of Sedimentologists. Oxford, London and Edinburgh: Blackwell Scientific.

Barnett, J.A.M., Mortimer, J., Rippon, J.H. and Walsh, J.J. (1987). Displacement in the volume containing a single normal fault. *Bulletin of the American Association of Petroleum Geologists*, **21**, 925–937.

Bernet, M., Brandon, M., Garver, J., *et al.* (2009). Exhuming the Alps through time: clues from detrital zircon fission-track thermochronology. *Basin Research*, doi:10.1111/1365–2117.2009.00400. x

Burbank, D.W. and Beck, R.A. (1991). Rapid, long-term rates of denudation. *Geology*, **19**, 1169–1172.

Carracedo, J.C., Rodriquez Badiola, E., Guillou, H. *et al.* (2007). Eruption and structural history of Teide and volcanic and rift zones of Tenerife, Canary Islands. *Geological Society of America Bulletin*, **119**, 1027–1051.

Fowler, C.M.R. (1990). *The Solid Earth*. Cambridge: Cambridge University Press.

Najman, Y. (2006). The sediment record of orogenic evolution: a review of approaches and techniques used in the Himalaya. *Earth Science Reviews*, **74**, 1–72.

Naylor, M. and Sinclair, H.D. (2007). Punctuated thrust deformation in the context of doubly vergent thrust wedges: implications for the localization of uplift and exhumation. *Geology*, **35**, 559–562.

Richards, A., Argles, T., Harris, N. *et al.* (2005). Himalayan architecture constrained by isotopic tracers from clastic sediments. *Earth and Planetary Science Letters*, **236**, 773–796.

Searle, M.P (1996). Cooling history, erosion, exhumation, and kinematics of the Himalayan-Karakoram-Tibet orogenic belt. In: *The Tectonics of Asia*, ed. An Yin and T.M. Harrison. Cambridge: Cambridge University Press.

Spiegel, C., Kuhlemann, J., Dunkl, I. *et al.* (2000). The erosion history of Central Alps: evidence from zircon fission track data from foreland basin sediments. *Nova*, **12**, 163–170.

Spiegel, C., Siebel, W., Kuhlemann, J. and Frisch, W. (2004). Toward a comprehensive provenance analysis: A multi-method approach and its implications for the evolution of the Central Alps. In: *Detrital Thermochronology-provenance Analysis, Exhumation and Landscape Evolution of Mountain Belts*, ed. M. Bernet, and C. Spiegel. *Geological Society of America Special Paper*, **378**, 37–50.

Stephenson, B.J., Searle, M.P., Waters, D.J. and Rex, D.C. (2001). Structure of the Main Central thrust zone and extrusion of the High Himalayan wedge, Kishtwar-Zanskar Himalaya. *Journal of the Geological Society of London*, **158**, 637–652.

Szulc, A.G., Najman, Y., Sinclair, H.D. *et al.* (2006). Tectonic evolution of thc Himalaya constrained by detrital Ar-40-Ar-39, Sm-Nd and petrographic data from the Siwalik foreland basin succession, SW Nepal. *Basin Research*, **18**, 375–391.

Van der Beek, P., Robert, X., Mugnier, J.L. *et al.* (2006). Late Miocene-Recent exhumation of the central Himalaya and recycling in the foreland basin assessed by apatite fission track thermochronology of Siwalik sediments, Nepal. *Basin Research*, **18**, 413–434.

Whipple, K.X. (2009). The influence of climate on the tectonic evolution of mountain belts. *Nature Geoscience*, **2**, 97–104.

Zeitler, P.K. and 19 others (2001). Crustal reworking at Nanga Parbat, Pakistan: metamorphic consequences of thermal-mechanical coupling facilitated by erosion. *Tectonics*, **20**, 712.

9 Sedimentary history of the foredeep basins

A familiar feature of orogens is the presence of a sedimentary basin, termed a foredeep basin, situated ahead of the orogen. The foredeep basins are filled in the main by the erosional detritus from the adjacent orogen, and good examples are the Siwalik basin of the Himalaya and the molasse basins of the Alps and Andes. The driving force for the erosion is mountain uplift. Before considering examples of foredeep basins we must first look at the controls on basin formation.

Isostasy and Bouguer anomalies

A full understanding of the physical relations between a mountain chain and the adjacent foreland basins calls for some knowledge of the gravitational factors involved in mountain building. The famous experiments in the Peruvian Andes by Bouguer in 1735 and 1745 established that there is a mass deficiency in mountain belts. Bouguer demonstrated this by measuring the deflection of a plumb line towards the Andes and showing that it was much less than expected from the huge bulk of the mountains. The gravity anomalies thus demonstrated are now named after him. Thus under mountains there is a negative Bouguer anomaly (Figs. 9.1, 9.2).

The concept of mass deficiency in mountains led to the idea of a mountain root with continents forming a buoyant crust floating on a denser mantle. The depth of the root is determined by the relative densities of the crust and mantle as well as the amount of convergence. From this discovery there followed the concept of isostasy (see Chapter 3) in which mountains may be viewed as floating bodies in a denser substratum, the mantle lithosphere and the asthenosphere. In the Alps, the Himalaya and the Andes the root extends downwards from the surface to 60–80 km; thus as with icebergs most is subsurface. The image of a mountain root has changed over the years. Originally it was a simple symmetrical down buckle of the crust whereas, as inspection of the cross-sections of the Alps in Chapter 5 will show, it can be more wedge-shaped and contain a stack of thrust sheets. In the Himalaya the gently north-dipping detachment separates a thrust pile of Indian mid/upper crust above from Indian lower crust and mantle below. In Tibet the root is 60–80 km deep and, as discussed in Chapter 5, it may be hybrid, partly Asian lithosphere and partially underthrust Indian lithosphere.

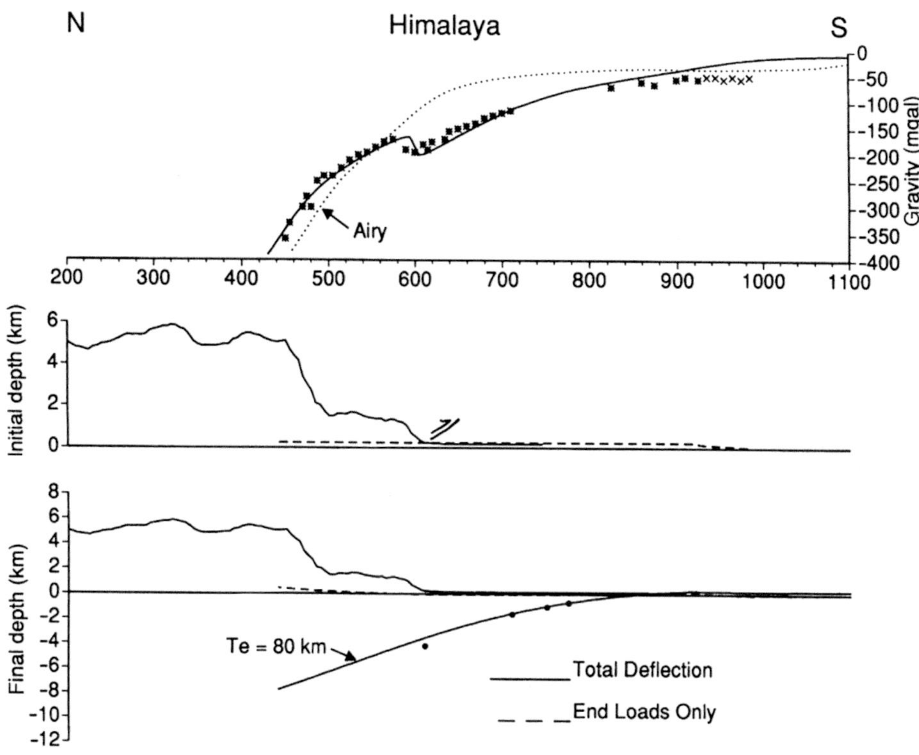

Figure 9.1 Bouguer anomalies across the Himalaya. Top – solid symbols show observed anomalies used to construct the flexural profile, solid line shows computed Bouguer anomaly, dotted line shows computed anomalies if Airy isostasy applies. Middle – shows present topography (solid) and preflexural topography (dashed line). Lower – observed depth of the foredeep. Te – the effective elastic plate. From Royden (1993), published with permission.

Loading the lithosphere

The existence of foredeep basins reminds us that continental lithosphere, because it rests on weak asthenosphere, is sensitive to the addition or removal of even quite small loads, as is evident from the isostatic rise or rebound in northern Europe after the disappearance of the recent ice sheets. In a similar manner the mass of the thickened lithosphere in an orogenic belt causes an isostatic depression of the Moho. However, gravity anomalies show that mountain belts and the adjacent basins are generally not in isostatic equilibrium. The 4–8 km High Himalaya are underlain by <70 km of crust, and the subsidence created by this load flexes the adjacent unthickened crust in the foreland. The crust is depressed in the hinterland as well and the depressions are known as fore (land) deeps and hinterland basins.

The theory of lithospheric flexure, whether in response to loading by ice sheets or mountain belts, is discussed by many authors (Fig. 9.3). The load causes a depression of the lithosphere and a bending moment is applied to the underlying plate which is subducted beneath the orogen. The plate can be viewed as a beam, and the load of the mountain belt

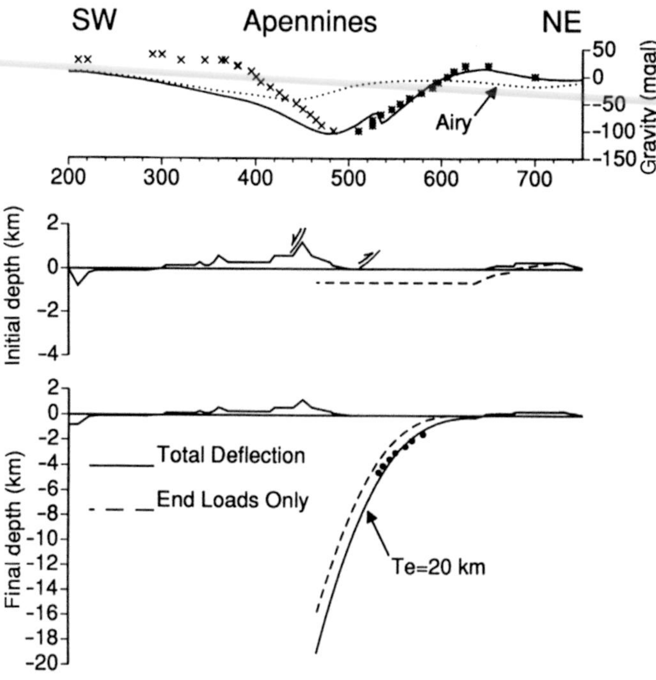

Figure 9.2 Bouguer anomalies across the Apennines. Top – measured anomalies with dotted line for anomalies expected with Airy isostasy. Middle – present topography with positions of thrust front and extensional front. Lower – depth of foredeep basin: solid curve shows best fitting profile and dashed line shows component of flexural profile due to subsurface subduction loads. Te – the effective elastic plate. From Royden, as Fig. 9.1, published with permission.

exerts a bending moment to the end of a plate which can be flexed up and down as the load varies. It is like a diving board which flexes downwards when there is a diver on the end of it. As the underthrusting plate bends downwards the outer arc of the elastic portion will undergo extension giving normal faulting. The collision of India and Asia applied a bending moment to the subducted Indian plate that is not only flexed but also contributes to the support of the Himalaya.

The shape of the flexure, that is the amplitude and wavelength of the foreland bulge and depression, depends on the flexural properties of the crust. The flexural rigidity, which denotes the ability of the lithosphere to bend, is defined by Young's modulus and Poisson's ratio of the rocks involved. The way the system works means that the amplitude of the basin is determined by the flexural rigidity of the plate. Young, hot crust has a low flexural rigidity and tends to form narrow but deep basins, whereas the higher flexural rigidity of cooler older crust, as exists in India, produces a wide foredeep basin and a more distinct foreland bulge. The lithosphere is not likely to be of a uniform composition along the mountain front, and therefore flexural rigidities will vary along the belt. A good illustration of this is the northern Alpine flexural basins. In France these basins are narrow because the flexural rigidity of the lithosphere is low there. However, in Germany and Austria where basement is made of relatively strong Variscan crust which possesses a high flexural rigidity, the basin is wider as a result. But it narrows again into Bohemia and then widens in Poland.

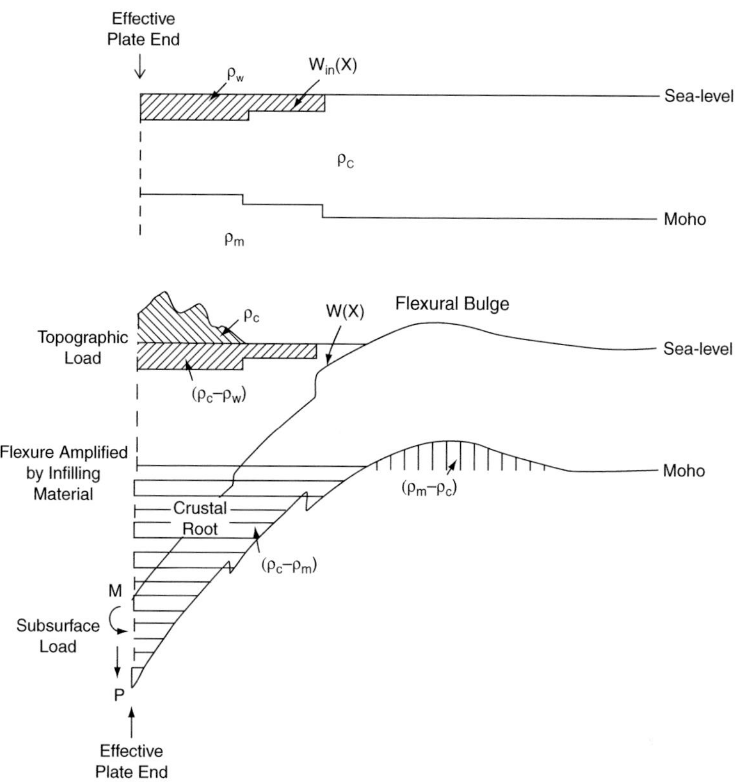

Figure 9.3 Plate geometry before and after flexure. Note the flexural bulge. Prior to flexure the foreland lithosphere is assumed to have Airy isostasy with an initial bathymetry of $w_{in}(x)$. ρ_w – density of water, ρ_c – density of crust, ρ_m – density of mantle. After flexure and thrusting, the depth to the top of the plate is $w(x)$. Flexural modelling of the deflection $w(x) - w_{in}(x)$ can be used to examine the flexural strength of the underlying lithosphere and the loads that act to give the deflection. Loads acting on the foreland lithosphere are shown by diagonal lines. The topographic load (load added to the plate) has density ρ_c above sea level and density $(\rho_c - \rho_m)$ below sea level. The subsurface load is represented by the bending moment M and a vertical shear force P at the effective plate end. Load is also due to the material which must fill the space below $w_{in}(x)$ and above $w(x)$, with density ρ_c. The only contributions to the Bouguer anomaly are the positive mass anomaly with density contrast $(\rho_m - \rho_c)$ corresponding to the crustal root beneath the mountains. From Royden (1993), published with permission.

A classification of foreland basins

As shown in previous chapters, the physical properties of a mountain belt represent a balance between tectonically induced uplift and erosional downwearing within an orogenic wedge. Erosion controls the net efflux of material from a mountain belt and this may be compared with the net influx of rock by deformation at the edge of the mountains. Therefore there is a mass balance in mountains of material being added to the orogen by for example accretion, and volume loss by erosion. The principal factors influencing sediment supply in the foredeep are tectonics, thrusting, drainage and climate (see Chapter 8).

Climate is a key control on erosion, hence it is expected that climatic change will exert an influence on the mechanics of orogenesis. This aspect has already been touched upon in discussing the history of orogenic wedges. The link is complex, owing perhaps to climatic variability added to which there is also a tectonic variability. Climatic factors and variation may produce a punctuated activity on thrusts which reflects the natural variability of frontal accretion and internal thickening of wedges. Furthermore, foredeep basins are compensated by topographic loads that are equal to the elevation of the thrust pile. But the situation is dynamic because the thrust pile not only advances and spreads but also thickens with time. Therefore, as the thrust propagates, the width and depth of the foreland basin will increase with time.

Royden (1993 and see Chapter 2, Figs. 2.1, 2.2) has set out the main characters of foreland basins. The basins are partly or completely filled by the detritus, maybe reaching a thickness of several kilometres, eroded off the mountains and their depth increases gradually towards the mountain, so that the maximum depth is found close to the mountain front. Sedimentary units in foredeeps thin out away from the adjacent mountains and are coarse-grained near the mountain front but finer-grained away from it. The sediments near the mountains may be involved in thrusting as the mountain front gradually advances laterally so as to incorporate the foredeep sediments into the orogen. For example, the Himalayan foredeep shows deformed and thrust rocks in the Main Frontal Thrust sheet, and the growth of blind thrusts within it. As the sediments are mostly derived from the mountains, the rate of flux of sediment into the foredeep is determined by the uplift rate in the mountains. In this respect the foredeep provides a record of exhumation. In the Siwalik Group of the Himalayan foredeep, magnetostratigraphy provides a remarkable record of the pulse rate for the uplift. Furthermore, since foredeep basins are compensated by topographic loads that are equal to the elevation of the thrust pile, it would be expected that the depth of the basin and its width will increase with time as the thrusts propagate.

It is important to remember that in some orogens much of the drainage in the foreland basin is parallel to the mountain fronts as seen in the course of the Ganges or in rivers flowing off the Pyrenees, and this means that sources for the sediments must be sought along strike. In the Alps, the present courses of the Rhone and Rhine rivers are roughly orthogonal to the Alpine front, but it is possible that the drainage during the formation of the Molasse basin was parallel to it.

Foredeeps and advancing thrust sheets

From a study of the different ages of sediments in foreland basins and their incorporation in the thrust belt, the rates of advance and the rates of displacement of the thrust belt can be measured. Where the foredeep basin has been carried in new thrusts which have grown underneath it, these are called piggy-back basins. Examples are the Ebro basin in the Pyrenees and the North Alpine Foreland Basin (this chapter and Chapter 5). Lyon-Caen and

Molnar (1985) have shown that the succession of depocentres (main centres of deposition) in the Himalayan foredeep marks the gradual southwards advance of the thrusts during the Himalayan orogeny (Fig. 9.4a,b). Using the pinch-outs (i.e. southern limits) of the dated sedimentary units, which show a progressive widening of the foredeep basin with time, it is possible to determine a rate of migration of the basin of 1–1.5 cm/a over the past 20 Ma. This is considered to be an approximation to the rate of thrusting and convergence between India and Asia. Molasse basins of the Western Alps indicate the same feature, and they suggest a convergence rate between Africa and Europe of *c.*0.7–0.8 km/a.

Basin evolution

Manifestly, a complete understanding of the evolution of an orogen requires knowledge of the exhumation history and the mechanisms involved. A good example of the link and interplay between tectonics and erosion is shown by the channel model for the Himalaya (see Chapter 6) which predicts the presence of an erosion front at the southern end of the channel, the material loss by erosion at that front serving as a driving force for motion in the channel. But that interplay is obviously of general importance in orogens. One of the lessons gained from the study of old orogens, such as the Appalachians or especially the Proterozoic and Archaean belts which exhibit most of the features described earlier for Cenozoic belts, is that, with important exceptions, they are no longer topographical mountains. But we cannot know whether the removal of cover resulted from erosion alone, as assumed in the past, or whether it was accomplished by low-angled faulting and orogenic collapse, i.e. tectonic denudation. As mentioned earlier the sedimentary record is very important in study of the decay of mountains and has been much studied in recent years in attempts to answer the question: when were the high-grade internal zones of orogens such as the Greater Himalayan Sequence and the Pennine Alps first exposed at the surface? This topic has already been touched on with respect to the Alps in Chapter 5, and the following provides further discussion on the topic.

As already stated, the reason for the foredeep basin is the flexural loading of the crust by the thrust sheets of the orogen. Sinclair and Allen (1992) gave a clear analysis of the process with reference to the North Alpine Foreland Basin (see Chapters 5 and 8). Using the wedge model they explored the mechanics of basin evolution and the accommodation (for sediments) space which is generated. Roughly, the accommodation space is the product of the thrust front advance and the deflection of the basin at the thrust front. The extent to which a basin is undersupplied or oversupplied with sediment is determined by the ratio F, which is calculated as the exhumation rate $\times$ the width of the thrust front, all divided by the thrust advance rate $\times$ the deflection at the thrust front. When $F > 1$ the basin tends to overfill, but when $F < 1$ the basin tends to underfill. The Alpine basin initially had a deep water submarine phase with fast thrust front advance and slow exhumation rates; Sinclair and Allen call this the accretionary wedge phase. But in the mid Oligocene exhumation rates increased and frontal advance rates were slowing down. In consequence the basin was filled to sea level ($F >> 1$) by shallow water marine and continental deposits. They call this the continental wedge phase. The thrust pile has

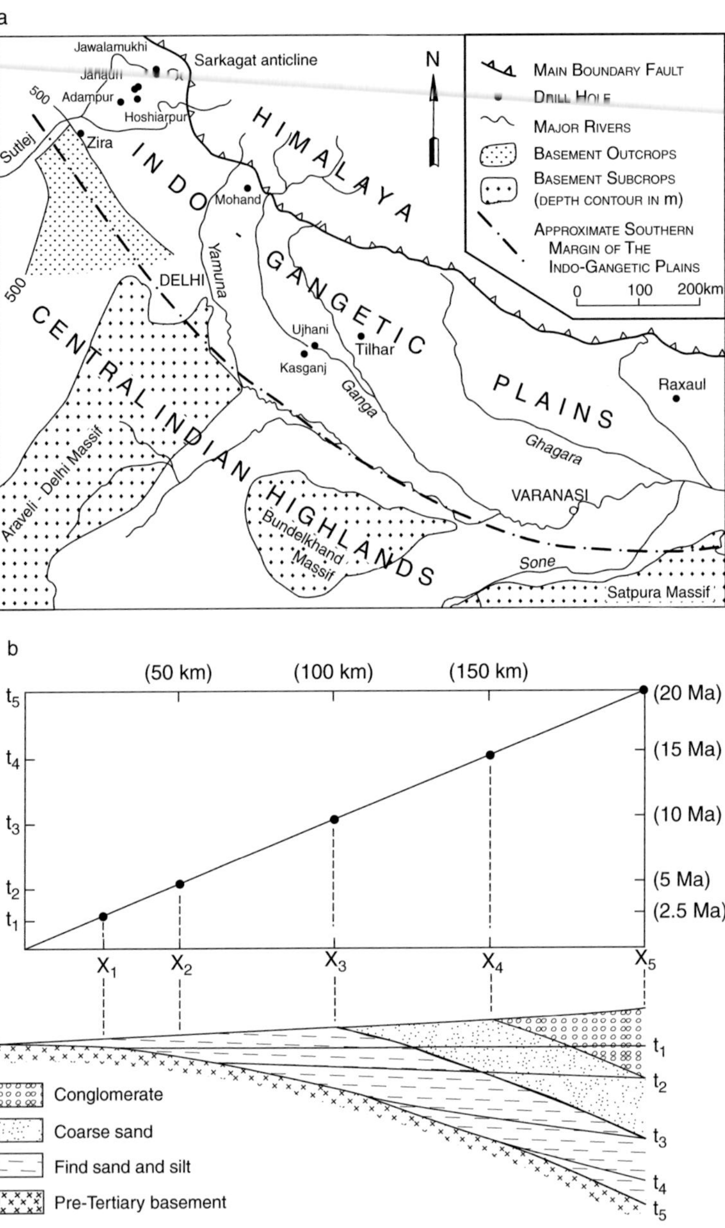

Figure 9.4 The Himalayan foredeep. a, Location of foredeep in the Indo-Gangetic plains and wells, basement outcrops and the edge of the Ganges basin. b, Onlap of foreland sediments. The Himalaya, from which coarse debris is shed into the basin, are to the right. Plot shows expected basement ages vs. distance from the southwest edge of the basin. The basement bends down and moves to the right under the advancing Himalayan thrust sheets. As the basement moves to the right, that is underthrust, it is buried by younger coarser material. Facies boundaries in the foreland basin are time transgressive so that age horizons, given by t_1 to t_5, cross such boundaries. The rate of convergence is the ratio of the distance to the edge of the basin divided by the age of the oldest foreland basin sediment overlying the basement. From Lyon-Caen and Molnar (1985), published with permission.

resulted in a doubling of crustal thickness. This recognition suggests that it should be possible to predict the shape of the foreland basin from a knowledge of the physical factors involved. The factors are the elastic plate thickness, the flexural rigidity, the bending moment at the plate end, and the vertical shear force at the plate end. From this it has been possible to construct profiles of basin geometry and depth which often agree well with the known form of any given basin.

Isostatic adjustments in foredeeps

As shown above, foredeep basins record the isostatic compensation of mountains, and in consequence the basins subside during orogenic thickening but rebound when the mountains are eroded and the load reduced. Although the consensus is that the erosion approached a steady state, Kuhlemann *et al.* (2002) and Cederbom *et al.* (2004) suggested that in the North Alpine foreland (molasse) basin in Switzerland there was a steep increase in erosion rate, with about 1–3 km of erosion, soon after 5 Ma, dated by the cooling ages of minerals determined by fission track methods. At this time deformation in the Alps had declined and there was a simultaneous increase in sedimentation rates in depocentres nearby. Accelerated unroofing and mass loss of the Alps triggered an isostatic rebound and erosion in the foredeep after 5 Ma. The isostatic rebound in the Swiss Alps appears to have caused at least 6.5 km erosion in parts of the Alps. This accelerated erosion is attributed to an increase of atmospheric moisture due to intensification of the Gulf Stream in the north Atlantic at 4.6 Ma. Thus increased erosion due to the climate triggered late orogenic and post-orogenic mass loss and isostatic rebound of both the Alps and the adjacent foreland basin. Flexural theory predicts that the rock uplift within the mountains is five times more than the value for the foreland basin, i.e. a minimum of 6.5 km. This calculation is consistent with the estimated uplift of the basin of 1.3–3.1 km.

Plate theory permits another prediction, which is that the type of sediment filling foredeep basins situated ahead of convergent orogens with advancing subduction boundaries (see Chapter 2) is characteristically molassic, that is arenaceous, whereas those in retreating subduction zone boundaries (see Chapter 2) are flysch, siltstone or greywacke. Royden (1993) considered that in orogens with retreating subduction zone boundaries, such as the Apennines and the Carpathians, the compensation for the foredeep basin is dominated by subduction loads acting on the downgoing dense ocean lithosphere which provide an additional load other than that expected from the weight of the mountains above. Such loads represent the bending moment of the slab and the vertical shear at the effective plate end. In other words gravity acts on the slab and sometimes, as in the Carpathian foredeep near the eastern extension of the Alpine orogen in Europe, this action may have produced a force necessary for the development of a foredeep. The comparative narrowness and depth of foredeeps related to retreating boundary orogens are controlled by the downward force of gravity acting on dense subducted lithosphere. Therefore the sedimentary facies (deep water flysch) reflects this factor.

Thermochronology using detrital minerals

The datings of detrital minerals in the foredeeps, such as muscovite and zircon, have been important in determining the growth and decay of the adjacent mountains.

Thermochronological studies on the sediments in foreland basins use (a) $^{40}Ar/^{39}Ar$ on detrital white micas, (b) low-temperature chronology given by fission track zircon and apatite analyses and (c) Nd analyses which are used for provenance studies; see the example from Taiwan in Chapter 4. The closure temperatures for zircon fission track are in a range of 225–175 °C depending on the condition of the zircon. The apatite fission track closure temperature is 45–120 °C. Another isotope system is (U–Th)/He in zircon which has a closure temperature of 40 °C. In provenance studies Nd ratios, for example in detrital epidote, are used to determine the age of rocks in a source area. The half-life of the parent ^{147}Sm is short enough to produce measurable differences in ^{143}Nd over a period of several million years, and thus Nd values indicate the age of the provenance from which the Nd is derived.

For good reviews of thermochronology with examples, see Reiners and Brandon (2006), Garver *et al.* (1999) and Armstrong (2005). The aim of these studies is to track the unroofing of an orogen and to determine the provenance of the detrital grains. This is clearly a complicated process, not surprisingly in view of the diverse and complex nature of most orogens. Thus foredeep sediments will have mixed signatures reflecting different parts of the orogen which is undergoing erosion. For full tracking of the erosion of an orogen it is desirable to use minerals with different closure temperatures and to undertake other types of geochronological studies. Another prerogative is to avoid the possibility that detrital ages have been reset, for example by igneous intrusion. Whereas zircon U–Pb isotope systems are regarded as resistant to resetting at temperatures prevailing in orogens, zircon fission track can be reset at temperatures above 180 °C.

Erosion acts to remove mountains from the top downwards; therefore the ages of detrital minerals in the foredeep should increase in age as we go down the structural section (Plate 39). This is because during uplift the minerals near the top of the mountains cooled first and thus give older ages than the deeper parts where temperatures are high enough for there to be open systems for a longer period. The assumption is that the degree of metamorphism increases when proceeding downwards deeper into an orogen.

One feature of the exhumation and deposition process is the time taken between closure of the isotope system of a mineral at depth, the removal of cover and the deposition of the mineral in the foredeep. This is called the lag time. Obviously we need to know the depth at which a mineral has grown, and this is calculated using the closure temperature and an assumed and appropriate geothermal gradient. The lag time can be surprisingly short, most of the time being taken up in reaching the surface, after which transport into the foredeep is fast. Thermochronological studies on, for example, fission track data from zircons provide a remarkable tracking of the process. In a particularly illuminating study, N. M. White *et al.* (2001a, 2002) determined the chronology of part of the Himalayan foredeep and considered a section about 4 km thick of the Dharamsala Group and the overlying Lower

Siwalik. The depositional age, 21–12 Ma, is known from the magnetostratigraphy. The disparity or lag time between the depositional age and the youngest mica ages varies, and depends on the exhumation rate. Before *c.*17 Ma rapid exhumation is indicated by the closeness (1–3 Ma) of these two ages. But after 17 Ma the youngest mica grains are a few Ma older than the depositional age, indicating a slowing of the exhumation rate in the Greater Himalaya with more material coming from the Lesser Himalaya as thrusts propagated southwards. The deposition of the Dharamsala Group was roughly synchronous with sillimanite metamorphism and melting in the Greater Himalaya. The minimum cooling rate was 60–40 °C/Ma for the period from 30–20 Ma.

Another excellent example is provided by Kirstein *et al.* (2009) in Taiwan (see Chapter 4). As shown by the cooling rate and thermal structure, in Taiwan *c.*3–9 km of cover has been removed. It is known from biostratigraphy and magnetostratigraphy that sediment deposition in the foredeep occurred between 4.1 and 0.8 Ma. From the time of closure of the isotope systems to the date of deposition in the foredeep gives the best estimate of the lag time as 0.4 to 1.3 Ma.

A slower lag time of 7.4±1.3 Ma or even 16 Ma is reported by Bernet *et al.* (2009) between the Oligocene and the present in the Alps.

Carrapa (2009) has related exhumation and lag times in the Alps to wedge theory (see Chapter 4). Subcritical wedges, where the taper angle is less than the critical value, involve increased hinterland exhumation and reflect rapid plate convergence. Supercritical wedges, where the taper angle is greater than the critical value, reflect decreased hinterland exhumation and slow plate convergence. Lag times decrease up-section in foreland basins, reflecting increased exhumation at 38–36 Ma. Lag times which increase up-section reflect decreased exhumation at 16–8 Ma. A counter-argument that steady-state exhumation prevailed during the post-Oligocene history of the Alps has been put forward by Bernet *et al.* (2009).

Exhumation of orogens

Two classic examples of foredeep basins are the Alps and the Himalaya.

The exhumation and uplift of the Alps

According to Spiegel *et al.* (2000), detrital thermochronometry suggests that coarse molassic sedimentation started in the late Repulian, *c.*31 Ma and ended in the Langhian–Serravallian, *c.*15 Ma. Thus the Central Alps were uplifted before *c.*30 Ma. First to be exposed were the Austro-Alpines, then at 21–20 Ma Pennine units were exposed. The exhumation of Penninic units at the Oligocene–Miocene boundary is related to orogen-parallel extension in the East and Central Alps, operating during continuing northward continental convergence. The extension is related to escape tectonics and the formation of the Pannonian basin. Steady-state exhumation occurred only since 14 Ma; until then cooling rates varied from 20 °C/Ma in the Lower Miocene to 40 °C/Ma in the mid Miocene. Eynatten *et al.* (1999) discussed the exhumation of the Central Alps using $^{40}Ar/^{39}Ar$ dates on white micas in the Northern molasse basin. Between 31 and 26 Ma they suggested that the Austro-Alps and Pennine nappes were being exhumed, and between 20 and 14 Ma the Lepontine dome reached the surface by slip on the Simplon detachment fault.

Alpine exhumation is linked to the foreland basins such as the North Alpine Foreland basin of Oligocene/mid Miocene age in Switzerland and, on the Adria plate, the Miocene Po basin, both of which are the result of erosion of the uplifted rocks affected by the Lepontine tectonometamorphic event. In the North Alpine Foreland basin coarse molasse sediments were deposited from *c.*31 Ma. This date is surprisingly close to the proposed dates for UHP metamorphism and later peak Barrovian metamorphism described in Chapter 5. The proposal that the Europe–Adria collision, UHP metamorphism, subduction of continental crust, and Lepontine metamorphism all occurred in a very short time span needs to be confirmed.

Carrapa *et al.* (2003) used $^{40}Ar/^{39}Ar$ ages on detrital minerals to study the evolution of the Piedmont basin, which is situated in North Italy and south of the Alps. The Piedmont basin contains *c.*4 km of clastic sediments, deposited after the late Eocene and late Oligocene, but sediments derived from the Western Alps only appear in lowest Miocene times, *c.*23 Ma. By mid to late Miocene times, detritals from Variscan crystallines of the Alps appear in the basin. Compared with the North Alpine Foreland Basin the arrival of sediments derived from the Western Alps is delayed, perhaps reflecting the prevailing drainage patterns.

Unroofing the Himalaya

In the Himalaya, Johnson (1994) and others calculated that some 25–35 km of cover has been removed, mostly in the past 20 Ma; indeed all the geochronological and sedimentary evidence points to the Miocene as being a time of great erosion, at least periodically, of the Himalaya. For example, in the Shivling area of the central GHS there is evidence of huge denudation between 23 and 21 Ma when an unroofing of some 20 km of overburden occurred (Searle *et al.*, 1999a). This event was broadly synchronous with the initiation and emplacement of the MCT.

Less than half of the Himalayan erosional detritus, probably amounting to *c.*2×10^7 km^3 in volume, resides in the foredeep but most has been carried off-shore into the huge Bengal fan and the Indus fan. The Himalayan foredeep is in two parts – the older foredeep of Eocene age and the later foredeep of early Miocene to present. The rocks of the older foredeep are exposed in thrust sheets in the footwall of the MBT and they are also present beneath the Ganges alluvial basin. In addition there are continental sedimentary deposits, mainly arenites and conglomerates, near the Indus Tsangpo suture that mark the final collision between India and Asia.

France-Lanord *et al.* (1993) showed by analysis of isotopes of Nd, O, Sr and H in the Bengal fan that after 17 Ma the main source was the Greater Himalaya, which was at a high elevation from that time, when the MCT was reaching the end of its main displacement.

The Himalaya has an average elevation of *c.*5 km and the foredeep basin has a maximum depth of *c.*5 km. On the outer margin of the foredeep is a forebulge, a good example of which is seen in north India on the southern margin of the Indo-Gangetic plain. The River Ganges alluvium represents the most recent addition to the foredeep. As first pointed out by Wadia in his famous book of 1953, the courses of the Indus, Ganges and Brahmaputra rivers of north India have changed over time. However, the nature of river capture remains controversial. An Yin (2006) suggested a significant stage in Miocene times when the headwaters of the Ganges captured those of the Indus, thereby reducing the catchment area of the Indus. However, Clift and Blusztaign (2005) proposed that at 5 Ma there was a rerouting of the major rivers of the Punjab into the Indus river, and before that date these rivers flowed into the Ganges. The evidence for this conclusion is faster erosion and high sediment accumulation rates in the Indus Fan during the Pleistocene. The faster erosion was linked to intensification of the monsoon in the last 4 Ma.

According to An Yin (2006) the river capture in northern India is the result of west-to-east increase in topography due to the diachronous collision of India and Asia and the eastward increase in shortening and convergence. The present drainage divide between the Ganges and Indus catchment area was probably established *c.*20 Ma and is due to a N–S ridge, formed by an uplift of basement rocks, which is possibly one of several such ridges that have determined the positions of the drainage divide over time. In contrast, Burbank *et al.* (1996) suggested that axial rivers may have flowed southeastwards from Pakistan during the early and middle Miocene.

The Himalaya is famous for antecedent drainage, rivers that existed before the rise of the mountains. The Yarlung–Tsangpo–Brahmaputra and the Indus are examples; they all rise on the Tibetan Plateau and flow across the Himalaya. Other possible antecedent rivers are the Sutlej and the Arun. Again this is controversial because river capture may be involved.

The older Himalayan foredeep consists of Upper Palaeocene to mid Eocene (Subathu Formation, *c.*650 m thick in India and the Gazij Group in Pakistan) marine sediments. They are thought to mark an early stage in the India–Asia collision when loading of the Indian plate caused a flexural downbend (see Najman, 2006). Basic igneous input in the Subathu indicates an ophiolitic source. In India the early foredeep rocks are overlain unconformably by non-marine clastic sediments (the Dagshai, Lower and Upper Dharamsala and Kasauli formations), constituting part of the younger foredeep. The unconformity is dated at <37 Ma in Pakistan (Najman *et al.*, 2001) and younger than 31 Ma in NW India, probably younger than *c.*22 Ma in central India and deposited between that date and *c.*12 Ma. The comparable formation in Pakistan is the continental clastic Murree Formation, which at 37 Ma is older than the first clastics in the Indian foredeep indicating diachronous uplift along the belt. The great change in sedimentary facies denotes an important stage in the evolution of the foredeep, the arrival of clastic sediments indicating, if not the first uplift of the mountain belt, then at least a major change in the uplift rate. The Dagshai and Kasauli formations contain fragments of low–medium grade metamorphic rocks which suggests that the lower-grade part of the Greater Himalayan Sequence was being unroofed at the time. Some garnet and staurolite appear in the lower part of the Dharmsala Group but kyanite did not appear until deposition of the mid and upper Siwalik – that is, after 11 Ma. In Nepal the Dumri Formation is thought to be synchronous with the Dagshai Formation and is younger than 17–20 Ma.

The bulk of the younger foredeep consists of the dominantly clastic Siwalik Group, which is *c.*5 km in thickness, and the alluvium of the Indo-Gangetic plain. These rocks contain high metamorphic grade fragments with garnet and staurolite first appearing in the Lower Siwalik sub-group and followed by kyanite at 13 Ma and sillimanite at 8 Ma. The implication is that unroofing of low- and medium-grade rocks of the Greater Himalaya Sequence (GHS) started at roughly 20 Ma and was continued and deepened by 13 Ma. The GHS may have been experiencing rapid exhumation until *c.*10 Ma, after which the rate decreased.

As already mentioned, volume estimates indicate that the foreland basins may contain a relatively small proportion of the detritus removed from the Himalayan orogen. Johnson (1994) calculated that *c.*1.9×10^7 km^3 has probably been eroded from the Himalaya and its associated belts. As already noted, less than half of the volume is found in the foredeep basins on the Indian continent. The bulk resides in the off-shore Indus and Bengal fans. The Indus river drains the western Himalaya and the Karakoram, together with part of southern Tibet. The Indus fan has an area of 1.1×10^6 km^2 and a maximum thickness of *c.*9 km. It is likely that the Indus fan contains a large amount of material from the Karakoram and the Pakistan Himalaya. Clift and Plumb (2008) described seismic plus dating evidence which shows that the Indus fan contains post-Lower Miocene turbidity flows, especially in the upper part. There was an increase in the sediment flux after 24 Ma with a peak at 16 Ma (mid Miocene) and less in the late Miocene to Pliocene, then another peak in the Pleistocene. It is proposed that enhanced chemical weathering occurred in the late Miocene to early Pliocene and it was caused by the arid climate in South Asia. The other fan system, the Bengal fan, is much larger and includes the onshore Bengal basin. It is truly enormous, 3000 km long and up to 21 km thick at its proximal end. Imagine Europe between London and Athens being covered in sediment as thick as that! This fan received detritus not only from the Himalaya but also from southern Tibet, Indo-Burma, Shillong and the Indian craton. Thus the volume of Himalayan-derived material quoted above in the Bengal fan may be considerably exaggerated. Most of our knowledge of the fan comes from deep drilling from the distal part of the fan which shows an early Miocene to recent history. After 15 Ma, the fan received low- to medium-grade metamorphic minerals probably derived from the GHS, which therefore must have been at the surface by then as has been concluded already from the younger foredeep. It would appear that most of the fan is pre-Miocene and, as indicated, possibly derived from outside the Himalayan regime. Neodymium isotopes in the Bengal fan are constant since 17 Ma, suggesting that there was a constant sediment flux from the Greater Himalaya. Therefore the source region for the Bengal fan did not change after 17 Ma and may have been relatively constant since 30 Ma. The river capture of several rivers flowing out of the Himalaya by the Ganges obviously altered the type of source material going into the Bengal fan.

To summarise, the available data indicate that at times exhumation was very rapid in the Himalaya, for example the extreme rates associated with the exhumation of the eclogites. But at other times exhumation and erosion rates were much slower, between 0.1 and 1 mm/a. This episodic nature of exhumation is a very important conclusion. The unroofing of the

GHS probably started in the early Miocene when there was rapid exhumation which lasted until late Miocene times. This is shown by the geochronological data and the dating of foreland sediments, in particular the accretionary, erosional and sedimentary flux in the orogenic system.

Sediment budgets in SE Asia

On a larger scale, Clift and Plumb (2008) have discussed sediment budgets in SE Asia as a whole and give a remarkably vivid account of the rapid uplift of the eastern part of the Tibetan Plateau occurring, from rather scanty geochronological data, in the late Miocene to the present. Their argument is that if erosion in the Himalaya was controlled by the monsoon then the effect should be seen in Indo-China and the Indo-China Sea. They describe evidence from the Red and Mekong rivers where $^{40}Ar/^{39}Ar$, U–Pb and fission track dates show that the main sources of detritals entering these river systems were crystalline rocks and sedimentary basins situated on the flanks of the Tibetan Plateau. On most of the plateau there is internal drainage, but on the eastern flank the Yangtse, Mekong and Red rivers flow off the Plateau. The courses of these rivers are bunched together east of the eastern syntaxis, and this may be due to the northeastward propagation of the syntaxis and the eastward growth of the plateau (Clark *et al.*, 2004). The evolution of these river courses involved a series of river capture events that happened prior to or coeval with the *c.*2 km uplift of the plateau in the late Miocene. Thus the ancestral Red River has lost drainage in the Cenozoic by drainage capture of its headwaters by the Yangtse. The Nd record, used as a provenance detector, shows that in the Eocene the Red River was not the same as now but that it extended much further north. The Red River delta shows this record in huge south-migrating foresets. The Tsangpo was originally a tributary of the palaeo Red River and has been captured by the Irrawaddy River. Finally, the river courses may be linked to activity on the Longmen Shan and Yalong thrust belts.

10

Deep structure, mountain support and phase changes

This chapter addresses the question of mountain support, isostasy and the tectonometamorphic and rheological processes operating in the deep structure of mountains that greatly influence their evolution. Referring to the outer part of the Earth, Barrell (1914) coined the terms lithosphere, the rocky part, and asthenosphere, the weaker lower part. This perceptive observation together with the concept of isostasy forms the basis for most of our thinking about mountains and their support. Clearly the mass of a mountain range or high plateau standing high above sea level requires a mechanism for its support. The creation of an elevated terrane means that work is done against gravity. The Tibetan Plateau could not stand at its present height without the support of a horizontal deviatoric stress, in this case the continuing plate convergence between India and Eurasia. Removal of this support would mean that the plateau would flow away – this is called orogenic collapse. It has been pointed out that the paradox in the Himalaya–Tibet example is that although the plate convergence continues steadily and the horizontal stress is maintained, the plateau is undergoing normal faulting which for many scientists implies that it is collapsing. This observation led to the idea that the lower part of the mantle lithosphere below Tibet has been delaminated and hence the Tibetan lithosphere acquired buoyancy.

Before going into the details of the geophysics of orogens it is necessary to consider models which show strength profiles through the lithosphere. There is ongoing debate between the 'jelly sandwich' model for the lithospheric rheology and the newer 'crème brûlée' model (Fig. 10.1). The jelly sandwich model is well established and might be regarded as the standard model for the rheology of the lithosphere; it is so called because it postulates a strong upper crust and upper mantle and a weak lower crust, whereas the more recent crème brûlée model, as proposed by Jackson (2002) and others, invokes a strong upper layer made up of the entire crust above a weak mantle.

Does the crème brûlée model contradict recent interpretations of the deep rheology of orogens? For example, channel flow is precluded in the crème brûlée model because it is based on the concept of a weak mid or lower crust which is capable of lateral flow. Likewise the several postulations (Royden *et al.*, 1997, and others) of the lateral eastward extrusion or expulsion, operating since the mid Miocene, of Tibetan mid–lower crust (by Poiseuille flow) are in conflict with the crème brûlée model.

Rock densities and isostasy

Continental lithosphere is in a delicate state of isostatic equilibrium reflecting contrasting rock densities. Lighter crust with a mean density of $2.8–2.9 \times 10^3$ kg/m^3 overlies a mantle

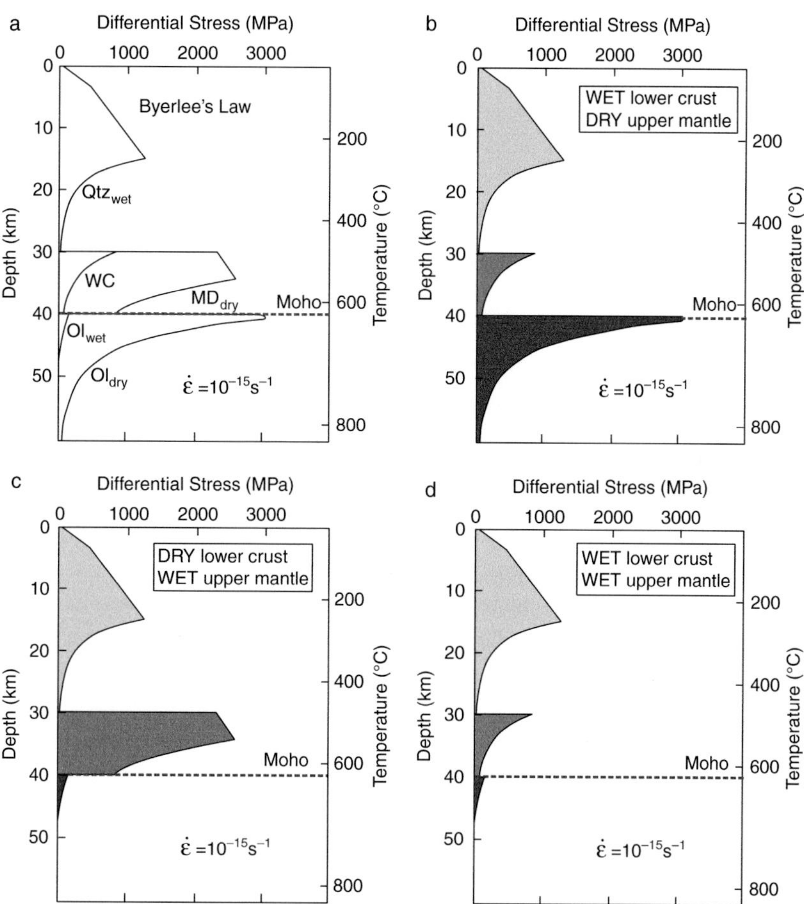

Figure 10.1 Support for mountains: experimental strength envelopes of differential stress (rock strength) versus depth for various continental conditions illustrating possible effects of water. a, Experimental results where lower crust is represented by dry diabase. b, Wet lower crust and dry upper mantle – this is the jelly sandwich model. c, Dry lower crust and wet upper mantle. d, Wet lower crust and wet upper mantle. Both are weak and nearly all strength resides in the seismogenic upper crust. This may be typical of most continental regions. From Jackson (2002), published by permission of the Geological Society of America.

with a density of 3.2–3.3 $\times$ 10^3 kg/m^3. Mantle lithosphere is at a lower temperature than asthenosphere and is therefore denser. Mountain ranges can be isostatically compensated and buoyed up by a thick crustal root like blocks of wood of different heights floating in water; the taller blocks stand higher. This is called Airy isostasy and it explains the equilibrium state existing between crustal blocks of different thicknesses. Alternatively, the mountain range can be buoyed up by relatively hot upper mantle, as in Pratt isostasy (see Chapter 3). As we will see later, an equilibrium state cannot be assumed to remain unchanged during orogenesis because of repeated changes in crustal thickness, density and thermal structure. The process of lithospheric thickening in a convergent plate margin often involves the upper mantle, and the thickened lithospheric root of cold mantle rocks should be denser than the hotter normal mantle of the asthenosphere at the same depth. Therefore the lithospheric mantle root provides a negative buoyancy that reduces the isostatic uplift of the thickened crust. The implications of these observations are pursued in later sections.

Support for the load of orogens

What is the support for the column of thickened lithosphere in an orogen, and is the deep structure capable of carrying the load? Jackson *et al.* (2004) have answered this question by postulating that in the Himalaya the thickened crust is supported by a strong 'root' composed of granulite facies rocks of the underthrust Indian Shield rocks (Fig. 10.2). However, the geophysical work summarised below indicates that this support may be transient because of metamorphic changes in the roots of mountains.

How can we identify the nature and composition of the deep unexposed parts of orogenic belts? Clearly not by direct observation if we are considering a recent orogen. Rarely, in some places, the nature of the deep lithosphere can be studied at the surface where the rocks have been exhumed such as in the Ivrea zone of the Alps (see Chapter 5) or perhaps by examining xenoliths which have been carried up from the lower crust. Deeply eroded Archaean rocks represent the deep parts of ancient orogens, and therefore provide a proxy answer, which is that the deep parts are composed of high-grade metamorphic rocks such as granulites. Fortunately, modern geophysical techniques increasingly fill in the gaps in our knowledge and are able to identify the nature of the lower crust and mantle in recent orogens. These techniques involve the analysis of the seismic record in order to determine the composition, likely temperature and rheology of the deeply buried parts of an orogen. The velocities of P- and S-waves are higher in the mantle than in the mid/upper crust. The main theme of this chapter is density, in particular how changes in the densities of rocks at depth in orogens have an important control on orogenic evolution, and how density differences in the lithosphere may control plate motions. It is early days in the employment of geophysics in the investigation of deep structure, and what follows here perhaps reflects the uncertainties at this time.

Geophysical studies in the Himalaya–Tibet

A much quoted cross-section of the Himalayan–Tibetan orogen shows underthrusting of Tibet by Indian lower crust and mantle (Fig. 5.9). This has been built up from a variety of geophysical studies. As an example of the geophysical approach, we can start with Molnar (1988) on the deep structure of the Himalaya and Tibet. He showed that seismic data are consistent with the proposed partial melting of the mantle in northern Tibet where recent volcanism is seen. Thus despite the rather constant elevation across the plateau it seems that the mantle beneath the plateau is heterogeneous. The results also indicate

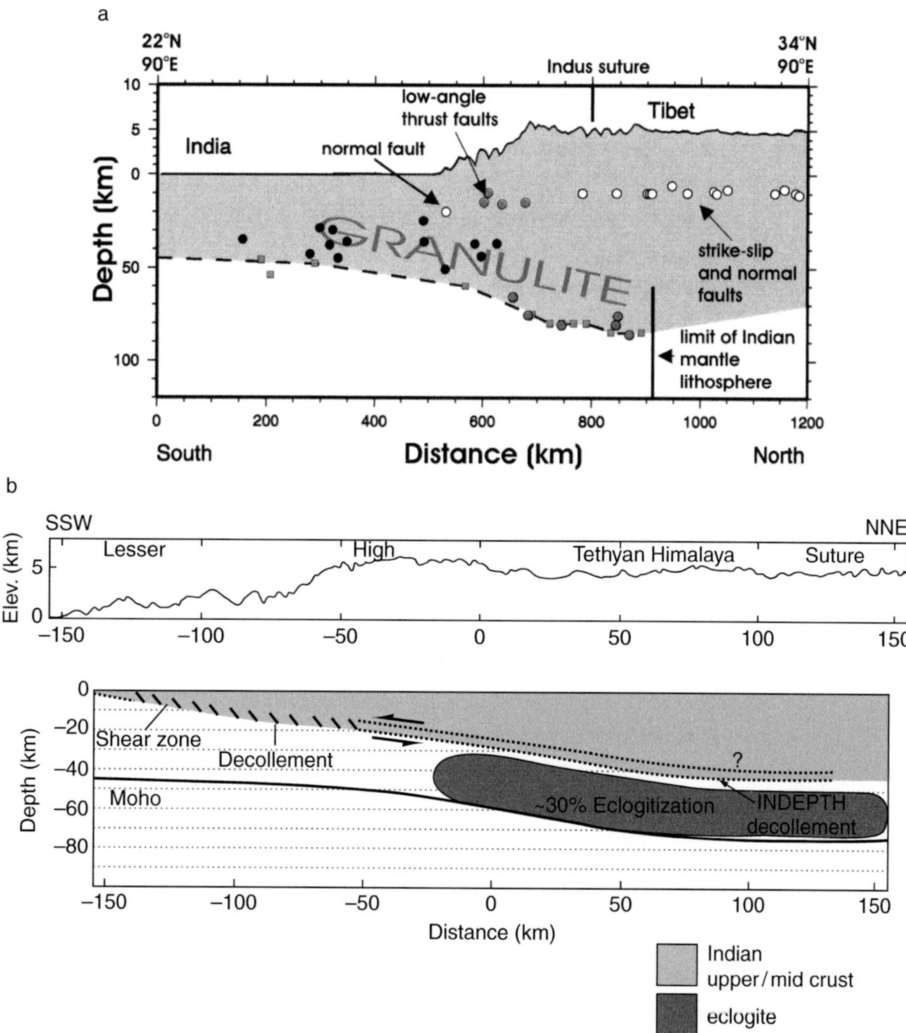

Figure 10.2 a, Granulite facies rocks provide a strong support for the Himalaya. Squares are Moho depths, circles are earthquakes from below the Himalaya. Vertical line at depth is the northern limit of isotropic S-wave signature on the Indian Shield. The lower crust beneath India and the Himalaya is thought to consist of dry granulite. From Jackson *et al.* (2004), published by permission of the Geological Society of America. b, Receiver function analysis in a profile across the Himalaya and southern Tibet to show the deep structure inferred from the data. Note the zone of eclogitisation at depth beneath the Tethyan Himalaya and southern Tibet. The INDEPTH profile is marked, as is the seismicity. From Schulte-Pelkum *et al.* (2005), published by permission of *Nature*.

a slightly thinner than normal (by about 10 km) crust in northern Tibet. Turning to the burning question of whether or not Tibet is underlain by underthrust Indian Shield, the upper mantle of the Indian craton has distinctive features indicated by P- and S-waves: these are not found under Tibet. More precisely, Molnar stated that Pn and Sn velocities under Tibet are relatively high and some may be consistent with a Shield-like structure under Tibet. Of course this observation alone does not prove underthrusting of Indian Shield. Molnar's conclusion is that if Tibet is underlain by Shield-like rocks they must have been modified by metamorphic changes. In the Himalaya, Molnar considered that several observations seem to permit the conclusion that the Indian Shield with its high velocity layer has been thrust under the Himalaya. In summary of Molnar's article, the deep structure of the Himalaya and Tibet suggests that Indian Shield has been underthrust under the Himalaya and perhaps under the southern part of the Tibetan Plateau for distances of 100–300 km. Later studies tend to support the higher figure (see Chapter 5). This conclusion supports Jackson *et al.* (2004): i.e. the weight of the Himalayan–Tibetan orogen is supported by a strong plate composed of granulite facies rocks.

Recent geophysical work on the deep structure of Tibet

Do recent geophysical developments confirm Jackson's contention? Caution is needed here. Molnar is clear about the difference between evidence being consistent with something as opposed to actually proving it. This follows from the obvious limitations of the geophysical approach arising from two points. Firstly, the geophysical data are a snap shot of the present-day features of the deep structure and say nothing about its history; secondly, identifying Shield-like rocks in the deeper part of the Tibetan lithosphere does not establish that they are Indian Shield rather than Asian Shield.

There has been important geophysical work done in the past few years on the deep structure of Tibet. Tilmann and Ni (2003) used tomographic imagery of the upper mantle beneath central Tibet to postulate a subvertical high velocity zone at between 100 and 400 km depth situated south of the Banggong suture. This is interpreted as downwelling Indian lithosphere. The amount of downwelling is supposed to account for the total shortening in the Himalaya and Tibet. To the north of the downwelling zone there is asthenosphere which is the heat source for the warm mantle under north and central Tibet (*cf.* the section of Owens and Zandt, 1997; Fig. 5.9). Hetényi *et al.* (2007) employed geophysical techniques to suggest that there must be considerable eclogite under Tibet (see below for discussion of paper by Schulte-Pelkum *et al.*, 2003). Chan *et al.* (2009) have analysed crustal xenoliths which have been carried up in an ultrapotassic dyke, and they concluded that the basement of south Tibet had reached 80 km depth by 17–14 Ma and has stayed at that thickness until the present day. This would appear to dispute the model, which involves convective removal of the lower mantle lithosphere under Tibet, put forward by England and Houseman (1986, 1988). Priestley *et al.* (2007, 2008) used seismic wave velocities to study the lithospheric mantle in Tibet. The speeds are slow down to *c.*130 km, but below that there is a fast speed down to *c.*250 km. The high-speed part merges with high-speed zones under India and SE China. Their interpretation is that the lower mantle of Tibet is intact and that delamination as proposed by England and Houseman has not occurred. The lithosphere has been underthrust by high-speed Indian mantle from the south and from the north by high-speed Asian mantle. Priestley *et al.* (2008) proposed that granulite facies Indian Shield has been underthrust under southern Tibet.

Receiver function analysis and shear wave splitting detect the deep structure

More recently the use of receiver function analysis (RFA) has given new insights into the deep structure of orogens. RFA determines the time delay between direct P-waves, which travel in a near-vertical direction, and converted S-waves which travel in subsurface interfaces with a velocity contrast. Delays between the direct and converted waves depend on the depth of the interface and on the transmission velocities along their paths. The amplitude of the converted arrival depends on the magnitude and sign of the velocity contrast.

Another technique permits the recognition of anisotropies in subsurface rocks. This technique refers to the conversion of P-waves into S-waves and the observation that S-waves are faster in one direction than another ('shear wave splitting') owing to the presence of a planar fabric in the rocks transmitting the earthquake waves. The planar fabric was initially interpreted as micro-cracks, but more recently as being produced by foliation due to the alignment of minerals in metamorphic rocks.

Schulte-Pelkum *et al.* (2003) used RFA (Fig. 10.2b) and resolved the dilemma of the conflicting rheological models for the lithosphere by postulating one model (crème brûlée) for the Himalayan Foreland lithosphere but another (jelly sandwich) for the Himalayan orogen, notably that part beneath the Greater Himalaya. In the Greater Himalayan lithosphere the jelly sandwich contains a strong granulite facies Indian lower crust which has been subducted under the Himalaya and a weak mid crust. However, the granulite facies rocks beneath the Tethyan zone of the Himalaya have undergone phase changes to eclogite. This eclogitisation does not extend much further north than the Indus–Tsangpo suture.

The supposed northern limit to eclogitisation under the Himalayan–Tibetan orogen suggests that most of the high plateau of Tibet is supported by strong granulite facies rocks. The model of Schulte-Pelkum *et al.* finds support in the demonstration by Chen and Yang (2004) that deep earthquakes occur under southern Tibet at depths of at least 100 km, indicating that the mantle there is strong. Earthquake focal depths from the Himalaya and Tibet indicate a strong upper crust down to depths of *c.*15 km. A weak, hot and partially melted mid-lower crust is indicated by seismic anisotropy occurring deeper than the low velocity zone shown on the INDEPTH profile. Thus from two approaches there seems to be support for the jelly sandwich model under the Himalayan–Tibet orogen.

The rheology of the lithosphere beneath the mid/lower crust (the jelly in the sandwich) was also discussed by Jackson *et al.* (2004) whose Fig. 2 shows that the S-waves indicate an anisotropic mantle under much of Tibet, with the change from isotropic mantle (Shield-type) occurring at latitude 30–32°, which is about 100 km north of the Indus Tsangpo Suture. The anisotropy reflects different seismic velocities in different directions, probably caused by preferred orientations of minerals such as mica. Therefore if Shield rocks do exist under Tibet then they must be modified by metamorphic transformations as were hinted at by Molnar. In addition, Jackson *et al.* interpreted the sub-crustal earthquakes under south Tibet, mentioned above, as possibly reflecting brittle behaviour in the upper mantle associated with the start of eclogitisation, possibly initiated by dewatering of the subducted Indian lithosphere (see below).

In summary, the recent geophysical work appears to confirm the view that India has underthrust Tibet for at least 300 km. It also disfavours the England and Houseman model. However, there are conflicting interpretations of the lithospheric thickness in northern Tibet.

Phase changes in the deep structure

The phase change from granulite to eclogite facies causes a rheological weakening and is attributed to the influx of hot water which acts as a trigger to metamorphic phase change. Similar phase changes have been described in the deep structure of the Norwegian Caledonides (see below), and they are important because a change to denser eclogite increases subsidence of the orogen. Conversely a phase change from eclogite to granulite facies decreases the density and permits uplift. Manifestly, if such phase changes exist in the deep structure of orogens, then their role in orogenic evolution is extremely important.

The prospect that the lower crust and mantle beneath Tibet may have undergone profound metamorphic alteration is supported by Le Pichon *et al.* (1999), who postulated an eclogite to granulite transformation under Tibet, with associated decrease in the density from 3.28 t/m^3 to 2.96 t/m^3. The resultant increase in buoyancy could be an explanation for the uplift of Tibet without the need for delamination or convective removal of the lower part of the mantle (*cf.* England and Houseman, 1986, 1988). A similar phase change could be applied to the problem of the raising of the Altiplano Plateau in the Andes. In agreement with Le Pichon *et al.* (1999), Schulte-Pelkum *et al.* (2003) speculated that the eclogite zone once extended under Tibet but unlike the situation under the Greater Himalaya, the lower crust has heated up north of the Indus Tsangpo Suture so as to promote the eclogite to granulite transition. What is lacking in these various suggestions of phase changes under the Himalaya and Tibet is a date for the initiation of the changes.

One example of the importance of phase changes could be their role in the rise and fall of the Tibetan Plateau. The collapse of the plateau is supposed to be marked by normal faulting. The proposition that the collapse is due to eclogitisation at depth appears to be in conflict with evidence that eclogitisation is confined to the south of Tibet. However, the evidence presented by Schulte-Pelkum *et al.* (2003) is based on the present-day situation and tells us nothing about the metamorphic composition of the lower crust a few Ma ago. In any case, the interpretation of the normal faults as collapse structures is disputed by McCaffry and Nabelek (1998) who explained them as resulting from basal drag due to the subduction of Indian crust under Asia. Lastly, the existence of widespread eclogite beneath Tibet conflicts with the proposal from Nelson *et al.* (1996) that the INDEPTH profile shows that the Tibetan Plateau is supported by Indian granulite facies lower crust. Unfortunately we are in the realm of speculation; perhaps a deep bore-hole is called for!

The deep structure of the Himalaya may be less obscure than that of Tibet. A strong mantle root under the Greater Himalayan Sequence and the Tethyan Sequence is shown by sub-crustal earthquakes, reaching 40 km below the Moho. Jackson *et al.* (2004) proposed a strong granulite facies root for the whole Himalaya, but this interpretation appears to be in conflict with at least part of Schulte-Pelkum's interpretation that eclogitisation is present at least in the lower crust under the Tethyan Himalaya.

The Tethyan Sequence is composed of low-grade sediments with small amounts of granite. Although the elevation in the Tethyan zone is considerably less than in the high peaks of the High Himalaya, it still reaches about 5 km which is the average height of the Himalaya. Is this elevation supported at present by a partially eclogitised root instead of a strong granulite facies root, as suggested by Jackson and co-workers? The consequence of eclogitisation of granulite facies rocks is an increase in density with the expectation that evidence of collapse with extensional tectonics should exist in the Tethyan Himalaya. Such evidence may be provided by N–S normal faults such as the Thakkola graben which is dated at *c.*14 Ma. If so, a recent date for eclogitisation can be ruled out. In addition, the NE–SW trending Yardong graben, which traverses the entire Tethyan Himalaya, suggests extension parallel to the strike of the belt, as do the North Himalayan gneiss domes which have been interpreted as metamorphic core complexes related to extension. Thus there is evidence of extensional collapse of the Tethyan Himalaya, but more work is required to establish any connection with eclogitisation at depth.

Eclogitisation and the 'weighting of orogens'

The studies reviewed above, though speculative, emphasise the importance of deep-seated metamorphic phase changes in the orogenic process. The concepts that follow are derived from the suggestion of Richardson and England in 1979 that eclogitisation of lower continental crust decreases crustal buoyancy by 'weighting' in orogens. This concept has been pursued by Dewey *et al.* (1993) who have shown, in the Caledonian orogen in southern Norway, that eclogites have formed from granulites as a result of the infiltration of hot water. This infiltration and eclogitisation caused weakening of the deep structure. Subsequent transformation of eclogite to amphibolite produced a decrease in density and led to the collapse of the mountains, not by erosion but by tectonic denudation.

The C_z/l_z ratio

The exploration of the role of eclogites in the evolution of orogens by Dewey *et al.* (1993) centres on the C_z/l_z ratio, where C_z is crustal thickness and l_z is lithospheric thickness. In their view the crust forms part of the lithospheric boundary conduction layer with thickness l_z, and a density ρ_c, the other part being mantle with density ρ_m. The whole lithosphere is compensated to the asthenospheric mantle with density ρ_a. Low-density crust tends to 'float' whereas the lithospheric mantle acts to 'sink' the lithosphere and so a mountain belt can be compared to an ice-berg floating in water.

The elevation or surface level in an orogen depends on the C_Z/l_Z ratio. In Chapter 9 it was shown that the cooler part of the lithospheric boundary layer acts as a strong flexural beam which supports loads. The elevation of a mountain is proportional to the vertical density distribution of the lithosphere. The condition of Airy isostasy and equilibrium is violated if the loads are bigger than allowed in the Airy model. This model calls for compensation level or depth at which the weight of various columns of rock with the same cross-sectional area must be the same. In the Airy model compensation is attained by variation in the thickness of the columns, and for this reason the relatively light mountains have deep roots.

So how does the C_Z/l_Z ratio operate in the orogenic process? A reasonable estimate from regions outside orogens gives values of $C_Z = 32$ km and $l_Z = 120$ km, giving a ratio of about 0.25. According to the calculations of Dewey *et al.*, this ratio limits crustal dimensions within an orogen; in particular it limits the amount of vertical stretching of the crust to *c.*70 km (slightly more than doubling the thickness) below a surface elevation of 3 km. Thus the ratio makes the orogen too buoyant and as a result volume is lost by erosion. This is a consequence of the balance that must exist between the vertical compression caused by isostatic compensation in a thickened crust and the horizontal compression due to plate convergence. As it happens, 70 km is a common estimate of crustal thickness in orogens, indicating that this thickness reflects the density and thickness of the crust.

Another implication of the ratio is that elevations of more than 3 km can only be attained by changing the density distribution in the lithosphere, either by crustal underplating or by thinning or delamination (essentially a peeling-off) of the lithosphere *without thinning the crust*. Both these processes should increase elevation because they have the effect of removing part of the dense mantle sink. Dewey *et al.* (1993) cited Tibet as an example of a delamination mechanism by convective removal of the lower mantle lithosphere in order to explain a possible sudden and recent increase in the elevation of the Tibetan plateau to *c.*5 km. However, as pointed out in Chapter 5, there are objections to this hypothesis.

A change in the deformation mechanics may occur if crustal thickening is buffered at 70 km. For example, the axes of stress may rotate, and the intermediate axis of compression rotate to become vertical which means that a wrench regime is established. If convergence continues then the orogen, now under a wrench fault regime, must spread laterally parallel to the minimum axis of compression. As described earlier, Dewey *et al.* explained the Cenozoic history of Tibet in these terms, i.e. the control of crustal thickness on the deformation mechanisms.

Accepting this analysis, how can horizontal orogenic shortening beyond 50% be explained? In the Himalaya, the calculated shortening is *c.*70%. The proposition is that this is not possible if we start with crust of normal thickness. However, shortening beyond 50% can be permitted either by substantial erosion occurring *during* shortening or by the existence of an initially (at the start of orogenesis) thinned crust, as happens when a rifted, thinned, continental margin is involved in the orogen. For example, the Alps and the Himalaya were built from crust that had undergone pre-Cenozoic rifting. In consequence the C_Z/l_Z ratio is reduced. These are ways of decreasing the buoyancy factor arising from the volume of continental crust in an orogen. In the Himalaya the average surface elevation reaches to *c.*5 km. At least 20 km has been lost by erosion.

However, these processes of thinning or delamination apparently do not work in the Caledonides of south Norway (Fig. 10.3). The presumption, given an initial ratio of 0.25,

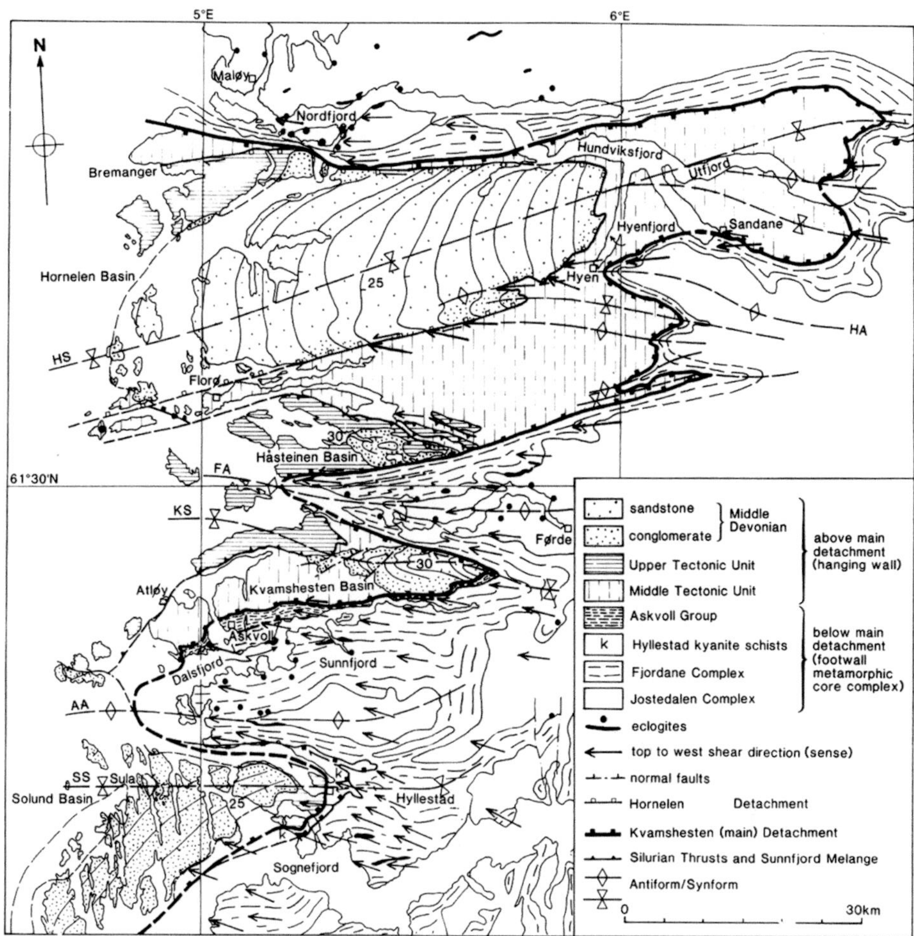

Figure 10.3 Map of the Caledonides in southern Norway to show the eclogites, the Kvamshesten detachment and the top to west shear direction on extensional structures. From Dewey *et al.* (1993), published by permission of the Geological Society of London.

that 70 km is the limit to crustal thickening is contradicted in the Norwegian example by the metamorphic assemblages found in the crustal rocks. These show that during orogenesis, exceptional pressures of 39 or 40 kbar were attained, which indicate a C_z of 120 km for the orogenically thickened crust. Indeed, crustal thickness could increase to more than 150 km during orogenic shortening. The very dense eclogite decreases crustal buoyancy and enhances subsidence, therefore permitting excessive shortening. In Norway, C_z and l_z thickening have worked in opposition owing to their different densities: C_z causes uplift whereas the denser l_z leads to subsidence. In Norway an advective removal of the mantle root, such as has been proposed in Tibet by England and Houseman (1986), leads to lithospheric thinning, uplift and the transformation of eclogites into lower-grade amphibolite.

On the assumptions set out above, crustal thicknesses in excess of 70 km cannot be achieved without the involvement of some other factor. Similarly in the Austrian and the

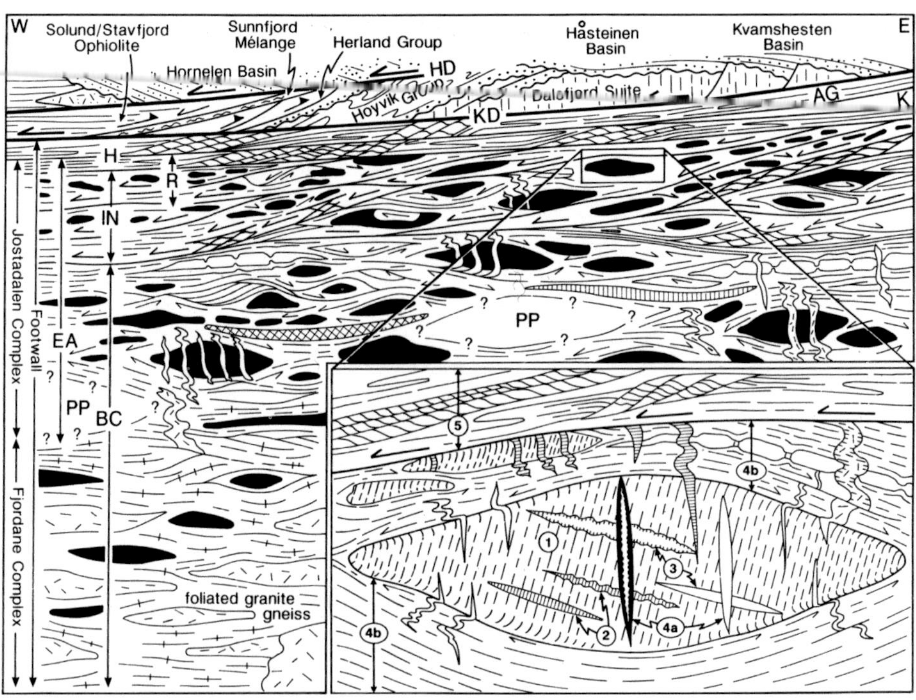

Figure 10.4 Schematic E–W section through the region shown in Fig. 10.3 to show the bulk coaxial extension affecting eclogites (in black) which are deformed to lensoid shapes ('halibuts'). Key to relevant lettering: BC, bulk coaxial deformation; EA, zone of preserved eclogite bodies in matrix of eclogite retrograded to amphibolite during coaxial extension; H, zone of homogeneous penetrative down to west non-coaxial deformation; Kv, Kvamshesten Detachment; R, zone of heavily retrograded eclogites. Inset: kinematic sequence preserved eclogite bodies. Encircled numbers: 1, L to L>S tectonites; 2, eclogite facies extensional veins; 3, early amphibolite facies extensional veins; 4a, amphibolite extensional veins in eclogite bodies; 4b, amphibolite facies bulk coaxial extensional veins; 5, superposed top down to the west non-coaxial deformation. From Dewey *et al.* (1993), published by permission of the Geological Society of London.

Swiss Alps ultrahigh pressures were attained during peak metamorphism (Chapter 6), yet the high-pressure rocks that are now at the surface as a result of extensive erosion have *c*.50 km of crust below them. Therefore the conclusion is that the C_z at the time of peak metamorphism was about 100 km. The problem is how to get the rocks down to depths of over 100 km during shortening and then back up to the surface. For this exhumation process, extensional collapse or crustal thinning is preferred because erosion is too slow given the time constraints.

In Norway, 10–20 Ma of Scandian crustal shortening occurred with lower crustal eclogite metamorphism (Fig. 10.3). The crucial point is that the eclogite transformation enhances the subsidence role of the mantle. This was followed by 20 Ma of vertical shortening and crustal extension with transformation of eclogites to amphibolites or greenschist facies. This transformation involved a decrease in density. Advective removal of the negatively buoyant lithospheric mantle led to lithospheric thinning, massive increase in the geothermal gradient and high heat flow. The process is called extensional collapse. Finally, thermal recovery involves lithospheric thickening and slowing subsidence.

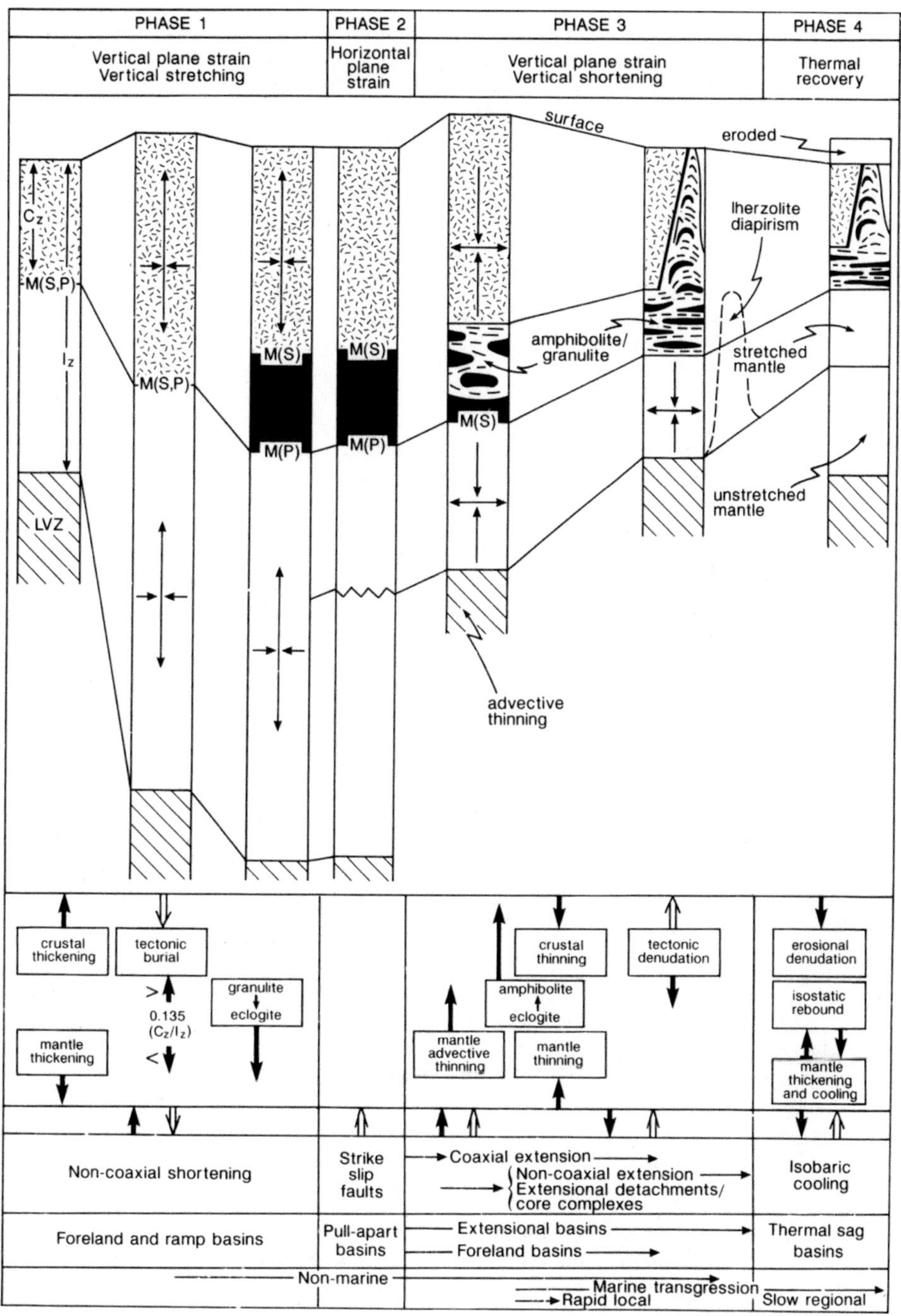

Figure 10.5 Eclogitisation and mountain uplift and subsidence. This figure shows an evolution for the Caledonides of South Norway. Phase 1 shows vertical plane strain and stretching. Phase 2 shows horizontal plane strain. Phase 3 shows vertical plane strain and shortening and Phase 4 is thermal recovery. C_Z, crustal thickness; LVZ, low velocity zone; l_z, thickness of the lithosphere; M, the Moho. Black is eclogite; horizontal broken lines shows horizontal extension; random ornament, continental crust. Solid arrows show the motion of rocks at the surface, open arrows show motion of rocks with respect to the surface. From Dewey *et al.* (1993), published by permission of the Geological Society of London.

The notable features of this model are, first, that C_Z of <150 km is possible because of the eclogite allowing shortening to exceed 50% and reach 80%. Furthermore, the mountains were only *c*.3 km in elevation, not the absurdly 10 km high mountains that are predictable in the absence of eclogite.

Extensional collapse is an important concept in understanding how the lithosphere in orogens returns to normal thickness. With extensional collapse, erosion is not the main mechanism, and it allows for a more rapid collapse of mountains than is likely with erosion acting alone. Extensional collapse also explains the fact that volumes of molasse in some orogens is insufficient to account for the loss of cover. Another point is that the volume of crust in a collision zone may be underestimated if the lower crust is eclogitised. Consequently, Dewey *et al.* (1993) do not accept the suggestion that crustal material may be lost in the mantle because the evidence from Norway suggests that eclogitised crust is transformed back into crustal densities by amphibolitisation and is thus 'returned' to the crust.

Eclogitisation and exhumation and collapse in other orogens

There is evidence that the presence or absence of eclogites exerts a control in other orogens apart from the Himalaya–Tibet orogen discussed earlier in this chapter. As described in Chapter 5, the Alps were formed by the southwards subduction of Europe beneath Adria. Continent–continent collision occurred at >50 Ma. The present convergence is at 8 mm/a, but earlier it was 4 mm/a. As a result of decoupling, the upper crust is shortened to produce some of the famous nappes of the Swiss Alps. But lower crust and mantle are subducted without undergoing deformation. Subduction is at an angle of 15° until a depth of 70 km is reached, when it steepens.

At depths of 50–60 km or more, eclogitisation of European lower crust occurs and the eclogite forms a high-density root underlying the axial zone (Penninic) of the Alps. According to Bousquet *et al.*(1997), this dense root is the reason why the Alps are limited to a 2.5 km maximum average elevation. But this fails to account for the metamorphic changes during the Lepontine metamorphism during which eclogites would presumably have been converted to amphibolite of greenschist facies rocks.

Deep-seated eclogitisation has been put forward to account for the low elevation (<2.5 km) of Alpine belts like the Zagros Mountains which are situated between the Alps and the Himalaya. In the Andes the uplift of the Altiplano may be due to increased buoyancy consequent upon an eclogite to granulite transformation.

The rapid phase change from granulite to eclogite facies and vice versa

In conclusion, although a strong granulite facies root is needed to support high mountains like the Himalaya, the evidence is that it may be modified by eclogitisation of the granulite facies rocks brought about by the influx of hot fluids. This results in a weakened, less buoyant mountain root and therefore in mountain collapse. Remarkably, Jackson *et al.* (2004) emphasised that the phase changes brought about by hydrous fluid are very fast. Fluid activity is shown by pseudotachylyte veins in which the eclogite minerals have a fine-grained skeletal form showing very rapid growth, even lasting only a few seconds. Water changes the reaction rate by a factor of 10^6, and it is the on–off switch controlling granulite to eclogite transitions. This fast reaction comes as a great surprise to tectonicians who are used to long time spans for events but now find that the eclogite transformation may occur in the short time span of 100–1000 years. A confirmation of this remarkable timescale is given by Camacho *et al.* (2005) who used geochronological techniques in order to demonstrate that eclogitisation can be very rapid. They argue that the total duration of the eclogite event in the Caledonides of south Norway was only *c.*20 ka, which is a surprise indeed!

Basement control, reactivation and reworking

Another feature to be considered here arises from the complexity of continental lithosphere, which in contrast to oceanic lithosphere is highly anisotropic, being made up of orogens of different ages which are woven into a complex pattern or 'grain' on the continents. Therefore the question arises: to what extent do pre-existing heterogeneities, imprinted by repeated orogenesis, in continental lithosphere influence its response to subsequent deformation? Fault zones are inherently weak and the continental lithosphere is riddled with them, so it is reasonable to suppose that subsequent deformation would utilise them. The deep structure or basement of an orogen may control and influence the patterns of structures formed in the higher-level cover structures. Inherited structure is evident at all structural levels: for example, the reactivation of normal faults as thrusts during basin inversion is widely recognised in the upper crustal levels – e.g. the North Sea, the Sunda Arc and the Rocky Mountains in Canada. Again, the South Tibetan Detachment of the Himalaya is a thrust which has been reactivated as a normal fault.

Two processes can be recognised: reactivation and reworking. Reactivation is rejuvenation of discrete structures without significant change of volume or orientation, while reworking is a focussing of metamorphism, deformation and magmatism on a piece of crust which has undergone previous orogeny. Reactivation is not the same as episodicity in orogenesis because the different events in reactivation must be perhaps millions of years apart. Reactivation is best shown in long-lived fault zones in which cataclasis and fluid flow in the lower part of the frictional regime assist retrograde metamorphism and changes in the deformation regime. Fault rocks are abundant in the geological record and are important as sites for economic resources and for seismic activity.

Depth may be the main difference between the two processes: reactivation is shallow and reworking is deep. Unlike reactivation, in continental reworking on an orogenic scale the older structures do not necessarily control the style, orientation and scale of the later structures. Therefore basement control is more difficult to demonstrate, especially in orogens where intense reworking has occurred. The terms 'basement' and 'cover' come from studies in the Swiss Alps, and the Penninic nappes show a core of basement of up to high-grade metamorphic Variscan rocks and a cover of Mesozoic and Cenozoic rocks. So what was the influence of the old structures inherited from the Variscan orogeny? It is difficult to identify any such influence, and this may be because of the intense reworking of the basement during the Alpine orogeny. A similar observation applies to the NW Caledonides in the Scottish Highlands where the orientation of the structures in the Moines, including the Moine and Sgurr Beag thrusts, completely ignored the structural pattern seen in the basement Lewisian rocks. Again, the Caledonian belts in the northeast USA cut across the strike of the late Proterozoic Grenville belt.

It should be remembered that orogenic belts are long linear or curvilinear zones which impress a 'grain' upon the crust. Looking at a map of the Archaean to Proterozoic orogens in the African continent, it is manifest that each pursues its own course and later ones cross-cut the earlier ones. In other words, there is little evidence that the grain of the basements influences that of the younger orogens.

Rather than looking for evidence of basement control in orogenic belts, it might be more profitable to consider non-orogenic regions where unmetamorphosed cover rests on a basement. In this example, it might be expected that patterns of faulting in the cover would be strongly influenced by the faults in the strong basement. However, even here in the example of the North Sea graben the roughly north–south alignment of the faults is in sharp contrast to the Caledonian trend of the Basement.

Reactivation is focussed on the rejuvenation of older structures instead of starting new structures. Thus old structures may control the location and pattern of lithospheric deformation in faults, orogens and rifted basins. Large pre-existing structures form a significant mechanical anisotropy in the lithosphere and therefore are likely to localise deformation. Many fault zones are long-lived, but some are not reactivated at all or only after long quiescence. Changes in internal fault zone strength may account for long-term reactivation of faults. The nature of fault zones is important if they are channels for hydrous fluids and magmas.

Reactivated faults in metamorphic basements show a long history of exhumation, and fault rocks are formed at different depths and times. Thus study of them gives insight into the evolution of a fault. The history is mainly retrograde so high-temperature mineral assemblages are partially or wholly overprinted by cataclastic structures. Frictional behaviour on faults results from different slip mechanisms, e.g. cataclastic and pressure solution, related to strain rate and pressure. Fluids are important.

Miller *et al.* (2001) have discussed the Great Glen Fault zone in the Scottish Highlands which cuts across the Scandian and Grampian orogenic belts. This fault shows structures formed near the brittle–ductile transition, and it has been reactivated over hundreds of million years. The fault zone is up to about 3 km wide and extends down to *c.*40 km. The main episodes of slip on the Great Glen Fault are (a) sinistral slip of several hundred kilometres between 425 and 390 Ma; (b) dextral slip of 20 km in the Devonian to Carboniferous; and (c) 10 km dextral slip in the Cenozoic. Historical earthquakes indicate recent slip. The fault zone shows fabrics formed at *c.*10 km depth during the main phase of sinistral slip which are semi-brittle, such as breccias and cataclasites, retrograding metamorphic assemblages. But in addition there has been growth of phyllosilicates resulting from fluid channelling, along with diffusive mass transfer by pressure solution. Elsewhere in the fault zone there is crystal plasticity in quartz and growth of foliated white mica and chlorite, indicating temperatures of 250–450 °C in low greenschist facies. These higher-grade fabrics were formed at 8–15 km depth near the brittle–ductile transition. The interpretation of different fabrics is that uplift of the fault zone has occurred or that there was shear heating.

The influx into the fault zone of a chemically active hydrous fluid phase caused retrogression of metamorphic fabrics belonging to the country rocks. Cataclastic deformation in the deeper part of the fault zone took place in a frictional regime plus near the brittle–ductile transition and led to pervasive fluid influx. On a long timescale there was a switch in deformational regime from cataclastic to fluid-assisted reaction softening and diffusion creep. Softening was due to growth of the phyllosilicates. This affected grain-scale fluid–rock interaction which begins to dominate the rheological behaviour. In the evolution there was progressive shallowing of the domain of viscous flow in the high strained region and a narrowing of the brittle–ductile transition.

Underhill and Patterson (1998) have used seismic data and onshore evidence to investigate structures associated with the E–W Purbeck/Isle of Wight Disturbance in southern England. Activity on the disturbance has resulted in tectonic inversion of structures in the Wessex basin which contains Permian–Cretaceous and Lower Cenozoic sediments resting on a Variscan basement. During activity on the disturbance, earlier normal faults of the basin became reactivated as thrusts, the motion on which has caused folding of strata overlying the faults into monoclines or asymmetric folds. The famous Lulworth Crumple in Dorset is one example of these folds. This crumple was formed by flexural slip and layer parallel shortening.

Further reading

Bilham, R. and Wu, F. (2005). Imaging the Indian subcontinent beneath the Himalaya. *Nature*, **450**, 1222–1225.

Butler, R.W.B. (2006). How weak are the continents? *Geoscientist*, **16**, 4–13.

Ding, L., Kapp, P., Yue, Y. and Lai, Q. (2007). Postcollisional calc-alkaline lavas and xenoliths from the southern Qiangtang terrane, central Tibet. *Earth and Planetary Science Letters*, **254**, 28–38.

McClay, K.R. and Buchanan, P.G. (1992). Thrust faults in inverted extensional basins. In: *Thrust and Nappe Tectonics*, ed. N.P. and K. McClay. *Geological Society of London, Special Publication*, **9**, 93–118.

11 Mountains and climate

The effects that mountain building may have on global climates and climate change have received considerable attention in recent years. However, the jury is still out on whether a direct causal link has been established between mountain building and climate or climate change, or at least the degree of influence of the one on the other. Many other factors are involved, not least the amount of CO_2 in the Earth's atmosphere. There is agreement that the proposition calls for not only high mountains but also ones covering a large area of the Earth's surface. Climatic modelling has emphasised that in order to influence climate a huge area of high ground is needed, and the modellers consider that the Alps or Himalaya are not big enough, although they may well disrupt a north–south air flow and thereby cause local climatic effects.

Mountains influence climate because they are obstacles to air circulation. Additionally, they are sources of elevated latent heat and they change the water exchanges between continental surfaces and the atmosphere. In this chapter there is no room for an expansive account of this topic, but we will highlight some features and in particular enter the discussion of the role of orogenesis in changing the climate in southern Asia. Most of this chapter is devoted to the ongoing controversies surrounding the hypothesis that the rise of the Tibetan Plateau influenced and strengthened the monsoon in the late Miocene.

Strengthened monsoons in different regions of South Asia need not bear the same relationship to the topography of the Tibetan Plateau and need not be simultaneous with the Indian monsoon. Some of the factors influencing the strength of the monsoon in eastern India may play a lesser role in the strength of the monsoon over the Arabian Sea, where the north–south trending African coastal mountain range exerts a strong effect on the Somali Jet (the eastward-flowing air stream off Somalia across the Indian Ocean) in summer.

Great interest centres on the Tibetan Plateau where many geoscientists have invoked a supposed connection between the monsoonal events in South Asia and the attainment of present elevation on the Tibetan Plateau. Tibet figures large in this discussion not only because of its height, *c.*5 km in elevation, but also because the plateau is more than twice the size of France, and is the largest area of high ground on the Earth. Clift and Plumb's book (2008) sets out a comprehensive discussion of the varied aspects of this topic and should be referred to if such an extended discussion is desired. Here we are primarily concerned with the tectonic aspects. Clift and Plumb referred to the importance of understanding the nature and cause of the monsoons because two-thirds of humanity are affected by them and the region is one of increasing socio-economic relevance. Monsoons are associated in our minds with SE Asia but in fact North and Central America, Australia and West Africa experience similar weather patterns. Monsoons may exist wherever there is a temperature contrast between land and ocean. Much attention has been given to a recent strengthening of the Asian monsoon, but in fact monsoons have fluctuated in intensity over longer time spans, even reaching millions of years.

Another point is that monsoon activity cannot be understood without taking account of other aspects of global ocean–atmosphere systems such as the Gulf Stream and El Niño. There also may be a correlation with glaciation: that is, summer monsoons appear to be stronger in warmer interglacial periods. This would appear to be the case with cycles of solar insolation known from cave records. The intensity of summer and winter monsoons has been affected by global climatic changes caused by variations in the Earth's orbit, the Milankovich cyclicity, on timescales of 20, 40 and 100 ka. The main concern of this chapter is to explore the longer-term variations in monsoonal strength imposed by tectonic activity.

It is widely thought that the existence of the Himalaya and Tibet has led to the development of the strongest monsoon on the planet. However, other possibilities must be considered, such as the retreat of the Paratethys Ocean which resulted in an increase in aridity in Asia (see below). Scale is important, but other elevated areas such as the Colorado Plateau may have influenced the climate of North America. Again, this is because they disrupt atmospheric circulation patterns and generate summer low-pressure systems that draw in moist air from the surrounding oceans.

Monsoons

The word monsoon comes from *mausim*, Arabic for seasons, and it carries different meanings in different areas and contexts. For maritime trade with East Africa the Arabs relied on seasonally reversing winds that blow from India to the Arabian Sea in winter and northeastward towards India in summer. The term monsoon refers to that weather pattern. In northern India, however, the summer monsoon is associated largely with the break in extreme summer heat brought about by heavy rain. The monsoon dominates the large-scale atmospheric circulation in Africa and South Asia. It arises through the thermal contrast between ocean and continent. In winter the continent is colder than the ocean and induces a northeasterly flow towards the Arabian Sea. In summer, overheated continents become warmer than the oceans and in consequence a low-pressure cell is generated over the continents and a high-pressure cell over the oceans. In addition, the albedo effect means the Sun's radiation is reflected back by the glaciers on the Tibetan Plateau. The thermal contrast between the slowly warming ocean and the rapidly warming land accelerates vertical air flow over the land and therefore drives the landward flow of moist air that is associated with the monsoon. Winds blowing northeastward, from the equatorial Indian Ocean, transport the moisture to be precipitated in the Indian subcontinent. Climatic records show that over time there have been marked variations in the intensity of the monsoon. Are these variations controlled by mountain growth?

The significance of the Tibetan Plateau

The air above Tibet is heated because the Earth's surface absorbs the Sun's radiation and then heats the nearby atmosphere so that during the summer months solar energy is transferred to the atmosphere above the plateau. The process is enhanced by the albedo. Therefore the Tibetan Plateau represents a huge heat source 5 km above sea level. As the warm air rises it spreads laterally and creates a low-pressure cell that draws in moist air from the Indian Ocean where there is a high-pressure cell. Only a plateau as big as Tibet can have this effect. The advection of moist air leads to heavy precipitation in the tropical belt from Africa to India and, most intensely, in Southeast Asia. The result is seasonal precipitation, the monsoon. But was this change in atmospheric circulation solely due to the Tibetan Plateau? To test this hypothesis calls for the answers to difficult questions. For instance, when did the Tibetan Plateau attain its peak elevation, and knowing that, are there clear indications of climate change at that time?

Fluteau *et al.* (1999) accepted that Tibetan topography was a cause of the intensification of the monsoon but also favour the plate tectonic effect of the closure of the Paratethys landlocked sea in northern Asia. Wide shallow seas covered large areas of central Asia in the early Cenozoic. These seas reduced the temperature contrast between land and ocean, thereby diminishing monsoonal intensity. This Paratethys Ocean shrank in size as the Tethys Ocean closed. The result of its disappearance in the late Miocene was a drying up of northern Asia with the establishment of a continental climate, presumably reinforcing the high-pressure belt in Asia.

Climate change in the Cenozoic

An enormous amount of data on climate change during the Cenozoic comes from the analysis of cores from the deep ocean. Oxygen isotope values, which can be measured in calcite in animal shells as well as rivers and oceans, are sensitive to ice volumes and ocean temperatures, hence can be used to reconstruct past climatic variations. There has been a long-term cooling of the Earth over the past 55 Ma, indicated by an increase of $\delta^{18}O$ over that time. During ice ages the evaporation–precipitation process operating in ice sheets causes a preferential uptake in continental ice sheets of the light isotope of oxygen, i.e. ^{16}O. Consequently the reduction of ^{16}O in the oceans means that there is an increase in $^{18}O/^{16}O$ ratios in the oceans during a glaciation. Along the way we have at *c.*36 Ma the first ice in Antarctica and possibly more ice cover in the mid Miocene to late Pliocene. Alaska had a tropical climate in the mid Oligocene, while fossil pollen shows that Antarctica and Australia were much warmer than now. Climate models show that the modern topography results in reduced precipitation in the Mediterranean and also arid regions such as the Sahara region which in the Oligocene–Miocene was grassland, rain forest and savanna.

There is some indication that global cooling is linked to plate tectonics. Thus Antarctica's ice may be due to its thermal isolation following separation from South

America and Australia, which happened about 36 Ma. Another possible link is shown by the rise of the complete isthmus of Panama at *c.*3 Ma. The isthmus is a volcanic arc situated above a subduction zone. Its rise cut off the continuous seaway between the Atlantic and Caribbean with the east Pacific. The timing is beautifully recorded in the observed changes in C and O isotopes in the two oceans. At *c.*3 Ma the isotopes tell us that separation had been achieved. Also, the salinity of the Caribbean increased. Mammals and other fauna were then able to migrate from North to South America. But more importantly for those of us who live at high northern latitudes, it resulted in a strengthening of the Gulf Stream. Before 3 Ma, the Atlantic North Equatorial current could reach the Pacific. After 3 Ma, the current was deflected northeastwards and there it reinforced the Gulf Stream.

The Gulf Stream increased in velocity in the late Miocene and reached a peak velocity in the mid Pliocene. Thus the gradual emergence of the Panama isthmus as a volcanic arc, from late Miocene times, eventually choked off the seaways through the gateway. The consequence of this was that the Gulf Stream became a source of moisture in high northern latitudes just at the time when the ice sheets of the Northern Hemisphere were starting to form. As a result, the circulation of ocean water was disrupted.

To summarise the global climatic changes: from the Cretaceous when climate was warmer than today, there has been a gradual cooling leading to the glaciations during the Quaternary. During the Miocene, climatic conditions were rather similar to today's but generally warmer and more humid. Boreal forests existed up to higher latitudes than today. Ocean circulation was similar to the present but, as the Panama isthmus was still open, there was an exchange of the higher-salinity water of the Atlantic and the lower-salinity waters of the Pacific. Micheels *et al.* (2007), using climatic modelling, assumed a Tibetan Plateau with about half its present height and a weaker monsoon in the late Miocene (11–7 Ma). They postulated a rapid uplift of Tibet and, as a result, a marked intensification of the monsoon, beginning *c.*8–7 Ma. The leeward effect of the rise of Tibet led to a drying of the Qaidam Basin and northern Pakistan. Although the authors mentioned above suggest a sudden intensification of the monsoon, they accept the evidence presented in Chapter 5 that the southern part of Tibet was already at its present elevation from 15 Ma. Rowley and Currie (2006) extended the date back to 35 Ma (see Chapter 5). Therefore they envisage a long-term intensifying of the monsoon over 30 Ma, reflecting the rise of the Tibetan Plateau. Later in this chapter, the point is made that for Tibet to make a major impact on climate the whole areal extent of the Tibetan Plateau at the present elevation must be involved (Chapter 5). The argument between the gradualists (rise of Tibet by slow long-term uplift) and the catastrophists (sudden uplift of the plateau in the late Miocene) still continues.

It is important to stress that it is likely that more than one factor may be involved in climate change. The amount of CO_2 in the atmosphere is related to temperature. This gas comes from outgassing of volcanoes, and we all are aware of mankind's contribution! Such factors provide the input into the atmosphere of CO_2; the output, withdrawal of CO_2 from the atmosphere, is by the chemical weathering of rocks. We know that during the Cretaceous CO_2 was high, possibly owing to volcanic action. Also, the higher sea levels at the time meant there was less land from which to chemically erode rocks. The inference is that higher temperatures prevailed in the Cretaceous.

Evidence for global climate change in the late Miocene and Pliocene

Global climate changed greatly about 3 to 2.4 Ma when glaciation in the Northern Hemisphere developed, and perhaps at about 15 Ma when a build-up in the Antarctic ice sheet occurred. Some have maintained that a dramatic change in the climate of southern Asia occurred at about 6–8 Ma, possibly linked to strengthened monsoon over the Arabian Sea, and northern Pakistan became more arid. The argument is that an intensification of the Asian monsoon caused seasonal fluctuations in rainfall that tended to favour grasses over trees. Contrary arguments are that at this time, upwelling off-shore from Oman led to a drier climate rather than increased rainfall, and the change to grasses can be interpreted otherwise (see Clift and Plumb, 2008, and later in this chapter).

Before discussing further the significance of the rise of the Tibetan Plateau it is appropriate to set out the evidence for a rapid climate change in the Indian Ocean and southern Asia about 6–9 Ma. The evidence is wide-ranging and comes from studies of changes in flora and fauna, the chemical composition of sediments etc. The following list follows Molnar *et al.* (1993):

1. Marine sediments in the northwestern Indian ocean record abrupt changes in microfossil assemblages that imply a strengthening of summer monsoon winds at about 8 Ma. Between 14.5 and 8 Ma, planktonic foraminifera in the Indian Ocean off the coast of Somalia were dominated by species that live in relatively infertile regions, where upwelling of nutrient-rich deeper water is not important. Upwelling occurs during summer monsoons because the strong wind from the southwest displaces surface water from the Arabian coast. The deficit is then replaced by cold, nutrient-rich water from intermediate depths. At about 8.5 Ma the relative abundance of the species *Globigerina bulloides* started to increase rapidly, from less than 5% of the fauna to more than 10% at 7.4 Ma. *Globigerina bulloides* currently thrives in subpolar oceans but is abundant also in tropical areas of upwelling. The concentration of *G. bulloides* in surface sediment in the western Indian ocean correlates strongly with present-day sea surface temperature: it comprises about 40% of plankton where the ocean temperature is 23 °C, but only 10% at 28 °C. It is the dominant species during the summer months when the monsoon is intense. Thus variations in microfossil abundance suggest that the monsoon wind strengthened considerably at about 8 Ma.
2. Abundances of siliceous microfossils from the Arabian Sea also show indications of late Miocene climate change. The species associated with oceanic upwelling either appeared for the first time or increased abruptly at *c.*8 Ma.
3. At *c.*11 Ma there was an increase in the rate of siliciclastic sedimentation followed by a major increase at 8–10 Ma, which also has been linked to a strong monsoon.
4. Forested areas in northern Pakistan yielded to grasslands between 7.4 and 6 Ma. This conclusion is deduced from a sharp increase in $\delta^{13}C$ in pedogenic carbonates (formed by leaching in soils) in the Siwalik Group. This implies an increased aridity.
5. An expansion in grasslands at *c.*6–8 Ma is indicated by changes in the fauna in Pakistan. In northern Pakistan at 9.5–7.4 Ma, browsing mammals gave way to those

dependent on grass. Between 9 and 7 Ma, rodent species also changed rapidly in northern Pakistan, apparently from forest dwellers to species more dependent on grass. There was an increase in aridity in Pakistan about 7–5 Ma. While it is reasonable to propose that an intensification of monsoonal rains at *c.*7 to 7.4 Ma would favour C4 grasses, there are, however, other explanations because the oldest evidence of the morphology of C4 photosynthesis, deduced from fossil grass in NW Kansas, dates from *c.*5–7 Ma. Although C4 photosynthesis may have developed in the Cretaceous, an abrupt change in $\delta^{13}C$ might be a result of natural selection that occurred without climate change. Again, the increases in C4 plants at 5–7 Ma in the Old and New World may reflect a decrease in atmospheric CO_2 levels.

6. Some pedogenic carbonates in northern Pakistan show a shift in $\delta^{13}C$ from $-10‰$ before 8.5 Ma to $-6.5‰$ after 7.5 Ma. This abrupt change also signals climate change or at least floral change. $\delta^{18}C$ ranges from about $-15‰$ to 0‰ in pedogenic carbonates younger than 10 ka.

   The geological record of $\delta^{18}O$ from pedogenic carbonates serves as a palaeothermometer for meteoric water, but there is a complication. Measurements of $\delta^{18}O$ in present-day precipitation during the Indian summer monsoon are depleted in $\delta^{18}O$ by as much as 6‰ compared with those precipitated elsewhere at the same temperature. The increase in $\delta^{18}O$ at 7–8 Ma in northern Pakistan reflects a change from largely winter rains, which would have been relatively cold, to mainly summer, and hence the warm rains that characterised the present-day monsoonal precipitation.
7. A change in mineralogy of sediment deposited in the Bengal fan may also result from climate change. Similarities of oxygen and strontium in sediment in the fan since 17 Ma with those in rocks in the Himalaya suggest that the sediment has been derived from the crystalline High Himalaya. Derry and France-Lanord (1996) showed that silts derived largely from a gneiss terrain were deposited. However, between 7 and 0.8 Ma, mud with clay minerals dominates the sequence. Both the mineralogy and the isotopic compositions imply derivation from a soil environment suggestive of more intense weathering than in the earlier or later periods. Glaciation, low temperatures and lower sea levels during glacial epochs may account for a reduced source of weathered material since 0.8 Ma and an increase in coarse material. An increase in weathering at about 7 Ma is consistent with both warming, suggested by the oxygen isotopes, and increased monsoonal precipitation. The reverse argument has been presented, namely, a wetter monsoon results in slower transport and more chemical weathering but less physical erosion. Clift and Plumb (2008) suggested that there was a switch to more chemical weathering in South China after 23 Ma.
8. Najman *et al.* (2004) suggested a climatic change from arid to more humid, shown by the red beds of the Dagshai Group giving way up succession to the caliche with logs, leaves and organic material of the Kasauli Group in the Himalayan foredeep. The date of this change is imprecise, but probably 20–15 Ma ago.

Any rapid strengthening of the monsoon at *c.*8 Ma may have altered the global climate. The change in $\delta^{13}C$ appears to be global and not confined to southern Asia, suggesting a worldwide expansion of C4 plants in the late Miocene which are better adapted to low concentrations of CO_2 than are C3 plants. Carbon dioxide in the atmosphere may have decreased by about 5–7 Ma. Thus the possible strengthening of the monsoon at *c.*8 Ma led to a decrease in one greenhouse gas, CO_2, and hence temperatures fell. The expectation would be that chemical weathering was reduced but, as we will see later, the opposite appears to be true.

In summary, observations of several types suggest major changes of climates of the northwestern Indian Ocean and southern Asia between 6 and 9 Ma, with different types of observations implying different climate change in the various regions, i.e. strengthening of upwelling presumably from a strengthening of seasonal winds in the Indian ocean, increasing aridity and warming in northern Pakistan, and perhaps increase both in temperature and in seasonal precipitation in the Indo-Gangetic plains just south of the Himalaya.

How does the Tibetan Plateau disrupt global circulation?

The following expands on earlier sections of this chapter. The Tibetan Plateau disrupts the prevailing west to east flow of air between Africa and southeast Asia, as well as flow from the southwest in summer. Moreover, it provides a source of heat displaced from the equator that should affect the meridional flow of air in summer. Heating over Asia disrupts the large-scale circulation of air from the tropics to roughly 35° N. As already stated, in summer, rising air over warming landmasses draws moist air in from the sea, and in winter, cooling of the landmass drives the opposite flow. The albedo effect also contributes to the pattern of air circulation over Tibet.

A localised heat source creates a pressure gradient, which induces winds. A further effect of the high plateau is a northward displacement of the high-pressure centre that commonly lies near 30°–35° N and fixing of a low-pressure centre over northern India in the summer monsoon. These pressure centres enhance summer winds from the Arabian Sea onto the Indian subcontinent. High pressure exists over the ocean and low pressure on land. The orographic barrier to northeast flow during the monsoon, in turn, causes the relatively moist boundary layer to precipitate its moisture along the eastern Himalaya. Pressure gradients are balanced by the Coriolis force, and winds blow nearly parallel to isobars, not perpendicular to them.

In classical Hadley symmetrical circulation, warm air rises in the tropics. The added latent heat gained by condensation and precipitation causes this air to rise to the top of the troposphere, where it then flows poleward. To conserve angular momentum, this poleward air must gain an increasing eastward component of velocity. At latitudes of about 30°–35° it flows essentially due east and sinks. Adiabatic heating dries the sinking air further, producing a high-pressure centre and a belt of deserts at this latitude. Tibet provides a heat source in the middle of the area where air otherwise would subside, and this heating further strengthens the northeast flow from sea to land.

Air moving from west to east, in the general circulation of the atmosphere, must either go over a high plateau or go round it. The effect of a plateau on such circulation depends not only on its height and north–south width but also on its latitude, because of conservation of momentum. The angular momentum of an air mass moving from west to east depends on its latitude, thickness and lateral extent. Conservation of angular momentum requires that a change in thickness of air mass, which can occur when it flows over a high terrane, be compensated by changes in latitude or lateral extent of the air mass. Eastward flow that is forced upwards or downwards over topography should also be deflected southwards or northwards.

A possible threshold height for the plateau

Theoretical global circulation modelling indicates a threshold height of a plateau, at *c.*1.6 km, that governs whether air is deflected horizontally around a plateau or flows over it. Modelling allows us to envisage the atmospheric patterns assuming a zero topography (no mountains) or with topography increasing up to the present elevation. With increase in the height of the plateau, the calculated locus of the mid-latitude high-pressure centre is displaced northward. So for a condition of no high terrane, the high-pressure belt has been calculated to lie at about 35° N over northern Pakistan, appropriate for normal Hadley circulation. For mean heights of roughly half those at present, the high-pressure centre has been calculated to lie about 45° N. Increasing the height more, even to mean heights roughly twice those at present, however, had little further effect on the location of the high-pressure centre. But modelling by Kitoh suggested a link with the East Asia monsoon and Tibet. The suggestion is that 60% of present height (*c.*3 km) triggers a marked change in East Asian circulations.

It has been shown that heating over Tibet causes air to ascend over the plateau. More important is the release of latent heat in the atmosphere near the plateau. Subtropical air below the trade cumulus boundary layer at about 2–3 km is relatively moist. Subsiding hot, dry air above it creates an inversion that is penetrated only occasionally by localised cumulus convection. As moist air in the lower troposphere rises near the plateau it cools adiabatically. As a result, water vapour condenses and precipitates, giving heavy rainfall over eastern India during the summer monsoon, mostly because of the orographic effect of the Himalaya on this moist air. If Tibet were lower than 2–3 km, this moist air could flow over the eastern Himalaya in summer without losing its moisture and releasing its latent heat.

Modelling of the Tibetan Plateau shows profound differences in temperature, precipitation and wind strengths in Asia to be associated with differences in mean elevations assigned to Tibet and its surroundings (Kutzbach *et al.*, 1989; Ruddiman and Kutzbach, 1989). For a smoothed elevation over this region half its present value, the calculated precipitation in Asia is roughly 20% lower than for present elevations. Calculated wind speeds are weaker by about 40%. For no plateau at all, the calculated precipitation in Asia is roughly 60% lower than for smoothed present elevations. Calculated southerly components of wind speeds are weaker by 70–100%, corresponding to a weak or non-existent monsoon. Apparently, however, neither theory nor resolution in global circulation models is adequate to evaluate the sensitivity of the monsoon to intermediate plateau heights or more localised heat sources. Thus, calculations with global circulation models have failed to confirm the proposed threshold in Tibet's height above which the monsoon should be strengthened in the dramatic manner suggested by the geological observations cited previously.

A causal relation between plateau uplift and mantle dynamics

The link between a possible late sudden uplift of Tibet and climatic change was developed by Molnar *et al.* (1993) into a remarkable account of how events happening in the mantle could have important consequences at the Earth's surface and in the atmosphere. They followed England and Houseman (see Chapter 5) in suggesting a sudden uplift at *c.*8 Ma which they explained by invoking a convective removal of the lower mantle, the so-called thermal boundary layer. Isostatic rebound was the result, and the Tibetan Plateau rose by 1 or more kilometres to its present elevation of *c.*5 km. The next step is the intensification of the monsoon. However, as we saw in Chapter 5, the timing of the uplift is far from solved. Indeed, recent work (see Chapter 5) suggests that at least parts of the plateau were at their present elevation before 8 Ma. But before we dismiss the Molnar *et al.* model we must consider that if thickening and uplift has migrated northwards across Tibet as the Indian indenter ploughed into Asia, then the uplift is progressive from south to north. The evidence that parts of the northern part of the plateau only attained their present elevation at *c.*8 Ma lends some support for this proposal. As already pointed out, modelling shows that not only high elevation but large area is called for. Therefore the possibility remains that attainment of maximum elevation for the *whole* plateau occurred at *c.*8 Ma. But this still does not support the dramatic uplift suggested by Molnar *et al.* (1993). Tests for their model call for more data on palaeoaltimetry and perhaps more datings of the normal faults. In addition, magnetostratigraphy on lake sediments in the graben may be interesting.

When did the plateau attain its present elevation?

The proposal has been made by Molnar *et al.* (1993) and others of a causal relationship between the climate change at 6–9 Ma and a nearly synchronous and sudden attainment of highest elevation of the Tibetan Plateau. There is general agreement that by 8 Ma there was an elevated Tibetan Plateau. What Molnar *et al.* maintain is the controversial idea that as late as the late Miocene the plateau was 1–2.5 km lower than at present. Evidence bearing on this proposal is scanty (see Chapter 5). All we know for sure is that at least parts of the plateau were below sea level at *c.*50 Ma and since then it has undergone uplift which could have occurred suddenly or slowly. The qualification in the last sentence is necessary because some have argued that Tibet underwent deformation and uplift in Cretaceous times and prior to the India–Asia collision. It has been argued that the present elevation seems to be necessary if Tibet has caused climate change. Otherwise, why did not a lower plateau elevation prior to 8 Ma result in strengthening of the monsoon? The implication is that the abrupt change in climate indicates that climate is sensitive to a threshold elevation and area for the Tibetan Plateau. All that can be deduced from the few estimates of palaeoaltimetry of the Tibetan Plateau (Chapter 5) is that the plateau may have had a smaller area but similar height to the present day before 8 Ma and maybe as early as 35 Ma.

Weakening of the monsoon at 8 Ma – a counter-argument

A radically different view to Molnar *et al.* of the history of the monsoon is given by Clift *et al.* (2008b) who presented weathering records from the South China Sea, Bay of Bengal and Arabian Sea that allow Asian monsoon climate to be identified back to the earliest Neogene. There is a correlation between Himalayan exhumation and monsoon intensity. A weakening of monsoon intensity is seen after 10–8 Ma. They propose that the Tibetan Plateau had reached a sufficient area and height by early Miocene times to permit channel flow and strengthening of the monsoon. This pattern was maintained through the Neogene, although temporal variations in monsoon intensity might have induced variations in the structural configuration. According to Clift and Plumb (2008), late Miocene weakening of the summer monsoon is responsible for the slower exhumation rates after *c.*10 Ma, possibly linked to global cooling. An acceleration of exhumation and erosion after 4 Ma correlates with the strengthening of the monsoon then.

An *et al.* (2001) proposed three stages of evolution of Asian climates: (1) at 9–8 Ma more aridity of the Asian interior, giving loess deposits and onset of Indian and East Asian monsoons; (2) continued intensification of East Asian summer and winter monsoons at 3.6–2.6 Ma; more dust supplied into the North Pacific; (3) variation and possible weakening of Indian and East Asian summer monsoon and continued strengthening of winter monsoon after 2.6 Ma.

The suggestion that erosion rates in the Himalaya increased at about 7–8 Ma has also been denied by Burbank *et al.* (1993) who considered that sediment accumulation rates in the Indo-Gangetic Foredeep and in the Bengal fan, as indicated by a change to finer-grained sediments and reduced volumes of sediment, declined at *c.*8 Ma despite the probability that the monsoon intensified then. They attributed this to reduced tectonic activity at that time, or decreased Himalayan glaciation or slope stabilisation due to dense plant cover. Later we will see that there are other grounds for inferring reduced erosion rates after 8 Ma. Burbank *et al.* (1993) suggested a stronger monsoon at 8 Ma and attributed the reduced erosion from the Ganges–Brahmaputra rivers to warming and moistening of the air over South Asia. This reduced the erosive power of glaciers and increased vegetation cover which in turn reduced run-off. But the present erosion is greatest in the eastern Himalaya which currently supplies the highest amount of clastics to rivers in the world. Are continental basins a true record of the flux? Is the Bengal fan a better indicator?

Barros *et al.* (2006) and others have given a different perspective. Certainly the existence of the Tibetan Plateau reinforces the land–sea contrasts and enhances the monsoon. But they point out the importance of the paths taken by the monsoonal storms which are exerting a strong control on precipitation amounts. They postulate a monsoon trough or low-pressure belt in which low-pressure systems are developed. The monsoon activity on the south-facing slopes of the Himalaya is associated with the development and westward propagation of monsoonal depressions in the Bay of Bengal. The paths of the depressions are determined by the Coriolis force giving a north deflection, also by the origin and strength of monsoon systems and the interactions with land mass as they travel over north India. The important conclusion is that mountain belts do not have an influence on whether or not monsoons occur. However, they do contribute to the intensity and lengthening of monsoons.

To sum up the present position on the controversial link between the rise of the Tibetan Plateau and the Asian monsoon, it is perhaps important to distinguish between two propositions. First, did the rise of the Plateau exert an important influence on the climate of SE Asia? Second, was there an intensification of the monsoon at around 8 Ma ago? As to the first proposition, it is undeniable that the elevation of the Himalayan barrier acted as a forcer of precipitation, and also the high plateau of Tibet forms a low-pressure area which draws in winds from the Indian Ocean. Climate models predict a strong response of the SW monsoon to the high plateau. Thus there is consensus amongst modellers that the monsoon must have responded to the uplift (see papers by Prell and Kutzbach, 1992, and others). However, the second proposition is much more challenging and calls for a rough synchroneity at *c.*8 Ma of peak elevation of the plateau, intensification of the monsoon and, in the hypothesis of Molnar *et al.* (1993), convective removal of the lower mantle lithosphere. Apart from doubts about the latter mechanism, the evidence from palaeoaltimetry is not strong enough to enable us to regard this as other than an interesting speculation. One could even say that the current sources of evidence give conflicting results. Perhaps the reader or others could be encouraged to find other sources of evidence to help resolve this conundrum – if it can be resolved.

Monsoon precipitation as a tectonic forcer

In Tibet there is also correspondence between topography, active faulting and precipitation. The plateau is arid and in the rain shadow of the Himalaya. Peak precipitation occurs along the Himalayan front and less so in north Tibet.

Did rains and erosion cause uplift? In Chapter 8 we mentioned the interesting point that stronger erosion in valleys removes material and unloads the crust, and as a result isostasy causes uplift of the peaks. A good example is the Nanga Parbat area where the Indus River removes detritus and promotes the remarkable uplift of the mountain (Plate 40, Koons and Zeitler, 2002). However, Whipple (2009) has shown that not much uplift results from this process (Clift and Plumb, 2008).

In Chapter 8, the study by Wobus *et al.* (2003) in the Marysanda river of central Nepal was mentioned. This concluded that a knick in the river profile separates a region of rapid uplift to the north composed of the Greater Himalayan Sequence, from the Lesser Himalaya to the south with lesser uplift. This poses the question: did monsoon-driven erosion cause the faulting? Burbank *et al.* (2003) presented a different interpretation based on a study of a 20 km wide zone in the Greater Himalaya where erosion rates deduced from apatite fission track dating show no difference in erosion, suggesting that variations in monsoonal precipitation do not correlate with erosion. Burbank, in contrast to those who suggest that climate and precipitation control exhumation, favoured the view that exhumation is controlled by rock uplift driven by tectonic forces. Thus a ramp in the MCT is the explanation for the uplift described by Hodges *et al.* (2004). Clift and Plumb (2008) think the conflict of interpretations may be explained by different timescales: for Hodges, short periods, and for Burbank, long periods.

Apparently in support of the argument used by Hodges, Thiede *et al.* (2004) in the Sutlej river showed a control by climate on orogenic structure. Heaviest precipitation occurs at the breaks in topography marked by the MCT and MBT. Fission track ages are younger in zones of high precipitation, perhaps indicating a link between precipitation and exhumation. Landslides correlate with monsoon intensity over the past 40 ka.

Monsoon controls erosion in the Himalaya

Clift and Plumb (2008) have discussed the climatic controls on erosion in SE Asia and use sediment budgets to show that erosion increased rapidly after 45 Ma, with a sharp rise after 33 Ma with a peak at 14 Ma and then decrease to a low in the late Miocene and early Pliocene. Their proposal is the monsoon controls erosion in the Himalaya and in Indo-China and determines the observed intermittency of erosion rates. During monsoons, heavy summer rainfall occurs and hence strong erosion occurs. There was worldwide rapid erosion in the Plio-Pleistocene linked to increasing glaciation. The influence of glacial periods on monsoon intensity is revealed by many studies which tend to show a close correspondence between stronger summer monsoons during the warmer periods in the glacial–interglacial cycles and weaker monsoons prevailing in glacial periods. Examination of the mineral contents of deep-sea cores from the Bengal fan shows that more smectite and kaolinite is present in interglacial times when the wet summer monsoon favours chemical weathering, whereas more illite and chlorite occur in glacial times.

Clift and Plumb (2008) argued that the monsoon is the main driver of erosion not only in SE Asia but elsewhere in the Earth. In the Alps, sediment accumulation shows a huge increase after 3–4 Ma with peaks at 22 and 17 Ma. This is not related to synchronous tectonic activity; instead it seems that global climate change increases monsoon activity and so peak erosion rates are linked to the monsoon. Continental erosion rates are linked to the summer monsoon. For example, in the Oman bajada alluvial basins, fast erosion in the Neogene was associated with the wet monsoon climate that prevailed in the latest Miocene–Pliocene and early Pleistocene in Arabia. Similarly, the Ganges–Brahmaputra delta shows fast erosion in the Holocene probably linked to a strong monsoon.

Clift *et al.* (2008b) argued that monsoon rains operating over the past 20–25 Ma are closely linked to, and indeed promote, Himalayan tectonic activity and erosion. They cite the lack of erosional detritus in the foredeep before the early Miocene as evidence that there was minimal topography and tectonic activity before *c.*20 Ma. In other words, without the monsoon there was no tectonic activity. Evidence of climatic control over Himalayan exhumation is the rapid cooling at 23 Ma in the Greater Himalaya probably related to the initiation of the MCT at 22–23 Ma. Surprisingly, there is no sedimentary record of exhumation before 23 Ma. The Dagshai–Dumri and Kasauli groups, deposited after *c.*22 Ma, show that exhumation increased during the period of deposition and reached down to garnet grade by the time of the deposition of the Kasauli Group.

In attempting to explain why Himalayan uplift was delayed for so long after collision, Clift and Plumb (2008) pointed out that 23 Ma was the onset of monsoon seen in the South China Sea and in the climatic zonation of China. But this argument neglects the evidence of pre-MCT thrusting (that is, before 23 Ma) of the Tethyan Sequence including the major EoHimalayan thrust of Eocene or Oligocene age (see Chapter 5). This thrusting was responsible for large crustal thickening accompanied by uplift, before *c.*35 Ma, but the likely erosional detritus has not been identified.

There is suspicion that chicken and egg arguments are being employed here and perhaps the prime causes – tectonic activity – are being overlooked. Finally, another possible link is suggested by Kitoh's (2004) modelling which showed that heavy rains only reached the Himalayan front after the Tibetan Plateau had reached 60% of its present height. At that time there was a link between a high Tibet, stronger monsoon and faster Himalayan exhumation.

Himalayan uplift and the chemical composition of the oceans

While the proposals on the role of the Tibetan Plateau in climate change are controversial, there is less controversy about another aspect of the link between mountain building and climates. Mechanisms of erosion affect climate on a global scale through influences on the carbon cycle. Weathering of Mg and Ca silicate rocks provides the essential buffer that balances the introduction of CO_2 into the atmosphere and the locking-up of that CO_2 in carbonate rocks. For Ca silicates, the chemical equation is:

$$CaSiO_3 + CO_2 = CaCO_3 + SiO_2$$

Erosion also promotes a drawdown of CO_2 by harvesting of biomass and burial of this material in sedimentary basins. Thus erosion mechanisms result in stabilisation of the Earth's climate. Galy *et al.* (2007) have shown that the Bengal fan contains and effectively buries 10–20% of the total terrestrial organic carbon, and this reflects the high erosion rates in the Himalaya and the consequent high sedimentation rates. It also reflects the low availability of oxygen in the Bay of Bengal.

Raymo, Ruddiman and Froelich (1988) proposed that increased chemical weathering and river fluxes, associated with mountain building especially in the Andes, Himalaya and

Tibet, may have drawn down CO_2 from the atmosphere since the late Miocene. Such a reduction in atmospheric CO_2 may have led to a lowering of temperature and the onset of the most recent Ice Age. In the late Miocene there were increased uplift rates in southwest USA, in the Swiss Alps and in the Southern Alps in New Zealand. However, the greatest loads of dissolved and particulate material are carried by the rivers draining the Himalayan–Tibetan–Andean orogens. They are the Yangtze, Amazon and Ganges–Brahmaputra–Indus rivers. The increase in the dissolved and particulate material that is carried by these rivers has resulted in increases in Sr and carbonate sedimentation in the oceans in late Miocene–Pliocene times.

Both accelerated erosion of higher terranes and increased rainfall associated with a strengthened monsoon may have conspired to increase weathering rates in the Himalaya. It is remarkable that the big rivers of the Indian subcontinent, the Ganges, Indus and Brahmaputra, contribute about 25% of all the dissolved material in the oceans and yet this is derived from a mere 5% of the Earth's land surface. This means that chemical weathering is largely concentrated in the Himalayan–Tibetan region. As explained above, the chemical weathering has caused a reduction or drawdown of CO_2 from the atmosphere. According to Prell and Kutzbach (1992), atmospheric concentrations of CO_2 were much higher in Cenozoic times with possibly twice present values existing at *c.*15 Ma ago. The warmer conditions associated with increased CO_2 result in greater wind velocities in the Arabian Sea and in an intensification of the monsoon. The cooling of climate over the past few million years may be linked to a decrease in atmospheric CO_2. But the monsoon has intensified since the Pliocene.

The $^{87}Sr/^{86}Sr$ plot

One of the best-known examples of the change in the composition of the oceans as a result of orogeny is the $^{87}Sr/^{86}Sr$ content of marine carbonates (Plate 41). It has increased markedly over the past 40 Ma. The plot of the ratios $^{87}Sr/^{86}Sr$ versus time is linked to the monsoon and reflects inputs from ocean hydrothermal activity and the dissolution of marine carbonates, but mainly from continental run-off. The strontium comes from the weathering of metamorphic carbonates in the Himalaya. The Ganges–Brahmaputra river system is the largest source of radiogenic ^{87}Sr in the oceans. The plot of the $^{87}Sr/^{86}Sr$ ratio versus time slightly flattens temporarily at *c.*16 Ma and at 3 Ma, indicating a reduced input from the Ganges–Brahmaputra rivers, but it is difficult to find any obvious link with changes in tectonic activity at those times.

Raymo, Ruddiman and Froelich (1988) suggested that the pronounced increase in oceanic $^{87}Sr/^{86}Sr$ ratios can only be explained by an increased delivery of radiogenic Sr from the land, the most likely explanation being an increase in the rate of chemical weathering over the Cenozoic. Geochemical studies by Hawthorne and James (2006), using Li/Ca and Li isotopes in the Pacific and North Atlantic Oceans, and by Piotrowski *et al.* (2000), using Hf isotopes in the North Atlantic and Indian Oceans, indicate some modifications to this statement. Variations in Li isotopes were interpreted as showing that silicate weathering and weathering intensity decreased between *c.*16 and *c.*8 Ma.

As noted there is a break in the slope of the $^{87}Sr/^{86}Sr$ plot at *c.*16 Ma. However, the river flux of Li has increased since *c.*8 Ma.

Can the strontium plot be entirely due to the erosion of the Himalaya? The upward swing in the plot occurs at about 40 Ma but, as Clift and Plumb (2008) emphasised, there is no evidence of erosion of the Himalaya before *c.*20 Ma. Therefore it may be that the part of the curve between 40 and 20 Ma reflects a strontium contribution into the oceans by Cenozoic orogens other than the Himalaya, for example the Alps or Andes.

Derry and France-Lanord (1996) have reinterpreted the famous strontium curve. In the Ganges–Brahmaputra basin they report a change in erosional and weathering regimes in the Himalaya near 7.4 Ma, from erosion dominated by mechanical erosion to one with significant chemical weathering. Thus we have a paradox of a reduction in the rate of erosion in the Himalaya coinciding with an increase in the monsoon rains, a conclusion shared with France-Lanord *et al.* (1993). The reduced erosion and increased monsoon can be attributed to hydrothermal activity on the sea floor but more probably it reflects an increase in the delivery of radiogenic Sr to the oceans from the land. Physically dominated erosion dissolves Sr from carbonates whereas chemically dominated erosion dissolves Sr from radiogenic silicates. The roles of these differing types of erosion were reversed at *c.*1 Ma.

In the chemically dominated period, smectite replaced illite as the dominant clay derived from the weathering of high-grade metamorphic rocks, while sediment accumulation decreased. Despite the intensifying of the monsoon there was a decrease in physical erosion in the Himalaya. The change in weathering increases the isotopic composition of the riverine Sr in the Ganges–Brahmaputra river system, and this increase requires a significant decrease in the flux of Sr to the oceans in order to fit the Sr plot.

Transport of calcium

Riverine transport of calcium can be used as a measure of weathering, and a possible approach to this topic is to study the calcite compensation depth. Below the compensation depth, $CaCO_3$ does not accumulate because deposition and dissolution are balanced. The compensation depth in oceans varies with time and depends on temperature and pressure and such factors as supply of carbonate and oceanic circulation, but is usually at several kilometres below sea level. The deepening of the calcite compensation depth in the oceans during the late Cenozoic can be interpreted as evidence for an increase in the flux of Ca to the oceans, and also in the rate of riverine transport of calcium to the oceans since 5 Ma. The calcite compensation depth has lowered and the carbonate mass accumulation rates in the Indian ocean have increased since 8–9 Ma. Increase in the rate of riverine transport is shown by increase of Ca^{2+} supplied to the oceans since 5 Ma. Raymo, Ruddiman and Froelich (1988) suggested an increase in Ca^{2+} of 4×10^{12} mol/a or 2.25×10^{11} kg/a of CaO. What is the possible rate of erosion of the Himalaya that is consistent with this increase? This has been calculated by Molnar *et al.* (1993), using certain assumptions, namely that CaO accounts for 10% of crustal rock and the rate of weathering increased linearly with time, and the assumption that all the Ca flux came from within the Himalaya.

They estimated that the Himalayan belt is roughly 100 km wide and 2500 km long, therefore the average rate of erosion since about 8 Ma was *c.*1.5 mm/a. This estimate is likely to be too high because they assume too small a width for the Himalayan belt, but evidently the calculated erosion rates are consistent with the proposed increase in CaO. For example, the cooling ages of minerals give average exhumation rates of 1–5 mm/a, and in the Bengal fan the stratigraphic ages of deposition also imply rapid erosion approaching, if not exceeding, 1–3 mm/a.

Drawdown of CO_2

Over the long term, weathering of silicate rock subtracts CO_2 from the atmosphere. Why not carbonates as well? Although the dissolution and aqueous transport of each Ca or Mg atom from a carbonate rock requires an extra molecule of CO_2 (to make two bicarbonate ions) this extra CO_2 molecule is released to the ocean–atmosphere system upon reprecipitation of carbonate in the ocean. However, the weathering of calcium from silicate minerals subtracts one mole of CO_2 from the atmosphere for each mole of Ca from the silicate. Whereas increasing the weathering of either silicates or carbonates might cause the calcite compensation depth to deepen, only weathering of silicates affects CO_2 in the atmosphere. Because most of the rock weathered in the Himalaya since 8 Ma has been silicate not carbonate, the change in erosion and weathering rates considered above might have caused a large increase in the rate of removal of atmospheric CO_2. Because CO_2 is a greenhouse gas, clearly this process must affect global climates, notably the initiation of glaciations.

Such a drawdown of CO_2 from the atmosphere must have been partially compensated by another process or processes. Even if only 10% of the increased calcium flux to the oceans was by weathering of silicates, a large amount would have been removed from the atmosphere as a result of an increase in the weathering rate. Molnar *et al.* (1993), assuming that the late Miocene atmospheric CO_2 was similar to the present, point out that the rate of removal proposed by Raymo *et al.* (1988) of CO_2 at 4×10^{11} molecules per year would have exhausted the CO_2 of the entire atmosphere in 100 ka.

Previously we saw that Molnar *et al.* (1993) considered that the two factors thought to affect erosion rates most, relief and precipitation, seem to have increased in or near the Himalaya at about 8 Ma. However, Clift and Plumb (2008) indicated a drying at 8 Ma which is about the time of initiation of loess deposition in Asia. Relief and precipitation should have led to increased weathering, one by creating new surface area of minerals to weather, and the other by increasing the flux of aqueous solutions to dissolve and transport weathered material to the oceans. Unless the drawdown of atmospheric CO_2 associated with apparently increased weathering rate was balanced by input from another source over this interval, it seems inescapable that some lowering of atmospheric CO_2 followed the strengthening of the monsoon. If such reduction in atmospheric CO_2 did occur, it may have lowered temperatures sufficiently to have caused the onset of the Northern Hemisphere glaciation.

Oxygen isotopes

The oxygen isotope composition of calcite, which is sensitive to ice volume and ocean temperature, is a valuable tool in reconstructing past climates. A pronounced enrichment in $^{18}O/^{16}O$ ratio in the oceans occurs during periods of glaciation. Thus study of the isotopes of the $CaCO_3$ shells of marine organisms from the deep-sea cores gives a record of past ocean temperatures. Oxygen isotope data from the past 55 Ma show an increase in $\delta^{18}O$ levels indicating long-term climatic cooling. Sharp increases at 36 Ma and in the late Miocene represent increases in ice volume in the Antarctica. The formation of the Panama isthmus seems to pre-date glaciation in the Northern Hemisphere by <2 Ma.

Proposed link between the Himalaya and chemical erosion

As mentioned above, Raymo, Ruddiman and Froelich in 1988 indicated an increase in dissolved fluxes in rivers over the past 5 Ma which they attributed to increased chemical erosion caused by uplifts in the Himalaya, the Andes and the Tibetan Plateau. They extended this link, notably the increase in fluxes and the associated reduction in atmospheric CO_2, to explain the growth of ice sheets in the Cenozoic. Northern Hemisphere glaciation may have been related to mountain uplift through the impact of high plateau and mountain ranges on atmospheric circulation and the impact of enhanced rates of chemical erosion on the oceanic carbon cycle and atmospheric CO_2. The increase in chemical weathering may also have influenced atmospheric CO_2, either by consumption of CO_2 during weathering reactions or by influencing the carbon cycle. If relief is an important factor in continental weathering, then atmospheric CO_2 levels may have been controlled not only by the balance between sea-floor spreading rates, land and sea distributions, and atmospheric CO_2 feedbacks on weathering rate, but also by relative rates of sea-floor spreading (CO_2 production) and continental orogeny (CO_2 consumption).

Erosion in channel flow

In Chapter 6, discussing channel flow in the Himalaya, we mentioned the idea put forward by Beaumont and his associates that climate may have acted as a control on the rate of channel flow. Thus influx of material in the channel was balanced by the efflux by erosion at the mountain front. Naylor and Sinclair (2007) have investigated control on a smaller scale by modelling the influence of exhumation on thrusting, notably whether exhumation causes the intermittent slip on thrusts. It is well known that all faults move in jerks punctuated by periods of quiescence. Their conclusion is that thrust activity is influenced by rates of surface uplift, frontal accretion and exhumation.

Further reading

Avouac, J.P. and Burov, E.B. (1996). Erosion as a driving mechanism of intracontinental mountain growth. *Journal of Geophysical Research*, **101**, B8, 17,747–17,769.

Clift, P.D., Giosan, L., Blusztajn, J., Campbell, I.H., Allen, C., Pringle, M., Tabrez, A., Danish, M., Rabani, M.M., Alizai, A., Carter, A. and Lückge, A. (2008a). Holocene erosion of the Lesser Himalaya triggered by intensified summer monsoon. *Geology*, **36**, 79–82.

12 Secular change in orogeny

Previous chapters have outlined the principles underpinning the understanding and analysis of orogenic belts in the modern world, with examples and applications to the Alps, the Himalaya, the Andes and western North America. We have also used examples from earlier times in Earth history to illuminate the issues surrounding, for example, the processes of thrust tectonics, metamorphism at extreme conditions, variability in exhumation rates, and contrasts in the amounts of magmatism in collisional and accretionary orogens. In doing this we have presented evidence from Palaeozoic orogens, such as the Variscan/Hercynian and Caledonian, Neoproterozoic orogens including the 'Pan African' belts of former Gondwanan fragments in India, Africa and Antarctica, and Mesoproterozoic orogens such as the Grenville of northeastern America. This begs the simple question – can we look back in time and see similarities in orogeny and the making of mountains throughout Earth history, or has mountain building changed in subtle or even dramatic ways since the earliest records of continental crust, deformation and metamorphism? In other words, has there been secular change in orogeny? Within the context of the recognised diversity in Phanerozoic orogenies, were Archaean and Proterozoic orogenies broadly the same, or different?

The question of secular change in mountain building has been the subject of considerable debate and a burgeoning literature. It forms part of the broader issue of the extent to which uniformitarianism can be applied far back in time – whether 'the present is the key to the past'. Recent major volumes on the subject include *Ancient Orogens and Modern Analogues* (Murphy *et al.*, 2009) and *When did Plate Tectonics Begin on Planet Earth*? (Condie & Pease, 2008). These build on an extensive literature in which the temporal evolution of the Earth's tectonic regime has been assessed or speculated upon using thermal and geodynamic modelling, geological observations from ancient and modern terrains, and geochemical and isotopic constraints on rates of continental growth through time (e.g. Hawkesworth & Kemp, 2006a, b; Kemp *et al.*, 2007) (see Fig. 12.5 later).

It is, unfortunately, beyond the scope of this contribution to examine all of the evidence and lines of argument used to support or deny the operation of plate tectonics back in time to the Archaean. We will instead focus on those lines of evidence and reasoning that apply directly to orogeny and its secular evolution. As shown from the preceding chapters, in the modern world orogeny is intimately and strongly linked with plate processes at convergent zones, whether orthogonal or oblique, from subduction and arc–back-arc development to subduction–accretion and thence, in many circumstances, to collision (Chapter 5). It is also conditioned, it would appear, by whether subduction advance or retreat dominates in the convergence history and then how this is accommodated in any terminal suturing and collision (Chapter 2). The subsequent evolution of the orogenic system then depends on the coupled thermal and physical properties and architecture of the resultant crust and resilience or destiny of its subjacent lithospheric mantle. It is almost inevitable that these will have changed through time. The reason for this is the Earth's declining heat production, referred to in Chapter 3, and its consequences for the thermal budget of the Earth and the potential temperature of the Earth's convecting upper mantle (Richter, 1985; Christensen, 1985; Sandiford, 1989a).

This chapter will begin with a brief review of heat production over time that establishes the essential reasoning for the argument that the Earth has cooled since at least the late Archaean and that tectonic processes in deep time had to be somewhat different from the Phanerozoic. We will then describe recent thermomechanical modelling that bears on the issue of the initiation of subduction and hence the formation of orogenic belts similar in most respects to those dwelt on in this work, and consider how this modelling may inform our understanding of orogenic processes through time. The metamorphic and structural evidence and interpretations from orogens, including deep continental crust now exposed at the Earth's surface, that point to similarities and differences in the thermal-tectonic regimes of orogeny back to the Archaean will then be evaluated using examples from the time period over which subduction is generally considered to have become prevalent on Earth.

Secular change in heat production and the Earth's thermal budget

The present Earth can be regarded as a heat engine losing heat from its surface through conduction through the lithosphere, plate construction and destruction processes, and heat advection from within-plate magmatism. The principal sources of heat being produced at depth, as noted in Chapter 3, are the decay of radionuclides and heat from exothermic processes in core formation and growth.

The present thermal budget of the Earth arises from the balance (or imbalance) between all the heat loss mechanisms, integrated over the entire Earth, and those producing heat. It is widely recognised, based on measurements of surface heat flow, that flux of heat *from* the Earth is greater than its present rate of heat production (e.g. Bickle, 1978; Christensen, 1985; Richter, 1985; Davies, 1992). Therefore, the present Earth is cooling. This simple boundary condition, coupled with basic geochemical constraints on the amounts of past heat production, implies that 'steady-state' temperatures at depth in the Earth, for example in the convecting upper mantle, were higher in the past. This is commonly expressed in terms of the potential temperature (T_p) of the mantle being higher than the presently accepted range for contemporary mantle (1280–1350 °C: England & Bickle, 1984; Davies, 1992). Typical estimates of the increase in T_p going back in time lie in the range 150–200 °C at the end of the Archaean (2500 Ma) and 250–300 °C in the middle Archaean (*c.*3000–3300 Ma) (Christensen, 1985; Nisbet *et al.*, 1993; Smithies *et al.*, 2003). Given that recent studies of the rates of continental growth through time broadly agree that this time period, 3300–2500 Ma, saw the production of between 40 and 60% of the Earth's

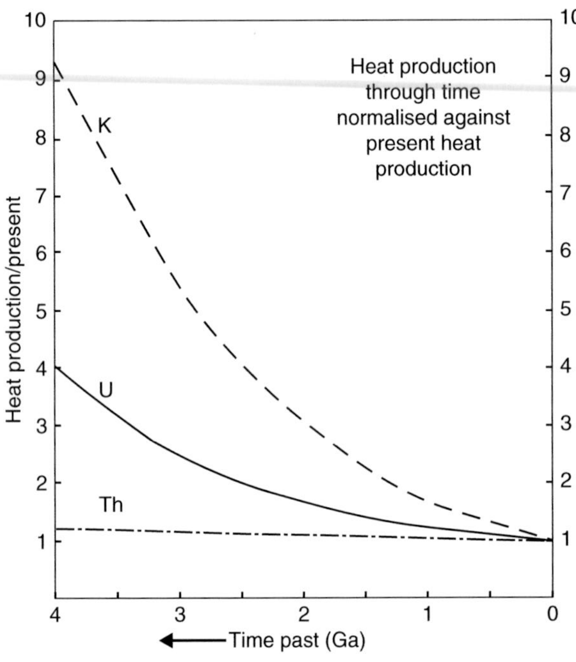

Figure 12.1 Heat production of the elements K, U and Th, calculated back in time and normalised to present heat production in each case. The much higher heat productions associated with K and U at 4 Ga (4000 Ma) are due to the significantly higher amounts of ^{40}K and ^{235}U, the minor isotopes of each of these elements. To convert this type of calculation to heat productions appropriate for individual rocks and Earth layers it is necessary to know the concentration of each element in the rock/layer, and the density of the material.

continental crust (Hawkesworth & Kemp, 2006a, b) it is apparent that any analysis of how continents form and grow must do so in the context of a hotter Earth with T_p some 200 °C higher than at present.

The theoretical basis for evaluating the impact of increased heat production in the past is both elegant and straightforward. A key assumption is that the Earth has not changed significantly in its bulk composition since at least the early Archaean (we will not examine the Hadean), particularly in terms of the total inventory of elements such as K, U, Th and Rb. In other words, large amounts of these elements have not been added to the Earth in its recent history. Given a fixed inventory of the main heat-producing elements it is straightforward, using the decay equations of the radioactive isotopes of these elements, to back-calculate their heat productions in the past relative to the present. Figure 12.1 shows an example of such a set of calculations, from which it can be seen that heat contributed by the decay of K (^{40}K) and U (^{238}U and especially ^{235}U) was markedly higher in the Archaean. Further assumptions are that the geochemical behaviour of elements, for example element incompatibility and partitioning between minerals (or residua) and melts, have not changed as they are governed by thermodynamics. Secular changes in the compositions of specific Earth layers can be accounted for given the two assumptions above and given knowledge of how elemental partitioning varies with mineral and melt compositions, and with temperature.

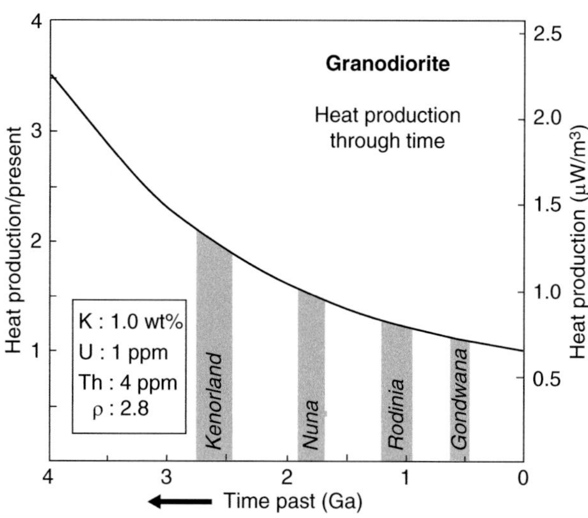

Figure 12.2 Time-dependent heat production in a model granodiorite, calculated using the concentrations listed in the inset key, and for a density of 2.8 g/cm^3. The vertical axis on the left gives the heat production normalised to present, and that on the right side the calculated heat productions in μW/m^3. Four proposed episodes of major crustal production and supercontinent formation (Brown, 2007; Hawkesworth & Kemp, 2006a, b; Kemp *et al.*, 2006) are shown. Addition of substantial amounts of granodiorite to the crust in the late Archaean (Kenorland episode) would have had a much greater impact on heat production than such additions in events leading to Rodinia and Gondwana.

An example of the calculated heat production for a given rock composition is presented in Fig. 12.2. In this case we have chosen a typical granodiorite with present-day heat production of 0.65 μW/m^3. Given its density, bulk composition and relative amounts of K, U, Th and Rb, the type of back-calculation performed in Fig. 12.1 yields the relative heat production curve of Fig. 12.2. At the end of the Archaean the same rock type, with *equivalent* bulk K, U and Th contents, would have about twice the heat production seen today. This result principally arises because more of the K was ^{40}K and more of the U was ^{235}U. This granodiorite is not an analogue for the continental crust as a whole as it is more felsic than the andesitic or mafic andesite compositions preferred by Rudnick and Gao (2003). The model granodiorite also has lower K (1 wt% *cf.* 1.5 wt%), U (1 ppm *cf.* 1.3 ppm) and Th (4 ppm *cf.* 5.6 ppm), and so is not as 'hot' either now or in the past as average crust (Rudnick and Gao, 2003).

The result shown in Fig. 12.2 can be generalised to the Earth as a whole, with further assumptions including the chosen composition of the bulk silicate Earth and its layers, and negligible radiogenic heat production from the core. This yields the heat production curve shown in Fig. 12.3, which is normalised relative to present-day global heat *loss*. In the absence of major changes in bulk silicate Earth composition it is inescapable that global heat production in the Archaean was 2.5 to 6 times greater than present-day heat loss (Bickle, 1978; Vlaar *et al.*, 1994).

Also illustrated in Fig. 12.3 are two curves for heat loss from the Earth, normalised against present-day heat loss. The heat loss curve showing no change with time (i.e. at

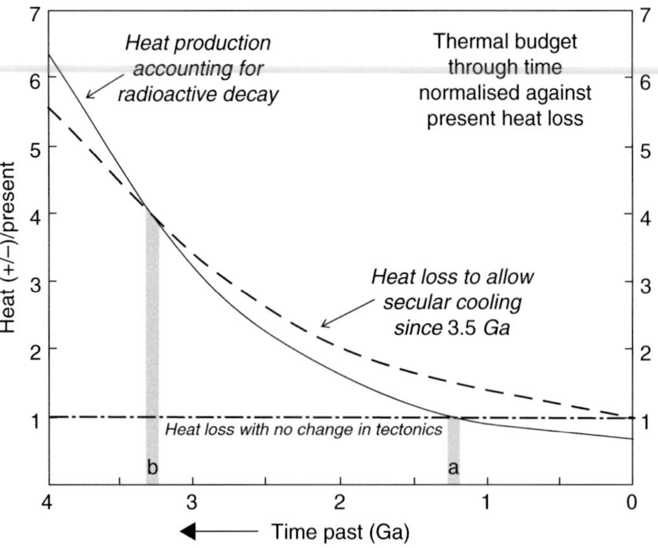

Figure 12.3 Time-dependent heat production calculated for the silicate Earth based on element concentrations given in Table 3.2 and following the approaches described by Bickle (1978), England and Bickle (1984) and Vlaar *et al.* (1994), referenced against present-day heat loss computed from surface heat flow integrated over the Earth (Bickle, 1978). Present-day heat production is 0.7–0.8 present heat loss. Heat loss back in time with no change in tectonics (e.g. rates of plate production and destruction, mantle temperature and material distributions) is simplified as equal to 1 (dash-dot line). For this situation past heat production becomes greater than heat loss in the time interval denoted by the shaded field a, at 1.3–1.2 Ga. The heat loss from the Earth that is required to allow secular cooling since 3.5–3.3 Ga is represented by the schematic dashed line discussed in the text.

normalised heat = 1) assumes no changes in the rates of tectonics and conductive heat loss. This is a simplification, as even with a fully uniformitarian approach and similar rates of plate production and destruction the involvement of a hotter mantle will lead to some increase in heat loss through advection, and thermal conductivities will also vary with temperature. Nevertheless, these changes are likely to be minor in comparison with those required to allow secular cooling from beyond the Mesoproterozoic. The hypothetical heat loss curve that allows for continuous secular cooling since 3.5 Ga and hence a decrease in T_p by 200–300 °C since that time (Fig. 12.3) requires that there have been major changes in the rates and styles of tectonic process on Earth over time. It is likely that those changes will have had an impact on the character and behaviour of orogens and mountain belts.

In order to evaluate the impact of higher heat production in the past, it is necessary to transform the heat production data into heat flow curves and thence into geotherms (Bickle, 1978). As explained in Chapter 3, in order to do this for the crust and thermal lithosphere it is necessary to make assumptions about the distribution of heat production in the crust and the magnitude of the heat flux into the crust from the mantle with time. For conductive geotherms associated with self-heating and thermal relaxation to steady-state conditions it is also necessary to assume or model the thermal conductivities. Figure 12.4a shows a 40 km thick model crust in which most heat production is contained in a 10 km thick

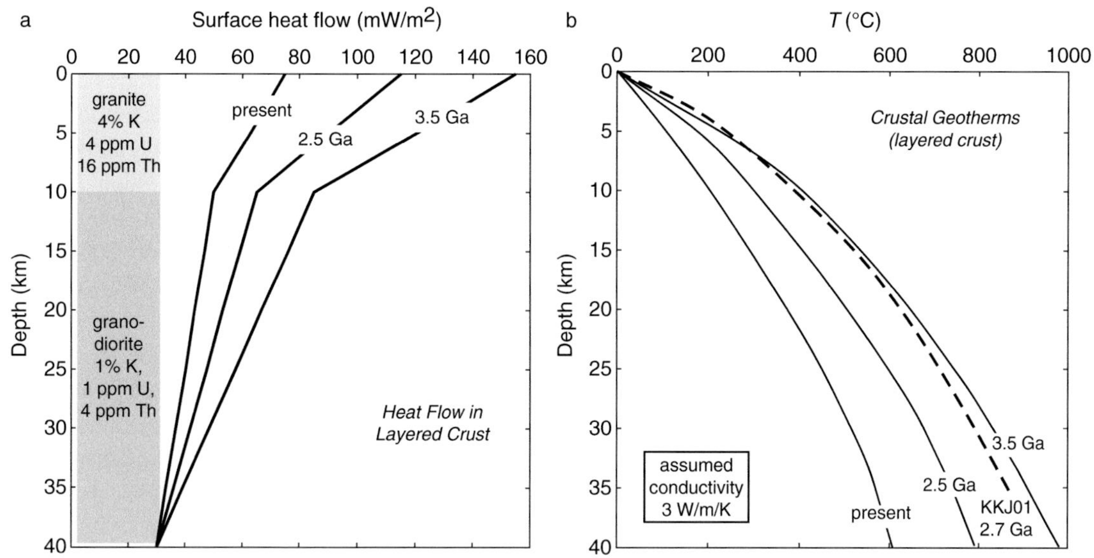

Figure 12.4 a, Surface heat flow in a model crust consisting of a lower layer, 30 km thick, composed of granodiorite, and an upper layer, 10 km thick, made up of granite. The effects of higher heat productions from the crustal layers in the past are shown by the three labelled curves. Surface heat flow for this crustal configuration is doubled at 3.5 Ga. In this simplified illustrative model the mantle heat flow into the crust is fixed at 20 mW/m^2 for all times. Note that depth increases downwards in this diagram. b, Crustal geotherms calculated for the present and at 2.5 Ga and 3.5 Ga for the model crust of (a) assuming a constant thermal conductivity of 3 W/(m K). For a 40 km thick crust the temperature at the Moho at 2.5 Ga is nearly 200 °C higher than the present calculated temperature (600 °C). At 3.5 Ga the Moho temperature is nearly 1000 °C in this simple model example. The geotherm curve labelled KKJ01 is that calculated by Kramers *et al.* (2001) for the Northern Marginal Zone of the Limpopo Belt of southern Africa using measured heat-production values for the exposed rocks and estimates of their volumes. This geotherm has been generated for the heat productions back-calculated to 2.7 Ga, approximating the time of major metamorphism, magmatism and tectonism in the region. The 2.7 Ga KKJ01 conductive geotherm lies at *T*–depth conditions similar to the metamorphic *P*–*T* conditions recorded in granulites from the Northern Marginal Zone (8–9 kbar, 850–900 °C).

granitic upper crust that overlies 30 km of granodiorite. The basal heat flux into the crust from the mantle is conservatively maintained at 20 mW/m^2 at all times, but the crustal heat production is allowed to increase back in time in accordance with the principles of radioactive decay (Fig. 12.1, 12.2). This crustal configuration has a present-day surface heat flow of 74 mW/m^2, but in the Archaean this would have been much higher (even for the same mantle heat flux) – between 110 and 150 mW/m^2 depending on age. Assuming a fixed thermal conductivity of 3 W/(K m) and applying the diffusion equations presented in Chapter 3, we can convert the heat flow versus depth relations of Fig. 12.4a into their corresponding steady-state geotherms for each chosen age (Fig. 12.4b).

The calculations show that conductive geotherms for the model crust, in which the highest heat productions are concentrated near the surface, lead to Moho temperatures exceeding 800 °C in the Archaean, even with mantle heat flux held constant. A 2.7 Ga geotherm calculated based on the present-day measured heat-production values of late Archaean charnockites and 8–10 kbar granulites from the Northern Marginal Zone of the Limpopo Belt of southern Africa (Kramers *et al.*, 2001) is shown in Fig. 12.4b for comparison. The steady-state geotherm in this case overlaps the peak metamorphic *P–T* conditions experienced by those rocks, and would be consistent with the suggestion that radiogenic self-heating in the Archaean could account for repeated low-*P*/high-*T* metamorphism, long-lived (2750–2580 Ma) high-*T* magmatism, and decompression-cooling (geotherm-parallel) *P–T* paths in the Northern Marginal Zone (Kramers *et al.*, 2001).

The discussion above demonstrates that increased heat production in the past will have resulted in hotter conductive geotherms in continental crust, unless that crust was on average and over much of its thickness depleted in K, U and Th relative to modern continents and estimates of average crustal composition (Rudnick & Gao, 2003). Examination of the voluminous Archaean TTG (tonalite–trondhjemite–granodiorite) suites, however, shows that this is not the case, at least for K. The higher geothermal gradients will weaken the crust (Marshak, 1999) and, for Moho temperatures of >800 °C, thermally weaken the mantle immediately beneath the Moho (Rey & Housemann, 2006), with consequences for the long-term support of thickened crust. Moreover, the higher mantle potential temperature is likely to increase the amount of melting in the mantle (Nisbet *et al.*, 1993; Vlaar *et al.*, 1994) and thereby enhance lithospheric weakening during convergence (Sizova *et al.*, 2010). This acts to inhibit subduction, which requires a strong lithosphere, and so provides a secular control on the onset of subduction and hence the onset of the 'modern' style of plate tectonics (Davies, 1992, 2006; Brown, 2006, 2007; Condie & Kroner, 2008; Sizova *et al.*, 2010).

Modelling the onset of subduction and its consequences

The onset of one-sided subduction, with a downgoing oceanic lithosphere slab converging with and descending beneath an adjacent plate so that a wedge of sub-lithospheric mantle occurs laterally and at depth between the two, is generally considered to be the episode in Earth history leading to modern plate tectonics characterised by calc-alkaline island arc magmatism and deformation that is strongly partitioned into narrow plate margin zones (Davies, 1992; Condie & Kroner, 2008). Estimates of the buoyancy of young, hot but thick oceanic crust and the vigour of convection in a hotter mantle have been widely used to argue that slab-pull would be diminished in a hotter early Earth, that as a consequence subduction would not be possible in the early Archaean (Davies, 1992), and that convergence may instead have been accommodated by the internal deformation of melt-weakened lithosphere. Geochemical signatures of TTG suite rocks, including their very low (depleted) heavy rare-earth element (HREE) and yttrium (Y) contents, high Sr/Y and lack of Eu depletion (Drummond & Defant, 1990), have been coupled with experimental work on melting of amphibolite and eclogite to infer the generation of voluminous TTG from the deep-level (>40–50 km) melting of mafic rocks with limited or no involvement of melts derived from the peridotitic mantle (Foley *et al.*, 2002;

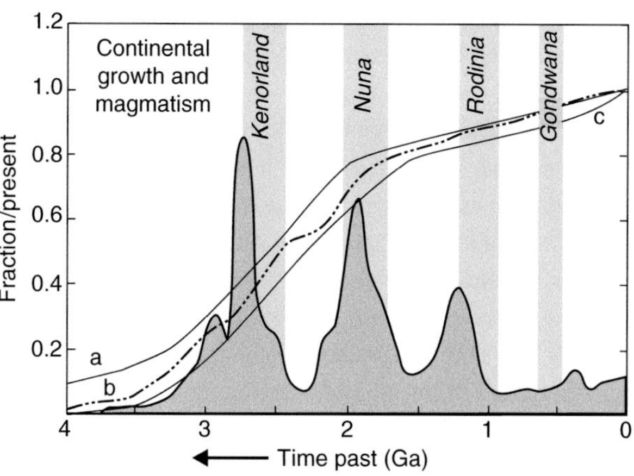

Figure 12.5 Estimates of the growth of continental crust (Rudnick & Gao, 2003; Condie, 2007) compared with proposed supercontinent events and relative abundance histogram of zircons from juvenile granitoid rocks (TTG, calc-alkaline magmas) inferred to reflect new continental crust (Kemp & Hawkesworth, 2003; Hawkesworth & Kemp, 2006a, b). Curves a and c are representative of the range of crustal growth estimates (a – up to 80% by 2 Ga; c – minimal before 3.5 Ga and 60% by 2 Ga). Curve b lies between the two and incorporates the possibility of episodic growth, particularly between 3 Ga and 1.8 Ga (Hawkesworth & Kemp, 2006a, b).

Rapp *et al.*, 2003). It has generally been argued that this melting, whether of high-*P* amphibolite or amphibole-bearing eclogite, could represent 'slab melting' in which the metamorphosed oceanic crust of a lower plate undergoes direct melting as it underthrusts an overlying plate in a flat style of 'hot' subduction or without any intervening mantle wedge (Foley *et al.*, 2003). Hawkesworth and Kemp (2006b) have compared the estimate of an overall intermediate composition for the continental crust with the more basic composition of magmas potentially added to the crust at arcs or in ancient flat subduction melting regimes. From this they have calculated that the prodigious crustal growth evident in the late Archaean (Fig. 12.5) from juvenile zircon Hf isotope signatures in TTG and other magmatic rocks (Hawkesworth & Kemp, 2006a; Kemp *et al.*, 2006) required the production and delamination of an eclogitic residue from melting that is volumetrically as great as the amount of continental crust generated and stabilised through further differentiation.

Sizova *et al.* (2010) have recently applied two-dimensional thermomechanical modelling of layered lithospheres to systematically evaluate the onset of subduction in Earth history. They investigated oceanic subduction in which an incoming oceanic plate converges with a continental one in a retreating subduction mode, with a wedge of marginal sediments present. They considered a range of parameters, and varied those that are believed to be dependent on age and thermal budget: crustal radiogenic heat production (normalised against the present) and mantle temperature (effectively T_p). They conducted a series of numerical experiments in which crustal radiogenic heat production was increased up to 4 times present heat production, and mantle temperature by up to 400 °C above a present value, taken as 1360 °C at 70 km depth, that reflects the present-day T_p. Mantle hydration and serpentinisation, partial melting of the mantle and crust, volcanism and sedimentation are all incorporated into these models, and related secondary variables such as oceanic crust thickness considered for subsets of the models.

The results indicate that, for the retreating subduction/convergent zone model, the style of tectonics changes from modern subduction (the starting boundary condition) through to 'no subduction' via a transitional, mantle-melt weakened, 'pre-subduction' regime that is asymmetric but lacking in any substantial mantle wedge above the lower plate or slab (Fig. 12.6 and Plate 42). For a range of radiogenic heat productions greater than present, the key parameter dictating the change in style is the mantle temperature, or T_p. In these models no subduction is possible if mantle temperature is increased by 200–250 °C. This result arises because of the marked internal deformation and loss of strength of the 'lithosphere', facilitated by the presence of abundant, high-percentage sub-lithospheric melts. A transitional 'pre-subduction' regime occurs when mantle temperatures are 175–250 °C above present T_p. In this regime the weak incoming plate underthrusts the other lithosphere at a shallow angle, without an obvious or persistent mantle wedge, and asymmetric (one-sided) subduction is not sustained. The oceanic plate underthrusts laterally and then further weakens by interaction with mantle melts. The buoyant ascent of melts at higher T_p conditions leads to slab/plate retreat and a form of 'back-arc' development to accommodate the melt-driven extension. It is pointed out that the evolution in the way convergence is accommodated, in particular the modelled geometries of the earlier regimes, can only be indicative as the boundary conditions imposed from the present types of forces acting on plates to some extent condition the model response. Hence the evolution depicted in the cartoon sections presented in Fig. 12.6 (modified from Sizova *et al.*, 2010) should be regarded with some caution.

The metamorphic and orogenic implications of this modelled evolution from a no-subduction to pre-subduction and then subduction regime are significant.

In the 'no-subduction' regime it is not possible to generate HP–UHP metamorphism, as for all time slices examined in the models the temperatures are too high at depths of 40–160 km in the initial weak zone of convergence. No zone of focussed deformation and vertical transfer of material is established, so even where oceanic crust is taken down to >50 km it is unable to be exhumed at low temperature, if at all. Thickening of the continental 'plate' is negligible, even where a deforming oceanic plate underlies it, so E-HPG metamorphism would not be expected. On the other hand, low-*P*/high-*T* metamorphism may occur over short timescales (a few Ma) as the continental crust thins, Moho temperatures increase to above 700 °C and melt-bearing asthenosphere rises to shallow levels. Despite the telescoping of isotherms to yield high-*T* at shallow levels, UHT metamorphism would not be seen in the model continental rocks as these are not thickened and might only experience $T > 700$ °C during thinning and short-lived spreading associated with asthenospheric decompression (Fig. 12.6). If G-UHT conditions were to be achieved, they would be contemporaneous with voluminous basaltic to basaltic komatiite magmatism. The proportion of pre-existing continental crust involved in the resulting G-UHT belt, which might only be exhumed 'accidentally' through the imposition of a younger and unrelated tectonic event, would be minor.

In the pre-subduction and transitional subduction regime, deformation is accommodated across a wide zone involving both plates, with the oceanic one undergoing buckling. The

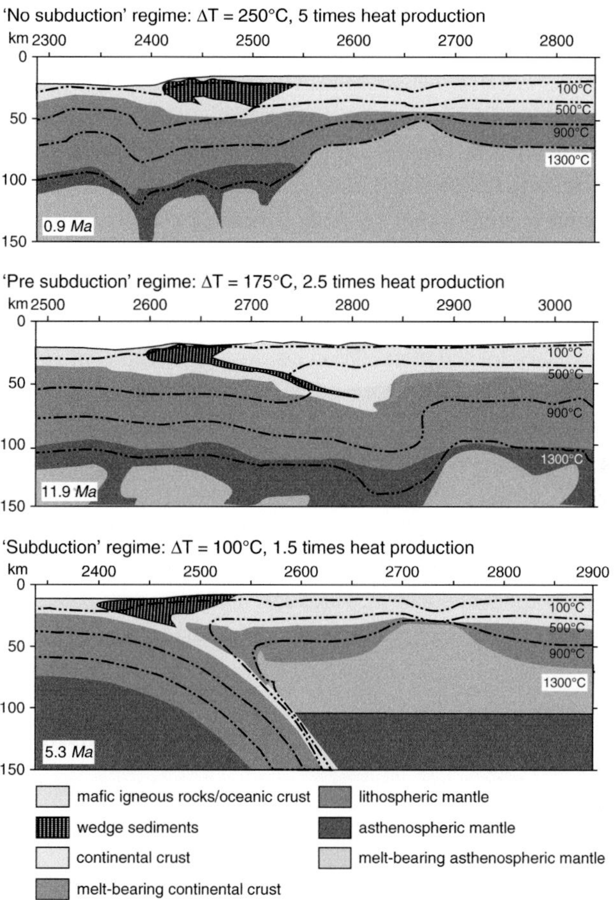

Figure 12.6 Selected 'snapshots' of lithospheric sections produced in the modelling of Sizova *et al.* (2010). Three cases are illustrated. The 'no subduction' case has a mantle temperature higher by 25 °C and crustal heat production 5 times present, and might represent a tectonic scenario for ages >3.3 Ga. The 'pre-subduction' regime may represent a late-Archaean situation, and the 'subduction' regime with higher mantle temperatures and crustal heat production than present may represent the situation since the Neoproterozoic. In each case the vertical axis is depth in kilometres, and the horizontal scale distance from the distal boundary of the model grid in kilometres. Isotherms for 100 °C, 500 °C, 900 °C and 1300 °C are shown by dash-dot lines in each case. See also colour plates section, Plate 42. Redrawn and published with the permission of Elsevier.

initial weak zone forms a low-angle ramp, perhaps 400 km or more in lateral extent, above which up to 50% thickening of the continental crust of the overlying slab may occur (Fig. 12.6). This region of thickening would contain medium-P/T and E-HPG rocks of both continental and oceanic affinities, metamorphosed at 600–800 °C for model depths of 50–70 km. Outside this zone, continental rocks at 30–40 km depths would attain temperatures of 700 °C, or more if TTG magmas derived from the melting of amphibolitic or eclogitic oceanic crust intruded and ponded in the deep crust. G-UHT metamorphism could result from such magmatic contributions to the lower crust. A testable implication of this is that the G-UHT metamorphism would be coeval with abundant juvenile magmatism at deep crustal levels. The involvement of pre-existing continental crust, especially if it undergoes partial melting, would limit the maximum *T* attainable. As in the no-subduction case, because the crust is not thickened the exhumation of the G-UHT deep crust would require a later tectonic event or the operation of complementary processes, such as gravitational redistribution of layered crust (e.g. Gerya *et al.*, 2000, 2001), during convergence.

With the onset of 'modern style' subduction the partitioning of deformation into the subduction zone and establishment of an overlying mantle wedge provides the geometry necessary for the marked downward perturbations of isotherms to deep levels required for the 'cold' HP–UHP metamorphism of subducting oceanic crust, the dehydration of which can fertilise the mantle wedge and initiate melting leading to calc-alkaline magma genesis. In addition, for the retreating subduction model, the upward perturbation of isotherms above decompressing asthenosphere associated with back-arc extension may facilitate low-*P*/high-*T* amphibolite to granulite metamorphism (e.g. Hidaka: Kemp *et al.*, 2007) and, with variations in slab retreat and advance, the development of low-*P*/high-*T* accretionary orogenic belts in which *P*–*T* paths are dominated by large excursions in temperature without much (i.e. greater than a few kbar) change in pressure.

Secular change in metamorphism in orogenic belts

The modelling outlined above emphasises the importance of past mantle temperatures as the primary control on the character of tectonics and metamorphism in convergent settings over time. The metamorphic record from past regional metamorphic belts therefore provides an important line of evidence for consideration of the extents to which ancient orogenies differ from those of the Phanerozoic. This metamorphic perspective on secular change has been examined by several workers. For example, Sandiford (1989a, b) noted that much of the Precambrian metamorphic record is dominated by low-*P*/high-*T* amphibolite to granulite terranes. He recognised the absence of blueschists and proposed secular change consequent on a decrease in T_p by about 200 °C since the late Archaean. The impacts of this were investigated by consideration of differing amounts of thickening of the crust versus the whole continental lithosphere during continental collision, on the assumption that such collisions occurred at least as early as the late Archaean (<3000 Ma). Sandiford (1989a) argued that the lithospheric mantle thermal boundary layer in the Archaean would be broader and more unstable, with a greater propensity to be removed because of more vigorous mantle convection. Removal of the thermal boundary layer during or following thickening would result in significant heating of the crust over a short timescale, leading to G-UHT metamorphism of the deepest rocks in the doubly thickened crust (60–80 km depths). As in the models later developed by Platt and England (1993), the over-thick and hot buoyant crust would undergo gravitational collapse, manifested as sub-horizontal G-UHT fabrics reflecting extension (Sandiford, 1989b). In this model for orogenesis the G-UHT metamorphism

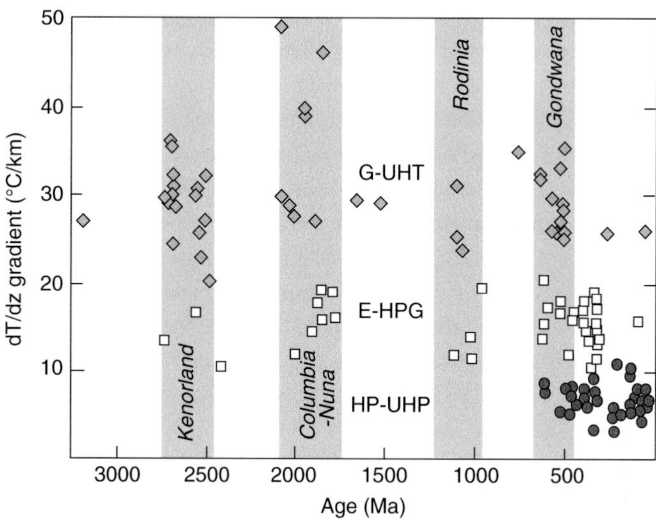

Figure 12.7 Apparent thermal gradient at peak metamorphism, d*T*/d*z*, plotted against age of metamorphism for the three divisions of metamorphism defined by Brown (2006, 2007) and discussed in Chapter 7. Shaded diamonds: G-UHT (low *P*/*T* facies series) metamorphic belts; open squares: E-HPG (medium *P*/*T* facies series) metamorphic belts; dark filled circles: HP–UHP (high *P*/*T* facies series) metamorphic belts. Vertical shaded fields represent proposed crust formation and supercontinent assembly episodes as shown in Fig. 12.5 and discussed in the text. Figure based on Brown (2007), modified to incorporate recent data on UHT metamorphic belts (Harley, 2008; Kelsey, 2008), published with the permission of the Geological Society of America.

event would be short-lived, associated with the extensional structures and terminated by cooling following rapid decompression. As mentioned in Chapter 7, the *P*–*T*–time record of the Cenozoic Betic–Rif orogen has been interpreted in terms of rapid near-isothermal decompression following mantle lithosphere removal.

Brown (2006, 2007) compiled *P*–*T* information from over 200 metamorphic belts or regions and on the basis of their d*T*/d*z* characteristics, using recorded peak conditions, divided them into the HP–UHP, E-HPG and G-UHT groups described in Chapter 7. In order to examine whether there have been secular trends in metamorphism, and hence orogeny, he plotted these groups in terms of the ages of peak metamorphism, insofar as those can be established from geochronology (Chapter 3). The results of his analysis, modified and complemented by new data for some of the belts, are presented here in Fig. 12.7.

Brown (2007) attributed the formation of G-UHT rocks to the closure and thickening of continental back arcs, with high geothermal gradients imposed by the abundance of accreted juvenile crust and shallow-level mantle-derived heat. In this situation where crustal thickening is limited, the *P*–*T* paths would be tight clockwise ones. E-HPG metamorphism is subduction-related but reflects subduction-to-collision orogeny, with 'warm' crustal material choking the subduction zone at moderate depths rather than being transported deep with a 'cold' subduction channel. Brown (2006, 2007) proposed that this 'duality' of thermal regimes has existed back to the late Archaean, and that this duality is the imprint or hallmark of plate tectonics in the sense that subduction and ocean-floor spreading have existed since that time. The duality is between the low-*P*/high-*T* (G-UHT) and medium-*P*/*T* to E-HPG groups, which are recorded in terrains as old as 2700 Ma and at least 2500 Ma respectively (Brown, 2007). The third type of metamorphism, high-*P*/low-*T* or HP–UHP metamorphism, is restricted to the Phanerozoic and Neoproterozoic, even though subduction with the involvement of a mantle wedge is modelled (Sizova *et al.*, 2010), and considered on metamorphic (Brown, 2006) and geochemical (Condie, 2007; Condie & Kroner, 2008) grounds, to have occurred since at least the late Archaean. Brown (2007) suggests that this reflects a change in the typical subduction zone thermal regime to 'cold' subduction and is related to the strengthening of the mantle lithosphere typically involved in plate convergence rather than to any first-order change in the nature of plate tectonics. The presence of strong lithospheric mantle is thought to facilitate the formation of subduction channels, processing of down-going crust to UHP depths, and focussed vertical displacement of rock packages at convergent plate boundaries (Gerya *et al.*, 2008).

The central observation in the analysis by Brown (2006, 2007) is that E-HPG and G-UHT metamorphic belts have been forming on Earth since at least the mid to late Archaean. He, and workers tracing the growth of continental crust (Hawkesworth & Kemp, 2006a; Kemp *et al.*, 2006) using zircon U–Pb and Hf isotopes, have linked the formation and preservation of these metamorphic belts with the assembly of past supercontinents through subduction-to-collision orogenesis and the associated amalgamation and stabilisation of previously distinct crustal blocks and cratons. Four proposed supercontinent assembly episodes (Kenorland, Nuna, Rodinia, Gondwana), some of which span several hundred million years, are compared with the metamorphic information compiled by Brown (2006, 2007) in Fig. 12.7. They are also shown in comparison to proposed growth curves for continental crust and summary histograms of zircon ages (Hawkesworth and Kemp, 2006a) in Fig. 12.5. Whilst there is a good correlation between the ages of E-HPG and G-UHT metamorphic belts and the assembly of supercontinents this is to some extent inevitable as the construction and definition of the pre-Gondwana supercontinents mostly rely on the metamorphic, structural and isotopic data from the exposed G-UHT and E-HPG belts themselves in the absence of sea-floor spreading records and ambiguities with palaeolongitudes for ancient crustal blocks. On the other hand, the marked differences in proportions of juvenile crust added to the continents during the episodes correlated with the dualistic metamorphic belts (Fig. 12.5: compare Kenorland and Nuna with Rodinia and Gondwana) suggest that even if assembly and amalgamation of different crustal domains through accretion and collision occurred at these times there had to be differences in how those domains formed and evolved prior to and possibly during assembly.

Accretion, amalgamation and collision from the Palaeoproterozoic to NeoArchaean

The following section presents selected evidence from ancient metamorphic belts that amalgamation and collision were central to orogenic processes back in time at least into the late Archaean. In order to determine the complex geological histories involved it is necessary to apply criteria derived from cross-cutting relationships, zircon and other accessory mineral U–Pb age data, and isotopic signatures such as the Hf isotopic compositions of zircon and Nd isotopic compositions of bulk rocks that enable the role of reworking to be evaluated and distinct crustal provinces to be recognised. The key criteria and evidence used to distinguish the different crustal domains, blocks or terranes that have later been amalgamated or 'stitched' together include distinctions in protolith ages of TTG suites; differences in the ages of early metamorphic and deformation events; and the presence versus absence of specific or uniquely 'fingerprinted' rock units. The key lines of evidence used to constrain the timing of amalgamation of the crustal blocks or terranes include the presence in two or more terranes of rock units of equivalent ages and character, and the presence of shared metamorphic events, with similar style and grades (Nutman *et al.*, 1993, 1996; Friend & Kinny, 1995).

The Trans-Hudson Orogeny of Canada

St-Onge *et al.* (2006) have synthesised the large-scale crustal architecture and evolution of the 1885–1795 Ma Trans-Hudson Orogen and compared its *c.*90 Ma record of terrane accretion and amalgamation followed by collision and post-collisional leucogranite magmatism to the Cenozoic history of the western Himalaya. The Trans-Hudson Orogen is a large orogen, some 4600 km long and up to 800 km in width, formed through 1820–1795 Ma collision of the Superior Craton, interpreted as a lower plate akin to India in the modern system, with a composite upper plate that was assembled through a series of pre-collisional accretionary and amalgamation events. These occurred over the time interval 1885–1845 Ma in the northern Quebec to Baffin Island sector of the orogen, where the composite upper plate, termed the Churchill plate (St-Onge *et al.*, 2006), was formed in at least two stages, firstly by accretion of a microcontinent to the Rae Craton at 1883–1865 Ma and secondly by accretion of an intra-oceanic island arc to this combined province at 1845–1842 Ma. Similar types of events occurred to the southwest where the Superior Craton collided against other cratons (Wyoming) or accreted belts (Hearne domain). In the reworked Superior Craton, 1820–1790 Ma lower plate metamorphic events associated with the collision were of medium-P/T character, with early kyanite–sillimanite grades in pelites. Upper plate metamorphism related to collision occurred under medium-P/T amphibolite conditions, and was retrograde with respect to earlier 1850–1835 Ma granulite assemblages associated with Andean-style magmatism that followed arc accretion. This well-documented sequential deformational and metamorphic history supported by detailed geochronology demonstrates the existence of distinct pre-collisional events in different domains, later metamorphism and deformation in the Superior Craton contemporaneous with that in the adjacent but distinct crustal blocks, and late-stage emplacement of leucogranites formed by crustal melting. These features strongly support the interpretation of the Trans-Hudson Orogen as a collisional orogenic belt involving continent–continent collision similar in style and character to the present day. St-Onge *et al.* (2006) argue that these observations extend the applicability of tectonics similar in most respects to Phanerozoic plate tectonics at least back to 1900 Ma.

The Lewisian of northwest Scotland

Geochronological studies principally utilising mineral (zircon, monazite, titanite) U–Pb SHRIMP methods (Friend & Kinny, 1995, 2001; Kinny & Friend, 1997; Corfu *et al.*, 1998; Kinny *et al.*, 2005; Love *et al.*, 2010) have resulted in the development and elaboration of a terrane model for the Lewisian of northwest Scotland. This complex is now considered to comprise a number of formerly distinct crustal blocks or terranes amalgamated in events younger than 2490 Ma and in most cases between 1870 Ma and 1670 Ma. The number of separate former terranes or distinct blocks is the subject of considerable debate as much of the evidence relies on the interpretation of complex and 'smeared out' zircon U–Pb age data and absence of zircon age records of specific events (Corfu *et al.*, 1998; Friend & Kinny, 2001). In addition, because zircon may be unresponsive to events in the deep crust that do not involve extensive fluid interaction or melt production (Harley *et al.*, 2007) it is not straightforward to evaluate the regional extents of those events that appear to be missing from some areas but present in others. Given these caveats, the rather different TTG protolith ages and distinct ages of early metamorphic events in the granulite facies Assynt and amphibolite facies Rhiconich terranes of the northern part of the mainland Lewisian strongly suggests that they were distinct crustal blocks prior to an amalgamation episode which must pre-date 1740 Ma, the age of a shared late-stage metamorphic overprint (Kinny *et al.*, 2005). The Assynt terrane is proposed to be distinct from the Gruinard terrane to its south by Kinny *et al.* (2005) on the basis of the preservation of 2730 Ma metamorphism and absence of 2490 Ma zircon rims in the latter, though the presence of comparable cross-cutting mafic dyke swarms, the famous Scourie dykes, in both terranes points to their amalgamation prior to dyke emplacement, which probably occurred by 2000 Ma. At least two suites of oceanic sediments and a 1890–1875 Ma Palaeoproterozoic arc complex (Baba, 1998), the South Harris Igneous Complex, were accreted to a similar Archaean crust block on the Outer Hebrides by 1700–1660 Ma (Kelly *et al.*, 2008), producing near-UHT granulites (11 kbar, 880 °C) with post-peak ITD P–T paths (Baba, 1998; Hollis *et al.*, 2006) that reflect the exhumation of the accretionary zone as collision ensued.

The Archaean Gneiss Complex of southwest Greenland

The Archaean block of southwest Greenland, part of the North Atlantic craton, is famous for the presence of some of Earth's oldest rocks, including the dominantly 3650–3700 Ma Amitsoq and similar TTG orthogneisses derived from intrusive magmatic precursors, and the oldest rocks of sedimentary-volcanic origin, the 3700 Ma and 3800–3870 Ma Isua Supracrustals and Akilia association (Whitehouse *et al.*, 1999; Friend & Nutman, 2005a; Whitehouse & Kamber, 2005) (Plate 43). However, it is the well-documented late Archaean geological evolution that is of significance here for the evaluation of orogeny associated with accretion and collision beyond the Proterozoic.

Within what had previously been considered to be one large but complex gneiss region, Nutman *et al.* (1993, 1996) identified tectonic boundaries, associated with high-strain zones, separating lithological units with distinctive age relations and, in some instances, metamorphic histories and grades. Although there are complexities in the details of the number of distinct crustal blocks in the region, and the number and timing of the tectonic boundaries between them, it is generally agreed that at least three major boundaries separate the region into at least four separate terranes, each of which had distinct histories prior to their amalgamation under high-grade metamorphic conditions. There were at least two amalgamation episodes relatively late in the overall history of the area, culminating in events at 2720 Ma and through 2650 Ma. A central terrane complex incorporating early Archaean supracrustals and orthogneisses, for example the Isua area and the 3800–3300 Ma Færingehavn terrane, and late Archaean 2825 Ma felsic gneisses (Tre Brødre terrane) was amalgamated and deformed into a nappe complex by 2720 Ma (Plate 44), contemporaneous with barrovian staurolite–kyanite to sillimanite-grade metamorphism in overriding and adjacent terranes (Friend & Nutman, 2005b). The Akia terrane, dominated by 3070–3000 Ma orthogneisses that experienced their initial metamorphism by 2970 Ma, and the broadly similar 3100 Ma Kapsillik terrane, were most likely amalgamated and in part interleaved with the combined terranes mentioned above at 2720 Ma, though it is likely from metamorphic ages obtained on monazites and zircons that terrane juxtaposition or reactivation continued on to 2650 Ma. The mainly late Archaean (2960–2840 Ma) Tasiusarsuaq terrane, which experienced granulite metamorphism at 2795 Ma, was thrusted over and accreted to the combined Færingehavn and Tre Brødre terranes, also at 2720–2700 Ma. On the basis of the evidence from the Nuuk region as a whole the *c.*2720 Ma tectonothermal event reflects collisional orogeny in the late Archaean. Subsequent metamorphism and deformation recorded in some of the terranes at 2650 Ma reflect post-amalgamation reactivation, shearing and refolding focussed along the previously established high-strain zones that bound the terranes.

Weak lithospheres and 'ultrahot' orogens

The examples briefly described in the preceding section demonstrate the likelihood of potentially modern-style collisional orogeny back into the late Archaean at least (Plates 45, 46). This is compatible with the modelling of the onset of subduction in late Archaean times (Sizova *et al.*, 2010) and with the synthesis of metamorphic thermal gradient data presented by Brown (2006, 2007). Equally, this does not exclude the possibility that some orogenic belts were rather different in character from the orogens that dominate the past 1000 Ma of Earth history, given the likely secular change in T_p and its consequences for lithospheric strength. The evidence for, and conceptual modelling of, low-topography orogens that may have been more common on an Earth with weaker lithospheres is examined and discussed in terms of this secular change in this concluding section.

The deformational pattern of the amalgamation-dominated belts described in the previous section is one of strain partitioned into relatively narrow shear zones in thick- and thin-skinned thrust complexes, leading to high-strain zones that may trace out deformed boundaries between very distinct crustal units and blocks, as in west Greenland.

This pattern is not shared by all ancient orogens. Distributed deformation patterns have been observed to typify a number of Precambrian regional metamorphic belts and Archaean granite–greenstone terrains (Bouhallier *et al.*, 1993; Choukroune *et al.*, 1995, 1997; Chardon *et al.*, 1996, 2009). It has further been proposed that such distributed deformation is a characteristic of orogenesis involving hot, weak lithospheres – 'ultrahot orogens' (UHO: Chardon *et al.*, 2009; Gapais *et al.*, 2009) characterised by rapid accretionary crustal growth, juvenile magmatism and high geothermal gradients. Higher T_p and geothermal gradient conditions would suggest that weak lithospheres were more common in the Archaean (Marshak, 1999; Rey & Houseman, 2006), and it could be inferred that UHO might be more common too. This is not to exclude the possibility that UHO may have formed in the Phanerozoic: for example it has been suggested that low-*P* accretionary orogenesis in the Lachlan Fold Belt is a type of UHO (Chardon *et al.*, 2009). In the Archaean, weaker lithosphere may have approached low viscosities consistent with the 'crème brûlée' model of Rey and Houseman (2006). If initial Moho temperatures exceeded 800 °C this may have facilitated distributed deformation, and for even higher Moho temperatures (>900 °C: Chardon *et al.*, 2009) an UHO would result, characterised by a mountain topography suppressed to elevations of 1.5–2 km, and accommodation of crustal accretion and growth by combined thickening and lateral spreading *during* convergence.

Large-scale flow with both horizontal and vertical flow components, and homogeneous thickening involving sagduction, diapirism and gravitational overturn or redistribution would be expected in this type of orogen (Chardon *et al.*, 2009). Based on observations from the Dharwar Craton of India, the Limpopo Belt, and some Archaean granite-greenstone belts, Cagnard *et al.* (2006) and Chardon *et al.* (2009) suggest that UHO structures are dominated by anastomosing transpressive shear zones that bound lensoidal domains which preserve either flat or steep fabrics (Fig. 12.8). These lensoidal regions may be on scales of kilometres to hundreds of kilometres. The shear zones may preserve vertical shear perpendicular to the main structural grain of the lenses, though oblique slip may also occur. Highly connected shear domains, especially in the upper crust, may show conjugate shearing in the horizontal plane. This is inferred to reflect orogen-parallel stretching synchronous or alternating with convergence (Fig. 12.8). Internal sagduction may occur in the lower parts of the upper crust of the UHO, stimulated by dense, possibly melt-depleted, supracrustals sinking into the weak layer beneath (Chardon *et al.*, 2009). This may be driven by inverted density stratification (e.g. Gerya *et al.*, 2002), perhaps acting in combination with coeval horizontal shortening that drives upward (migmatite dome) and downward (supracrustal keel) extrusion. In the lower crust, composite fabrics, sharing a sub-horizontal stretching lineation, predominate in the lenses between the steep high-strain zones. All such lower crustal structures would form contemporaneously with partial melting.

In the UHO model proposed by Chardon *et al.* (2009), the upper and lower crustal deformation regimes are connected across an attachment zone in the middle crust (12–20 km depths: Fig. 12.8) that accommodates the change in strain patterns. The attachment zone partitions strain between the upper and lower crustal domains but is not permanent. For very weak and buoyant deep crust the structures above the attachment zone may not pass through it, in which case the deep crust will be dominated by sub-horizontal planar

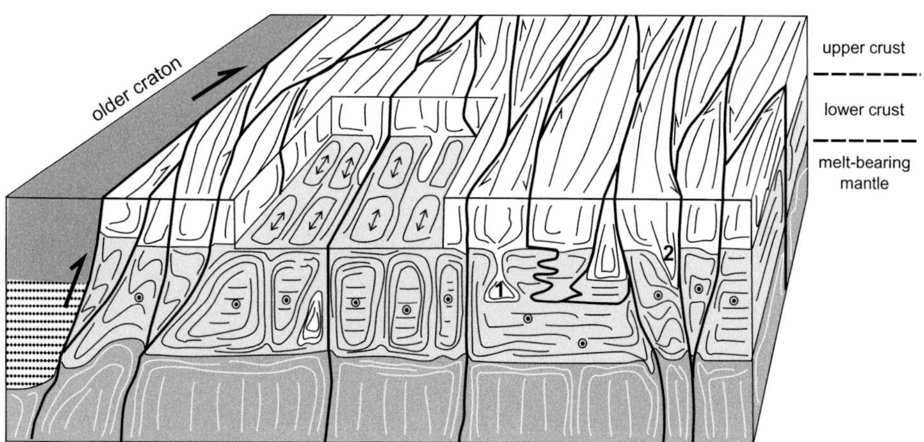

Figure 12.8 Schematic block diagram of fabric patterns and kinematics in an accretionary ultrahot orogen (UHO) as envisaged by Chardon *et al.* (2009). In this model, deformation in the upper crust is linked to that in the lower crust via the sub-vertical high-strain zones that may penetrate into the uppermost mantle, though this linkage may not be present in cases with weaker lower crust and mantle. Local sagduction is shown as occurring both localised along shear zones and by sagging of upper crust into weak lower crust independent of these shears. Lower crustal regions that are not penetrated by sub-vertical shears will be dominated over large regions by sub-horizontal foliations and mineral elongation lineations. Melt-rich mantle is located immediately beneath the weak lower crust, and it is assumed that there is little strength contrast between them for Moho temperatures in excess of 900 °C. *P–T* paths for points 1 and 2 are explained in Fig. 12.9. Figure modified from Chardon *et al.* (2009). See also colour plates section, Plate 47. Published by permission of Elsevier.

and linear fabrics. If the deep crust is not quite so weak it is argued that some structures may pass through the attachment and so partition lower crust deformation transiently into cells bounded by zones of steep orogen-parallel flow, in which internal vertical mass transfer (e.g. sagduction and diapirism) can occur independently of that in other cells. This is the situation depicted for much of the deep crust shown in Fig. 12.8.

In terms of their metamorphism the UHO would be typified by large areas of low-*P*/high-*T* metamorphism, and isograds sub-parallel to the land surface. The metamorphism is long-lived or involves groups of shorter-lived events within a long-lived overall evolution (e.g. 80–100 Ma) in which metamorphism is broadly contemporaneous with magmatic accretion. Metamorphic *P–T* paths in UHO will vary depending on the evolution of the deformation cell (Fig. 12.8) – whether the rock package stays in the deep crust or is extruded into the attachment zone and vertically upwards. If the rocks stay deep they will eventually undergo IBC. If they enter the attachment then IBC and decompression-cooling will occur (depending on extrusion, or incorporation into shear zone walls). Looped *P–T* paths characterised by small changes in *P* through at most a few kbars and by excursions to high-*T* and UHT will result in this case (Fig. 12.9). Gapais *et al.* (2009) argue that rocks staying within their crustal panel or cell will track out post-peak *P–T* paths sub-parallel to geotherms, or decompression-cooling paths. The small loop in *P* in the model *P–T* paths shown in Fig. 12.9 arises as a result of syn-collisional accretion being coupled with spreading at all stages of orogen growth.

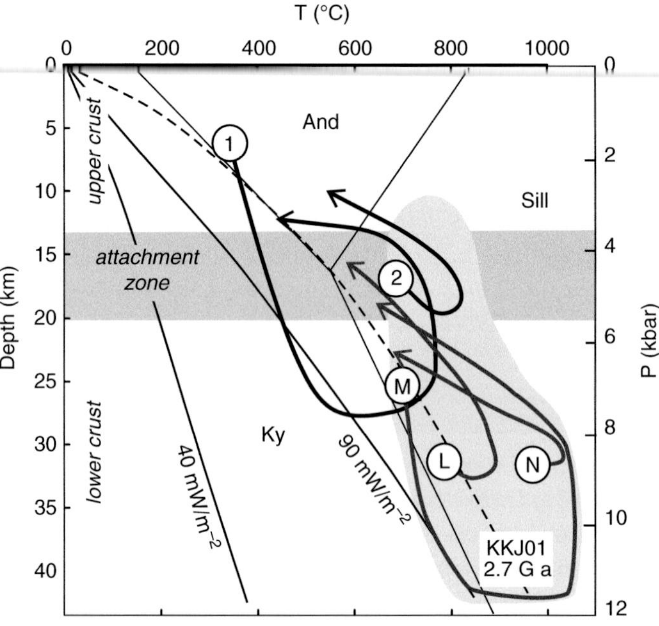

Figure 12.9 Temperature–depth diagram illustrating the *P–T* conditions and histories that may be developed in an ultrahot orogen (UHO) similar to that depicted in Fig. 12.8. *P–T* path 1 is that of a supracrustal sequence from the upper crust descending by sagduction into the lower crust, through the attachment zone (point 1 in Fig. 12.8), followed by upward extrusion, perhaps by incorporation in a subvertical high-strain zone. *P–T* path 2 is a geotherm-parallel path (Cagnard *et al.*, 2006) that might be recorded by material that is buried but does not go beneath the attachment zone. *P–T* paths for three selected HT-UHT metamorphic regions are shown for comparison with these UHO paths. M: 590 Ma UHT-ITD granulites of the Mather Supracrustals, Antarctica (Harley, 1998a, 2008); L: 2700–2650 Ma HT-ITD granulites from the Northern Marginal Zone (NMZ) of the Limpopo Belt (Kramers *et al.*, 2001); N: 2590–2510 Ma UHT-IBC granulites of the Napier Complex, Antarctica (Harley, 1998a, 2008; Kelly & Harley, 2005). Stippled field depicts the *P–T* conditions determined for Proterozoic and late Archaean HT–UHT metamorphic regions based on the compilations of Harley (1992), Brown (2007) and Kelsey (2008). Two geotherms characterised by their surface heat flows (mW/m^2) are added, along with the Limpopo Belt NMZ geotherm calculated by Kramers *et al.* (2001) [KKJ01].

The UHO concept bears some similarities to models for Archaean hot orogens that propose a major role for gravitational instability and overturn (Gerya *et al.*, 2001, 2002) as important in the exhumation of UHT granulites from the *P–T* conditions shown by the shaded field in Fig. 12.9. A key difference, however, is that in the latter models there is no mid-crustal attachment zone that acts as a limit on the amounts and extents of sagduction and vertical extrusion. Gerya *et al.* (2002) have examined the geothermal and rheological consequences of forming multi-layer crusts, stacked for example through accretion or volcanism, and allowing these to self-heat. Where lower-density felsic rocks underlie denser mafics and supracrustals, the geometry eventually becomes gravitationally unstable and ripe for reorganisation or overturn, especially when Moho temperatures are high. Gerya *et al.* (2002) show that Archaean heat production parameters lead to high transient geotherms that may reduce the viscosities of crustal rocks enough to enable density-driven sagduction and complementary diapirism. This model of gravitational redistribution was applied to the Limpopo Belt and argued to be an effective mechanism for production of large-scale domes and keels, and relatively rapid exhumation of medium-*P* granulites.

Gerya *et al.* (2002) have examined the *P–T* path consequences of gravitational redistribution on near-normal thickness crust. They show that geotherm-parallel retrograde *P–T* paths involving decompression-cooling or a combination of this with IBC can be generated, with exhumation at 0.25 cm/a (25 km in 10 Ma) and cooling through 400–500 °C over a time interval of only 9 Ma from peak conditions of 7.5–8.5 kbar and 800–900 °C. The ascent of granulite diapirs and adjacent down-sagging of overlying greenstones gives rise to lateral heat exchange leading to this relatively rapid cooling with decompression in the granulites.

Notwithstanding the conceptual and numerical modelling that supports the general proposition that hot orogens may occur, and have been more common in the Archaean because of the higher geothermal gradients, it is not obvious that the hottest UHT metamorphic belt on Earth, the Napier Complex of East Antarctica, can be interpreted in terms of the UHO and related gravitational instability models.

The Napier Complex preserves UHT at >1000 °C on a regional scale and over a significant syn-metamorphic thickness of crust (*c.*15 km) at pressures corresponding to the deep crust (7–11 kbar). Metamorphism is syn-deformational, and that deformation is dominated by strong flat-lying fabrics with layer-parallel extension or flattening, mineral elongation lineations and lineation-parallel boudinage, chocolate-tablet boudinage, and parallelism of observed partial melt veins with layering. Supracrustal sequences occur interleaved with orthogneisses. It preserves a post-peak *P–T* path dominated by near-isobaric cooling (IBC: Fig. 12.9) following minor near-peak decompression, and was exhumed by orogenies that are much younger than and unrelated to that which formed the UHT granulites (Harley, 1998b, 2008). In terms of timing and overall timescale of metamorphism, the UHT event and subsequent IBC are long-lived. UHT conditions at $T > 900$ °C lasted for at least 80 Ma, from 2590 Ma to 2510 Ma, and the granulites only cooled below 800 °C by 2480 Ma. Combining the zircon age data with lower-*T* thermochronometers leads to an estimate of protracted slow cooling for 200 Ma at only 2 °C/Ma (Kelly & Harley, 2005; Harley, 2008).

The features briefly described above are reconcilable with the UHO model of Chardon *et al.* (2009) applied to rocks that reside in the deep crust beneath the attachment zone, especially in the case where bulk plane strain dominates and acts on a buoyant, weak lower crust. However, unlike the UHO model, the Napier Complex has little juvenile crust at the time of tectonism, its crustal section being dominated by 2840 Ma and older (3850 Ma, 3300 Ma and 3000 Ma) TTG orthogneisses. Whilst 2610–2590 Ma magmatic protoliths occur, these appear to be very minor components of the Napier Complex and are not juvenile crustal contributions in terms of their isotope geochemistry. In short, although the UHT Napier Complex does have some of the properties of a UHO, not least in terms of the geothermal gradient and temperatures at 30–40 km in the crust, it does not correspond to an accretionary UHO in the sense proposed by Chardon *et al.* (2009) as it lacks the long-lived juvenile magmatism and marked acceleration in crustal growth that is regarded as characteristic of UHO. Indeed, given the minimal crustal growth that appears to be associated with this outstanding UHT metamorphic belt, the Napier Complex may instead represent the oldest, and hottest, example of the deep crust trapped within a collisional 'hot orogen', albeit one that had low elevation compared with present orogenic plateaux.

Further reading

Baldwin, J.A., Bowring, S.A., Williams, W.L. and Williams, I.S. (2004). Eclogites of the Snowbird tectonic zone: petrological and U-Pb geochronological evidence for Paleoproterozoic high-pressure metamorphism in the western Canadian Shield. *Contributions to Mineralogy and Petrology*, **147**, 528–548.

Bohlen, S.R. (1991). On the formation of granulites. *Journal of Metamorphic Geology*, **9**, 223–229.

Collins, W.J. and Vernon, R.H. (1991). Orogeny associated with anticlockwise *P–T–t* paths: evidence from low-*P*, high-*T* metamorphic terranes in the Arunta inlier, central Australia. *Geology*, **19**, 835–838.

Collins, W.J., Van Kranendonk, M.J. and Teyssier, C. (1998). Partial convective overturn of Archaean crust in the East Pilbara craton, Western Australia: driving mechanisms and tectonic implications. *Journal of Structural Geology*, **20**, 1405–1424.

Duclaux, G., Rey, P., Guillot, S. and Ménot, R.-P. (2007). Orogen-parallel flow during continental convergence: numerical experiments and Archean field examples. *Geology*, **35**, 715–718.

Goscombe, B.D. and Gray, D.R. (2008). Structure and strain variation at mid-crustal levels in a transpressional orogen: a review of Kaoko Belt structure and the character of West Gondwana amalgamation and dispersal. *Gondwana Research*, **13**, 45–85.

Goscombe, B.D., Gray, D. and Hand, M. (2005). Extrusional tectonics in the core of a transpressional orogen; the Kaoko Belt, Namibia. *Journal of Petrology*, **46**, 1203–1241.

Martelat, J.-E., Lardeaux, J.-M., Nicollet, C. and Rakotondrazafy, R. (2000). Strain pattern and late Precambrian deformation history in Southern Madagascar. *Precambrian Research*, **102**, 1–20.

Rey, P., Vanderhaeghe, O. and Teyssier, C. (2001). Gravitational collapse of the continental crust: definition, regimes, and modes. *Tectonophysics*, **342**, 435–449.

Sandiford, M. and Powell, R. (1990). Some isostatic and thermal consequences of the vertical strain geometry in convergent orogens. *Earth and Planetary Science Letters*, **98**, 154–165.

Santosh, M. and Sajeev, K. (2006). Anticlockwise evolution of ultrahigh-temperature granulites within a continental collision zone in southern India. *Lithos*, **92**, 447–464.

Santosh, M. and Kusky, T. (2010) Origin of paired high pressure-ultrahigh-temperature orogens: a ridge subduction and slab window model. *Terra Nova*, **22**, 35–42.

Warren, R.G. and Ellis, D.J. (1996). Mantle underplating, granite tectonics, and metamorphic *P–T–t* paths. *Geology*, **24**, 663–666.

Windley, B.F. (1992). Proterozoic collisional and accretionary orogens. In: *Proterozoic Crustal Evolution*, ed. K.C. Condie. Amsterdam: Elsevier, pp. 419–446.

b

a

Plate 1 a. The Moine thrust at Knockan Crag, Assynt, Scotland. Moine schists have been placed on top of Cambro-Ordovician dolomites. This is probably the most visited exposure of the Moine thrust and there is a visitor centre, but the thrust here is a late brittle fault with breccias of mylonitic rocks. b. The ductile Moine thrust (arrow) at the Stack of Glencoul, north Assynt. Dark Moine mylonites rest on highly deformed Cambrian Pipe Rock in which the Skolithos Pipes have rotated into near parallelism with the bedding. Photographs by Michael Johnson.

Plate 2 Beinn Eighe in the Moine thrust zone, Kinlochewe, NW Scotland. Red Torridonian rocks (Neoproterozoic) thrust over white Cambrian Quartzite in a duplex or imbricate zone. Photograph by Michael Johnson.

Plate 3 Heart Mountain in the Canadian Rocky Mountain fold and thrust belt. The Exshaw thrust, shown by a heavy line, has been rotated through the vertical by late movements from an original dip to the west (right side on the photograph). Photograph by Michael Johnson.

Plate 4 Mount Yamnuska in the Canadian Rocky Mountain fold and thrust belt, showing a large ramp-like structure in the McConnell thrust. The hanging wall is Cambrian carbonate, the footwall late Mesozoic shales etc. Photograph by Michael Johnson.

Plate 5 Mount Kidd in the Canadian Rocky Mountain fold and thrust belt showing the termination of the Lewis thrust which runs along the lower part of the mountain on the left hand side of the photograph. Across the valley the Lewis thrust is absent and is replaced by chevron folds (the "ductile bead" of Elliott). Photograph by Michael Johnson.

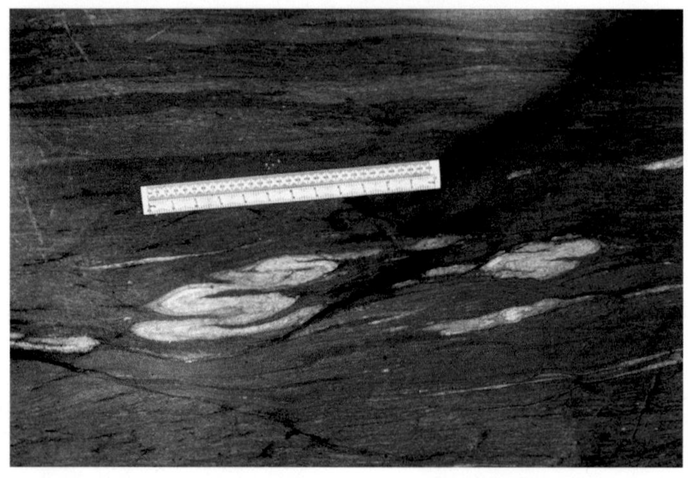

Plate 6 Intrafolial folds in mylonite: axes are near-parallel to the stretching direction. Vogogna, Insubric Line. Photograph by Michael Johnson.

Plate 7 Mylonite-Gneiss, S-L tectonite, from the Main Central Thrust Zone, Joshimath, India. Photograph by Michael Johnson.

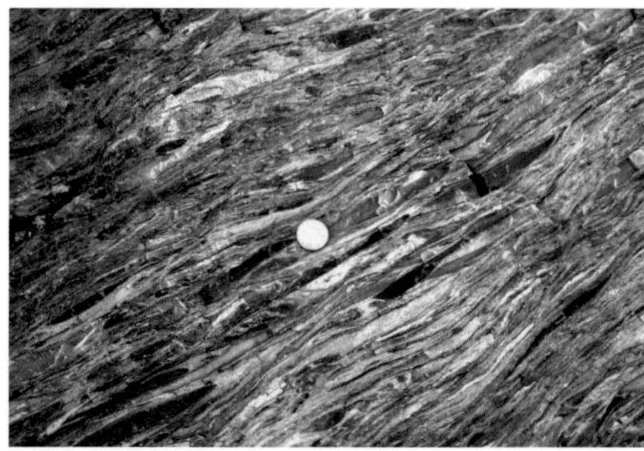

Plate 8 Bygdin Conglomerate, footwall of the Jotunheim thrust, Jotunheim, Norwegian Caledonides. The pebbles are flattened into cake-like (oblate ellipsoids) objects or into rods (prolate ellipsoids). Photograph by Michael Johnson.

Plate 9 Canadian Rockies, foothills showing the McConnell thrust (arrow) which places Lower Palaeozoic onto Cretaceous foredeep deposits. The thrust runs roughly at the base of the cliffs. Photograph by Michael Johnson.

Plate 10 View northwards across the Lesser Himalaya of the High Himalaya showing the steep topographic front. The Main Central Thrust is situated roughly towards the base of the front. The distance between the viewpoint and the High Himalaya is *c.*80 km. Photograph by Michael Johnson.

Plate 11 View northwards showing part of the Annapurna range in Nepal. The Main Central Thrust is situated in the lower slopes. The difference of elevation between the viewpoint and the summits is *c.*6–7 km. Photograph by Michael Johnson.

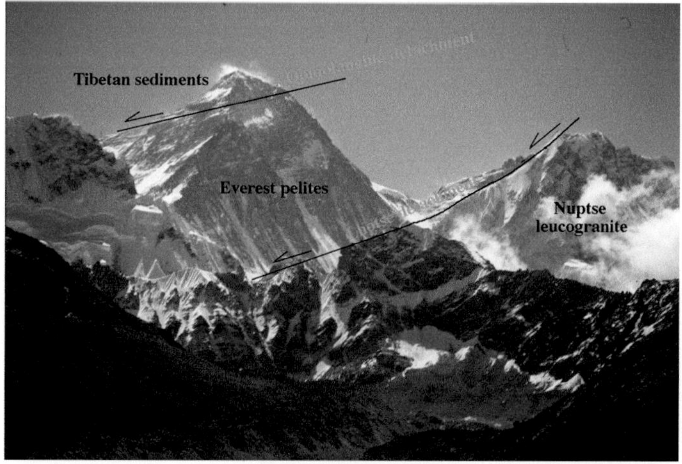

Plate 12 Everest in the eastern Himalaya showing two north-dipping detachments along which extension has occurred. The upper Quomolongmo Detachment above the yellow band of Ordovician age, may be equivalent to the South Tibetan Detachment seen in most of the Himalaya. Photograph by M.P. Searle.

Plate 13 Satellite view looking northwest of the Himalaya and the Tibetan Plateau. The along strike distance is *c*.2000 km. The Indus–Tsangpo suture is marked by the Indus and Tsangpo rivers to the north of the high peaks. Note the late N.S. grabens in the Himalaya and Tibet (one occupied by a lake). To the south of the Himalaya is the Ganges basin of the Foreland. NASA photograph.

Plate 14 Aerial view of the Dent de Morcles in the Helvetic Alps, Switzerland. The large northwesterly verging fold is in Mesozoic rocks in the Morcles Nappe. The Morcles thrust is in the lower part of the photograph. Photograph by J.G. Ramsay who has kindly given permission to use it.

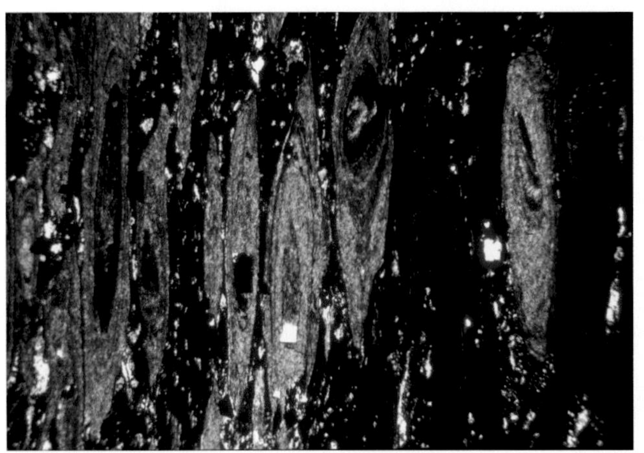

Plate 15 Deformed chamosite ooids from the Helvetic Alps, Windgallen, Switzerland. The longest dimension of the ooids (X axis of the deformation ellipsoid) is roughly perpendicular to the fold axes. See Figure 5.20. Photograph by Michael Johnson.

Plate 16 The Mythen in the external zone of the Swiss Alps. PreAlpine klippe with Upper Trias to Cretaceous sediments, mainly carbonates, exposed in the cliffs resting on Eocene flysch of the Helvetic nappes, in the tree-and grass-covered lower slopes. It is a far travelled nappe derived from the internal zone (Pennine Alps) Panoramic image from 16 frames created by Hannes Rost, courtesy of Wikimedia Commons.

Plate 17 The Gibraltar orocline looking to the east, with the Betic Cordillera of southern Spain to the north of the Straits of Gibraltar and the Rif mountains of Morocco to the south. Photograph by NASA.

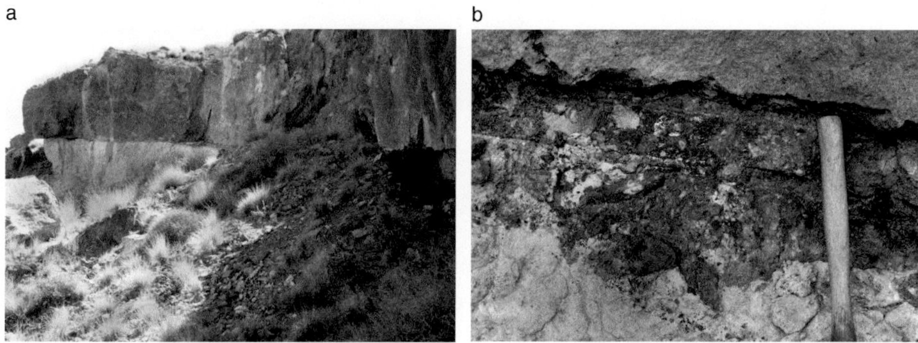

Plate 19 a and b The Keystone thrust in southern Nevada, USA. The thrust is part of the Sevier orogenic belt, and it has emerged at the Earth's surface and travelled across river gravels. Cambrian rocks of the hanging wall (dark), thrust over Mesozoic rocks (yellow coloured). b. shows gouge along the thrust. Photographs by Michael Johnson.

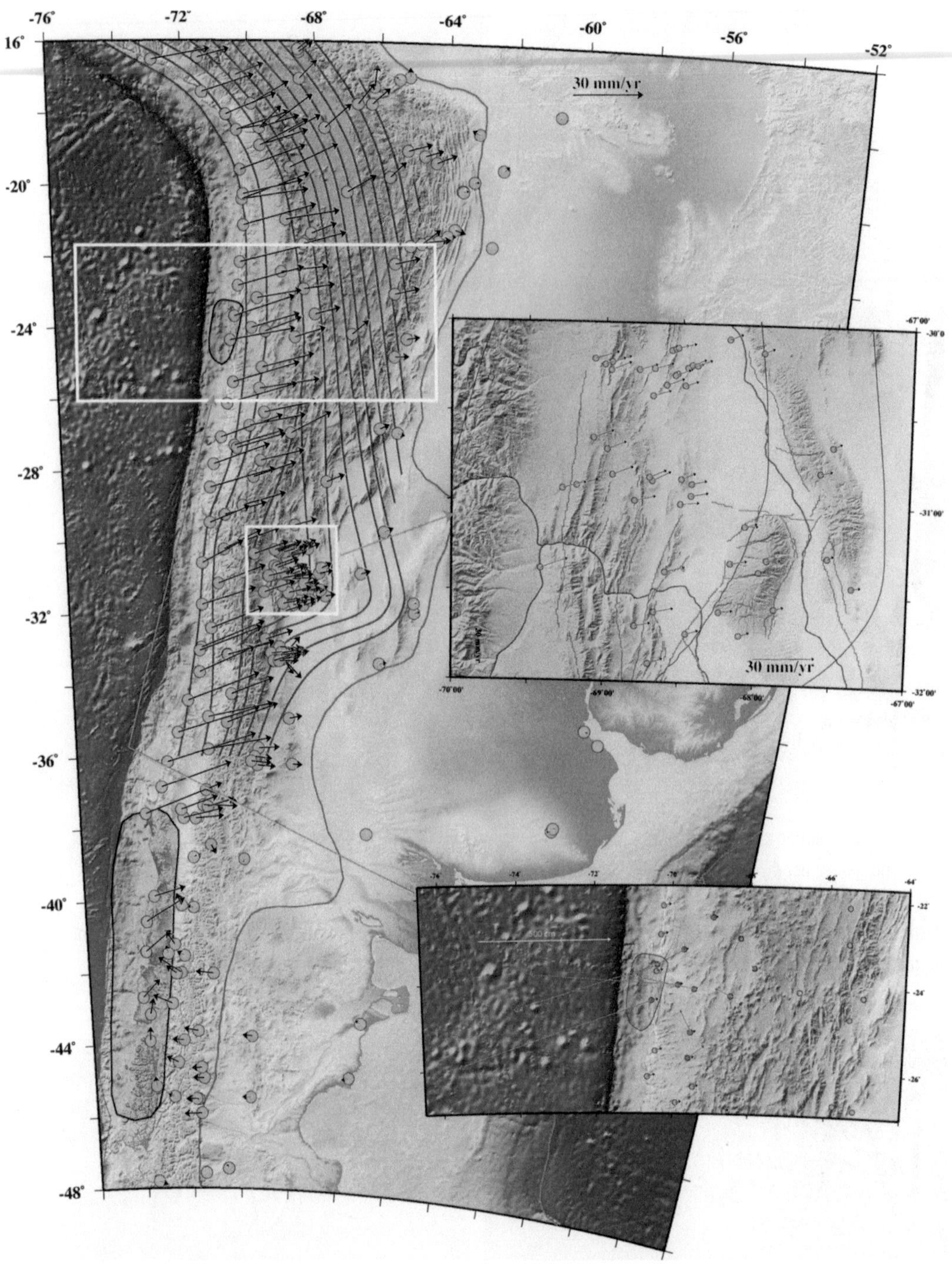

Plate 18 GPS data showing the present-day eastwards decrease in strain rate across the Andes. Compiled by Prof. R. Smalley.

Plate 20 The Keystone thrust: Cambrian rocks of the hanging wall (dark), Mesozoic rocks (yellow coloured). Photograph by Michael Johnson.

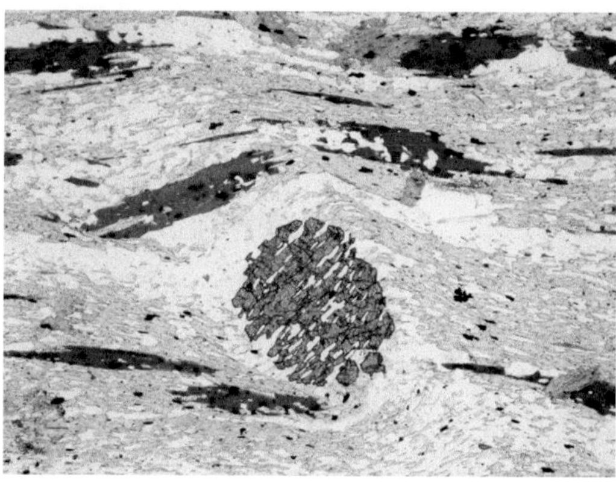

Plate 21 Garnet zone schist from Barrow's zones of Scotland. The garnet porphyroblast contains a rotated internal fabric that is continuous in this case with the external quartz-muscovite foliation. Field of view 7.4 mm across. Photograph by Simon Harley.

Plate 22 Kyanite + staurolite + garnet + biotite mica schist from Barrow's zones of Scotland. This view shows kyanite occurring as platy porphyroblasts and within the biotite + muscovite + quartz foliation. Field of view 5 mm across. Photograph by Simon Harley.

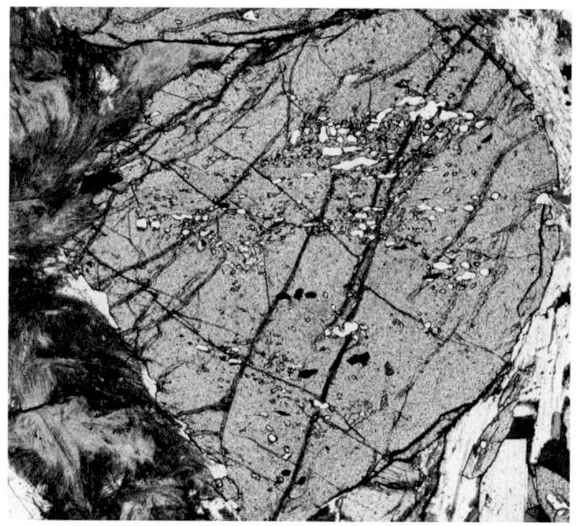

Plate 23 Garnet porphyroblast in the kyanite zone schist of Plate 22. Mats of fibrous sillimanite are seen on garnet on the left hand side of the field of view. Field of view 2.5 mm across. Photograph by Simon Harley.

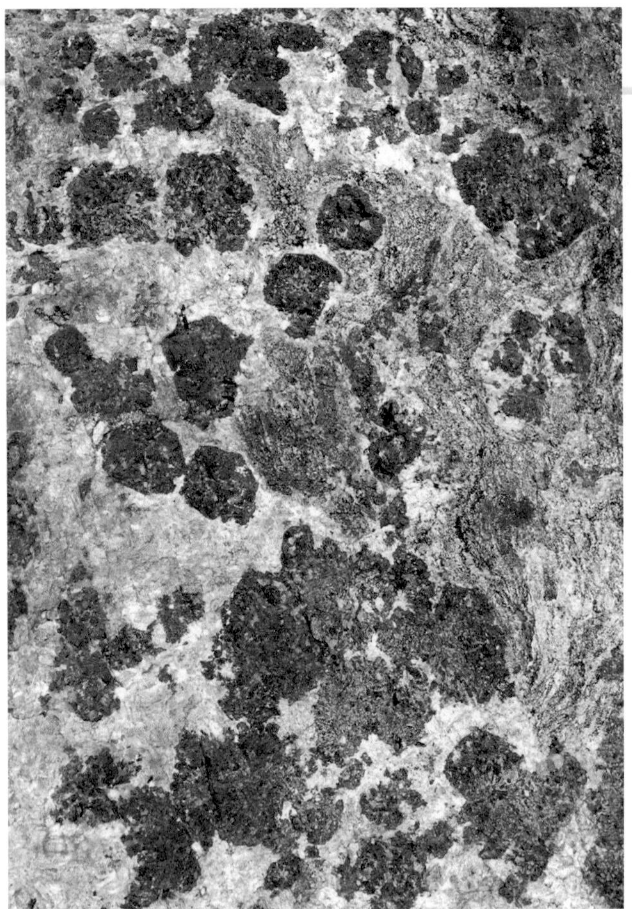

Plate 24 Coarse-grained patch migmatite in granulite facies pelite, Larsemann Hills, East Antarctica. The leucosome contains garnet porphyroblasts up to 10 cm in diameter. The portion of the rock that retains a foliation (a mesosome) consists of cordierite + sillimanite + plagioclase + quartz and local spinel. Field of view 0.5 metres across. Photograph by Simon Harley.

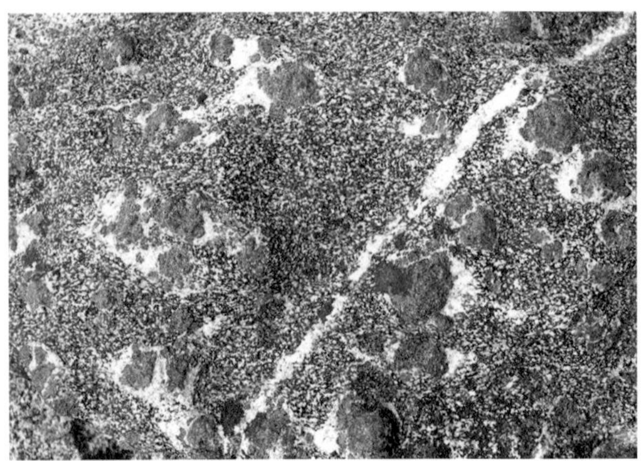

Plate 25 Localised patch leucosomes in an amphibolite facies metabasic gneiss, Nuuk region, south west Greenland. Patches consisting of garnet, plagioclase, quartz and clinopyroxene are interpreted to be the crystalline products of melting. Some melt has been extracted along vein systems. Field of view 0.3 metres across. Photograph by Simon Harley.

Plate 26 High-degree partial melting in granulite facies gneisses, Brattstrand Bluffs, East Antarctica. The upper part of the cliff is comprised by locally discordant garnet leucogranite sheets that are deformed along with their garnet + sillimanite + cordierite gneiss hosts. Reclined folds sole downwards into a 10 metre thick sub-horizontal granulite facies high-strain zone (the figure is standing in this zone). Whilst the ages of melting and deformation events here are difficult to determine from accessory mineral geochronology, the available data suggest migmatisation at 930 Ma and reworking at 530 Ma. Photograph by Simon Harley.

Plate 27 Detail of field relationships involved in melting and migmatisation, and later deformation, from part of the area shown in plate 30. This view of the Brattstrand Bluffs site is taken adjacent to the upper of the two figures in Plate 26. Leucogranite sheets cut and anastomose within the early gneissic foliation. Field of view 3 metres across. Photograph by Simon Harley.

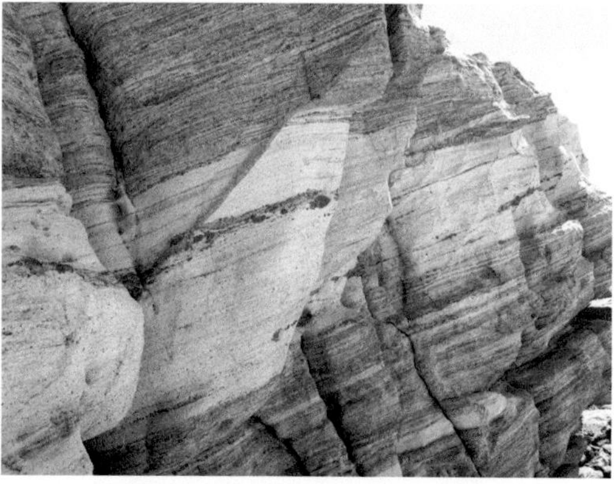

Plate 28 Locally discordant leucogranite in garnet + cordierite + sillimanite pelitic migmatite gneisses from within the high-strain zone beneath the lower platform shown in Plate 26. Coarse garnet (pink), sillimanite, spinel (black) and cordierite grain trails are strung out as a consequence of overprinting deformation. Outcrop section is 7 metres long. Photograph by Simon Harley.

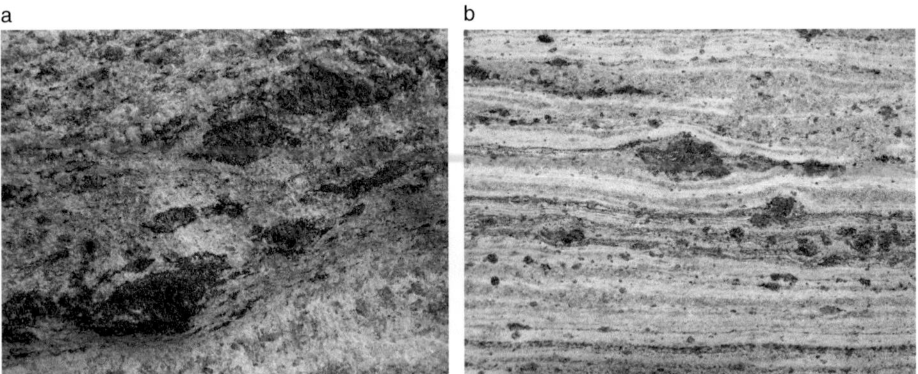

Plate 29 a, b Intense strain associated with Cambrian reworking of granulite facies migmatites, Brattstrand Bluffs, East Antarctica. a) shows flattening of garnet in a garnet + sillimanite migmatitic gneiss. Dark coronas on garnet consist of spinel (black) + cordierite (rare, brown-black) + ilmenite. These have formed through syn-deformational garnet breakdown. Field of view 0.2 metres across. b) shows even more intense flattening in a leucosome-rich migmatitic granulite. Garnet (pink) + spinel (black) + cordierite (local brown discoloured areas) layers are attenuated into stringers. Field of view 0.4 metres across. Photographs by Simon Harley.

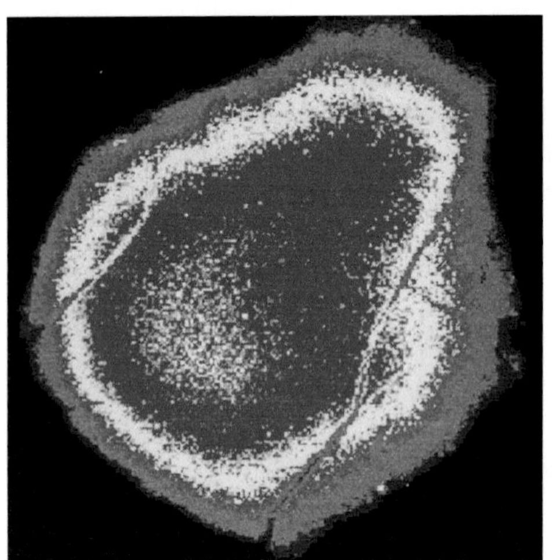

Plate 30 Prodigious growth zoning in garnet. The false colours reflect a change in garnet Ca content from 22–20 mole% grossular in cores (red) to 4–2 mole% grossular on rims (paler and darker blue). This example is from a E-HPG garnet leucogranulite from the Bohemian Massif, Czech Republic. Diameter of the garnet is only 180 microns. Image by Simon Harley.

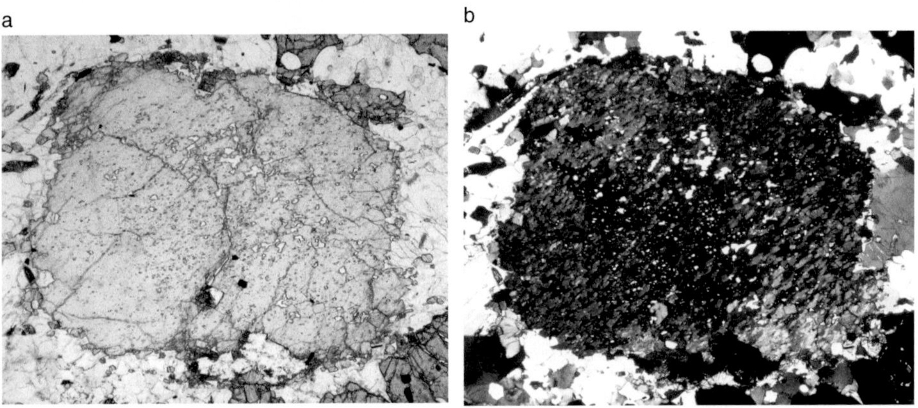

Plate 31 a, b Coarse kyanite pseudomorphed by sillimanite in a migmatitic granulite, Rauer Islands, East Antarctica; a) is plane-polarised light, b) is in crossed polars. Sillimanite forms a fine oriented texture composed of subgrains with diamond-shaped cross-sections. Electron back-scatter diffraction imaging of these confirms their consistent spatial arrangement, with lattice misorientations of only a few degrees between grains. Field of view 0.6 mm across. Photographs by Simon Harley.

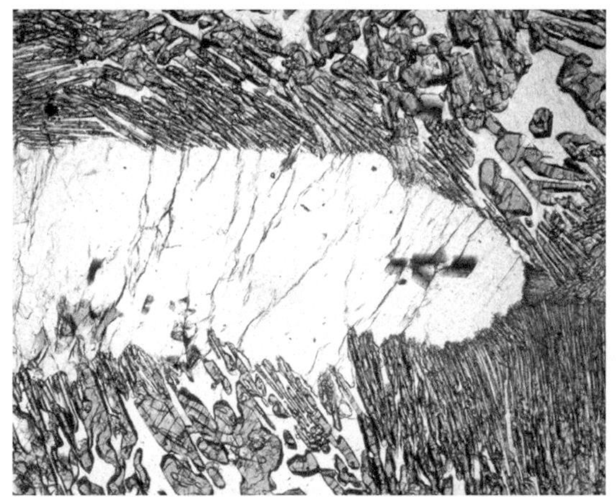

Plate 32 UHT reaction texture in granulite facies pelite, Rauer Islands, East Antarctica. Sillimanite (and garnet that is outside the field of view) is partially replaced by a structured symplectitic corona consisting of sapphirine (blue) and cordierite. The texture reflects the reaction garnet + sillimanite = sapphirine + cordierite, traversed on decompression at temperatures of >900 °C as pressures changed from 10 kbar to 7 kbar. Field of view 1.2 mm across. Photograph by Simon Harley.

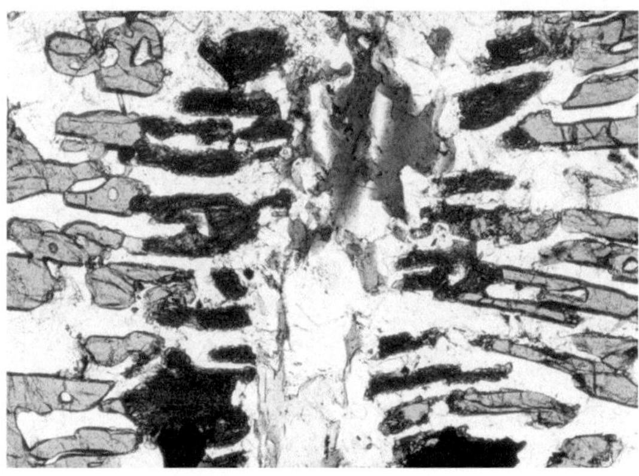

Plate 33 Two-stage reaction texture, Rauer Islands, East Antarctica. The initial symplectite involves sapphirine (blue) and cordierite, which have replaced garnet + sillimanite (see plate 26). This intergrowth is then partially replaced by a lower-P/T mineral assemblage of spinel + cordierite + biotite or orthopyroxene. Field of view 0.3 mm across. Photograph by Simon Harley.

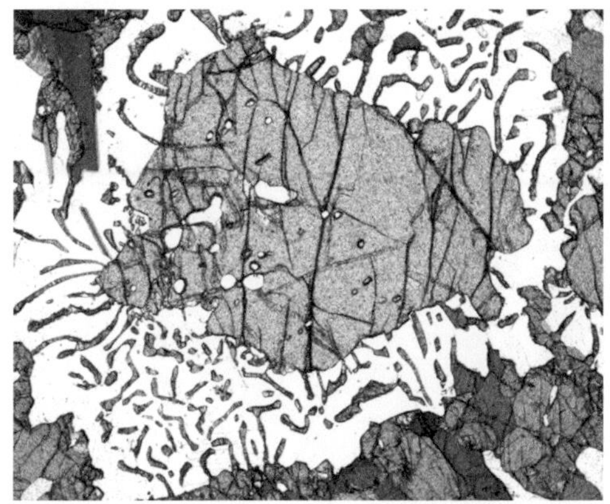

Plate 34 (34x). Reaction texture in a garnet-bearing mafic granulite, Rauer Islands, East Antarctica. The garnet is partially resorbed to produce a radial symplectite of orthopyroxene + plagioclase that merges with blocky orthopyroxene rinds on nearby clinopyroxene (pale green) and hornblende (darker green / brown). Field of view 5 mm across. Photograph by Simon Harley.

Plate 35 (35x). Garnet + omphacite + glaucophane + epidote + rutile + quartz blueschist, Sesia zone, northern Italy. Garnet porphyroblasts show near-rim retrograde alteration to chlorite (green) + albite. Glaucophane (pale blue-purple), epidote (pale, high relief) and omphacite (coarse, elongate pale-green grain), forms a foliation that wraps the garnet. Rutile occurs throughout. Field of view 2.5 mm across. Photograph by Simon Harley.

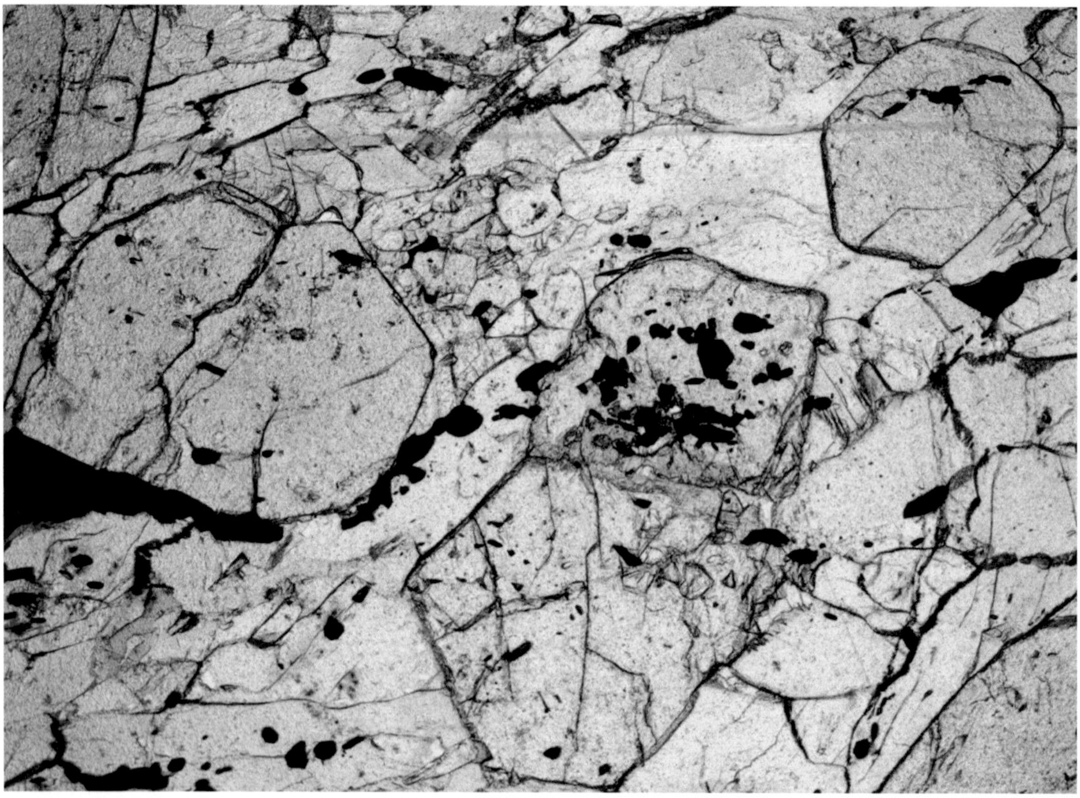

Plate 36 (36x). High-grade epidote eclogite, Naustdal, Western Norway. Euhedral garnet porphyroblasts are strongly zoned in colour, reflecting core to rim growth zoning (Mn and Fe decreasing rimwards). Rutile occurs throughout the rock. Epidote occurs as pale blocky grains forming a foliation along with omphacite. The omphacite is strongly and discontinuously colour zoned, from pale green to colourless on rims. Some garnets show near-rim alteration to chlorite (green) and blue-green amphibole, reflecting minor greenschist retrogression. Field of view 2.5 mm across. Photograph by Simon Harley.

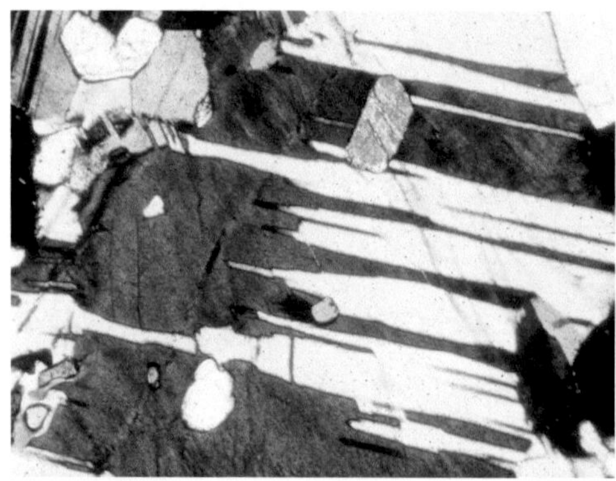

Plate 37 Twinning in cordierite from a garnet + sillimanite + sapphirine + cordierite granulite from the UHT Napier Complex, East Antarctica. Field of view 0.35 mm across. Photograph by Simon Harley.

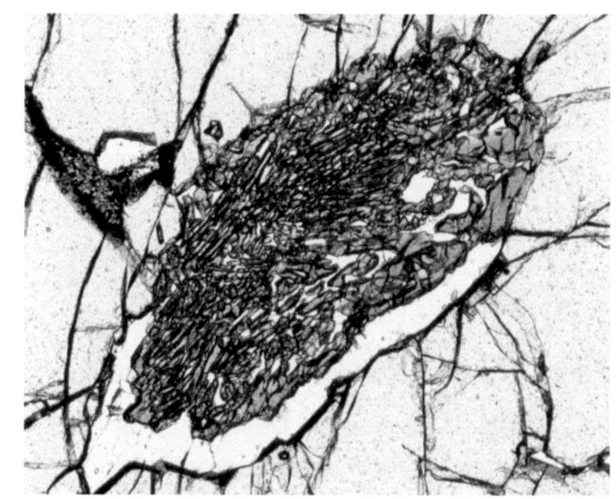

Plate 38 Sapphirine (pale blue) + quartz intergrowth within magnesian garnet, Rauer Islands, East Antarctica. This UHT intergrowth has replaced earlier kyanite and sillimanite in the garnet. The inferred 'UHT-ITD' P–T path is late Neoproterozoic in age. Field of view 0.3 mm across. Photograph by Simon Harley.

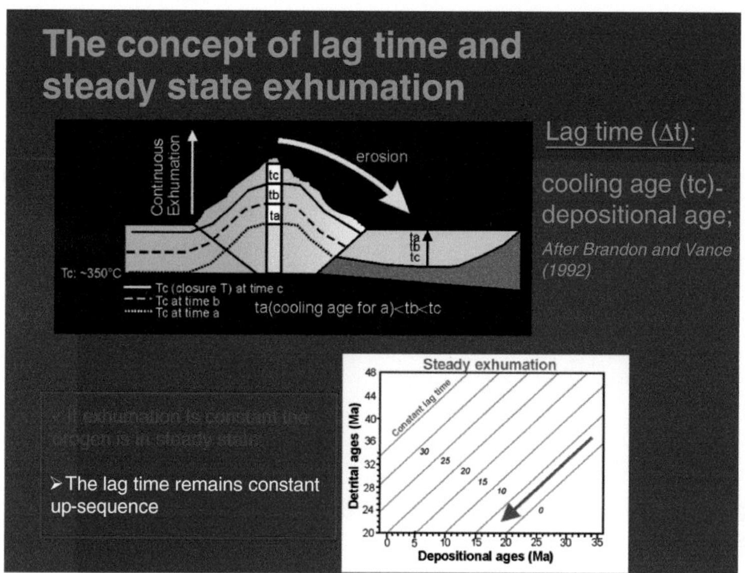

Plate 39 The uplift-cooling model for the interpretation of age sequences in an orogen and the adjacent foreland basin. Whereas in the orogen the cooling ages of minerals decrease in age from the top of the orogen downwards, this sequence is reversed in the basin. Image by B. Carrapa.

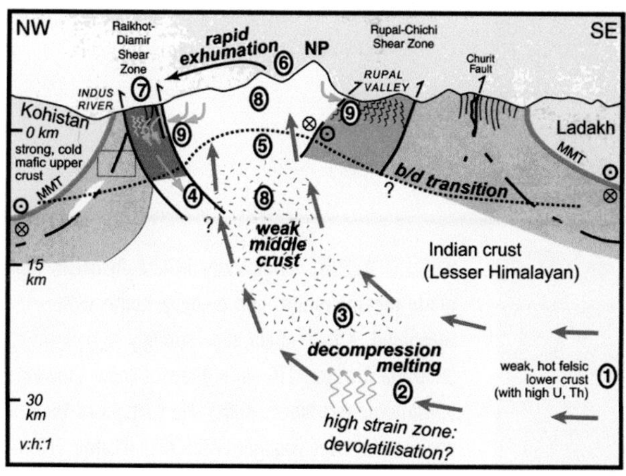

Plate 40 At Nanga Parbat (western syntaxis, Himalaya) coupling of erosion and rheology is manifested as "tectonic aneurysm". Incision (at rates of *c*.12 mm/a) of a deep Indus river gorge causes a positive feedback to develop as crust undergoing active anatexis and metamorphism flows into the weak zone (1,2,3). Decompression melting occurs (4). Rapid advection of material builds a mountain localised over the weak zone (5,6) as Nanga Parbat is unroofed (7) and the river continues to remove material and expose migmatites (8). A strong meteoric circulation is generated (9). From Zeitler *et al.* (2001).

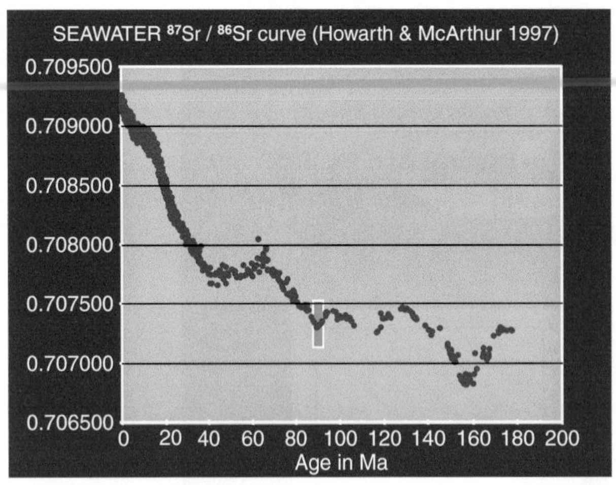

Plate 41 The Sr^{87}/Sr^{86} plot for the oceans showing the famous steepening at *c*. 40 Ma which seems to be linked with the rise of the Alps and Himalaya. Image based on Howarth and McArthur (1997).

Plate 43 Complexity in early Archaean high grade metamorphism and orogeny. 3650 Ma Amitsoq TTG suite gneiss, complexly migmatised and folded by *c*.3500 Ma (Friend and Nutman, 2005a). This example is from Itilleq Fjord, south west Greenland. Outcrop section is 2.5 metres across. Photograph by Simon Harley.

Plate 44 Complexity in late Archaean high grade metamorphism and orogeny, south west Greenland. A high-strain zone strongly re-deforms 3650 Ma Amitsoq TTG suite gneisses (now showing asymmetric recumbent folds) and juxtaposes them against and over younger (2825 Ma) Ikkatoq TTG gneisses of the Tre Brødre terrane (Friend and Nutman, 2005b). The amphibolite facies tectonothermal event responsible for amalgamation and interleaving of the distinct gneiss suites occurred at 2720 Ma. Norsanna, near Ameralik Fjord, south west Greenland. Photograph by Simon Harley.

'No subduction' regime: ΔT = 250 °C, 5 times heat production

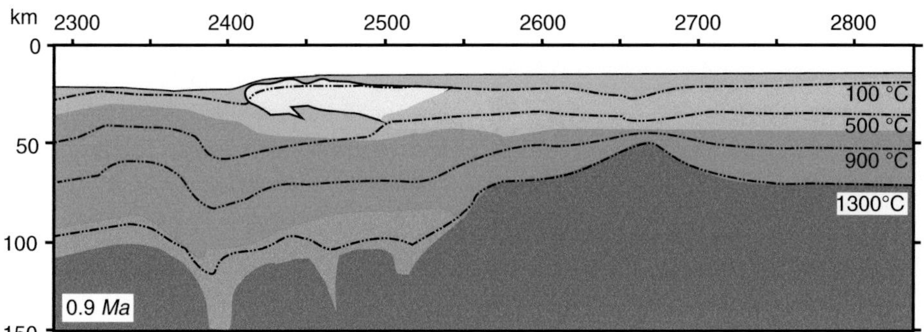

'Pre subduction' regime: ΔT = 175 °C, 2.5 times heat production

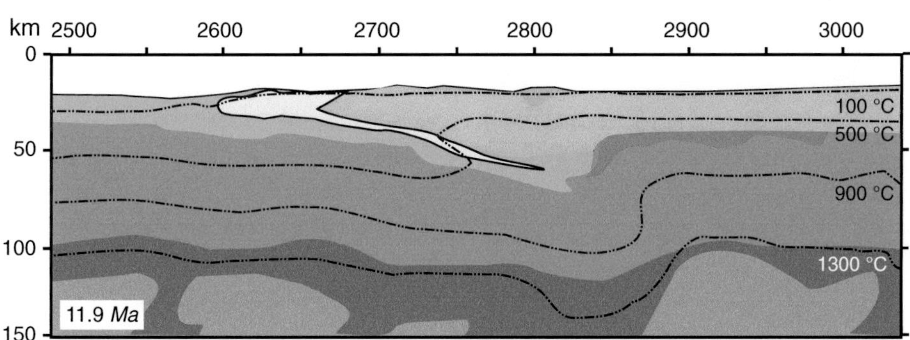

'Subduction' regime: ΔT = 100 °C, 1.5 times heat production

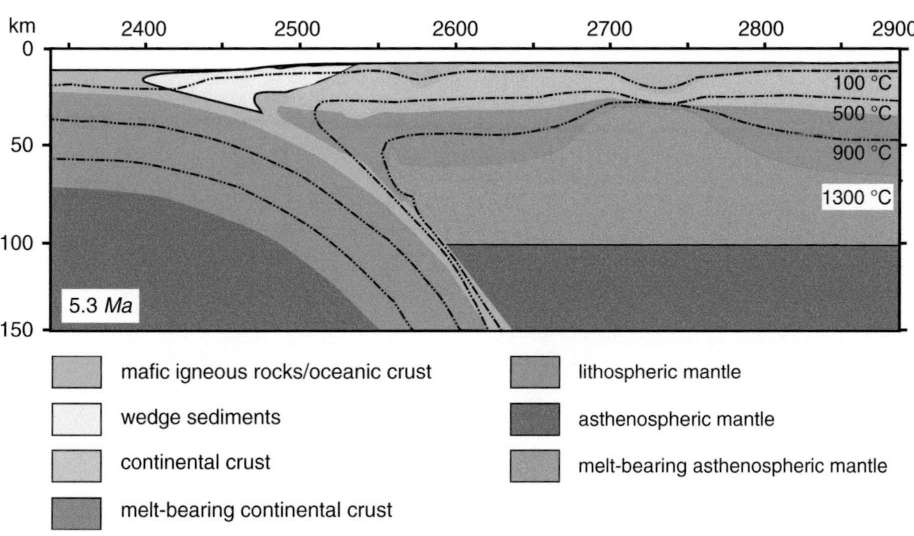

Plate 42 Colour version of Fig. 12.6.

Plate 45 Complexity in Archaean high grade metamorphism and orogeny, Rauer Islands, East Antarctica. A c. 2800 Ma TTG suite orthogneiss sheet, itself subsequently deformed and metamorphosed under granulite facies conditions, cutting an older composite layered and magmatitic gneiss basement. The composite gneisses have protolith ages of up to 3500 Ma (Simon Harley., unpubl.). Torckler Island, Rauer Group. Photograph by Simon Harley.

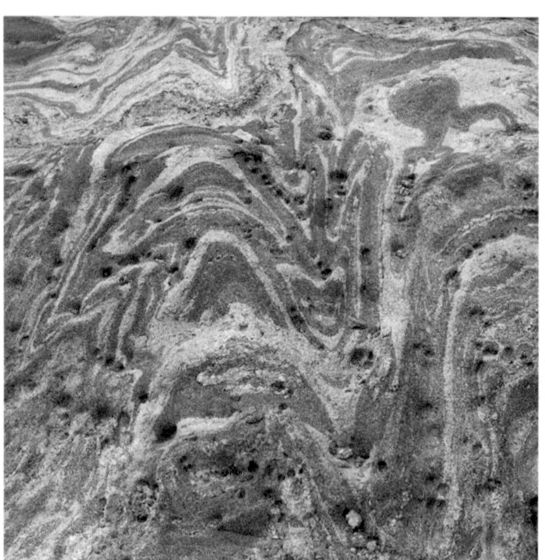

Plate 46 Complexity in Archaean high grade metamorphism and orogeny, Rauer Islands, East Antarctica. Composite layered gneisses displaying complex interference folds. Early folded mafic layers are cut by thin TTG sheets, and hence are >2800 Ma old. The composite layering produced is then refolded about two fold axis orientations to produce close folds trending across the field of view that are then re-oriented about near-upright close to open folds with axes trending towards the reader. Torckler Island, Rauer Group. Field of view 1.5 metres across. Photograph by Simon Harley.

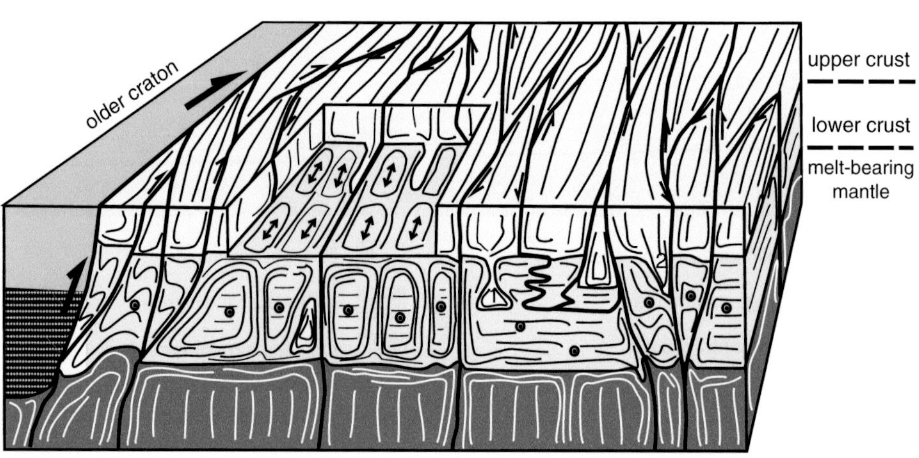

Plate 47 Colour version of Fig. 12.8.

References

Adams, C.J. (1981). Uplift rates and thermal structure in the Alpine Fault Zone and Alpine Schists, Southern Alps, New Zealand. In: *Thrust and Nappe Tectonic*, ed. K.R. McClay and N.J. Price. *Geological Society of London Special Publication*, **9**, 211–222.

Allen, P.A., Crampton, S.L. and Sinclair, H.D. (1991). The inception and early evolution of the North Alpine Foreland Basin, Switzerland. *Basin Research*, **3**, 143–163.

Almendinger, R.W. and Gubbels, T. (1996). Pure and simple shear plateau uplift, Altiplano-Puna, Argentina, Bolivia. *Tectonophysics*, **259**, 1–13.

Alsop, G.I. and Holdsworth, R.E. (1997). Flow perturbation in shear zones. In: *Deformation of the Continental Crust*, ed. A.C. Ries, R.W.H. Butler and R.H. Graham. *Geological Society of London Special Publication*, **272**, 75–101.

Amato, J.M., Johnson, C.M., Baumgartner, L.P. and Beard, B. (1999). Rapid exhumation of the Zermatt–Saas ohiolite deduced from high-precision Sm–Nd and Rb–Sr geochronology. *Earth and Planetary Science Letters*, **171**(3), 425–438.

Andersen, T.B. and Jamtveit, B. (1990). Uplift of deep crust during orogenic extensional collapse – a model based on field studies in the Sogn-Sunnfjord region of western Norway. *Tectonics*, **9**, 1097–1111.

Angiboust, S., Agard, P., Jolivet, L. and Beyssac, O. (2009). The Zermatt-Saas ophiolite: the largest (60-km wide) and deepest (*c.* 70–80 km) continuous slice of oceanic lithosphere detached from a subduction zone? *Terra Nova*, **21**, 171–180.

An Yin, (2006). Cenozoic tectonic evolution of the Himalayan orogen as constrained by along-strike variation of structural geometry, exhumation history, and foreland sedimentation. *Earth Science Reviews*, **76**, 1–131.

An Zhisheng, Kutzbach, J.E., Prell, W.L. and Porter, S.C. (2001). Evolution of Asian monsoon and phased uplift of the Himalaya since late Miocene times. *Nature*, **411**, 62–66.

Argand, E. (1924). La tectonique de l'Asie. *Comptes-rendu du 13th International Geological Congress*, **7**, 171–372.

Armstrong, P.A. (2005). Thermochronometers in sedimentary basins. *Reviews of Mineralogy and Geochemistry*, **58**, 1–29.

Avouac, J.P. and Burov, E.B. (1996). Erosion as a driving mechanism of intracontinental mountain growth. *Journal of Geophysical Research*, **101**, B8, 17747–17769.

Baba, S. (1998). Proterozoic anticlockwise *P–T* path of the Lewisian Complex of South Harris, Outer Hebrides, NW Scotland. *Journal of Metamorphic Geology*, **16**, 819–841.

Bailey, E.B. (1935). *Tectonic Essays Mainly Alpine*. Oxford: Clarendon Press.

Baldwin, J.A. and Brown, M. (2008). Age and duration of ultrahigh temperature metamorphism in the Anápolis–Itauçu Complex, Southern Brasilia Belt, central Brazil – constraints from U–Pb geochronology, mineral rare earth element chemistry and trace-element thermometry. *Journal of Metamorphic Geology*, **26**, 213–233.

Baldwin, J.A., Brown, M. and Schmitz, M.D. (2007). First application of titanium-in-zircon thermometry to ultrahigh-temperature metamorphism. *Geology*, **35**, 295–298.

Baldwin, J.A., Powell, R., Brown, M., Moraes, R. and Fuck, R.A. (2005). Modelling of mineral equilibria in ultrahigh-temperature metamorphic rocks from the Anápolis–Itauçu Complex, central Brazil. *Journal of Metamorphic Geology*, **23**, 511–531.

Baldwin, S.L., Monteleone, B.D., Webb, L.E. *et al.* (2004). Pliocene exhumation at plate tectonic rates in eastern Papua New Guinea. *Nature*, **431**, 263–267.

Barazangi, M. and Isacks, B.L. (1976). Spatial distribution of earthquakes and subduction of the Nazca plate beneath South America. *Geology*, **4**, 686–692.

Barrell, J. (1914). The strength of the Earth's crust. *Geology*, **22**, 28–48.

Barros, A.P., Chiao, S., Lang, T.J., Burbank, D. and Putkonen, J. (2006). From weather to climate – seasonal interannual variability of storms and implications for erosion processes in the Himalaya. In: *Tectonics, Climate, and Landscape Evolution*, ed. S.D. Willett, N. Hovius, M.T. Brandon and D.M. Fisher. *Geological Society of America, Special Paper*, **398**, 17–38.

Barrow, G. (1893). On an intrusion of muscovite biotite gneiss in the southeastern Highlands of Scotland, and its accompanying metamorphism. *Quarterly Journal of the Geological Society of London*, **49**, 330–358.

Barrow, G. (1912). On the geology of lower Deeside and the southern Highland border. *Proceedings of the Geologists Association*, **23**, 268–284.

Barry, R.G. (1992). *Mountain Weather and Climate*. Routledge.
This gives an account of the perturbation of airflow over mountains and cloud formation and precipitation resulting from mountain barriers. It includes theory and examples from around the world.

Beaumont, C., Jamieson, R.A., Butler, J.P. and Warren, C.J. (2009). Crustal structure: a key constraint on the mechanism of ultra-high-pressure rock exhumation. *Earth and Planetary Science Letters*, **287**, 116–129.

Beaumont, C., Jamieson, R.A. and Nguyen, M.H. (2010). Models of large hot orogens containing a collage of reworked and accreted terranes. *Canadian Journal of Earth Sciences*, **47**, 485–515.

Beaumont, C., Jamieson, R.A., Nguyen, M.H. and Lee, B. (2001). Himalayan tectonics explained by extrusion of a low-viscosity crustal channel coupled to focused denudation. *Nature*, **414**, 738–742.

Beaumont, C., Jamieson, R.A., Nguyen, M.H. and Medvedev, S. (2004). Crustal channel flows: 1. Numerical models with applications to the tectonics of the Himalayan–Tibetan orogen. *Journal of Geophysical Research*, **109**, B06406.

Beaumont, C., Nguyen, M.H., Jamieson, R.A. and Ellis, S. (2006). Crustal flow modes in large hot orogens. In: *Channel Flow, Ductile Extrusion, and Exhumation of Lower-mid Crust in Continental Collision Zones*, ed. R.D. Law, L. Godin and M.P. Searle. *Geological Society of London Special Publication*, **268**, 91–145.

Beck, S.L. and Zandt, G. (2002). The nature of orogenic crust in the central Andes. *Journal of Geophysical Research*, **107**, B10, 2230, ESE 71–16.

Beltrando, M., Compagnoni, R. and Lombardo, B. (2010). (Ultra-) High-pressure metamorphism and orogenesis: an Alpine perspective. *Gondwana Research*, **18**, 147–166.

Bendick, R. and Flesch, L. (2007). Reconciling lithospheric deformation and lower crustal flow beneath central Tibet. *Geology*, **35**, 895–898.

Berger, A. and Bousquet, R. (2008). *Subduction-related Metamorphism in the Alps: Review of Isotope Ages Based on Petrology and their Dynamic Consequences. Geological Society of London Special Publication*, **298**, 117–144.

Bernet, M., Brandon, M., Garver, J. *et al.* (2009). Exhuming the Alps through time: clues from detrital zircon fission-track thermochronology. *Basin Research*, doi:10.1111/1365–2117. 2009.00400. x

Bertrand, M. (1883–84). Rapports de structure des Alpes de Glaris et du bassin houiller du Nord. *Géologie Société de France Bulletin*, 3rd ser. **12**, 318.

Bickle, M.J. (1978). Heat loss from the Earth: a constraint on Archaean tectonics from the relation between geothermal gradients and the rate of plate production. *Earth and Planetary Science Letters*, **40**, 301–324.

Blisniuk, P.M., Hacker, B.R., Glodny, J. *et al.* (2001). Normal faulting in central Tibet since at least 13.5 myrs ago. *Nature*, **412**, 628–632.

Bohlen, S.R. (1991). On the formation of granulites. *Journal of Metamorphic Geology*, **9**, 223–229.

Booth, A.L., Zeitler, P.K., Kidd, W. *et al.* (2004). U–Pb constraints on the tectonic evolution of southeastern Tibet and Namche Barwa areas. *American Journal of Science*, **304**, 889–929.

Bouhallier, H., Choukroune, P. and Ballèvre, M. (1993). Diapirism, bulk homogeneous shortening and transcurrent shearing in the Archaean Dharwar craton: the Holenarsipur area, southern India. *Precambrian Research*, **63**, 43–58.

Bousquet, R., Goffée, B., Henry, P., Le Pichon, X. and Chopin, C. (1997). Kinematic and petrological model of the Central Alps: Lepontine metamorphism in the upper crust and eclogitisation in the lower crust. *Tectonophysics*, **273**, 105–127.

Bowen, N.L. (1940). Progressive metamorphism of siliceous limestone and dolomite. *Journal of Geology*, **48**, 225–274.

Boyer, S.E. (1992). Geometric evidence for synchronous thrusting in the southern Alberta and northwest Montana thrust belts. In: *Thrust Tectonics*, ed. K.R. McClay. London and NewYork: Chapman and Hall, 377–390.

Boyer, S.E. and Elliott, D. (1982). Thrust systems. *American Association of Petroleum Geologists Bulletin*, **66**, 1196–1230.

Braun, I., Montel, J.M. and Nicollet, C. (1998). Electron microprobe dating monazites from high-grade gneisses and pegmatites of the Kerala Khondalite Belt, southern India. *Chemical Geology*, **146**, 65–85.

Braun, J. and Shaw, R. (2001). A thin plate model of Palaeozoic deformation of the Australian lithosphere: implications for understanding the dynamics of intracratonic deformation. In: *Continental Reactivation and Reworking*, ed. J.A. Miller, R.E. Holdsworth, I.S. Buick and M. Hand. *Geological Society of London Special Publication*, **184**, 165–193.

Brocard, G.Y. and van der Beek, P.A. (2006). Influence of incision rates, rock strength and bedrock river gradients and valley flat widths: Field based evidence and calibration from Alpine rivers (south east France). In: *Tectonics, Climate and Landscape Evolution*, ed. S.D. Willett, N. Hovius, M.T. Brandon and D.M. Fisher. *Geological Society of America Special Paper* **398**, 101–126.

Brouwer, F.M. (2000). Thermal evolution of high pressure metamorphic rocks in the Alps. *Geologica Ultraiectina*, **199**, 221pp.

Brown, M. (2004). The mechanism of melt extraction from lower continental crust of orogens. *Transactions of the Royal Society of Edinburgh*, **95**, 35–48.

Brown, M. (2006). A duality of thermal regimes is the distinctive characteristic of plate tectonics since the Neoarchean. *Geology*, **34**, 961–964.

Brown, M. (2007). Metamorphic conditions in orogenic belts: a record of secular change. *International Geological Review*, **49**, 193–234.

Brown, M. (2009). Metamorphic patterns in orogenic systems and the geological record. *Geological Society of London Special Publications*, **318**, 37–74.

Brown, R.L. and Gibson, H.D. (2006). An argument for channel flow in the southern Canadian Cordillera and comparison with Himalayan tectonics. In: *Channel Flow, Ductile Extrusion and Exhumation in Continental Collision Zones*, ed. R.D. Law, M.P. Searle and L. Godin, *Geological Society of London Special Publication*, **268**, 543–559.

Bull, J.M. and Scrutton, R.A. (1992). Seismic images of intraplate deformation, central Indian ocean and their tectonic significance. *Journal of the Geological Society of London*, **149**, 955–966.

Burbank, D.W. and Beck, R.A. (1991). Rapid, long-term rates of denudation. *Geology*, **19**, 1169–1172.

Burbank, D.W., Beck, R.A. and Mulder, T. (1996). The Himalayan foreland basins. In: *The Tectonic Evolution of Asia*, ed. An Yin and T.M. Harrison. Cambridge University Press, 149–188.

Burbank, D.W., Blythe, A.E., Putkonen, J.L. *et al.* (2003). Decoupling erosion and precipitation in the Himalaya. *Nature*, **426**, 652–655.

Burbank, D.W., Derry, L.A. and France-Lanord, C. (1993). Reduced Himalayan sediment production 8 Ma ago despite an intensified monsoon. *Nature*, **364**, 48–54.

Burg, J.-P. and Gerya, T.V. (2005). The role of viscous heating in Barrovian metamorphism of collisional orogens: thermomechanical models and application to the Lepontine Dome in the Central Alps. *Journal of Metamorphic Geology*, **23**, 75–95.

Buxtorf, A. (1907). Zur tektonik des klettenjura. Ber Versammlung oberhaupt. *Geologische Verhandlungen*, **30/40**, 29–38.

Cadell, H.M. (1989). Experimental researches in mountain building. *Transactions of the Royal Society of Edinburgh*, **35**, 337–357.

Cagnard, F., Durrieu, N., Gapais, D., Brun, J.-P. and Ehlers, C. (2006). Crustal thickening and lateral flow during compression of hot lithospheres, with particular reference to Precambrian times. *Terra Nova*, **18**, 72–78.

Cahill, T. and Isacks, B.L. (1992). Seismicity and shape of the subducted Nazca plate. *Journal of Geophysical Research*, **97**, 17503–17529.

Camacho, A., Lee, K.W., Hensen, B.J. and Braun, J. (2005). Short-lived orogenic cycles and the eclogitisation of cold crust by spasmodic fluids. *Nature*, **435**, 1191–1196.

Capitanio, F.A., Morra, G., Goes, S., Weinberg, R.F. and Moresi, L. (2010). India–Asia convergence driven by the subduction of the Greater Indian continent. *Nature Geoscience*, **3**, 136–139.

Carlson, W.D. (2006). Rates of Fe, Mg, Mn, and Ca diffusion in garnet. *The American Mineralogist*, **91**, 1–11.

Carmichael, D.M. (1969). On the mechanism of prograde metamorphic reactions in quartz-bearing pelitic rocks. *Contributions to Mineralogy and Petrology*, **20**, 244–267.

Carrapa, B. (2009). Tracing exhumation and orogenic wedge dynamics in the European Alps with detrital thermochronology. *Geology*, **37**, 1127–1130.

Carrapa, B., Wijbrans, J. and Bertotti, G. (2003). Episodic exhumation in the Western Alps. *Geology*, **31**, 601–604.

Carrington, D.P. and Harley, S.L. (1995). Partial melting and phase relations in high-grade metapelites: an experimental petrogenetic grid in the KFMASH system. *Contributions to Mineralogy and Petrology*, **120**, 270–291.

Carswell, D.A., Brueckner, H.K., Cuthbert, S.J., Mehta, K. and O'Brien, P.J. (2003). The timing of stabilisation and the exhumation rate for ultra-high pressure rocks in the Western Gneiss Region of Norway. *Journal of Metamorphic Geology*, **21**, 601–612.

Cederbom, C.E., Sinclair, H.D., Schlunegger, F. and Rahn, M.K. (2004). Climate-induced rebound and exhumation of the European Alps. *Geology*, **32**, 709–712.

Chan, G.H.N., Waters, D.J., Searle, M.P. *et al.* (2009). Probing the basement of southern Tibet: evidence from crustal xenoliths entrained in Miocene ultrapotassic dyke. *Journal of the Geological Society of London*, **166**, 45–52.

Chardon, D., Choukroune, P. and Jayananda, M. (1996). Strain patterns, décollement and incipient sagducted greenstone terrains in the Archaean Dharwar craton (South India). *Journal of Structural Geology*, **18**, 991–1004.

Chardon, D., Gapais, D. and Cagnard, F. (2009) Flow of ultra-hot orogens: A view from the Precambrian, clues for the Phanerozoic. *Tectonophysics*, **477**, 105–118.

Chemenda, A.I., Mattauer, M., Malavieille, J. and Bokun, A.N. (1995). A mechanism for syncollisional rock exhumation and associated normal faulting; results from physical modeling. *Earth and Planetary Science Letters*, **132**, 225–232.

Chemenda, I., Burg, J.-P. and Mattauer, M. (2000). Evolutionary model of the Himalayan–Tibet system: geopoem based on new modelling, geological and geophysical data. *Earth and Planetary Science Letters*, **174**, 397–409.

Chen, W.-P. and Yang, Z. (2004). Earthquakes beneath the Himalaya and Tibet: evidence for a strong lithospheric mantle. *Science*, **304**, 1949–1952.

Chopin, C. (1984). Coesite and pure pyrope in highgrade blueschists of the western Alps: a first record and some consequences. *Contributions to Mineralogy and Petrology*, **86**, 107–118.

Chopin, C. (2003). Ultrahigh-pressure metamorphism; tracing continental crust into the mantle. *Earth and Planetary Science Letters*, **212**, 1–14.

Choukroune, P., Bouhallier, H. and Arndt, N.T. (1995). Soft lithosphere during periods of Archean crustal growth or crustal reworking. In: *Early Precambrian Processes*, ed. M.P. Coward, A.C. Ries. *Geological Society of London Special Publication*, **95**, 67–86.

Choukroune, P., Ludden, J.N., Chardon, D., Calvert, A.J. and Bouhallier, H. (1997). Archaean crustal growth and tectonic processes: a comparison of the Superior Province, Canada and the Dharwar craton, India. In: *Orogeny through Time*, ed. J.-P. and M. Ford. *Geological Society of London Special Publication*, **121**, 63–98.

Christensen, U.R. (1985). Thermal evolution models for the Earth. *Journal of Geophysical Research*, **90**, 2995–3007.

Chung, S.-L., Mei-Fei Chu, Jianqing, Ji *et al.* (2009). The nature and timing of crustal thickening in Southern Tibet: Geochemical and zircon Hf constraints from postcollisional adakites. *Tectonophysics*, **477**, 36–48.

Clark, M.C. and Royden, L.H. (2000). Topographic ooze: building the eastern margin of Tibet by lower crustal flow. *Geology*, **28**, 703–706.

Clark, M.K., Schoenbohm, L.M. and Royden, L.H. (2004). Surface uplift, tectonics and erosion of eastern Tibet from large-scale drainage patterns. *Tectonics*, **23**, TC 1006, doi:10.1029/2002 TC 001402.

Clark, M.K., Schoenbohm, L., Royden, R.H. *et al.* (2004). Surface uplift, tectonism in the evolution of eastern Tibet from large-scale drainage patterns. *Tectonics*, **23**, doi:10. 1029z/2002 TC001402.

Clemens, J. and Vielzeuf, D. (1987). Constraints on melting and magma production in the crust. *Earth and Planetary Science Letters*, **86**, 207–306.

Clift, P.D. and Blusztaign, J. (2005). Reorganisation of the western Himalaya river system after five millions years ago. *Nature*, **438**, 1001–1003.

Clift, P.D., Giosan, L., Blusztajn, J. *et al.* (2008a). Holocene erosion of the Lesser Himalaya triggered by intensified summer monsoon. *Geology*, **36**, 79–82.

Clift, P.D., Hodges, K.V., Heslop, I. *et al.* (2008b). Correlation of Himalayan exhumation rates and Asian monsoon intensity. *Nature Geoscience*, **1**, 875–880.

Clift, P.D. and Plumb, R.A. (2008). *The Asian Monsoon*. Cambridge University Press.

Cobbold, P.R. and Quinquis, H. (1980). Development of sheath folds in shear regimes. *Journal of Structural Geology*, **2**, 119–126.

Coleman, M. and Hodges, K.V. (1995). Evidence for Tibetan Plateau uplift before 14 myr ago from a new mineral age for east–west extension. *Nature*, **374**, 49–52.

Collins, W.J. and Vernon, R.H. (1991). Orogeny associated with anticlockwise *P–T–t* paths: evidence from low-P, high-T metamorphic terranes in the Arunta inlier, central Australia. *Geology*, **19**, 835–838.

Collins, W.J. and Vernon, R.H. (1992). Palaeozoic arc growth, deformation and migration across the Lachlan Fold Belt, southeastern Australia. *Tectonophysics*, **214**, 381–400.

Condie, K.C. (2007). Accretionary orogens in space and time. *Geological Society of America Special Paper*, **200**, 145–158.

Condie, K.C. and Kroner, A. (2008). When did plate tectonics begin? Evidence from the geological record. In: *When Did Plate Tectonics Begin on Planet Earth?* ed. K.C. Condie and V. Pease. *Geological Society of America Special Paper*, **440**, 281–294.

Condie, K.C. and Pease, V. (2008) *When Did Plate Tectonics Begin on Planet Earth? Geological Society of America Special Paper* **440**.

Coney, P.J. 1980. Cordilleran Metamorphic Core Complexes: An overview. In: *Cordilleran Metamorphic Core Complexes*, ed. M.D. Crittenden, P.J. Coney and G.H. Davis. *Geological Society of America Memoir*, **53**, 463–483.

Corfu, F., Crane, A., Moser, D. and Rogers, G. (1998). U–Pb zircon systematics at Gruinard Bay, northwest Scotland: implications for the early orogenic evolution of the Lewisian Complex. *Contributions to Mineralogy and Petrology*, **133**, 329–345.

Coward, M. (1994). Continental collision. In: *Continental Deformation*, ed. P.L. Hancock. London: Pergamon Press, 264–288.

Crespo-Blanc, A. (2007). Superimposed folding and oblique structures in the palaeomargin-derived units of the Central Betics (SW Spain). *Geological Society of London Journal*, **164**, 621–636.

Crouzet, C., Menard, G. and Rochette, P. (2001). Cooling history of the Dauphinois zone (Western Alps, France) deduced from the thermopalaeomagnetic record: geodynamic implications. *Tectonophysics*, **340**, 79–93.

Currie, B.S., Rowley, D.B. and Tabor, N.J. (2005). Mid Miocene Palaeoaltimetry of southern Tibet implications for the role of mantle thickening and delamination in the Himalayan orogen. *Geology*, **33**, 181–184.

Dahlen, F.A., Suppe, J. and Davis, D. (1984). Mechanics of fold-and-thrust belts and accretionary wedges: cohesive Coulomb theory. *Journal of Geophysical Research*, **89**, 10087–10101.

Dahlstrom, C.D.A. (1969). Balanced cross sections. *Canadian Journal of Earth Science*, **6**, 743–757.

Dahlstrom, C.D.A. (1970). Structural geology in the eastern margin of the Canadian Rockies. *Canadian Petroleum Geology Bulletin*, **18**, 332–406.

Davies, G.F. (1992). On the emergence of plate tectonics. *Geology*, **20**, 963–966.

Davies, G.F. (2006). Gravitational depletion of the early Earth's upper mantle and the viability of early plate tectonics. *Earth and Planetary Science Letters*, **243**, 376, 382.

Davis, D., Suppe, J. and Dahlen, F.A. (1983). Mechanics of fold-and-thrust belts and accretionary wedges. *Journal of Geophysical Research*, **88**(B2), 1153–1172.

DeCelles, P.G., Gehrels, G.E., Najman, Y., Martin, A.J. and Garzanti, E. (2004). Detrital geochronology and geochemistry of Cretaceous–Early Miocene strata Nepal: implications for timing and diachroneity of initial Himalayan orogenesis. *Earth and Planetary Science Letters*, **227**, 313–330.

DeCelles, P.G., Quade, J., Kapp, P. *et al.* (2007). High and dry in central Tibet during the Late Oligocene. *Earth and Planetary Science Letters*, **253**, 389–401.

DeCelles, P.G., Robinson, D.M. and Zandt, G. (2002). Implications of shortening in the Himalayan fold-thrust belt for uplift of the Tibetan Plateau. *Tectonics*, **21**(6), 1062.

De Paoli, M.C., Clarke, G.L., Klepeis, K.A., Allibone, A.H. and Turnbull, I.M. (2009). The eclogite–granulite transition: mafic and intermediate assemblages at Breaksea Sound, New Zealand. *Journal of Petrology*, **50**, 2307–2343.

de Roever, W.P. (1955). Some remarks concerning the origin of glaucophane in the North Berkeley Hills, California. *American Journal of Science*, **253**, 240–244.

de Roever, W.P. (1964). On the cause of the preferential distribution of certain metamorphic minerals in orogenic belts of different age. *Geologische Rundschau*, **53**, 324–336.

Derry, L.A. and France-Lanord, C. (1996). Neogene Himalayan history and river ^{87}Sr/^{86}Sr: impact on the marine Sr record. *Earth and Planetary Science Letters*, **142**, 59–74.

de Sigoyer, J., Chavagnac, V., Blichert-Toft, J. *et al.* (2000). Dating the Indian continental subduction and collisional thickening in the northwest Himalaya: Multichronology of the Tso Morari eclogites. *Geology*, **28**, 487–490.

de Sigoyer, J., Guillot, S. and Dick, P. (2004). Exhumation of the ultrahigh pressure Tso Morari Unit in eastern Ladakh (NW Himalaya); a case study. *Tectonics*, **23**, 18, doi:10.1029/2002TC001492.

Dewey, J.F. and Bird, J.M. (1970). Mountain belts and the new global tectonics. *Journal of Geophysical Research*, **75**, 2625–2647.

Dewey, J.F. and Lamb, S.H. (1992). Active tectonics of the Andes. *Tectonophysics*, **205**, 79–89.

Dewey, J.F., Pitman, W.C., Ryan, W.B.F. and Bonnin, J. 1973. Plate tectonics and evolution of the Alpine system. *Geological Society of America Bulletin*, **84**, 3137–3180.

Dewey, J.F., Ryan, P.D. and Andersen, T.B. (1993). Orogenic uplift and collapse, crustal thickness, fabric and metamorphic phase changes: the role of eclogites. In: *Magmatic Processes and Plate Tectonics*, ed. H.M. Pritchard, T. Alabaster, N.B.W. Harris and C.R. Neary. *Geological Society Special Publication*, **76**, 325–343.

Dewey, J.F. and Shackleton, R.M. (1984). A model for the evolution of the Grampian tract in the early Caledonides and Appalachians. *Nature*, **312**, 115–121.

Dewey, J.F., Shackleton, R.S., Chang Chengfa and Sun Yivin (1988). The tectonic evolution of the Tibetan Plateau. In: *The Geological Evolution of Tibet*, ed. R.S., Shackleton, J.F, Dewey and Jixiang Yan. *Proceedings of the Royal Society Series A*, **327**, 365–413.

Dewey, J.F. and Strachan, R.A. (2003). Changing Silurian-Devonian relative plate motion in the Caledonides: sinistral transpression to sinistral transtension. *Journal of the Geological Society of London*, **160**, 219–229.

Dickinson, W.R. (2009). Anatomy and global context of the North American Cordillera. In: *Backbone of the Americas: Shallow Subduction, Plateau, and Ridge and Terrane Collision*, ed. S.M. Kay, V.A. Ramos and W.R. Dickinson. *Geological Society of America Memoir*, **204**, 1–29.

Dinarès, J., McCelland, E. and Santanach, P. (1992). Contrasting rotations within thrust sheets and kinematics of thrust tectonics as derived from palaeomagnetic data: an example from the Southern Pyrenees. In: *Thrust Tectonics*, ed. K.R. McClay. London and New York: Chapman and Hall, 265–275.

Ding, L., Kapp, P., Yue, Y. and Lai, Q. (2007). Postcollisional calc-alkaline lavas and xenoliths from the southern Qiangtang terrane, central Tibet. *Earth and Planetary Science Letters*, **254**, 28–38.

Dodson, M.H. (1973). Closure temperature in cooling geochronological and petrological systems. *Contributions to Mineralogy and Petrology*, **40**, 259–274.

Drummond, M.S. and Defant, M.J. (1990). A model for trondhjemite-tonalite-dacire genesis and crustal growth via slab melting: Archean to modern comparisons. *Journal of Geophysical Research*, **95**, 21503–21521.

Dyksterhuis, S. and Müller, R.D. (2007). Cause and evolution of intraplate orogeny in Australia. *Geology*, **36**, 495–498.

Elger, K., Oncken, O. and Glodney, J. (2005). Plateau-style accumulation of deformation: southern Altiplano. *Tectonics*, **24**, TC4020.

Elliott, D. (1976). The energy balance and deformation mechanisms of thrust sheets. *Philosophical Transactions of the Royal Society, London*, A**283**, 289–312.

Elliott, D. and Johnson, M.R.W. (1980). Structural evolution of the northern part of the Moine thrust belt, NW Scotland. *Transactions of the Royal Society of Edinburgh: Earth Sciences*, **71**, 69–96.

Ellis, D.J. (1987). Origin and evolution of granulites in normal and thickened crust. *Geology*, **15**, 167–170.

Ellis, D.J. and Green, D.H. (1979). An experimental study of the effect of Ca upon garnet–clinopyroxene Fe–Mg exchange equilibria. *Contributions to Mineralogy and Petrology*, **71**, 13–22.

Ellis, D.J., Sheraton, J.W., England, R.N. and Dallwitz, W.B. (1980). Osumilite–sapphirine–quartz granulites from Enderby Land, Antarctica: mineral assemblages and reactions. *Contributions to Mineralogy and Petrology*, **90**, 123–143.

England, P.C. (1982). Some numerical investigations of large-scale continental deformation. In: *Mountain Building Processes*, ed. K.J. Hsu. New York: Academic Press.

England, P.C. and Bickle, M.J. (1984). Continental thermal and tectonic regimes in the Archaean. *Journal of Geology*, **92**, 353–367.

England, P.C. and Houseman, G.A. (1986). Finite strain calculations of continental deformation: 2. Comparison with the India–Asia collision zone. *Journal of Geophysical Research*, **91**, 3664–3676.

England, P.C. and Houseman, G.A. (1988). The mechanics of the Tibetan Plateau. *Philosophical Transactions of the Royal Society, London*, A**326**, 301–320.

England, P.C. and Richardson, S.W. (1977). The influence of erosion upon mineral facies of rocks from different metamorphic environments. *Journal of the Geological Society, London*, **134**, 201–213.

England, P.C. and Thompson, A.B. (1984) Pressure–temperature–time paths of regional metamorphism, Part I: Heat transfer during the evolution of regions of thickened continental crust. *Journal of Petrology*, **25**, 894–928.

Ernst, W.G. (1971). Do mineral parageneses reflect unusually high-pressure conditions of Franciscan metamorphism? *American Journal of Science*, **270**, 81–108.

Ernst, W.G. (1973). Blueschist metamorphism and *P–T* regimes in active subduction zones. *Tectonophysics*, **17**, 255–272.

Ernst, W.G. (1975). Systematics of large-scale tectonics and age progressions in Alpine and circum-Pacific blueschist belts. *Tectonophysics*, **26**, 229–246.

Esher, A., Masson, H. and Steck., A. (1993). Nappe geometry in the western Swiss Alps. *Journal of Structural Geology*, **15**, 501–509.

Eskola, P. (1915). On the relations between the chemical and mineralogical composition in the metamorphic rocks of the Orijarvi region. *Bulletin de la Commission Geologique de Finlande*, **44**, 109–143.

Eskola, P. (1920). The mineral facies of rocks. *Norsk Geologisk Tidsskrifl*, **6**, 143–194.

Eskola, P. (1929). Om mineral facies. *Geologiske Förening i Stockholm Förhandlingar*, **51**, 157–173.

Eynatten, H., Schlumegger, F., Gaupp, R. and Wijbrans, J.R. (1999). Exhumation of the Central Alps: evidence from ^{40}Ar/^{39}Ar dating of detrital white micas from the Swiss Molasse Basin. *Terra Nova*, **11**, 284–289.

Faccenda, M., Gerya, T.V. and Chakraborty, S. (2008). Styles of post-subduction collisional orogeny: influence of convergence velocity, crustal rheology and radiogenic heat production. *Lithos*, **103**, 257–287.

Faccenda, M., Minelli, G. and Gerya, T.V. (2009). Coupled and decoupled regimes of continental collision: numerical modelling. *Earth and Planetary Science Letters*, **278**, 337–349.

Ferry, J.M. (1976). P, T, f_{CO2} and f_{H2O} during metamorphism of calcareous sediments in the Waterville–Vassalboro area, south-central Maine. *Contributions to Mineralogy and Petrology*, **57**, 119–143.

Ferry, J.M. and Spear, F.S. (1978). Experimental calibration of the partitioning of Fe and Mg between biotite and garnet. *Contributions to Mineralogy and Petrology*, **66**, 113–117.

Fluteau, F., Ramstein, G. and Besse, J. (1999). Simulating the evolution of the Asian and African monsoons during the past 30 Myr using an atmospheric general circulation model. *Journal of Geophysical Research*, **104**, D10, 11995–12018.

Foley, S.F., Buhre, S. and Jacob, D.E. (2003). Evolution of the Archaean crust by delamination and shallow subduction. *Nature*, **421**, 249–252.

Foley, S.F., Tiepolo, M. and Vannucci, R. (2002). Growth of early continental crust controlled by melting of amphibolite in subduction zones. *Nature*, **417**, 837–840.

France-Lanord, C., Derry, L. and Michard, A. (1993). Evolution of the Himalaya since Miocene time: isotopic and sedimentological evidence from the Bengal Fan. In: *Himalayan Tectonics*, ed. P.J. Treloar and M.P. Searle. *Geological Society of London Special Publication*, **74**, 605–621.

Fraser, G., Worley, B. and Sandiford, M. (2000) High-precision geothermobarometry across the High Himalayan metamorphic sequence, Langtang Valley, Nepal. *Journal of Metamorphic Geology*, **18**, 665–681.

Freeman, S.R., Butler, R.W.H., Cliff, R.A. and Rex, D.H. (1998). Direct dating of mylonite evolution: a multidisciplinary study from the Moine Thrust Zone, NW Scotland. *Journal of the Geological Society of London*, **155**, 745–758.

Friend, C.R.L. and Kinny, P.D. (1995). New evidence for protolith ages of Lewisian granulites, northwest Scotland. *Geology*, **23**, 1027–1030.

Friend, C.R.L. and Kinny, P.D. (2001). A reappraisal of the Lewisian Gneiss Complex: geochronological evidence for its tectonic assembly from disparate terranes in the Proterozoic. *Contributions to Mineralogy and Petrology*, **142**, 198–218.

Friend, C.R.L. and Nutman, A.P. (2005a). Complex 3670–3500 Ma orogenic episodes superimposed on juvenile crust accreted between 3850 and 3690 Ma, Itsaq Gneiss Complex, southern West Greenland. *Journal of Geology*, **113**, 375–397.

Friend, C.R.L. and Nutman, A.P. (2005b). New pieces to the Archaean terrane jigsaw puzzle in the Nuuk region, southern West Greenland: steps in transforming a simple insight into a complex regional tectonothermal model. *Journal of the Geological Society of London*, **162**, 147–162.

Friend, C.R.L., Nutman, A.P., Baadsgaard, H., Kinny, P.D. and McGregor, V.R. (1996). Timing of late Archaean terrane assembly, crustal thickening and granite emplacement in the Nuuk region, southern West Greenland. *Earth and Planetary Science Letters*, **142**, 353–365.

Frost, B.R. and Chacko, T. (1989). The granulite uncertainty principle: limitations on thermobarometry in granulites. *Journal of Geology*, **97**, 435–450.

Galy, V., France-Lanord, C., Beyssac, O. *et al.* (2007). Efficient organic carbon burial in the Bengal Fan sustained by Himalayan erosional system. *Nature*, **450**, 407–410.

Gapais, D., Cagnard, F., Gueydan, F., Barbey, P. and Ballèvre, M. (2009). Mountain building and exhumation processes through time: inferences from nature and models. *Terra Nova*, **21**, 188–194.

Garver, J.I., Brandon, M.T., Roden-Tice, M. and Kamp, P. (1999). Exhumation history of orogenic highlands determined by detrital fission-track thermogeochronology. In: *Exhumation Processes: Normal Faulting, Ductile Flow and Erosion*, ed. U. Ring, M.T. Brandon, G.S. Lister and S.D. Willett. *Geological Society of London Special Publication*, **154**, 283–304.

Geisler, T., Schaltegger, U. and Tomaschek, F. (2007). Re-equilibration of zircon in aqueous fluids and melts. *Elements*, **3**, 45–51.

Gerbault, M., Martinod, J. and Hérail, G. (2005). Possible orogeny-parallel lower crustal flow and thickening in the Central Andes. *Tectonophysics*, **399**, 59–72.

Gerya, T.V., Maresch, W.V., Willner, A.P., Van Reenan, D.D. and Smit, C.A. (2001). Inherent gravitational instability of thickened continental crust with low- to medium-pressure granulite facies metamorphism. *Earth and Planetary Science Letters*, **190**, 221–235.

Gerya, T.V., Perchuk, L.L. and Burg, J.-P. (2008). Transient hot channels: perpetrating and regurgitating ultrahigh-pressure, high-temperature crust–mantle associations in collision belts. *Lithos*, **103**, 236–256.

Gerya, T.V., Perchuk, L.L., Maresch, W.V. *et al.* (2002). Thermal regime and gravitational instability of multi-layered continental crust: implications for the buoyant exhumation of high-grade metamorphic rocks. *European Journal of Mineralogy*, **14**, 687–699.

Gerya, T.V., Perchuk, L.L., Van Reenan, D.D. and Smit, C.A. (2000). Two dimensional numerical modelling of pressure–temperature–time paths for the exhumation of some granulite facies terrains in the Precambrian. *Journal of Geodynamics*, **30**, 17–35.

Gerya, T.V., Stöckhert, B. and Perchuk, A.L. (2002). Exhumation of high pressure metamorphic rocks in a subduction channel – a numerical simulation. *Tectonics*, **21**, 1–15.

Godin, L., Grujic, D., Law, R.D. and Searle, M.P. (2006). Channel flow, ductile extrusion and exhumation in continental collision zones. In: *Channel Flow, Ductile Extrusion in Continental Collision Zones*, ed. R.D. Law, M.P. Searle and L. Godin. *Geological Society of London Special Publication*, **268**, 1–23.

Godin, L., Parrish, R.R., Brown, R.L. and Hodges, K.V. (2001) Crustal thickening leading to exhumation of the Himalayan metamorphic core of central Nepal: insight from U–Pb geochronology and $^{40}Ar/^{39}Ar$ thermochronology. *Tectonics*, **20**, 729–774.

Goldschmidt, W.M. (1915). Geologisch-petrographische Studien im Hochgebirge des südlichen Norwegens. III. Die Kalksilikatgneise und Kalksilikatglimmerschiefer im Trondhjem-Gebiete. *Norsk Videnskapelig Selskap i Oslo Skrifter, Matematisk-Naturvidenskapelig Klasse*, **10**.

Goodenough, K.M., Millar, I., Strachan, R.A., Krabbendam, M. and Evans, J.A. (2011). Timing of regional deformation and development of the Moine Thrust Zone in the Scottish Caledonides: constraints from the U–Pb geochronology of alkaline intrusions. *Journal of the Geological Society of London*, **168**, 99–114.

Goscombe, B.D. and Gray, D.R. (2008). Structure and strain variation at mid crustal levels in a transpressional orogen: a review of Kaoko Belt structure and the character of West Gondwana amalgamation and dispersal. *Gondwana Research*, **13**, 45–85.

Goscombe, B., Gray, D. and Hand, M. (2006). Crustal architecture of the Himalayan metamorphic front in eastern Nepal. *Gondwana Research*, **10**, 232–255.

Green, D.H. and Ringwood, A.E. (1967). An experimental investigation of the gabbro-to-eclogite transformation and its petrological applications. *Geochimica et Cosmochimica Acta*, **31**, 767–833.

Greenwood, H.J. (1962). Metamorphic systems involving two volatile components. *Carnegie Institute Washington Yearbook*, **61**, 82–85.

Grubenmann, U. and Niggli, P. (1924). *Die Gesteinsmetamorphose*. Berlin: Gebrueder Borntriger.

Gutiérrez-Alonso, G., Fernández-Suárez, J. and Weil, A.B. (2004). Orocline triggered lithospheric delamination. In: *Orogenic Curvature: Integrating Palaeomagnetic and Structural Analyses*, ed. A.J. Sussman and A.B. Weil. *Geological Society of America Special Paper* **383**, 121–130.

Hacker, B.R. (2006). Pressures and temperatures of ultrahigh pressure metamorphism; implications for UHP tectonics and H_2O in subducting slabs. *International Geology Review*, **48**, 1053–1066.

Hacker, B.R., Calvert, A.J., Zhang, R.Y., Ernst, W.G. and Liou, J.G. (2003). Ultrarapid exhumation of ultrahigh-pressure diamond-bearing metasedimentary rocks of the Kokchetav Massif, Kazakhstan? *Lithos*, **70**, 61–75.

Hacker, B.R., Wallis, S.R., Ratschbacher, L.R., Grove, M. and Gehrels, G. (2006). High-temperature geochronology constraints on the tectonic history and architecture of the ultrahigh pressure Dabie-Sulu Orogen. *Tectonics*, **25**, TC5006.

Hanchar, J.M. and Miller, C.F. (1993). Zircon zonation patterns as revealed by cathodoluminescence and backscattered electron images: implications for interpretation of complex crustal histories. *Chemical Geology*, **110**, 1–13.

Harker, A. (1932). *Metamorphism: A Study of the Transformation of Rock Masses* (2nd edn 1939). London: Methuen.

Harley, S.L. (1989). The origins of granulites: a metamorphic perspective. *Geological Magazine*, **126**, 215–247.

Harley, S.L. (1992). Proterozoic granulite terranes. In: *Proterozoic Crustal Evolution*, ed. K.C. Condie. Amsterdam: Elsevier, 301–359.

Harley, S.L. (1998a). On the occurrence and characterisation of ultrahigh temperature crustal metamorphism. In: *What Drives Metamorphism and Metamorphic Reactions?*

ed. P.J. Treloar and P. O'Brien. *Geological Society of London Special Publication*, **138**, 75–101.

Harley, S.L. (1998b). On the occurrence and characterisation of ultrahigh temperature crustal metamorphism. In: *What Drives Metamorphism and Metamorphic Reactions?* ed. P.J. Treloar and P. O'Brien. *Geological Society of London Special Publication*, **138**, pp. 75–101.

Harley, S.L. (1998c). Ultrahigh temperature granulite metamorphism (1050 °C, 12 kbar) metamorphism and decompression in garnet (Mg70)-orthopyroxene-sillimanite gneisses from the Rauer Group, East Antarctica. *Journal of Metamorphic Geology*, **16**, 541–562.

Harley, S.L. (2004). Extending our understanding of ultrahigh temperature crustal metamorphism. *Journal of Mineralogical and Petrological Sciences*, **99**, 140–158.

Harley, S.L. (2008). Refining the *P–T* records of UHT crustal metamorphism. *Journal of Metamorphic Geology*, **26**, 125–154.

Harley, S.L. and Carrington, D.P. (2001). The distribution of H_2O between cordierite and granitic melt: H_2O incorporation in cordierite and its application to high-grade metamorphism and crustal anatexis. *Journal of Petrology*, **42**, 1595–1620.

Harley, S.L. and Green, D.H. (1982). Garnet–orthopyroxene barometry for granulites and peridotites. *Nature*, **300**, 697–701.

Harley, S.L., Kelly, N.M. and Moller, A. (2007). Zircon behaviour and the thermal history of mountain chains. *Elements*, **3**, 25–30.

Harris, N.B.W. (2007). Channel flow and the Himalyan–Tibet orogen: a review. *Journal of the Geological Society of London*, **164**, 511–523.

Harrison, T.M. (2009). The Hadean Crust: Evidence from >4 Ga zircons. *Annual Review of Earth and Planetary Sciences*, **37**, 479–505.

Hawkesworth, C.J. and Kemp, A.I.S. (2006a). Evolution of the continental crust. *Nature*, **443**, 811–817.

Hawkesworth, C.J. and Kemp, A.I.S. (2006b). The differentiation and rates of generation of the continental crust. *Chemical Geology*, **226**, 134–143.

Hawthorne, E.C. and James, R.H. (2006). Temporal record of lithium in seawater: a tracer for silicate weathering? *Earth and Planetary Science Letters*, **246**, 393–406.

Heim, A. (1922). *Geologie der Schweiz*. Leipzig: Tauchnitz.

Hensen, B.J. (1971). Theoretical phase relations involving cordierite and garnet in the system FeO-MgO-Al_2O_3-SiO_2. *Contributions to Mineralogy and Petrology*, **33**, 191–214.

Hensen, B.J. and Green, D.H. (1973). Experimental study of the stability of cordierite and garnet in pelitic compositions at high pressures and temperatures. III. Synthesis of experimental data and geological applications. *Contributions to Mineralogy and Petrology*, **38**, 151–166.

Hermann, J. and Rubatto, D. (2003). Relating zircon and monazite domains to garnet growth zones: age and duration of granulite facies metamorphism in the Val Malenco lower crust. *Journal of Metamorphic Geology*, **21**, 833–852.

Hermann, J., Rubatto, D., Korsakov, A. and Shatsky, V.S. (2001). Multiple zircon growth during fast exhumation of diamondiferous, deeply subducted continental crust (Kokchetav massif, Kazakhstan). *Contributions to Mineralogy and Petrology*, **141**, 66–82.

Hetényi, G., Cattin, R., Brunet, F. *et al.* (2007). Density distribution of the Indian plate beneath the Tibetan plate: geophysical and petrological constraints on the kinetics of lower crustal eclogitization. *Earth and Planetary Science Letters*, **264**, (1–2), 226–244.

Hodges, K.V., Wobus, C., Ruhl, K., Schildgen, T. and Whipple, K. (2004). Quaternary deformation, river steepening, and heavy precipitation at the front of the Higher Himalayan ranges. *Earth and Planetary Science Letters*, **220**, 379–389.

Hodges, K.V., Wobus, C., Ruhl, K., Schildgen, T. and Whipple, K. (2004). Quaternary deformation, river steepening, and heavy precipitation at the front of the Higher Himalayan ranges. *Earth and Planetary Science Letters*, **220**, 379–389.

Hokada, T. (2001). Feldspar thermometry in ultrahigh-temperature metamorphic rocks: Evidence of crustal metamorphism attaining ~1100 °C in the Archean Napier Complex, East Antarctica. *American Mineralogist*, **86**, 932–938.

Holdsworth, R.E., Stewart, M., Imber, J. and Strachan, R.A. (2001). The structure and rheological evolution of reactivated continental fault zones: a review and case study. In: *Continental Reactivation and Reworking*, ed. R.E. Holdsworth, I.S. Buick and M. Hand. Geological Society of London, Special Publication, **184**, 115–137.

Holland, T.J.B. and Powell, R. (1998). An internally consistent thermodynamic data set for phases of petrological interest. *Journal of Metamorphic Geology*, **16**, 309–343.

Hollis, J.A., Harley, S.L., White, R.W. and Clarke, G.L. (2006). Preservation of evidence for prograde metamorphism in ultrahigh-temperature, high-pressure kyanite-bearing granulites, South Harris, Scotland. *Journal of Metamorphic Geology*, **24**, 263–279.

Holmes, A. (1928). Radioactivity and continental drift. *Geological Magazine*, **65**, 236–238.

Holmes, A. (1929). Radioactivity and earth movements. *Transactions of the Geological Society of Glasgow*, **18**, 559–606.

Holmes, A. (1944). *Principles of Physical Geology*. London, Edinburgh, Paris, Melbourne, Toronto and New York: Thomas Nelson and Sons Ltd.

Hoskin, P.W.O. and Black, L.P. (2000). Metamorphic zircon formation by solid state recrystallisation of protolith igneous zircon. *Journal of Metamorphic Geology*, **18**, 423–439.

Hossack, J.R. (1968). Pebble deformation and thrusting in the Bygdin area (Southern Norway). *Tectonophysics*, **5**, 315–339.

Hossack, J.R. (1979). The use of balanced cross sections in the calculation of orogenic contraction – a review. *Journal of the Geological Society of London*, **136**, 705–711.

Hubbard, M.S. (1989). Thermobarometric constraints on the thermal history of the Main Central Thrust Zone and Tibetan slab in eastern Nepal Himalaya. *Journal of Metamorphic Geology*, **7**, 19–30.

Hubbert, M.K. and Rubey, W.W. (1959). Role of fluid pressure in mechanics of overthrust faulting. Parts 1 and 2. *Geological Society of America Bulletin*, **70**, 115–205.

Huiqi, L., McClay, K.R. and Powell, D. (1992). Physical models for thrust wedges. In: *Thrust Tectonics*, ed. K.R. McClay. London and New York: Chapman and Hall, 71–81.

Isacks, B.L. (1988). Uplift of the Central Andean Plateau and bending of the Bolivian orocline. *Journal of Geophysical Research*, **93**, No. 84, 3211–3231.

Jackson, J.A. (2002). Strength of the continental lithosphere: time to abandon the jelly sandwich model? *GSA Today*, September, 4–9.

Jackson, J.A., Austrheim, H., McKenzie, D. and Priestley, K. (2004). Metastability, mechanical strength, and the support of mountain belts. *Geology*, **32**, 625–628.

Jadoon, I.A.K., Lawrence, R.D. and Lillie, R.J. (1992). Balanced and retrodeformed geological cross-sections from the frontal Sulaiman Lobe, Pakistan. In: *Thrust Tectonics*, ed. K.R. McClay. London and New York: Chapman and Hall, 343–356.

Jamieson, R.A. and Beaumont, C. (2011). Coeval thrusting and extension during lower crustal ductile flow – implications for exhumation of high-grade metamorphic rocks. *Journal of Metamorphic Geology*, **29**, 33–51.

Jamieson, R.A., Beaumont, C., Medvedev, S. and Nguyen, M.H. (2004). Crustal channel flows: 2. Numerical models with implications for metamorphism in the Himalayan–Tibetan orogen. *Journal of Geophysical Research*, **109**, B06406.

Jamieson, R.A., Beaumont, C., Nguyen, M.H. and Grujic, D. (2006). Provenance of the Greater Himalayan Sequence and associated rocks: predictions of channel flow models. In: *Channel Flow, Ductile Extrusion, and Exhumation of Lower–Mid Crust in Continental Collision Zones*, ed. R.D. Law, L. Godin and M.P. Searle, *Geological Society of London Special Publication*, **268**, 165–182.

Jamieson, R.A., Beaumont, C., Warren, C.J. and Nguyen, M.H. (2010). The Grenville Orogen explained? Applications and limitations of integrating numerical models with geological and geophysical data. *Canadian Journal of Earth Sciences*, **47**, 517–539.

Janots, E., Engi, M., Berger, A., Rubatto, D. and Gregory, C. (2007). Heating rate in the northern Lepontine dome (Central Alps) from in situ isotope dating of allanite and monazite. *Geophysical Research Abstracts*, **9**, 08743.

Jiménez-Munt, I., Garcia-Castellanas, D. and Fernandez, M. (2005). Thin-plate modelling of lithospheric deformation and surface transport. *Tectonophysics*, **407**, 3–4, 239–255.

Johnson, M.R.W. (1981). The erosion factor in the emplacement of the Keystone thrust sheet (South Nevada) across a land surface. *Geological Magazine*, **118**, 501–507.

Johnson, M.R.W. (1994). Volume balance of erosional loss and sediment deposition related to Himalayan uplifts. *Journal of the Geological Society of London*, **151**, 217–220.

Kay, S.M. and Coira, B.L. (2009). Shallowing and steepening subduction zones, continental lithospheric loss, magmatism, and crustal flow under the Central Andean Altiplano-Puna Plateau. In: *Backbone of the Americas: Shallow Subduction, Plateau Uplift, and Ridge and Terrane Collision*, ed. S.M. Kay, V.A. Ramos and W.R. Dickinson. *Geological Society of America Memoir*, **204**, 229–259.

Kelly, N.M. and Harley, S.L. (2005). An integrated microtextural and chemical approach to zircon geochronology: refining the Archaean history of the Napier Complex, east Antarctica. *Contributions to Mineralogy and Petrology*, **149**, 57–84.

Kelly, N.M., Hinton, R.W., Harley, S.L. and Appleby, S. (2008). New SIMS U–Pb zircon ages from the Langavat Belt, South Harris, NW Scotland: implications for the Lewisian Terrane Model. *Journal of the Geological Society of London*, **165**, 967–981.

Kelsey, D.E. (2008). On ultrahigh-temperature crustal metamorphism. *Gondwana Research*, **13**, 1–29.

Kelsey, D.E., White, R.W., Holland, T.J.B. and Powell, R. (2004). Calculated phase equilibria in K_2O-MgO-FeO-Al_2O_3-SiO_2-H_2O for sapphirine-quartz-bearing mineral assemblages. *Journal of Metamorphic Geology*, **22**, 559–578.

Kelsey, D.E., White, R.W. and Powell, R. (2005). Calculated phase equilibria in K_2O-MgO-FeO-Al_2O_3-SiO_2-H_2O for silica-undersaturated sapphirine-bearing mineral assemblages. *Journal of Metamorphic Geology*, **23**, 217–239.

Kemp, A.I.S. and Hawkesworth, C.J. (2003) Granitic perspectives on the generation and secular evolution of the continental crust. In: *The Crust. Treatise on Geochemistry*, **3**. Amsterdam: Elsevier, 349–410.

Kemp, A.I.S., Hawkesworth, C.J., Paterson, B.A. and Kinny, P.D. (2006). Episodic growth of the Gondwana supercontinent from hafnium and oxygen isotopes in zircon. *Nature*, **439**, 580–583.

Kemp, A.I.S., Shimura, T., Hawkesworth, C.J. and EIMF (2007). Linking granulites, silicic magmatism, and crustal growth in arcs: ion microprobe (zircon) U–Pb ages from the Hidaka metamorphic belt, Japan. *Geology*, **35**, 807–810.

Kendrick, E.C., Bevis, M., Cifuentes, O. and Galban, F. (1999). Current rates of convergence across the Central Andes: estimates from continuous GPS observations. *Geophysical Research Letters*, **26**, 541–544.

Kinny, P.D. and Friend, C.R.L. (1997). U–Pb isotopic evidence for the accretion of different crustal blocks to form the Lewisian Complex of northwest Scotland. *Contributions to Mineralogy and Petrology*, **129**, 326–340.

Kinny, P.D., Friend, C.R.L. and Love, G.J. (2005). Proposal for a terrane-based nomenclature for the Lewisian Gneiss Complex of NW Scotland. *Journal of the Geological Society of London*, **162**, 175–186.

Kirstein, A., Fellin, M.G., Willett, S.D. *et al.* (2009). Pliocene onset of rapid exhumation in Taiwan during arc-continent collision: new insights from detrital thermochrononology. *Basin Research*, doi:10.1111/j.1365–2117. 2009. 00426.x

Kitoh, A. (2004). Effects of mountain uplift on east Asian summer climate investigated by a coupled atmosphere–ocean GCM. *Journal of Climatology*, **17**, 783–802.

Klemperer, S.L. (2006). Crustal flow in Tibet: geophysical evidence for the physical state of Tibetan lithosphere, and inferred patterns of active flow. In: *Channel Flow, Ductile Extrusion and Exhumation in Continental Collision Zones*, ed. R.D. Law, M.P. Searle and L. Godin. *Geological Society of London Special Publication*, **268**, 39–70.

Knipe, R.J. (1990). Microstructural analysis and tectonic evolution in thrust systems: examples from the Assynt region of the Moine thrust zone, Scotland. In: *Deformation Processes in Minerals, Ceramics and Rocks*, ed. D.J. Barber and P.G. Meredith. London: Unwin Hyman, 228–261.

Kohn, M.J. (2003). Geochemical zoning in metamorphic minerals. In: *Treatise on Geochemistry, The Crust* Volume **3**, ed. R. Rudnick. Amsterdam: Elsevier, 229–261.

Kohn, M.J. (2008). *P–T–t* data from central Nepal support critical taper and repudiate large-scale channel flow of the Greater Himalayan Sequence. *Geological Society of America Bulletin*, **120**, 259–273, doi:10.1130/B26252.

Kohn, M.J., Wieland, M.S., Parkinson, C.D. and Upreti, B.N. (2005). Five generations of monazite in Langtang gneisses: implications for chronology of the Himalayan metamorphic core. *Journal of Metamorphic Geology*, **23**, 399–406.

Koons, P.O. and Zeitler, P.K. (2002). Mechanical links between erosion and metamorphism in Nanga Parbat, Pakistan Himalaya. *American Journal of Science*, **302**, 499–773.

Korzhinskii, D.S. (1959). *Physicochemical Basis of the Analysis of the Paragenesis of Minerals*. New York: Consultant Bureau.

Krabbendam, M. (2001). When the Wilson Cycle breaks down: how orogens can produce strong lithosphere and inhibit their future reworking. In: *Continental Reactivation and Reworking*, ed. J.A. Miller, R.E. Holdsworth, I.S. Buick and M. Hand. *Geological Society of London Special Publication*, **184**, 57–75.

Kramers, J.D., Kreissig, K. and Jones, M.Q.W. (2001). Crustal heat production and style of metamorphism: a comparison between two Archaean high grade provinces in the Limpopo Belt, southern Africa. *Precambrian Research*, **112**, 149–163.

Kuhlemann, J., Frisch, W., Székely, B., Dunkl, I. and Kázmér, M. (2002). Post-collisional sediment budget history of the Alps: tectonic versus climate control. *International Journal of Earth Science*, **91**, 818–837.

Kutzbach, J.E., Guetter, P.J., Ruddiman, W.F. and Prell, W.L. (1989). The sensitivity of climate to late Cenozoic uplift in southern Asia and the American West. *Journal of Geophysical Research, Atmosphere*, **94**, 18,393–18,407.

Lapen, T.J., Johnson, C.M., Baumgartner, L.P. *et al.* (2007). Coupling of oceanic and continental crust during Eocene eclogite facies metamorphism: evidence from the Monte Rosa nappes, Western Alps. *Contributions to Mineralogy and Petrology*, **153**, 139–157.

Lapen, T.J., Johnson, C.M., Baumgartner, L.P. *et al.* (2003). Burial rates during prograde metamorphism of an ultra high pressure terrane: an example from Lago di Cignana, western Alps. *Earth and Planetary Science Letters*, **215**, 57–72.

Lapworth, C. (1883). The secret of the Highlands. *Geological Magazine*, **10**, 120–128.

Laubscher, H. (1992). The Alps – a transpressive pile of peels. In: *Thrust Tectonics*, ed. K. McClay. London and New York: Chapman and Hall, 277–285.

Law, R.D. (1987). Heterogeneous deformation and quartz crystallographic fabric transition: natural examples from the Moine thrust zone at the Stack of Glencoul northern Assynt. *Journal of Structural Geology*, **9**, 819–833.

Law, R.D., Casey, M. and Knipe, R.J. (1986). Kinematic and tectonic significance of microstructures and crystallographic fabrics within quartz mylonites from the Assynt and Eriboll regions of the Moine thrust zone, NW Scotland. *Transactions of the Royal Society of Edinburgh, Earth Sciences*, **77**, 9–123.

Lease, R.O., Burbank, D.W., Gehrels, G.E., Wang, Z. and Yuan, D. (2007). Signatures of mountain building: detrital U–Pb ages from northeastern Tibet. *Geology*, **35**, 239–242.

Leffler, L., Stein, S., Mao, A. *et al.* (2009). Constraints on present-day shortening rate across the Central Eastern Andes from GPS data. *Geophysical Research Letters*, **24**(9), 1031–1034.

Le Pichon, X., Henry, P. and Goffée, B. (1999). Uplift of Tibet from eclogites to granulites – implications for the Andean Plateau and the Variscan belt. *Tectonophysics*, **273**, 57–76.

Leslie, G. (2009). Border Skirmish. *Geoscientist*, **19**, (3), 16–20.

Liati, A. and Froitzheim, N. (2006). Assessing the Valais Ocean, Western Alps: U–Pb Shrimp zircon geochronology of eclogite in the Balma unit, on top of the Monte Rosa nappe. *European Journal of Mineralogy*, **18**, 299–308.

Lister, G.A., Forster, M.A. and Rawling, T.J. (2001). Episodicity in Orogenesis. In: *Continental Reactivation and Reworking*, ed. J.A. Miller, R.E. Holdsworth, I.S. Buick and M. Hand. *Geological Society of London Special Publication*, **184**, 89–113.

Lopez, A. (2006). Geological Society of America Abstracts, *Special meeting* no. **2**, p.30.

Love, G.J., Kinny, P.D. and Friend, C.R.L. (2010). Palaeoproterozoic terrane assembly in the Lewisian Gneiss Complex on the Scottish mainland, south of Gruinard Bay: SHRIMP U–Pb zircon evidence. *Precambrian Research*, **183**, 89–111.

Lui, H., McClay, K.R. and Powell, D. (1992). Physical modelling of thrust wedges. In: *Thrust Tectonics*, ed. K.R. McClay. London and New York: Chapman and Hall, 71–81.

Lui, M., Youqing, Y., Stein, S., Yuanqing, Z. and Engeln, J. (2000). Crustal shortening in the Andes: why do GPS rates differ from geological rates? *Geophysical Research Letters*, **27**(18), 3005–3008.

Luth, S.W. and Willingshofer, E. (2008). Mapping of the post-collisional cooling history of the Eastern Alps. *Swiss Journal of Geology*, **101**, 207–223.

Lyon-Caen, H. and Molnar, P. (1985). Gravity anomalies, flexure of the Indian plate, and the structure, support and evolution of the Himalaya and Ganga basin. *Tectonics*, **4**, 513–538.

Marshak, S. (1999). Deformation style way back when: thoughts on the contrasts between Archean/Paleoproterozoic and contemporary orogens. *Journal of Structural Geology*, **21**, 1175–1182.

Martin, A. (1986). Effect of steeper Archaean geothermal gradient on geochemistry of subduction zone magmas. *Geology*, **14**, 753–756.

Massey, J.A., Reddy, S.M., Harris, N.B.W. and Harmon, R.S. (1994). Correlation between melting, deformation and fluid interaction in the continental crust of the high Himalayas, Langtang Valley, Nepal. *Terra Nova*, **6**, 229–237.

McCaffrey, R. and Nabelek, J. (1998). Role of oblique convergence in the active deformation of the Himalayas and southern Tibet. *Geology*, **26**, 691–694.

McClay, K.R. (ed.) (1992). *Thrust Tectonics*. London and New York: Chapman and Hall.

McClelland, W.C. and Gilotti, J.A. (2003). Late-stage extensional exhumation of high-pressure granulites in the Greenland Caledonides. *Geology*, **31**, 259–262.

McKenzie, D. (1978). Active tectonics of the Alpine-Himalayan belt: the Aegean sea and surrounding regions. *Geophysical Journal of the Royal Astronomical Society*, **55**, 217–254.

McKenzie, D. and Priestley, K. (2008). The influence of lithospheric thickness variations on continental evolution. *Lithos*, **102**, 1–11.

McQuarrie, N. (2002). Initial plate geometry, shortening variations and evolution of the Bolivian Orocline. *Geology*, **30**, 867–870.

McQuarrie, N., Horton, B.R., Zandt, G., Beck, S. and DeCelles, P.G. (2005). Lithospheric evolution of the Andean fold-and-thrust belt, Bolivia, and the origin of the central Andean Plateau. *Tectonophysics*, **399**, 15–37.

Meffan-Main, S., Cliff, R.A., Baumgartner, L.P., Lombardo, B. and Campagnoni, R. (2004). A Tertiary age for Alpine high pressure metamorphism in Gran Paradiso Massif, Western Alps: a Rb–Sr microsampling study. *Journal of Metamorphic Geology*, **2**, 261–281.

Merle, O. (1986). Patterns of stress trajectories and strain rates in spreading gliding nappes. *Tectonophysics*, **124**, 211–222.

Merle, O. and Abidi, N. (1995). Approche expérimentale du fontionnement des ramps émergentes. *Société Géologue du France*, **166**, 5, 439–451.

Merschat, A.J., Hatcher, R.D. and Davis, T.L. (2005). The northern inner Piedmont, S. Appalachians, U.S.A.: kinematics of transpression and S–W directed mid-crustal flow. *Journal of Structural Geology*, **27**, 1252–1281.

Mezger, K., Rawnsley, C.M., Bohlen, S.R. and Hanson, G.N. (1991). U–Pb garnet, sphene, monazite, and rutile ages – implications for the duration of high-grade metamorphism and cooling histories, Adirondack Mountains, New York. *Journal of Geology*, **99**, 415–428.

Micheels, A., Bruch, A.A., Uhl, D., Utescher, T. and Mosbrugger, V. (2007). A late Miocene climate model simulation with ECHAM4/ML and its quantitative validation with terrestrial proxy data. *Palaeogeography, Palaeoclimatology, Palaeoecology*, **253**, 251–270.

Miller, J.A., Holdsworth, R.E., Buick, I.S. and Hand, M. (eds.) (2001). *Continental Reactivation and Reworking. Geological Society of London Special Publication*, **184**.

Milnes, A.G. and Koyi, H.A. (2000). Ductile rebound of an orogenic root: case study and numerical model of gravity tectonics in the Western Gneiss Complex, Caledonides, southern Norway. *Terra Nova*, **12**, 1–7.

Miyashiro, A. (1949). The stability relation of kyanite, sillimanite and andalusite, and the physical conditions of the metamorphic processes. *Journal of the Geological Society of Japan*, **55**, 218–223.

Miyashiro, A. (1961). Evolution of metamorphic belts. *Journal of Petrology*, **2**, 277–311.

Miyashiro, A. (1973). *Metamorphism and Metamorphic Belts*. London: George Allen and Unwin.

Möller, A., Hensen, B.J., Armstrong, R.A., Mezger, K. and Ballèvre, M. (2003). U–Pb zircon and monazite age constraints on granulite-facies metamorphism and deformation in the Strangways Metamorphic Complex (central Australia). *Contributions to Mineralogy and Petrology*, **145**, 406–423.

Molnar, P. (1988). A review of geophysical constraints on the deep structure of the Tibetan Plateau, the Himalaya and the Karakoram, and their tectonic implications. In: *Tectonic Evolution of the Himalayas and Tibet*, ed. R.M. Shackleton, J.F. Dewey and B.F. Windley. *Philosophical Transactions of the Royal Society, London*, A**326**, 3–16.

Molnar, P. (2001). Climate change, flooding in arid environments, and erosion rates. *Geology*, **29**, 1071–1074.

Molnar, P. and England, P.C. (1990). Late Cenozoic uplift of mountain ranges and global climate change: chicken or egg? *Nature*, **346**, 29–34.

Molnar, P., England, P.C. and Martinod, J. (1993). Mantle dynamics, uplift of the Tibetan Plateau and the Indian monsoon. *Reviews of Geophysics*, **31**, 357–396.

Molnar, P. and Garzione, C. (2007). Bounds on the viscosity coefficient of continental lithosphere from removal of mantle beneath Altiplano and Eastern Cordillera. *Tectonics*, **26**, TC 2013, doi:10.1029/ 2006 TC 001964.

Molnar, P., Houseman, G.A. and England, P.C. (2006). Palaeoaltimetry of Tibet. *Nature, Earth Sciences*, **444**, E4.

Molnar, P. and Stock, J.M. (2009). Slowing of Indian convergence with Asia since 20 Ma and its implications for Tibetan mantle dynamics. *Tectonics*, **28**, TC 3001, doi:10.1029/2008 TC2271.

Molnar, P. and Tapponnier, P. (1975). Cenozoic tectonics of Asia: effects of a continental collision. *Science*, **189**, 419–426.

Molnar, P. and Tapponnier, P. (1977). The relation of the tectonics of eastern Asia to the India–Asia collision: an application of slip line field theory to large scale continental tectonics. *Geology*, **5**, 212–216.

Moraes, R., Brown, M., Fuck, R.A., Camargo, M.A. and Lima, T.M. (2002). Characterization and *P–T* evolution of meltbearing ultrahigh-temperature granulites: an example from the Anápolis–Itauçu Complex of the Brasília fold belt, Brazil. *Journal of Petrology*, **43**, 1673–1705.

Morag, N., Avigad, D., Harlavan, Y., McWilliams, M.O. and Michard, A. (2008). Rapid exhumation and mountain building in the Western Alps: petrology and $^{40}Ar/^{39}Ar$ geochronology of detritus from Tertiary basins of southeastern France. *Tectonics*, **27**, TC 2004, doi:1029/2007.

Murphy, J.B., Keppie, J.D. and Hynes, J. (2009). Ancient orogens and modern analogues: an introduction. In: *Ancient Orogens and Modern Analogues*, ed. J.B. Murphy, J.D. Keppie and J. Hynes. *Geological Society of London Special Publication*, **327**, 1–8.

Murphy, M.A., An Yin, Harrison, T.M. *et al.* (1997). Did the Indo-Asian collision alone create the Tibetan Plateau? *Geology*, **25**, 719–722.

Najman, Y. (2006). The sediment record of orogenic evolution: a review of approaches and techniques used in the Himalaya. *Earth Science Reviews*, **74**, 1–72.

Najman, Y., Carter, A., Oliver, G.J.H. and Garzanti, E. (2005). Provenance of Eocene foreland basin sediments, Nepal: Constraints to the timing and diachroneity of early Himalayan orogenesis. *Geology*, **33**, 309–312.

Najman, Y., Johnson, K., White, N. and Oliver, G.J.H. (2004). Evolution of the Himalayan foredeep basin, NW India. *Basin Research*, **16**, 1–24.

Najman, Y., Pringle, M., Godin, L. and Oliver, G.J.H. (2001). Dating the oldest continental sediments from the Himalayan foredeep basin. *Nature*, **410**, 194–197.

Nance, D.R., Murphy, J.B., Strachan, R.A. *et al.* (2008). Neoproterozoic-early Palaeozoic tectonostratigraphy and palaeogeography of the peri-Gondwanan terranes: Amazonian v. West African connections. In: *The Boundaries of the West African Craton*, ed. N. Ennih and J-P. Liégeois. *Geological Society of London Special Publication*, **297**, 345–383.

Naylor, M. and Sinclair, H.D. (2007). Punctuated thrust deformation in the context of doubly vergent thrust wedges: Implications for the location of uplift and exhumation. *Geological Society of America Bulletin*, **35**, 599–562.

Nelson, K.D., Zhao, W., Brown, L.D. *et al.* (1996). Partially molten middle crust beneath Southern Tibet: synthesis of Project INDEPTH results. *Science*, **274**, 1684–1688.

Nisbet, E.G., Cheadle, M.J., Arndt, N.T. and Bickle, M.J. (1993). Constraining the potential temperature of the Archaean mantle – a review of the evidence from Komatiites. *Lithos*, **30**, 291–307.

Norris, R.J. and Cooper, A.F. (2001). Late Quaternary slip rates and slip partitioning on the Alpine fault, New Zealand. *Journal of Structural Geology*, **23**, 2–3, 507–552.

Nutman, A.P., Friend, C.R.L., Kinny, P.D. and McGregor, V.R. (1993). Anatomy of an early Archaean gneiss complex – 3900 to 3600 Ma crustal evolution in southwest Greenland. *Geology*, **21**, 415–418.

Nutman, A.P., McGregor, V.R., Friend, C.R.L., Bennett, V. and Kinny, P.D. (1996). The Itsaq Gneiss complex of southern West Greenland: the world's most extensive record of early crustal evolution. *Precambrian Research*, **78**, 1–39.

O'Brien, P.J. (2001). Subduction followed by collision: Alpine and Himalayan examples: *Physics of the Earth and Planetary Interiors*, **127**, 277–291.

O'Brien, P.J. (2008). Challenges in high-pressure granulite metamorphism in the era of pseudosections: reaction textures, compositional zoning and tectonic interpretation with examples from the Bohemian Massif. *Journal of Metamorphic Geology*, **26**, 235–251.

O'Brien, P.J. and Rötzler, J. (2003). High-pressure granulites: formation recovery of peak conditions and implications for tectonics. *Journal of Metamorphic Geology*, **21**, 3–20.

Oldow, J.S., Bally, A.W., Avé Lallemant, H.G. and Leeman, W.P. (1989). Phanerozoic evolution of the North American Cordillera: United States and Canada. In: *The Geology of North America – An Overview*, ed. Bally, A.W. & Palmer, A.R. Boulder: Geological Society of America, 139–232.

Osanai, Y., Nakano, N., Owada, M. *et al.* (2004). Permo-Triassic ultrahigh-temperature metamorphism in the Kontum massif, central Vietnam. *Journal of Mineralogical and Petrological Sciences*, **99**, 225–241.

Ouzegane, K. and Boumaza, S. (1996). An example of ultrahigh temperature metamorphism: orthopyroxene–sillimanite–garnet, sapphirine–quartz, and spinel–quartz parageneses in Al–Mg granulites from In Hihaou, In Ouzzal, Hoggar. *Journal of Metamorphic Geology*, **14**, 693–708.

Owens, T.J. and Zandt, G. (1997). Implications of crustal property variations for models of Tibetan Plateau evolution. *Nature*, **387**, 37–43.

Oxburgh, E.R. and Turcotte, D.L. (1970). Thermal structure of island arcs. *Geological Society of America Bulletin*, **81**, 1665–1688.

Parrish, R.R., Gough, S.J., Searle, M.P. and Waters, D.J.W. (2006). Plate velocity exhumation of ultra high pressure eclogites in Pakistan. *Geology*, **34**, 989–992.

Patiño Douce, A.E. and Johnston, A.D. (1991). Phase equilibria and melt productivity in the pelitic system: implications for the origin of peraluminous granitoids and aluminous granulites. *Contributions to Mineralogy and Petrology*, **107**, 202–218.

Pattison, D.R.M., Chacko, T., Farquhar, J. and McFarlane, C.R.M. (2003). Temperatures of granulite-facies metamorphism: constraints from experimental phase equilibria and thermobarometry corrected for retrograde exchange. *Journal of Petrology*, **44**, 867–900.

Peach, B.N., Horne, J., Gunn, W., Clough, C.T. and Hinxman, L.W. (1907). *The Geological Structure of the North-west Highlands of Scotland. Memoir of the Geological Survey of Great Britain.*

Peacock, S. (1992). Blueschist-facies metamorphism, shear heating, and *P–T–t* paths in subduction shear zones. *Journal of Geophysical Research*, **97**, 17693–17707.

Pecher, A. and Le Fort, P. (1986). The metamorphism in central Himalaya, its relation with the thrust tectonics. *Sciences de la Terre Memoire*, **47**, 285–309.

Perchuk, L.L. and Lavrent'eva, I.V. (1983). Experimental investigation of exchange equilibria in the system cordierite–garnet–biotite. In: *Kinetics and Equilibrium in Mineral Reactions*, ed. S.K. Saxena. New York: Springer-Verlag, 199–239.

Pfiffner, A. (1986). Evolution of the north Alpine foreland basin in the central Alps. In: *Foreland Basins*, ed. P.A. Allen and P. Homewood. Oxford: Blackwell Scientific Publications, 219–228.

Pfiffner, A. (1992). Alpine orogeny. In: *A Continent Revealed: The European Geotraverse*, ed. D. Blundell, R. Freeman, and S. Mueller. Cambridge: Cambridge University Press, 180–190.

Pfiffner, O.A., Ellis, S. and Beaumont, C. (2000). Collision tectonics in the Swiss Alps: insights from geodynamic modelling. *Tectonics*, **19**, 1065–1094.

Piotrowski, A.M., Lee, D.-C., Christensen, J.N. *et al.* (2000). Changes in erosion and ocean circulation recorded in Hf isotopic compositions of North Atlantic and Indian Ocean ferromanganese crusts. *Earth and Planetary Science Letters*, **181**, 315–325.

Pisarevsky, S.A. (2008). Neoproterozoic–early Palaeozoic tectonostratigraphy and palaeogeography of the peri-Gondwanan terranes: Amazonian v. West African connections. In: *The Boundaries of the West African Craton*, ed. N. Ennih and J-P. Liégeois. *Geological Society of London Special Publication*, **297**, 345–383.

Pisarevsky, S.A., Murphy, J.B., Cawood, P.A. and Collins, A.S. (2008). Late Neoproterozoic and Early Cambrian palaeogeography: models and problems. In: *West Gondwana: Pre-Cenozoic Correlations Across the South Atlantic Region*, ed. R.J. Pankhurst, R.A.J. Trouw, B.B. Brito Neves and M.J. De Wit. *Geological Society of London Special Publication*, **294**, 9–31.

Platt, J.P. (1986). Dynamics of orogenic wedges and the uplift of high-pressure metamorphic rocks. *Geological Society of America Bulletin*, **97**, 1037–1053.

Platt, J.P., Allerton, S., Kirker, A. *et al.* (2003a). The ultimate arc: differential displacement, oroclinal bending and vertical axis rotation in the External Betic–Rif arc. *Tectonics*, **22**(3), 1017, doi:10.1029/200ITC001321.

Platt, J.P. and England, P.C. (1993). Convective removal of lithosphere beneath mountain belts: thermal and metamorphic consequences. *American Journal of Science*, **293**, 307–336.

Platt, J.P., Soto, J.-I., Whitehouse, M.J., Hurford, A.J. and Kelley, S.P. (1998). Thermal evolution, rate of exhumation, and tectonic significance of metamorphic rocks from the floor of the Alboran extensional basin, western Mediterranean. *Tectonics*, **17**, 671–689.

Platt, J.P., Whitehouse, M.J., Kelley, S.P., Carter, A. and Hollick, L. (2003b). Simultaneous extensional exhumation across the Alboran Basin: implications for the causes of late orogenic extension. *Geology*, **31**, 251–254.

Powell, R. (1978). *Equilibrium Thermodynamics in Petrology*. Harper & Row, London.

Powell, R. and Holland, T.J.B. (1988). An internally consistent dataset with uncertainties and correlations: 3. Applications to geobarometry, worked examples and a computer program. *Journal of Metamorphic Geology*, **6**, 173–204.

Powell, R., Holland, T.J.B. and Worley, B. (1998). Calculating phase diagrams involving solid solutions via non-linear equations, with examples using THERMOCALC. *Journal of Metamorphic Geology*, **16**, 577–588.

Prell, W.L. and Kutzbach, J.E. (1992). Sensitivity of the Indian monsoon to forcing parameters and implications for its evolution. *Nature*, **360**, 647–651.

Price, R.A. (1981). The Cordilleran foreland thrust and fold belt in the southern Canadian Rockies. In: *Thrust and Nappe Tectonics*, ed. K. McClaym and N.J. Price. *Geological Society of London Special Publication*, **9**, 427–448.

Priestley, K., Debayle, E., McKenzie, D. and Barron, J. (2007). The upper mantle 3D Sv wave speed structure beneath Tibet. *American Geophysical Union, Abstract.*

Priestley, K., Jackson, J.A. and McKenzie, D. (2008). Lithospheric structure and deep earthquakes beneath India, the Himalaya and southern Tibet. *Geophysical Journal International*, **172**, 345–362.

Pyle, J.M. and Spear, F.S. (2003). Four generations of accessory-phase growth in low-pressure migmatites from SW New Hampshire. *American Mineralogist*, **88**, 338–351.

Raith, M., Karmakar, S. and Brown, M. (1997). Ultra-high-temperature metamorphism and multistage decompressional evolution of sapphirine granulites from the Palni Hill Ranges, southern India. *Journal of Metamorphic Geology*, **15**, 379–399.

Ramoz, V.A. (1999). Plate tectonic setting of the Andean Cordillera. *Episodes*, **22**, 183–190.

Ramsay, J.G. (1969). The measurement of strain and displacement in orogenic belts. In: *Time and Place in Orogeny*, ed. P.E. Kent, G.E. Satterthwaite and A.M. Spencer. *Geological Society of London Special Publication*, **3**, 43–79.

Ramsay, J.G. (1992). Some geometric problems of ramp-flat thrust models. In: *Thrust Tectonics*, ed. K.R. McClay. London and New York: Chapman and Hall, 191–200.

Ramsay, J.G., Casey, M. and Kligfield, R. (1983). The role of shear in the development of the Helvetic thrust belt of Switzerland. *Geology*, **11**, 439–442.

Ramsay, J.G. and Huber, M. (1983 and 1987). *The Techniques of Modern Structural Geology* Volume 1: *Strain Analysis;* Volume 2: *Folds and Fractures*. London, New York: Academic Press.

Ramsay, J.G. and Lisle, R. (2000). *The Techniques of Modern Structural Geology* Volume 3: *Applications of Continuum Mechanics in Structural Geology*. London, New York: Academic Press.

Rapp, R.P., Shimizu, N. and Norman, M.D. (2003). Growth of early continental crust by partial melting of eclogite. *Nature*, **425**, 605–609.

Ratschbacher, L., Frisch, W., Linzer, H-G. and Merle, O. (1991b). Lateral extrusion in the Eastern Alps, Part 2: Structural analysis. *Tectonics*, **40**, 339–350.

Ratschbacher, L., Merle, O., Davy, P. and Cobbold, P. (1991a). Lateral extrusion in the eastern Alps, Part 1: Boundary conditions and experiments scaled for gravity. *Tectonics*, **10**(2), 245–256.

Raymo, M.E. and Ruddiman, W.F. (1992). Tectonic forcing of late Cenozoic climate. *Nature*, **359**, 121–122.

Raymo, M.E., Ruddiman, W.F. and Froelich, P.N. (1988). Influence of late Cenozoic mountain building on ocean geochemical cycles. *Geology*, **16**, 649–653.

Reddy, S.M., Searle, M.P. and Massey, J.A. (1993). Structural evolution of the High Himalayan gneiss sequence of the Langtang Valley, Nepal. In: *Himalayan Tectonics*. ed. P.J. Treloar and M.P. Searle, *Geological Society of London Special Publication*, **74**, 375–390.

Reddy, S.M., Timms, N.E., Pantleon, W. and Trimby, P. (2007). Quantitative characterization of plastic deformation of zircon and geological implications. *Contributions to Mineralogy and Petrology*, **153**, 625–645.

Reiner, P.W. and Brandon, P.K. (2006). Using thermochronology to understand orogenic erosion. *Annual Reviews in Earth and Planetary Science*, **34**, 419–466.

Reiners, P.W. and Brandon, M.T. (2006). Using thermochronology to understand orogenic erosion. *Annual Reviews in Earth and Planetary Science*, **34**, 419–773.

Rey, P.F. and Houseman, G. (2006). Lithospheric scale gravitational flow: the impact of body forces on orogenic processes from Archaean to Phanerozoic. In: *Analogue and Numerical Modelling of Crustal-Scale Processes*, ed. S.J. Buiter and G. Schreurs. *Geological Society of London Special Publication*, **253**, 153–167.

Rey, P.F., Teyssier, C. and Whitney, D.L. (2009). The role of partial melting and extensional strain rates in the development of metamorphic core complexes. *Tectonophysics*, **477**, 135–144.

Rey, P., Vanderhaeghe, O. and Teyssier, C. (2001). Gravitational collapse of the continental crust: definition, regimes, and modes. *Tectonophysics*, **342**, 435–449.

Richardson, S.W. and England, P.C. (1979). Metamorphic consequences of crustal eclogite production in overthrust zones. *Earth and Planetary Science Letters*, **42**, 183–190.

Richardson, S.W., Gilbert, M.C. and Bell, P.M. (1969). Experimental determination of kyanite–andalusite and andalusite–sillimanite equilibria; the aluminium silicate triple point. *American Journal of Science*, **267**, 259–272.

Richter, F.N. (1985). Models for the Archaean thermal regime. *Earth and Planetary Science Letters*, **73**, 350–360.

Ring, U., Brandon, M.T., Lister, G.S. and Willett, S.D. (eds.) (1999). *Exhumation Processes: Normal Faulting, Ductile Flow and Erosion. Geological Society of London Special Publication*, **154**.

Rivers, T. (2009). The Grenville Province as a large hot long-duration collisional orogen: insights from the spatial and thermal evolution of its orogenic fronts. In: *Ancient Orogens and Modern Analogues*, ed. J.B. Murphy, J.D. Keppie and A.J. Hynes. *Geological Society of London Special Publication*, **327**, 405–444.

Roberts, E.A. and Houseman, G.A. (2001). Geodynamics of central Australia during the intraplate Alice Springs Orogeny: thin viscous models. In: *Continental Reactivation and Reworking*, ed. J.A. Miller, R.E. Holdsworth, I.S. Buick and M. Hand. *Geological Society of London Special Publication*, **184**, 139–164.

Rosenbaum, G. and Lister, G. (2005). The Western Alps from the Jurassic to Oligocene – spatio-temporal constraints and evolutionary reconstructions. *Earth Science Reviews*, **69** (3–4), 281–306.

Rosenbaum, G. and Lister, G.S. (2004). Formation of arcuate orogenic belts in the western Mediterranean region. In: *Orogenic Curvature: Integrating Paleomagnetic and Structural Analyses*, ed. A.J. Sussman and A.R. Weil. *Geological Society of America Special Paper*, **383**, 41–56.

Rowan, M.G. and Kligfield, R. (1992). Kinematics of large-scale asymmetric buckle folds in overthrust shear: an example from the Helvetic Alps. In: *Thrust Tectonics*, ed. K.R. McClay. London, New York: Chapman and Hall, 165–173.

Rowley, D.B. and Currie, B.S. (2006). Palaeoaltimetry of the late Eocene Miocene Lumpola basin, central Tibet. *Nature*, **439**, 677–681. See also discussion by Molnar, Houseman and England in the same issue.

Royden, L.H. (1993a). The tectonic expression slab pull at continental convergent boundaries. *Tectonics*, **12**, 303–325.

Royden, L.H. (1993b). The evolution of retreating subduction boundaries formed during continental collision. *Tectonics*, **12**, 629–638.

Royden, L.H., Burchfiel, B.C., King, R.W. *et al.* (1997). Surface deformation and lower crustal flow in eastern Tibet. *Science*, **276**, 788–790.

Rubatto, D. (2002). Zircon trace element geochemistry: partitioning with garnet and the link between U–Pb ages and metamorphism. *Chemical Geology*, **184**, 123–138.

Rubatto, D. and Hermann, J. (2001). Exhumation as fast as subduction? *Geology*, **29**, 3–6.

Ruddiman, W.F. and Kutzbach, J.E. (1989). Cenozoic northern hemisphere climate by uplift in southern Asia and the American West. *Journal of Geophysical Research, Atmosphere*, **94**, 18409–18427.

Rudnick, R.L. and Gao, S. (2003). The composition of the continental crust. In: *The Crust.* Treatise on Geochemistry, **3** (ed. R.L. Rudnick; series eds. H.D. Holland and K.K. Turekian). Oxford: Elsevier-Pergamon, 1–64.

Sandiford, M. (1989a). Secular trends in the thermal evolution of metamorphic terrains. *Earth and Planetary Science Letters*, **95**, 85–96.

Sandiford, M. (1989b). Horizontal structures in granulite terrains: a record of mountain building or mountain collapse? *Geology*, **17**, 449–452.

Sandiford, M., Hand, M. and McLaren, S. (1998). High geothermal gradient metamorphism during thermal subsidence. *Earth and Planetary Science Letters*, **163**, 149–165.

Sandiford, M., Hand, M. and McLaren, S. (2001). Tectonic feedback, intraplate orogeny and the geochemical structure of the crust: a central Australian perspective. In: *Continental Reactivation and Reworking*, ed. J.A. Miller, R.E. Holdsworth and S. McLaren. *Geological Society of London Special Publication*, **184**, 195–218.

Sandiford, M. and McLaren, S. (2002). Tectonic feedback and the ordering of heat producing elements within the continental lithosphere. *Earth and Planetary Science Letters*, **204**, 133–150.

Sandiford, M. and Powell, R. (1990). Some isostatic and thermal consequences of the vertical strain geometry in convergent orogens. *Earth and Planetary Science Letters*, **98**, 154–165.

Sandiford, M. and Powell, R. (1991). Some remarks on high-temperature – low pressure metamorphism in convergent orogens. *Journal of Metamorphic Geology*, **9**, 333–340.

Sarkarinejad, K. and Azizi, A. (2008). Slip partitioning and inclined dextral transport during the Zagros Thrust System, Iran. *Journal of Structural Geology*, **30/1**, 116–136.

Sawyer, E.W. (1994). Melt segregation in the continental crust. *Geology*, **22**, 1019–1022.

Sawyer, E.W. (1999). Criteria for the recognition of partial melt. *Physics and Chemistry of the Earth*, **24**, 269–279.

Sawyer, E.W. (2001). Melt segregation in the continental crust: Distribution and movement of melt in anatectic rocks. *Journal of Metamorphic Geology*, **18**, 291–309.

Schaltegger, U., Fanning, C.M., Günther, D. *et al.* (1999). Growth, annealing and recrystallisation of zircon and preservation of monazite in high-grade metamorphism: conventional and in-situ U–Pb isotope, cathodoluminescence and microchemical. *Contributions to Mineralogy and Petrology*, **134**, 186–201.

Schellart, W.P. and Lister, G.S. (2004). Tectonic models for the formation of arc-shaped convergent zones and backarc basins. In: *Orogenic Curvature: Integrating Paleomagnetic and Structural Analyses*, ed. A.J. Sussman and A.B. Weil. *Geological Society of America Special Paper*, **383**, 237–258.

Schertl, H.R., Schreyer, W. and Chopin, C. (1991). The pyrope-coesite rocks and their country rocks at Parigi, Dora Maira massif, western Alps: detailed petrography, mineral chemistry and PT-path. *Contributions to Mineralogy and Petrology*, **108**, 1–21.

Schlunegger, F. and Willet, S. (1999). Spatial and temporal variations in exhumation in the central Swiss Alps and implications for exhumation mechanisms. *Journal of the Geological Society of London*, **154**, 157–179.

Schmid, S.M. and Kissling, E. (2000). The arc of the western Alps in the light of geophysical data on deep crustal structure. *Tectonics*, **19**, 62–85.

Schoenbohm, L.M., Burchfiel, B.C. and Chen, L. (2006). Propagation of surface uplift, lower crustal flow, and Cenozoic tectonics of the southeast margin of the Tibetan Plateau. *Geology*, **34**, 813–816.

Schreyer, W. (1988). Experimental studies on metamorphism of crustal rocks under mantle pressures. *Mineralogical Magazine*, **52**, 1–26.

Schreyer, W. (1995). Ultradeep metamorphic rocks: the retrospective viewpoint. *Journal of Geophysical Research*, **100** B, 8353–8366.

Schulmann, K., Lexa, O., Stipska, P. *et al.* (2008). Vertical extrusion and horizontal channel flow of orogenic lower crust: key exhumation mechanisms in large hot orogens? *Journal of Metamorphic Geology*, **26**, 273–297.

Schulte-Pelkum, V., Monsalve, G., Sheehan, A. *et al.* (2003). Indepth 111 Seismic Team. Seismic imaging of the downwelling Indian lithosphere beneath Tibet. *Science*, **300**, 1424–1427.

Schulte-Pelkum, V., Monsalve, G., Sheehan, A. *et al.* (2005). Imaging the Indian subcontinent beneath the Himalaya. *Nature*, **435**, 1222–1225.

Searle, M.P. (1997). Shisha Pangma leucogranite, South Tibetan Himalaya: field relations, geochemistry, age, and emplacement. *The Journal of Geology*, **105**, 295–317.

Searle, M.P. (2007). Diagnostic features and processes in the construction and evolution of Oman-, Zagros-, Himalayan-, Karakoram-, and Tibetan-type orogenic belts. In: *4D Framework of Continental Crust*, ed. R.D. Hatcher, M.P. Carlson, J.H. McBride and J.R. Martínez-Catalán. *Geological Society of America Memoir*, **200**, 1–20, doi:10.1130/2007.1200 (04).

Searle, M.P., Law, R.D., Godin, L. *et al.* (2008). Defining the Himalayan Main Central Thrust in Nepal. *Journal of the Geological Society of London*, **165**, 523–534.

Searle, M.P., Law, R.D. and Jessup, M.J. (2006). Crustal structure, restoration and evolution of the Greater Himalaya in Nepal–south Tibet: implications for channel flow and ductile extrusion of the middle crust. In: *Channel Flow, Ductile Extrusion and Exhumation in Continental Collision Zones*, ed. R.D. Law, M.P. Searle and L. Godin. *Geological Society of London Special Publication*, **268**, 355–378.

Searle, M.P., Noble, R., Hurford, A.J. and Rex, D.C. (1999a). Age of crustal melting, emplacement and exhumation history of the Shivling Leucogranite, Garhwal Himalaya. *Geological Magazine*, **136**, 513–525.

Searle, M.P., Simpson, R.L., Law, R.D., Parrish, R.R. and Waters, D.J. (2003). The structural geometry, metamorphic and magmatic evolution of the Everest massif, High Himalaya on Nepal–south Tibet. *Journal of the Geological Society of London*, **160**, 345–366.

Searle, M.P., Stephenson, B., Walker, J. and Walker, C. (2007). Restoration of the Western Himalaya: implications for metamorphic protoliths, thrust and normal faulting, and channel flow models. *Episodes*, **30**(4), 242–257.

Searle, M.P., Waters, D.J., Rex, D.C. and Wilson, R.N. (1992). Pressure, temperature and time constraints on Himalayan metamorphism from eastern Kashmir and western Zanskar. *Journal of the Geological Society of London*, **149**, 753–773.

Searle, M.P., Waters, D.J.W., Dransfield, M.W. *et al.* (1999b). *Thermal and Mechanical Models for the Evolution of the Zanskar Himalaya. Geological Society of London Special Publication*, **364**, 139–156.

Sederholm, J.J. (1907). Om granit och gneis. *Bulletin de la Commission Geologique de Finlande*, **23**.

Seward, D. and Burg, J.-P. (2008). Growth of Namche Barwa syntaxis and associated evolution of the Tsangpo gorge: constraints from structural and thermogeochronologic data. *Tectonophysics*, **451**, 282–289.

Sibson, R. (1977). Fault rocks and fault mechanism. *Journal of the Geological Society of London*, **133**, 191–213.

Sinclair, H.D. and Allen, P.A. (1992). Vertical versus horizontal motions in the Alpine orogenic wedge: stratigraphic response in the foreland basin. *Basin Research*, **4**, 215–232.

Sinclair, H.D., Gibson, M., Naylor, M. and Morris, R.G. (2006). Asymmetric growth of the Pyrenees revealed through measurement and modelling of orogenic fluxes. *American Journal of Science*, **305**, 369–406.

Sizova, E., Gerya, T., Brown, M. and Perchuk, L.L. (2010). Subduction styles in the Precambrian: insight from numerical experiments. *Lithos*, **116**, 209–229.

Sleep, N.H. and Blanpied, M.L. (1992). Creep, compaction and the weak rheology of major faults. *Nature*, **359**, 687–692.

Smith, D.C. (1984). Coesite in clinopyroxene in the Caledonides and its implications for geodynamics. *Nature*, **310**, 641–644.

Smithies, R.H., Champion, D.C. and Cassidy, K.F. (2003). Formation of Earth's early Archaean continental crust. *Precambrian Research*, **127**, 89–101.

Sobolev, N.V. and Shatsky, V.S. (1990). Diamond inclusions in garnets from metamorphic rocks: a new environment for diamond formation. *Nature*, **343**, 742–746.

Spear, F.S. (1993). *Metamorphic Phase Equilibria and Pressure-Temperature Time Paths*. Washington: Mineralogical Society of America Monograph.

Spear, F.S., Kohn, M.J. and Cheney, J.T. (1999). *P–T* paths from anatectic pelites. *Contributions to Mineralogy and Petrology*, **134**, 17–32.

Spear, F.S. and Parrish, R.R. (1996). Petrology and cooling rates of the Valhalla complex, British Columbia, Canada. *Journal of Petrology*, **37**, 733–765.

Spear, F.S., Selverstone, J., Hickmott, D., Crowley, P. and Hodges, K.V. (1984). *P–T* paths from garnet zoning: a new technique for deciphering tectonic processes in crystalline terranes. *Geology*, **12**, 87–90.

Spiegel, C., Kuhlemann, J., Dunkl, I. *et al.* (2000). The erosion history of Central Alps: evidence from zircon fission track data from foreland basin sediments. *Nova*, **12**, 163–170.

Spicer, R.A., Harris, N.B.W. and Widdowson, M. (2003). Constant elevation of southern Tibet over the past 15 million years. *Nature*, **421**, 622–624.

Srivastava, P. and Mitra, G. (1994). Thrust geometries and deep structure of the outer and lesser Himalaya, Kumaon and Garhwal (India): implications for evolution of the Himalaya fold-and-thrust belt. *Tectonics*, **13**, 89–109.

Steck, A. (2003). Geology of the NW Indian Himalaya. *Eclogae Geologicae Helvetia*, **96**, 147–213.

Steppuhn, A., Micheels, A., Geiger, G. and Mosbrugger, V. (2006). Reconstructing the Late Miocene climate and oceanic flux using the AGCM ECHAM4 coupled to a mixed-layer ocean model with adjusted flux correction. *Palaeogeography, Palaeoclimatology, Palaeoecology*, **238**, 399–423.

Stevens, G. and Clemens, J.D. (1993). Fluid-absent melting and the roles of fluids in the lithosphere: a slanted summary? *Chemical Geology*, **108**, 1–17.

Stewart, R.J., Hallet, B., Zeitler, P.K. *et al.* (2008). Brahmaputra sediment flux dominated by highly localised rapid erosion from easternmost Himalaya. *Geology*, **36**, 711–714.

Stöckhert, B. and Gerya, T.V. (2005). Pre-collisional high pressure metamorphism and nappe tectonics at active continental margins: A numerical simulation. *Terra Nova*, **17**, 102–110.

Stockmal, G.S., Beaumont, C., Nguyen, M. and Lee, B. (2007). Mechanics of thin-skinned fold and thrust belts: insights from numerical models. *Geological Society of America Special Paper*, **433**, 63–98.

St-Onge, M.R., Searle, M.P. and Wodicka, N. (2006). Trans-Hudson Orogen of North America and Himalaya–Karakoram–Tibetan Orogen of Asia: structural and thermal characteristics of the lower and upper plates. *Tectonics*, **25**, TC 4006, 1–22.

Strachan, R.A. (2003). The Northern Highlands. In: *The Geology of Scotland* 4th edition, ed. N.H. Trewin. Geological Society of London.

Stüwe, K. (2002). *Geodynamics of the Lithosphere*. Springer.

Sussman, A.J. and Weil, A.R. (eds.) (2004). *Orogenic Curvature: Integrating Paleomagnetic and Structural Analyses. Geological Society of America Special Paper, **383***.

Suzuki, K. and Kato, T. (2008). CHIME dating of monazite, xenotime, zircon and polycrase: protocol, pitfalls and chemical criterion of possibly discordant age data. *Gondwana Research*, **14**, 569–586.

Sylvester, D. and Janecky, D.R. (1988). Structure and petrofabrics of quartzite and elongate pebbles at Sanviksfjell, Bergen, Norway. *Norsk Geologisk Tidsskrift*, **68**, 31–50.

Tanner, P.W.G. (1992). The duplex model: implications from a study of flexural-slip duplexes. In: *Thrust Tectonics*, ed. K.R. McClay. London and New York: Chapman and Hall, 201–220.

Tapponnier, P. and Molnar, P. (1979). Active faulting and Cenozoic tectonics of the Tien Shan, Mongolia and Bayka regions. *Journal of Geophysical Research*, **84**, 3425–3459.

Tapponnier, P., Peltzer, G. and Armijo, R. (1986). On the mechanics of the collision between India and Asia. In: *Collision Tectonics*, ed. M.P. Coward and A.C. Ries. *Geological Society of London Special Publication*, **19**, 115–157.

Thiede, R.C., Bookhagen, B., Arrowsmith, J.R., Sobel, E.R. and Strecker, M.R. (2004). Climatic control on rapid exhumation along the Southern Himalayan Front. *Earth and Planetary Science Letters*, **222**, 791–806.

Thompson, A.B. (1982). Dehydration melting of pelitic rocks and the generation of H_20-undersaturated granitic liquids. *American Journal of Science*, **282**, 1567–1595.

Thompson, A.B. and England, P.C. (1984). Pressure–temperature–time paths of regional metamorphism. II. *Journal of Petrology*, **25**, 929–955.

Thompson, J.B. Jr (1955). The thermodynamic basis for the mineral facies concept. *American Journal of Science*, **253**, 65–103.

Thompson, J.B. Jr (1957). The graphical analysis of mineral assemblages in pelitic schists. *American Mineralogist*, **42**, 842–858.

Tinker, M.A., Wallace, T.C., Beck, S.L., Myers, S. and Papanikolas, A. (1996). Geometry and state of stress in the Nazca plate beneath Bolivia and its implications for the evolution of the Bolivian orocline. *Geology*, **24**, 387–390.

Tomkins, H.S., Powell, R. and Ellis, D.J. (2007). The pressure dependence of the zirconium-in-rutile thermometer. *Journal of Metamorphic Geology*, **25**, 703–713.

Treloar, P.J. (1997). Thermal controls on early Tertiary short-lived rapid regional metamorphism in NW Himalaya, Pakistan. *Tectonophysics*, **273**, 77–104.

Treloar, P.J., Coward, M.P., Chambers, A.F., Izatt, C.N. and Jackson, K.C. (1992). Thrust geometries, interference and rotations in the Northwest Himalaya. In: *Thrust Tectonics*, ed. K.R. McClay. London and New York: Chapman and Hall, 325–341.

Treloar, P.J., O'Brien, P.J., Parrish, R.R. and Khan, M.A. (2003). Exhumation of early Tertiary coesite-bearing eclogite from Pakistan Himalaya. *Journal of the Geological Society of London*, **160**, 367–376.

Treloar, P.J. and Rex, D.C. (1990). Post-metamorphic cooling history of the Indian plate crystalline stack, Pakistan Himalaya. *Journal of the Geological Society of London*, **147**, 735–738.

Trumpy, R. (1960). Palaeotectonic evolution of the Central and Western Alps. *Geological Society of America Bulletin*, **71**, 843–907.

Turcotte, D.L. and Schubert, G. (1982). *Geodynamics: Applications of Continuum Physics to Geological Problems*. New York: John Wiley and Sons.

Turner, F.J. (1948). *Mineralogical and Structural Evolution of the Metamorphic Rocks. Geological Society of America Memoir*, **30**.

Turner, F.J. (1981). *Metamorphic Petrology: Mineralogical, Field and Tectonic Aspects*, 2nd edition. New York: McGraw-Hill.

Turner, F.J. and Verhoogen, J. (1951). *Igneous and Metamorphic Petrology* (2nd edition 1960). New York: McGraw-Hill.

Turner, S., Hawksworth, C., Jlaqi Lui *et al.* (1993). Timing of Tibetan uplift constrained by analysis of volcanic rocks. *Nature*, **364**, 50–54.

Underhill, J.R. and Patterson, S. (1998). Genesis of tectonic inversion structures: seismic evidence for the development of key structures along the Purbeck–Isle of Wight Disturbance. *Journal of the Geological Society of London*, **155**, 975–992.

Valley, J.W., Lackey, J.S., Cavosie, A.J. *et al.* (2005). 4.4 billion years of crustal maturation: oxygen isotope ratios of magmatic zircon. *Contributions to Mineralogy and Petrology*, **150**, 561–580.

Vanderhaeghe, O. (2009). Migmatites, granites and orogeny: Flow modes of partially-molten rocks and magmas associated with melt/solid segregation in orogenic belts. *Tectonophysics*, **477**, 119–134.

Vanderhaeghe, O., Medvedev, S., Fullsack, P., Beaumont, C. and Jamieson, R.A. (2003). Evolution of orogenic wedges and continental plateaux: insights from crustal thermal–mechanical models overlying subducting mantle lithosphere. *Geophysical Journal International*, **153**, 27–51.

Vanderhaeghe, O., Teyssier, C. and Wysoczanski, R. (1999). Structural and geochronologic constraints on the role of partial melting during the formation of the Shuswap metamorphic core complex at the latitude of the Thor-Odin dome, British Columbia. *Canadian Journal of Earth Sciences*, **36**, 917–943.

Vannay, J.-C. and Grasemann, B. (2001). Himalayan inverted metamorphism and syn-convergence extension as a consequence of a general shear extrusion. *Geological Magazine*, **138**, 253–276.

Vannay, J.-C. and Hodges, K.V. (1996). Tectonometamorphic evolution of the Himalayan metamorphic core between Annapurna and Dhaulagiri, central Nepal. *Journal of Metamorphic Geology*, **14**, 635–656.

Vernon, A.J., van der Beck, P.A., Persaw, C., Foeken, J. and Stuart, F.M. (2009). Variable late Neogene exhumation of the central European Alps: Low-temperature thermochronology from the Aar Massif, Switzerland and the Lepontine Dome, Italy. *Tectonics*, **28**, TC 5004, doi:10.1029/2008T Coo 2387.2009.

Vielzeuf, D. and Holloway, J.R. (1988). Experimental determination of the fluid absent melting relations in the pelitic system. *Contributions to Mineralogy and Petrology*, **98**, 257–276.

Vlaar, N.J., Vankeken, P.E. and Vandenberg, A.P. (1994). Cooling of the Earth in the Archean – consequences of pressure-release melting in a hotter mantle. *Earth and Planetary Science Letters*, **121**, 1–18.

Walsh, J.J., Bailey, W.R., Childs, C., Nicol, A. and Bonson, C.G. (2008). Formation of segmented faults: a 3D perspective. *Journal of Structural Geology*, **25**, 1251–1262.

Warren, C.J., Beaumont, C. and Jamieson, R.A. (2008). Formation and exhumation of ultra-high pressure rocks during continental collision: role of detachment in the subduction channel. *Geochemistry Geophysics Geosystems (G^3)*, **9**, Q04019.

Waters, D.J. (1988). Partial melting and the formation of granulite facies assemblages in Namaqualand, South Africa. *Journal of Metamorphic Geology*, **6**, 387–404.

Waters, D.J. (1991). Hercynite-quartz granulites: phase relations, and implications for crustal processes. *European Journal of Mineralogy*, **3**, 367–386.

Watson, E.B., Wark, D.A. and Thomas, J.B. (2006). Crystallization thermometers for zircon and rutile. *Contributions to Mineralogy and Petrology*, **151**, 413–433.

Wdowinski, S. and O'Connell, R.J. (1989). A continuum model of continental deformation above subduction zones: applications to the Andes and the Aegean. *Journal of Geophysical Research*, **94**, No. B8, 10,331–10,346.

Weil, A.B. and Sussman, A.J., eds. (2004). Oroclines; progressive arcs and primary arcs: integrating palaeomagnetism and structural analysis. *Geological Society of America Special Paper*, **383**.

Wernicke, B.P. and Burchfiel, B.C. (1983). Modes of extensional tectonics. *Journal of Structural Geology*, **4**, 105–115.

Whipple, K.X. (2009). The influence of climate on the tectonic evolution of mountain belts. *Nature Geoscience*, **2**, 97–104.

Whitehouse, M.J. and Kamber, B.S. (2005). Assigning dates to thin gneissic veins in high grade metamorphic terranes: a cautionary tale from Akilia, Southwest Greenland. *Journal of Petrology*, **46**, 291–318.

Whitehouse, M.J., Kamber, B.S. and Moorbath, S. (1999). Age significance of U–Th–Pb zircon data from early Archaean rocks of West Greenland: a reassessment based on combined ion-microprobe and imaging studies. *Chemical Geology*, **160**, 201–204.

Whitehouse, M.J. and Platt, J.P. (2003). Dating high-grade metamorphism – constraints from rare-earth elements in zircon and garnet. *Contributions to Mineralogy and Petrology*, **145**, 61–74.

White, N.M., Pringle, M., Garzanti, E. *et al.* (2002). Constraints on the exhumation and erosion of the High Himalaya slab, NW India, from foreland basin deposits. *Earth and Planetary Science Letters*, **195**, 29–44.

White, N., Parrish, R.R., Bickle, M.J. *et al.* (2001a). Metamorphism and exhumation of the NW Himalaya constrained by U-Th-Pb analyses of detrital monazite grains from early foreland basin sediments. *Journal of the Geological Society of London*, **158**, 625–635.

White, R.W., Powell, R. and Holland, T.J.B. (2001b). Calculation of partial melting equilibria in the system Na_2O-CaO-K_2O-FeO-MgO-Al_2O_3-SiO_2-H_2O (NCKFMASH). *Journal of Metamorphic Geology*, **19**, 139–153.

Whitney, D.L., Teyssier, C. and Vanderhaeghe, O. (2004). Gneiss domes and crustal flow. In: *Gneiss Domes in Orogeny*, ed. D.L. Whitney, C. Teyssier and C.S. Siddoway. *Geological Society of America Special Publication*, **380**, 15–33.

Whittington, A.G., Hofmeister, A.M. and Nabelek, P.I. (2009). Temperature dependent thermal diffusivity of the Earth's crust and implications for magmatism. *Nature*, **458**, 319–321.

Wilde, S.A., Valley, J.W., Peck, W.H. and Graham, C.M. (2001). Evidence from detrital zircons for the existence of continental crust and oceans on the Earth 4.4 Gyr ago. *Nature*, **409**, 175–178.

Willett, S.D. (1999). Orogeny and orthography: The effects of erosion on the structure of mountain belts. *Journal of Geophysical Research*, **104**, 28957–28981.

Willett, S.D., Beaumont, C. and Fullsack, P. (1993). Mechanical model for the tectonics of doubly vergent compressional orogens. *Geology*, **21**, 371–374.

Willett, S.D. and Brandon, M.T. (2002). On steady states in mountain belts. *Geology*, **30**, 175–178.

Willett, S.D., Hovius, N., Brandon, M.T. and Fisher, D.M. (2006). *Tectonics, Climate, and Landscape Evolution. Geological Society of America Special Paper*, **398**.

Williams, P.F. and Jiang, D. (2005). An investigation of lower crustal deformation: evidence for channel flow and its implications for tectonics and structural studies. *Journal of Structural Geology*, **27**, 1486–1504.

Wilson, C.W. and Stearns, R.G. (1958). The structure of the Cumberland Plateau, Tennessee. *Geological Society of America Bulletin*, **69**, 1283–1296.

Wobus, C.W., Hodges, K.P. and Whipple, K.X. (2003). Has focussed denudation sustained active thrusting at the Himalayan topographic front? *Geology*, **31**, 861–864.

Zack, T., Moraes, R. and Kronz, A. (2004). Temperature dependence of Zr in rutile: Empirical calibration of a rutile thermometer: *Contributions to Mineralogy and Petrology*, **148**, 471–488.

Zeitler, P.K. and 19 others (2001). Crustal reworking at Nanga Parbat, Pakistan: metamorphic consequences of thermal-mechanical coupling facilitated by erosion. *Tectonics*, **20**, 712.

Zen, E-An (1966). *Construction of Pressure–Temperature Diagrams for Multicomponent Systems After the Method of Schreinemakers – a Geometrical Approach. US Geological Survey Bulletin*, **1225**.

Zhang, P.Z., Zhenkang Shen, Min Wang *et al.* (2004). Continuous deformation of the Tibetan Plateau from global positioning system data. *Geology*, **32**, 809–812.

Zhang, R.Y., Liou, J.G. and Ernst, W.G. (2009). The Dabie–Sulu continental collision zone: a comprehensive review. *Gondwana Research*, **16**, 1–26.

Zhao, W.L. and Morgan, W.J. (1985). Uplift of the Tibetan Plateau. *Tectonics*, **4**, 359–369.

Zwart, H.J. (1962). On the determination of polymorphic mineral associations, and its application to the Bosost area (Central Pyrenees). *Geologische Rundschau*, **52**, 38–65.

Zwart, H.J. (1963). Some examples of the relations between deformation and metamorphism from the central Pyrenees. *Geologie en Mijnbouw*, **42**, 143–154.

Zwart, H.J. (1967). The duality of orogenic belts. *Geologie en Mijnbouw*, **46**, 283–309.

Zwart, H.J. (1969). Metamorphic facies series in the European orogenic belts and their bearing on the causes of orogeny. *Geological Association of Canada Special Paper*, **5**, 7–16.

Index

Printed in the United States
by Baker & Taylor Publisher Services